Internal and External Narrative Generation Based on Post-Narratology:

Emerging Research and Opportunities

Takashi Ogata
Iwate Prefectural University, Japan

A volume in the Advances in Linguistics and Communication Studies (ALCS) Book Series

IGI Global
DISSEMINATOR OF KNOWLEDGE

Published in the United States of America by
IGI Global
Information Science Reference (an imprint of IGI Global)
701 E. Chocolate Avenue
Hershey PA, USA 17033
Tel: 717-533-8845
Fax: 717-533-8661
E-mail: cust@igi-global.com
Web site: http://www.igi-global.com

Library of Congress Cataloging-in-Publication Data

Names: Ogata, Takashi, 1958- author.
Title: Internal and external narrative generation based on post-narratology : emerging research and opportunities / by Takashi Ogata.
Description: Hershey, PA : Information Science Reference, [2020]
Identifiers: LCCN 2019013140| ISBN 9781522599432 (hardcover) | ISBN 9781522599456 (ebook) | ISBN 9781522599449 (softcover)
Subjects: LCSH: Discourse analysis, Narrative. | Narration (Rhetoric)
Classification: LCC P302.7 O46 2020 | DDC 401/.41--dc23 LC record available at https://lccn.loc.gov/2019013140

This book is published in the IGI Global book series Advances in Linguistics and Communication Studies (ALCS) (ISSN: 2372-109X; eISSN: 2372-1111)

British Cataloguing in Publication Data
A Cataloguing in Publication record for this book is available from the British Library.

For electronic access to this publication, please contact: eresources@igi-global.com.

Advances in Linguistics and Communication Studies (ALCS) Book Series

ISSN:2372-109X
EISSN:2372-1111

Editor-in-Chief: Abigail G. Scheg Western Governors University, USA

MISSION

The scope of language and communication is constantly changing as society evolves, new modes of communication are developed through technological advancement, and novel words enter our lexicon as the result of cultural change. Understanding how we communicate and use language is crucial in all industries and updated research is necessary in order to promote further knowledge in this field.

The **Advances in Linguistics and Communication Studies (ALCS)** book series presents the latest research in diverse topics relating to language and communication. Interdisciplinary in its coverage, ALCS presents comprehensive research on the use of language and communication in various industries including business, education, government, and healthcare.

COVERAGE

- Language in the Media
- Language and Identity
- Computer-Mediated Communication
- Sociolinguistics
- Interpersonal Communication
- Discourse Analysis
- Language Acquisition
- Non-Verbal Communication
- Dialectology
- Computational Linguistics

IGI Global is currently accepting manuscripts for publication within this series. To submit a proposal for a volume in this series, please contact our Acquisition Editors at Acquisitions@igi-global.com or visit: http://www.igi-global.com/publish/.

The Advances in Linguistics and Communication Studies (ALCS) Book Series (ISSN 2372-109X) is published by IGI Global, 701 E. Chocolate Avenue, Hershey, PA 17033-1240, USA, www.igi-global.com. This series is composed of titles available for purchase individually; each title is edited to be contextually exclusive from any other title within the series. For pricing and ordering information please visit http://www.igi-global.com/book-series/advances-linguistics-communication-studies/78950. Postmaster: Send all address changes to above address.

Titles in this Series

For a list of additional titles in this series, please visit:
https://www.igi-global.com/book-series/advances-linguistics-communication-studies/78950

Enhancements and Limitations to ICT-Based Informal Language Learning Emerging Research and Opportunities
Rashad Ahmed (Miami University, USA) Abdu Al-kadi (Ibb University, Yemen) and Trenton Hagar (Utrecht University, The Netherands)
Information Science Reference • © 2020 • 201pp • H/C (ISBN: 9781799821168) • US $175.00

Cases on Communication Disorders in Culturally Diverse Populations
Diane M. Scott (North Carolina Central University, USA)
Information Science Reference • © 2020 • 300pp • H/C (ISBN: 9781799822615) • US $185.00

Redefining the Role of Language in a Globalized World
Ai-Ling Wang (Tamkang University, Taiwan)
Information Science Reference • © 2020 • 300pp • H/C (ISBN: 9781799828310) • US $195.00

International Approaches to Bridging the Language Gap
Cristina-Aránzazu Huertas-Abril (University of Córdoba, Spain) and María Elena Gómez-Parra (University of Córdoba, Spain)
Information Science Reference • © 2020 • 317pp • H/C (ISBN: 9781799812197) • US $195.00

Toward an Integrated Approach to Narrative Generation Emerging Research and Opportunities
Takashi Ogata (Iwate Prefectural University, Japan)
Information Science Reference • © 2020 • 432pp • H/C (ISBN: 9781522596936) • US $215.00

For an entire list of titles in this series, please visit:
https://www.igi-global.com/book-series/advances-linguistics-communication-studies/78950

IGI Global
DISSEMINATOR OF KNOWLEDGE

701 East Chocolate Avenue, Hershey, PA 17033, USA
Tel: 717-533-8845 x100 • Fax: 717-533-8661
E-Mail: cust@igi-global.com • www.igi-global.com

Table of Contents

Preface

The narrative (*monogatari* in Japanese) generation system study by the author has been planned and practiced under a systematic framework that covers the following: theoretical and philosophical thoughts of narratives and narrative generation, analytical research of related narrative genres and narrative works, the design and development of various types of narrative generation systems, external or social and internal or personal developments and practices of narrative analyses and narrative generation systems, and narrative works writing and creation using narrative generation systems. This study has also been conducted as a trans-disciplinary approach related to diverse research genres and fields, including human sciences, such as narratology, literary theories, and philosophy; social sciences, such as advertisement and sociology; and technologies and information sciences, such as artificial intelligence and cognitive science. For this new approach, the author has presented various terms, including "informatics of narratology," "informational narratology," "integrated approach to narrative generation," and "post-narratology." Although the author has published several authored and edited books using these terms (Ogata & Akimoto, 2016, 2019; Ogata & Asakawa, 2018; Ogata & Kanai, 2010; Ogata, Kawamura, & Kanai, 2018), these books have not necessarily had a consistent organization that could show the above research characteristics systematically. By contrast, the book *Internal and External Narrative Generation Based on Post-Narratology: Emergent Research and Opportunities* and the recent work and *Toward an Integrated Approach to Narrative Generation: Emergent Research and Opportunities* were written under a consistent and systematic organization. These two books comprise a series. The first book, *Toward an Integrated Approach to Narrative Generation*, focuses on topics regarding the concepts, thoughts, and philosophy supporting the narrative generation study (also delving into concrete narrative surveys). The second book, *Internal and External Narrative Generation Based on Post-Narratology*, presents the design and development of narrative generation systems and narrative analyses dependent on the conceptual and philosophical discussion in the first book. Moreover, *Internal and External Narrative Generation Based on Post-Narratology* partially refers to the themes regarding narrative work creation as arts and literature through narrative generation systems beyond narrative generation systems as a technology.

Going back to the first book, Chapter 1 [*What Are Narrative Generation Phenomena*] introduces an idea that deals with narrative phenomena as the integration between the individual level, or narrative generation and reception system, and social level, or narrative production and consumption system. This idea is called the "multiple narrative structures model." Chapter 2 [*Areas of Narratives or Narrative Genres*] presents a tentative and large categorized system of narrative genres, namely, a "narrative genre system." Chapter 3 [*Narratology and Post-Narratology*] describes the narratology or post-narratology that synthesizes and develops various narrative-related studies, including previous narrative research, narrative and narrative generation studies in the broad sense, and narratology and literary theories. In Chapter 4 [*Theoretical or Philosophical Considerations for an Integrated Narrative Generation Approach*], although narrative philosophy or thought are the fundamental concepts supporting the study, a point different from previous works relating to narrative philosophy is the aim to make products in the technological field of narrative generation systems, instead of philosophy or thought, about narrative itself. Additionally, the **Introduction** chapter and the **Conclusion** chapter also provide rich and deep information for reading the above four chapters.

CHAPTER ORGANIZATION

This book is composed of an introductory section, four chapters, and concluding section.

The main purposes of **Introduction** are to show the themes of the author's integrated narrative generation study and to outline the book's contents to the readers. To achieve these purposes, this part reviews four long papers and five books written by the author to consider themes in the integrated narrative generation study concretely. This book and the recent book *Toward an Integrated Approach to Narrative Generation*, which comprise a series, systematically collect a part of the author's previous results and expand them toward new directions, thereby constructing "post-narratology" and external or social and internal or private development of narrative generation study.

The main content of **Chapter 1 [An Integrated Narrative Generation System: Synthesis and Expansion]** is a presentation of the overall system architecture of the Integrated Narrative Generation System (INGS), which has been developed as a central part of the narrative generation study by the author. The INGS architecture has been described in previous books since 2016 (Ogata & Akimoto, 2016, 2019; Ogata & Asakawa, 2018; Ogata, Kawamura, & Kanai, 2018) (in addition, Ogata & Kanai (2010), which have given a more theoretical and detailed discussion for a comprehensive narrative generation architecture related to the current INGS.). Thus, this chapter reflects on the extended components of this system, which has been gradually extended through experimental execution. In particular, the author describes the design and development process of INGS and various components and mechanisms related to INGS, and then introduces related conceptual dictionaries in detail. Moreover, new execution examples are presented. An important point concerning INGS is that the system is positioned as a collection of themes through concrete system mechanisms as well as the *Geino* Information System (GIS), explained in Chapters 2 and 4 in detail. The post-narratology of the author will be realized through INGS (and GIS).

Chapter 2 [*Kabuki* as a Synthetic Narrative: Synthesis and Expansion] comprehensively introduces the present status regarding *kabuki* from the viewpoint of narrative generation. The importance of *kabuki* study for the author is closely related to GIS, which is one of the core systems supporting the integrated narrative study besides INGS. In the research framework, narrative generation is divided into a narrative generation–reception process and a narrative production–consumption process based on the multiple narrative structures model by the author. GIS is offered for the side of the narrative production–consumption process in contrast to INGS, which is conducted on the narrative generation–reception process. The *kabuki* study in Chapter 2 approaches the *kabuki* seen as a comprehensive narrative phenomenon, including creating and performing each play, from writing a scenario to constructing a synthetic work, to staging a play by actors having the same name in a historical story. Such narrative generation of *kabuki* contributes to the design of the GIS architecture. The main content of the first half of this chapter is an introduction and discussion of diverse topics, such as the general characteristics of *kabuki*, the multiple elements based on the author's research objective, and narrative techniques of *kabuki* using many examples. The second half shows an experimental system using several narrative techniques of *kabuki*. These systematized narrative techniques will be incorporated into GIS and INGS.

Chapter 3 [An Analysis of Advertisements for Narrative Generation] chiefly describes a narrative analysis of television commercials for an advertising narrative generation system as an application system of the core narrative generation systems, namely, INGS and GIS, partially. The narrative analysis of advertisements focuses on the main components of advertising scenarios, including persons, objects (products or brands and other objects), places or stages, times or seasons, and product manufacturing, distribution, and purchasing processes to investigate precisely these components from the perspective of normal or ordinary and abnormal or extraordinary states. A conclusion acquired from this study is that the basic method of advertising narrative generation is based on the defamiliarized or deviated rhetoric from normal and ordinary usage and appearance of a product or brand. Further, such defamiliarized rhetoric will be introduced into narrative techniques for general narrative generation.

Chapter 4 [Toward External and Internal Practices of Narrative Generation] bridges the previous research on integrated narrative generation to the future stage that aims at the social or "external" and personal or private (i.e., "internal") development or practice. In these two directions, the external direction is practically associated with the development of GIS. As GIS that corresponds to the aspect of the narrative production–consumption process is directly related to the social narrative development through diverse narrative production and consumption mechanisms, the author states the problem of the external practice of narrative generation through a conceptual and technical description of GIS. Meanwhile, the internal practice of narrative generation refers to the direction of the generation, or creation, of narrative works in the personal or private level through diverse potential possibilities of correlations with automated narrative generation mechanisms, namely, INGS, GIS, and other application systems. Apart from discussing the vision and directions of internal narrative generation, this chapter also attempts at the collection and organization of the author's personal information, aiming at systematization, as the materials for future narrative creation using automated narrative generation systems.

Lastly, the results and indications of the preceding chapters and future research directions are comprehensively presented in the final section, **Conclusion**. In the last part of this section, the author briefly shows two research directions for the entire book, which correspond to the external and internal practices mentioned in Chapter 4.

As shown in the above description, through the six chapters, including **Introduction** and **Conclusion**, of *Internal and External Narrative Generation*

Based on Post-Narratology: Emergent Research and Opportunities and the six chapters, including **Introduction** and **Conclusion**, of *Toward an Integrated Approach to Narrative Generation: Emergent Research and Opportunities*, the author presents the vision, plans, processes, results, and considerations of an integrated narrative generation approach and post-narratology.

In addition, the following groups will find this text useful in increasing their research exposure to pertinent and interdisciplinary topics and in furthering their own research efforts in this field: (1)Researchers and advanced-level students of information technologies and cognitive studies (including artificial intelligence, cognitive science, narrative or story generation, natural language comprehension and generation, computational creativity, communication studies, media studies, neuroscience, etc.), (2) Researchers and advanced-level students of literary studies (including narratology, literary theories, discourse theory, rhetoric, poetics, semiotics, linguistics, film theory, etc.), and (3) Designers, creators, developers, researchers, and advanced-level students of advanced digital content design and development (including advertisements, computer game, interactive storytelling, automated content generation, human interface, the Web design, etc.)

REFERENCES

Ogata, T., & Akimoto, T. (2016). *Computational and cognitive approaches to narratology*. Hershey, PA: IGI Global. doi:10.4018/978-1-5225-0432-0

Ogata, T., & Akimoto, T. (2019). *Post-narratology through computational and cognitive approaches*. Hershey, PA: IGI Global. doi:10.4018/978-1-5225-7979-3

Ogata, T., & Asakawa, S. (2018). *Content generation through narrative communication and simulation*. Hershey, PA: IGI Global. doi:10.4018/978-1-5225-4775-4

Ogata, T., & Kanai, A. (2010). *An introduction to informatics of narratology: Toward the thoughts and technologies of narrative generation*. Tokyo, Japan: Gakubunsha.

Ogata, T., Kawamura, Y., & Kanai, A. (2018). *Informational narratology: Artificial intelligence/cognition/social process and narrative generation.* Takashi Ogata Iwate Prefectural University.

Acknowledgment

I would like to acknowledge the supportive staff at IGI Global. Their expertise, ongoing support, and effective advice have made publishing of this project a success.

I would also like to give acknowledgment to this book's reviewers. They read the manuscripts and gave many constructive comments and advice. They were very helpful in this book's revision and completion.

Each chapter of this book was supported by JSPS KAKENHI Grant Number 18K18509, Kayamori Foundation of Information Science Advancement Research Grant, The Telecommunication Advancement Foundation Research Grant, and AOI TYO Holdings Inc. I would like to acknowledge these organizations.

Furthermore, I would like to thank the students and staff at the university for their help in this writing project's successful completion.

Finally, I would also like to thank my wife for her love and support throughout this project.

Takashi Ogata
Iwate Prefectural University, Japan

Introduction

TWO BOOKS THAT INTEGRATE A SERIES OF STUDIES

The author has an opinion regarding books as media. In contemporary academia, the main medium for the report of research results, especially in scientific and technological areas, is the academic journal. Papers submitted to academic journals play the role of evaluation standards for each researcher. However, as the entire systematic construction of all research is also emphasized in the academic areas of humanities and social sciences, research presented in the form of a book is still recognized as valuable. Although the author's main publication media are ordinarily academic journals and conferences, the author considered the presentation of this research in the form of a book to be important for the systematic and synthetic features of the narrative generation study. Books are a good medium for systematically and synthetically describing a study.

This opinion is closely related to the point that the author's "narratology of narrative generation" or "post-narratology," shown in the two books written based on a consistent plan, *Toward an Integrated Approach to Narrative Generation: Emerging Research and Opportunities* and *Internal and External Narrative Generation Based on Post-Narratology: Emerging Research and Opportunities*, is an interdisciplinary study linking broad academic fields. While the author's study covers the narrative generation field using artificial intelligence (AI) and cognitive science, at the same time, it also has elements of philosophy, folklore, and advertisement science. Thus, the author aims to establish a new research system or research genre that includes or relates to a variety of academic areas. The author defines the new research concept using words such as "narratology of narrative generation" or "post-narratology." For the systematic introduction of a new research field that transcends the existing academic barriers, the author thinks that the format of book as a synthesized medium is effective.

Building on this idea, the author has attempted to show this study, including the thought and philosophy, systematically through reviewing several books and relatively long papers. The section **SYNTHETIC APPROACHES TO NARRATIVE GENERATION THROUGH SYSTEM-ORIENTED BOOKS** introduces these books and papers to show the systematic and synthetic feature of the author's study. Here, the stories and narratives embody the concept of "an integrated approach to narrative generation." The next section, **THEMES OF THE BOOK SERIES AND CONFIGURATION OF THE BOOK**, shows all parts of the book that synthetically and systematically complete the parts of narrative generation study in the current stage.

This book and the recent book (Ogata, 2019d) are a series of systematic study. These books comprise the systematic writing of the past books and papers, building on the treatment of problems in them and the following topics. Ogata (2019d) dealt with the themes related to (1) and (2). This book treats the themes of (3), (4), (5), and (6):

1. **Basic standpoints—A unified consideration of narrative phenomena:** Scientific methods and narratological methods are two representative methods that articulate and order changeable worlds and societies and give them operational frameworks. Scientific methods are based on the analytical method that mainly anatomizes a world or object into the elements. In contrast, the narratological methods show a synthesized direction that comprehends the world or object as an organic synthesis of diverse elements. These take contrasting positions. The narratological method is a way or a set of ways that all human societies innately possess but that typically show unique characteristics of human brain and thought. This study pursues the narratological principle as a representation of the human brain's mechanism to extend results in various directions. Therefore, the narratives considered in these books cover broad cultural areas beyond the literary and artistic ranges to the areas of medicines related to the brain and psychology.
2. **The philosophy and thought of narrative or narrative generation**: The author presents several philosophical foundations that give direction for the development, application, and distribution of one or more narrative generation systems based on the above basic consideration. These include "post-narratology" (or "expanded literary theory," "informational narratology," "informatics of narratology," etc.), the multiple narrative structures model, narrative generation's circular control, narrative generation through norms and deviation, and narrative generation of

fluidity and fixation. These indicate higher-level or meta-level strategic mechanisms in a broad range, including the architecture of narrative generation systems, output information, and design and development processes. These always function with the execution processes of each narrative generation system, even if they are not explicitly described.

3. **The consideration of narrative generation mechanisms through narrative analyses:** Intending to use directly the practices of narratology, this study develops a narrative rhetorical system for diverse narrative genres. The author considers how narrative generation parallels the human brain and its mechanism and positions rhetoric as objects' representation in the world and developmental processes, namely the generation and transformation of semantic structures. Our pursuit of principles of the world can be realized through observation at various levels and the constructive exploration of the principles—that is, the rhetoric of generation and transformation. In this study, various narrative analyses are also conducted according to such direction. Basically, analyzing narrative texts or works in various genres and narrative generation processes in many genres from personal level to collective, organizational, and social levels makes it possible to acquire a unique rhetoric in each genre and systematically abstract the acquired rhetorical information. The author also presents constructive analyses of rhetorical information through computationally and cognitively re-constructing previous narratology, literary theories, and literary critiques. An essential reason that this study uses computational and cognitive methods is the experimental simulation of the knowledge and mechanisms at various narrative levels of narratology and cognitive science. Another reason is that they provide an effective means for considering mutual relationships among hierarchical levels.
4. **A fundamental narrative generation system:** A narrative generation system presented by Ogata (1995) was constructed as a general framework that integrates diverse types of knowledge related to the conceptual parts of narrative generation. This system architecture has the following three categories of narrative knowledge: "narrative techniques," for operating narrative structures; "narrative strategies," which are rule-based knowledge for controlling the use of the above narrative techniques; and "narrative conceptual structures" generated by narrative techniques and strategies. Although narrative strategies actually need to be executed though cognitive motivations and the desires of the narrator who controls the techniques and strategies, the experimental system in the

research stage more simply refers to several parameters that represent the characteristics and goals of the generated narrative. The narrative generation system introduced in these books and discussed intensively in Chapter 1 in this book, the "Integrated Narrative Generation System (INGS)," has been designed and developed based on the above concepts and a previous story generation system by Ogata (1992). INGS is not a static system; it is always in being refined and augmented. In each stage, INGS will be a bridge to later developments and applications.

5. **Applications and the social development and distribution of narrative generation:** The next part of the research is the process in which narrative rhetorical analyses and narrative generation systems are applied in several directions. The first application system that the author conducted was a marketing- and advertising-integrated support system (Ogata, Watanabe, Hori, & Ohsuga, 1995), and it was a narrative generation's application system for the consistent support of tasks from the targeting work of a product by marketers to the CM scenario creation by advertising creators. One of the objectives of the application system was to stimulate the thinking process of marketers, advertising creators, and other participants, to expand their perspectives, and to support their trial-and-error works using diverse and flexible narrative generation ability to stimulate idea creation. Later, the author's advertising narrative generation studies have continued to develop, and the author is currently developing a new INGS-related advertising project with the biggest advertisement production company in Japan (Ito, Sasaki, & Ogata, 2018; Ono, Sasaki, & Ogata, 2018, 2019a, 2019b; Ono, Sasaki, Ito, & Ogata, 2018). Another large application area was *geinō*, a Japanese word that includes dramas such as *kabuki*, folkloric performances, and modern entertainment or show business. Traditionally in Japan, the world of *geinō* has been an effective space of integrated narrative production and consumption with complex organizations and media mechanisms constructed by talented personal creators. The *geinō* can be regarded as a system that continues to produce narratives as works of various genres, including cross-media rumors and gossip by *geinōjin* (performers) and talents as characters. The author has analyzed and simulated the vast world of modern *geinō* from several perspectives such as the level of produced narratives; the level of organizations, media, and creators that produce narratives to bridge them to the design; and the development of social and technological systems that support or simulate the narrative production and consumption mechanism of *geinō*. The author is currently

designing *geinō* phenomena as the model and system of an integrated mechanism, "*Geinō* Information System (GIS)," that creates narratives of *geinōjins* through the organizational association of the following mechanisms: narrative production, narrative development, reception space, narrative reception, and narrative interpretation. GIS has developed a system that included or works with INGS beyond mere applications and intends to develop a fundamental mechanism for pushing narrative generation study into the world of social distribution and development.

6. **The production of literary and artistic works or literary and artistic organizations:** Finally, an objective of the author in using these mechanisms is to explore experimentally new narrative genres and works that have characteristics such as extemporaneity, mutability, immediacy. However, this plan does not yet have a concrete goal or a strategy for achieving that goal. However, the author considers that although a narrative writer's creativity is shown in the work and whether the texts are literary musical score, the writer's goals and intentions are not necessarily reflected directly in the presentation of the narrative work. Similar to the creation of a pottery, the narrative production process includes the image in which the narrative work is involved in the non-autonomic process in the furnace of computers and acquires diverse representation possibilities according the writing process. In addition, the narrative work is gradually produced through the continuous and mutual practices among various agents in the multiple narrative structure model, including the receivers. An important consideration is how to explore and develop various possibilities for narrative representation and creation that are impossible in current narrative genres, including novels. In the future, all the author's research and development will be organized from the above viewpoints. At the same time, the author plans in parallel, to explore the principle of literary or artistic movement beyond business in the narrow sense. The point where the principles of engineering and business are overcome or downfall mean that a mental principle may be a goal in the author's entire activity. In particular, produced narratives will appear in the final stages.

In particular, the recent book (Ogata, 2019d) described the following topics through the following chapters:

Chapter 1 [What Are Narrative Generation Phenomena?] introduced an idea that dealt with narrative phenomena as the integration between the individual level (narrative generation and reception system) and social level (narrative production and consumption system); this idea was called the "multiple narrative structures model." This chapter described the future image of a human-machine symbiosis system that includes narrators (senders) and narratees (receivers) as artificial intelligence. Furthermore, based on the concept of "visible narratives" and "invisible narratives," the author analyzed the narrative components or elements to consider methods for synthesizing the analyzed elements. This idea of the analysis and synthesis of various narrative elements is systematized in INGS that is the theme of Chapter 1 in this book.

In **Chapter 2 [Areas of Narratives or Narrative Genres]**, the author presented a tentative and large categorized system of narrative genres, i.e., a "narrative genre system." It was related to the division of the objects or materials of research and analysis in the synthetic narrative generation study based on computational methods. Although, throughout Ogata (2019d) and this book, the author consciously uses Japanese narratives that include both universal narrative characteristics and local or cultural features, this narrative genre system was basically constructed using Japanese narrative genres as concrete materials. As an overview, the narrative genre system included the following five narrative categories: (1) as a work in the narrow sense, (2) as a work in the broad sense, (3) as social and emergent phenomena, (4) invading the real phenomena, and (5) as human physiological and psychological phenomena. In each explanation, after the corresponding narrative genre category was defined and explained, a concrete genre under the large genre category was treated for discussing the characteristics.

Chapter 3 [Narratology and Post-Narratology] described the narratology or post-narratology that synthesized and developed various narrative-related studies, including previous narrative research, narrative and narrative generation studies in the broad sense, and previous narratology and literary theories. This chapter stated various narrative studies in the broad sense and then studied and surveyed narrative and narrative generation researches in more narrow sense. Further, dependent on these backgrounds, the author surveyed the fields of narratology and literary theories. On the other hand, as a cultural approach, this chapter referred especially to Japan's literature. In summary, dependent on the above topics, this chapter presented the concept of post-narratology, the expanded literary theory in the author's previous term.

In **Chapter 4 [Theoretical or Philosophical Considerations for an Integrated Narrative Generation Approach]**, although narrative philosophy or thought were the fundamental concepts supporting this study, a point different from the previous studies relating to narrative philosophy was that this study aims to make products in the technological field of narrative generation systems instead of philosophy or thought about narrative itself. From the viewpoint of philosophy or thought, the narrative generation systems were applications. In contrast, from the goal of narrative generation systems, philosophy or thought corresponded to a kind of strategic framework for establishing the vision, strategy, and direction. In particular, the first philosophical concept was "multiple narrative structures." Next, the author addressed the following three concepts: "circular narrative control," "norm and deviation," and "fluidity and fixation." They were not concepts that are respectively individual. These philosophical concepts built the dynamical characteristics of narrative generation through their interrelationships.

In contrast, this book, *Internal and External Narrative Generation Based on Post-Narratology: Emergent Research and Opportunities*, mainly describes more concrete topics regarding narrative analysis, system design and implementation, as well as social level system development and distribution. The ultimate theme in this book is the proposition of the concepts and methods of "internal and external narrative generation." As an overview, external narrative generation corresponds to the practices of social level development and distribution of narrative generation systems, in particular INGS and GIS. On the other hand, internal narrative generation implies narrative creation, including the author's personal elements and social and historical narratives. External narrative generation is at the level of mechanism or form. In contrast, internal narrative generation corresponds to the level of contents or narrative works. Furthermore, based on the level of internal narrative generation, the author's narrative generation study substantially covers the area of narrative creation and literature. In particular, this book includes four chapters: **Chapter 1 [An Integrated Narrative Generation Systems: Synthesis and Expansion]**, **Chapter 2 [*Kabuki* as a Synthetic Narrative: Synthesis and Expansion]**, **Chapter 3 [An Analysis of Advertisements for Narrative Generation]**, and **Chapter 4 [Toward External and Internal Practices of Narrative Generation]**. Additionally, the book has the **Conclusion** chapter corresponding to this **Introduction** chapter.

SYNTHETIC APPROACHES TO NARRATIVE GENERATION THROUGH SYSTEM-ORIENTED BOOKS

The author has strategically divided the previous narrative generation research into "paper" and "book." The book is a tool that will be utilized to describe the comprehensive results and plans of the research and development, whereas the paper will be used to provide a detailed description of the various research issues.

For showing the systematic and synthetic aspects of the author's narrative generation study, this section introduces several system-oriented books (including relatively long papers in a book) that the author published previously, which include a master dissertation, a doctoral dissertation, and five co-authored books with various authors. In the recent book (Ogata & Akimoto, 2019), the author used the word "post-narratology" as a comprehensive title that shows the integrated approach to narrative generation. On the other hand, as a more concrete topic, the author presents interdisciplinary frameworks or philosophical concepts leading to the author's narrative generation studies, which are primarily represented by the term "expanded literary theory" (Ogata, 2002, 2003a, 2003b, 2014, 2019h). If there is a difference between the field of post-narratology developed and expanded literary theory, although the former is a comprehensive concept centered on narrative generation systems, expanded literary theory is one aspect of the fundamental and philosophical thoughts that provide background support for the design and development of narrative generation systems. In the current theoretical system developed by the author, suffice it to say that expanded literary theory is one component of post-narratology. Through these books, the author also provided other terms, "the informatics of narratology," "informational narratology," and "computational and cognitive approaches to narratology." However, these terms show a related pioneering research field.

[A] Preliminary Studies

(A-1) Master's Dissertation

[Goals] The main objective of the master's dissertation by the author (Ogata, 1992), *Study on a Narrative Generation System Based on Explanation*, was the design and development of a narrative, in particular story, generation system using AI and cognitive science, based on concepts from the humanities, including Vladimir Propp's (1895–1970) (1968) literary theory, folktale theories, and folklore.

[Structure] This paper has main four parts. The first part describes the background research fields of the paper, narratology and literary theories, artificial intelligence, and narrative understanding and generation. The next part explains in detail the morphology of the folktale by Propp and shows the formalization of story grammar through Propp's theory and the design of an experimental system. It also describes a system implemented in the Common Lisp programming language; the generation results are described in the next part. Although the final part corresponds to the appendix of the paper, it surveys and introduces the defamiliarization theory in literary thought and several folkloric theories regarding social defamiliarization and discusses the possibilities and problems of the application to narrative generation of these approaches.

[Overview] Here, the author transforms the categorized knowledge regarding the basic components of stories described in Propp's folktale theory into a form of story grammar. For this transformation, the author describes the story grammar based on Propp's original theory more faithfully, without using the existing definitions of story grammar in cognitive science; Propp's theory relies more on rich literary knowledge than on oversimplified existing cognitive science models. Further, the developed system can generate many different stories using a simple sentence generation mechanism. At the same time, this study also uses the method of explanation-based learning proposed in AI to be able to generate stories according to fixed story patterns generated using rules and inferences. Although this method is not "learning" in the sense of acquiring new knowledge, it can gradually increase the efficiency of story generation by simplifying the method, particularly generation speed.

In addition to these technological propositions, this paper also refers to many narratological and folkloric concepts in the research field to explore richer story generation methods and describes several concepts concerning narratives and folktales, such as theories of foreigners, strangers, and tricksters. In this sense, this paper is the first systematic and synthetic result for the author's narrative generation study and the first collation of the proposed various techniques and knowledge, such as the literary theory by Propp, grammatical knowledge-formation of narratives and stories, learning methods by AI, and surveys in the wide range of humanities and cultural sciences, including narratology, literary theories, folklore, etc.

(A-2) Doctoral Dissertation

[Goals] In the author's doctoral dissertation, *Narrative Generation: An Approach Based on Techniques and Strategies for the Narrative*, Ogata (1995) was based on a fundamental view of the changing roles of information technologies due to their progression. In particular, one core idea in this dissertation was that computer technologies, which were previously used as only one very small part of human intelligence activities, were increasingly expected to play various roles in areas related to human emotion and feeling. This dissertation proposed a narrative generation system based on the concepts of narrative techniques and narrative strategies.

[Structure] This long paper consists of eight chapters. This dissertation presents the narrative generation system's study as one aspect of a new type of information processing aiming at advanced human interfaces, human sensitive (*kansei*) information processing, and various creativity aids. In Chapter 2, the author first described the problems of knowledge information processing and the new research directions towards solving them. The author also presented an approach focusing on human narrative generation abilities and situated it within the flow of knowledge information processing studies to demonstrate its significance. Further, the unique information characteristics in narrative and their relationships with new knowledge information processing were presented. Chapter 3 comprehensively showed a historical survey of the interdisciplinary research areas related to this dissertation. In particular, AI and cognitive science were described in detail, as they constitute the theoretical and technological backgrounds for the study of narrative generation systems, along with narrative and discourse studies, which included story grammar, narrative analysis, narrative generation, discourse analysis, and natural language processing. Further, several areas of narratology and literary theories, such as Propp's theory, were also referenced. These surveys showed one origin of the author's studies from the expansion of literary theory to post-narratology. Chapter 4 conducted a structural narrative analysis of a short story to extract and consider necessary knowledge for the generation of a computational narrative. Although this analysis described a narrative using a tree structure, it differentiated the narrative tree formed by the text of the narrative itself (the narrative tree of the plot or narrative discourse structure) from the narrative tree that represents the world described by the narrative (the narrative tree of the story). In this section, according to the basic structure of a narrative in narratology, the story was differentiated from the narrative discourse. Here, when a narrative was considered on the conceptual level, a kind of structural hierarchy emerged. Moreover, the author also analyzed the macro structure representing typical narrative patterns based on narratology. Chapter 5, based on the theory and results of the narrative analyses, considered a narrative generation process from a computational perspective. The chapter presented a narrative generation methodology based on the concepts of narrative technique and strategy, and presented types of narrative techniques and the parameters used by narrative strategies. In Chapter 6, the author developed an experimental narrative generation system based on the methods proposed in the fifth chapter and described the results of the experiments and evaluations. Knowledge bases, narrative techniques, and narrative strategies comprising the system were described in detail and examples of the usage of the system were shown. The narrative generation system presented here is positioned as a kind of prototype establishing the basic construction of the narrative generation mechanism created by the author to connect to the integrated narrative generation system (INGS). Additionally, methods or knowledge used in the *geinō* information system (GIS) were based on the ideas that used materials of folklore and *geinō* in Japan, although the first ideas were shown instead in the preceding master's dissertation instead of this doctoral dissertation. Chapter 7 discussed applications of narrative generation systems. From the viewpoints of new styles of human interfacing and the methods for aiding creative activities, this section discussed diverse and applicable possibilities of narrative generation systems, such as narrative creation tools and narrative representation media. In post-narratology, narrative generation's applications will reach the themes and topics of narrative production and creation. Although this dissertation did not directly refer to these themes and topics, it indicated these problems in the context of either narrative creation support or aid (Hori, 2007). In Chapter 8, the author stated the results from the proposed narrative generation system's application to the area of marketing and advertising creation. Based on the experiments and through discussion with several experts in a real advertising company, this chapter demonstrated the usefulness of the narrative generation system. In addition, although the research theme based on the application of narrative generation to advertising generation appeared in this first stage of research and has continued intermittently subsequently, this theme is currently reviving. In the current advertisement-related research, beyond the generation of the advertising in the narrow sense, the issue has been expanded to address the problem of the generalized representation of created narratives. As mentioned above, this dissertation was the first extensive manuscript that presented a framework of research and development to bridge directly to post-narratology in which both information-oriented researches centered on AI, cognitive science, and natural language processing, and human and social sciences, such as narratology, literary theories, and advertising studies, were unified in the design and development of narrative generation systems or the INGS and various narrative analyses.

[Overview] In the context described in **Structure**, this dissertation presented narrative generation systems as the study of novel information processing aimed at advanced human interfaces, *kansei* (sensitivity) information processing, and various types of creation aids. Overall, the key points of this research were the following: it systematically described the problems of knowledge information processing and new research directions as solutions; it presented an approach that was focused on human narrative generation ability situated in the context of knowledge information processing. In particular, this dissertation comprehensively surveyed areas of interdisciplinary research from the past to the present that were related to this narrative generation study and attempted a narrative structural analysis using a short story to extract and consider the knowledge that is needed by a computer for narrative generation. Based on the theories and the results of the narrative analysis, this dissertation considered the narrative generation process from the viewpoint of computer processing to present a methodology of narrative generation based on the following two concepts: "narrative techniques" and "narrative strategies." The types of narrative techniques and the parameters used for the narrative strategies were systematized. In the next stage of the research design, an experimental narrative generation system was developed and experiments and its evaluation were conducted. Moreover, from the viewpoint of new styles of human interface and methods of aiding human creativity, the dissertation discussed various applicable possibilities of narrative generation systems, including narrative creation tools and media related to narrative representation. The experimental systems applied the proposed narrative generation system to marketing and advertisement areas, which were also introduced. As mentioned above, this dissertation integrated informational approaches, such as AI, cognitive science, and natural language processing, with human and social scientific approaches, such as narratology, literary theories, and advertisement study, based on the axis of design and development of narrative generation systems and various narrative analyses for the systems. In this sense, it was the first extensive body of work that presented a framework of research and development connected to post-narratology. Additionally, the master's dissertation by the author (Ogata, 1992) was also a relatively long dissertation; it adopted a style that was the similar to the above doctoral dissertation as the main objective was the design and development of a story generation system using AI and cognitive science based on concepts from the humanities, such as Propp's literary theory, folktale theories, and folklore.

(A-3) Two Chapters in a Book

[Goals] Two relatively long papers (Ogata, 2003a, 2003b) in Yoshida (2003) describe systematically and comprehensively the entire narrative generation study of the author at this point.

[Structure] The book regarding complex systems, including economic and management areas, edited by Yoshida (2003) comprises two relatively long papers by the author as shown in **Overview**.

[Overview] The author's first paper (Chapter 5), "System Narratology," presents a new research concept that mixes computational sciences, such as cognitive science and artificial intelligence, and literary theories and narratology. Although the author no longer uses the term "system narratology," its intended use has morphed into the author's later terms for the integrated approach to narrative generation, including "informatics of narratology," "informational narratology," and "post-narratology." In particular, one of the author's objectives was to grasp narrative as a general concept showing human cognitive features instead of regarding it as mere a field of literature and, in so doing, to give a unified method to narrative-related studies that were previously conducted individually in different academic fields, according to the difference of genres and media. This chapter in the book discusses narrative as a complicated and diverse phenomena based on the multiple narrative structures model presented by the author and also deals with the technological methods that implement synthetically various narrative sides as simulation systems by computer and literary theories and narratology as a rich origin of narrative knowledge.

This paper also presents the concepts of narrative genres and "expanded literary theory." The expanded literary theory is an original concept by the author that aims to expand previous literary theories and narratology by introducing cognitive science, AI, and related informational methods. In this paper, the author also discusses literary theories as the generic name of literary research that pursues general literary principles instead of individual literary phenomena, such as each novelist or poet or each literary work or narrative. Similarly, the author describes narratology as the general principles of structures and functions of narratives as extreme cross-genre existences.

This recognition connects the systematic idea of narrative genres, that is "narrative genre system" stated in Chapter 2 of Ogata (2019d) (Ogata, 2019f).

Another, more technological purpose of the expanded literary theory is to introduce literary and narrative knowledge to the ontological knowledge bases in the author's narrative generation systems. Although previous literary theories and narratology have mainly proposed methods for reading and interpretation of literary texts, this chapter reconsiders them from the direction of narrative production based on the expanded literary theory. At the same time, previous literary theories and narratology were chiefly dependent on the proposition and verification of a concept through the application to real literary works and interpretation. However, the author's approach introduces systematic formality into the literary theories and narratology and tries to expand a theoretical concept to the forms in which the implementation by computer programming becomes possible. This introduces problems at the higher-level cognitive structure, and these have been studied through literary theories and narratology in cognitive science; narratology has important significance for establishing and extending the expanded literary theory.

In the second chapter (Chapter 11), the author explains several aspects of narrative generation using computer simulation and informational technologies. First, the second section, "Expanded Story Theory," considers the aspect of story in a narrative generation process by reconsidering the Propp-based story generation system that the author first studied for his master's dissertation. In this paper, the main modules of a narrative generation process are, at the broadest level, divided into story generation, narrative discourse generation, and narrative representation. These can be respectively treated as relatively individual mechanisms. In the Propp-based story generation system, special elements in Russian folktales were removed, and his functions theory, characters, and their action types were preferentially implemented because they were considered more universal and common elements. However, in the current view and in the opinion of the author, the special elements are probably elements related to the aspect of narrative contents, such as concrete characters, objects, and places. As Propp himself was a practical folklorist, he collected a variety of folkloric materials, including folktales. Narrative generation systems need to have and acquire this type of narrative content knowledge to be able to generate truly rich stories and narratives. According to such a viewpoint's transposition or alteration, the author currently undertakes such collections and definitions of story content knowledge (Fujiwara & Ogata, 2014; Fujiwara, Ono, & Ogata, 2015; Ito, Ono, & Ogata, 2018a, 2018b; Ogata, Fujiwara, Imabuchi, 2014; Ono, Ogata, & Ito, 2019), mainly using

Japanese folktale studies by Keigo Seki (1899–1990) and other researches (Ozawa, 1976, 1997, 1999; Seki, 1957, 1966, 1977, 1978a, 1978b, 1978c, 1978d, 1978e, 1979a, 1979b, 1979c, 1979d, 1980; Seki, Nomura, & Ōshima, 1979, 1980).

In next section, "Expanded Narrative Discourse Theory," the author reconsiders the narrative discourse theory by Gerald Genette (1930–2018), one of the most influential theorists in narratology, in view of the expanded literary theory. Genette's narrative discourse theory systematically classified various methods of how to narrate a story—i.e., narrative discourse techniques. The theory was based on the idea that the method of narration of the content of a story is an important theme in addition to what is being narrated. The narrative generation study by the author interprets the relation of the two aspects (story and narrative discourse) as the structural transformation from a story to narrative discourses to build a bridge to design and implement the computer program. This chapter deals with the problem of temporal order transformation and formalizes Genette's theory. Moreover, the author extends the part that Genette did not cover with a unique analysis and investigates the validity of this extended theory using computer simulation. This narrative analysis is performed using material from movie works. Although Genette's theory and other many narratological theories were mainly based on narratives in written language, especially novels, the author adds visual narratives into the analysis from the viewpoint of cross-media narrative concepts and also aims to expand the literary theories in this sense.

In the fourth section, "The Problems of Narrative Communication and Hypertext-type Narratives," the author treats the problem of hypertext novels in which a narrative is developed by the narratee's or reader's subjective selection as material for considering the idea that each narrative is generated through the mutual relations between the narrator and narratee. Although this study presents two experimental systems in the form of hypertext, the author introduces the thought of the narrative discourse theory as a theory for constructing on a computer a "hypercomic" that represents a *manga* work based on the hypertext method. The concept of hypertext has frequently been discussed in relation to the reception theory in literary theories. Moreover, this paper's strategy, which overlaps reception theory with narrative discourse theory, contributes to extend the range of discussion to the problem of the interface among different literary theories in the expanded literary theory. Further, the fifth section, "The Theory of Multiple Systems: The *Geinō* Information System," presents the concept of the *Geinō* Information System (GIS), an implementation image on a computer of the multiple narrative structures model. In this study, the author implements the experimental system of the part that adds life stories to the *geinō* characters in the GIS to model a different aspect from story and narrative discourse in the entire narrative generation framework.

As discussed above, two of the chapters synthetically consider many previous themes in preparation for later research development from the perspective of a research system of narrative generation.

[B] An Introduction to Informatics of Narratology

[Goals] *An Introduction to Informatics of Narratology: Around the Thoughts and Techniques in Narrative Generation Systems* (Ogata & Kanai, 2010) comprises a collection focused on the "informatics of narratology," which is an approach to narratology that is dependent on narrative generation and introduces AI and cognitive science.

[Structure] This book was co-authored with Akihito Kanai who is a cognitive scientist of image narrative and image rhetoric. Chapters 4, 5, 6 and 7 were papers by Ogata and Chapters 2 and 8 were co-authored papers. Further, this book is divides into two parts, and Part I (Chapter 1 to 4) and Part II (Chapter 5 to 8). They have respective individual introductory chapters.

[Overview] In these papers, Ogata thoroughly investigated the problems of consciousness and future vision in the context of the research and development of narrative generation systems, including philosophical elements. Generally, in contrast to ordinary journal papers and conference presentations, this book provided an in-depth description of philosophical thoughts, techniques, and technologies for narrative generation. The conceptual goal was to extend the concept of "narratology" based on informational methods and the technological vision was to develop narrative generation systems. In this book, the relatively long papers by Ogata primarily discussed the following contents: narrative generation systems, including an advertisement generation system and its narrative elements or components; the philosophical ideas of the type of design and development of the systems, such as "fluidity and fixation" and "multiple narrative structures"; and an analysis and discussion on the narrative generation phenomena of humans and societies based on the framework of a narrative generation system.

The introductory chapter to Part I, which describes research backgrounds through the writing process of this book in this sub-section referred to a basic issue relating to narratology and narrative generation. The issue is that the text of a narrative is divided into aspects of story and narrative discourse and previous information studies, such as AI and cognitive science, have emphasized the former. However, the author also focused on the latter. The second chapter (Chapter 2) comprehensively described the method for applying a narrative generation system to that of automatic advertising and referred to methods of concept description related to events, worlds, and narratives. At the same time, one of the objectives of this chapter was, through an explanation of the advertising narrative generation, to discuss a more general narrative generation system used as the engine of the advertising application. The central concept was that advertising narratives are required to represent both the aspects of story and brand. This chapter proposed "the rhetoric of cutting a story" and "the rhetoric of familiarization and de-familiarization" in the representation as its concrete methods. Chapter 4 consisted of a paper related to the author's deepest motivation for studying narrative generation systems as informatics and referred to many novels, stories, and other literary works that the author had read previously on the relationships by presenting the concepts of "fluidity and fixation." This was the notion that literature (further, all documents written and represented by humans) needed to be treated as two directions between the fluidity corresponding to human thought and representation processes and the fixation of books and documents. Moreover, this consideration indicated that future narrative generation systems and computational contents must be constructed according to the above two aspects.

The introduction to Part II states that different technologies and science fields also have unique and ambiguous aspects as a kind of thought and philosophy and, at the same time, such parts rather contribute toward determining the actual realization forms in technologies and sciences. This chapter asserted that the author focuses on the thinking and philosophical elements in the narrative generation researches, in addition to narrative generation technologies. Chapter 5 described the author's referential backgrounds and several basic ideas. First, the chapter showed that the narrative generation study by the author was formed as an interdisciplinary field, which included folklore, anthropology, history, philosophy. Further, it, extending beyond the disciplines, includes AI, cognitive science, and natural language processing. This chapter also presented several viewpoints, including a multiple narrative structures model, narratives as simulations of society and humans, and others. Moreover, although the chapter could not discuss their philosophical and literary contents in an in-depth manner, the author referenced two great authors in modern Japan, Kumagusu Minakata (1867–1941), a folklorist and biologist, and Kenji Miyazawa (1896–1933), a poet and story teller. They were thinkers and philosophers and their works covered diverse academic and literary or narrative genres. Narratology and narrative itself can embody interdisciplinary, transdisciplinary, and over-disciplinary natures. Chapter 6 explained, in a top-down manner, the entire process of the narrative generation system, which was connected to the current INGS (Integrated Narrative Generation System) that the author has been developing. The description corresponded the conceptual design of the system. The largest modules comprised story mechanism, narrative discourse mechanism, and narrative representation mechanism, and each of these included sub-modules. At the same time, these modules were generated in a free and flexible order instead of one that was fixed or sequential in its execution. Furthermore, this chapter also proposed a longer and broader plan for the "narratology of informatics" than what was originally written in the first place. Although, currently, the described contents did not realize it, the author was planning to write a system which consists of a broader narrative generation study in the following books. The plan was considered as works which lay the foundation for a longer book on the system of narrative generation. In Chapter 7, the author regarded the multiplicity as a key concept. The multiplicity has creative functions that allow for the development of new viewpoints regarding the world and human beings by providing new perspectives on reality and the world instead of maintaining a monotonic viewpoint. The ultimate objective of narrative generation systems is to develop machines that support such multiplicity using computers. Although Chapter 2 used advertising narrative generation as its material, Chapter 8 continuously stated concrete methods of advertising narrative generation. Advertisement is not merely a field in which narrative generation may be applied; it is also a dense narrative genre and involves many functions for narrative generation. Namely, advertising generation used as the material in this book was a sort of microcosmic version for narrative generation. Further, this chapter overviewed previous advertising studies that were divided into several areas from the viewpoint of generation.

[C] Computational and Cognitive Approaches to Narratology

[Goals] The second book by the author, *Computational and Cognitive Approaches to Narratology* (Ogata & Akimoto, 2016), simultaneously discusses the following issues: narrative-related information and computational technologies, and the cognitive mechanisms and analyses; and theories as the foundation of technology, particularly those from the humanities and social sciences.

[Structure] This book contains 16 chapters or respectively individual papers and the author's papers correspond to Chapters 1, 3, and 16. In more detailed structure, this book was composed of the following three sections. First, "Generation and Creation of Narratives" contains five chapters for discussion that were centered on the generative, creative, or productive aspects of narratives. Next, "Language and Narrative" comprises four papers that are primarily relevant to the fields of linguistics and natural language processing. Finally, "Cognition, Culture, and Narrative" explores topics on the aspects of reception and the cultural nature of narratives. One characteristics of the book is that several chapters include Japanese cultural elements.

[Overview] This book especially focused on the systems or analyses of the narrative or story generation process and its applications to advertisement, management, welfare, literature, and art (movie, *rakugo,* a form of Japanese verbal entertainment, and *kabuki*, a form of traditional Japanese drama with dance and music); the elements within its scope were narratology, literary theories, poetics, semiotics, rhetoric, philosophy, linguistics, psychology, marketing and advertisement. The media and genres that could be examined in the narrative analysis and used in its applications cover a very broad range. Further, this book's chapters included diverse studies with various backgrounds, objectives, and methodologies and were based on three processes in narrative phenomena: generation, text, and reception. In terms of social organizations, these processes correspond to production, content, and consumption. In the above broad areas, diverse narrative genres were explored. Moreover, the chapters were characterized by various objectives and approaches. Technological researches included those which focused on computational modeling and implementation. Conversely, the experiential researches of human narrative reception encapsulated cognitive or psychological studies, conceptual considerations, design, critiques of narrative phenomena, and qualitative and quantitative analyses. As stated above, although this was a co-authored book with other writers, two papers by Ogata dealt with the Integrated Narrative Generation System (INGS) and the narrative generation of *kabuki*. In particular, the former systematically focused on an actual computer program and provided a relatively detailed description of related diverse topics, such as the background and foundation of the system's design and development. Thus, this paper was equivalent to a synthetic summary of the research and development corresponding to post-narratology and the narratology of narrative generation by Ogata.

First, two chapters by the author are introduced. Chapter 1 by Ogata (2016a) examines and discussed the interdisciplinary approach that has been referenced using the terms, expanded literary theory, informatics of narratology, and post-narratology. The first part of the chapter provided comprehensive overviews of narratology and related literary theories, computational and cognitive theories regarding narratology and narrative generation, and narrative generation systems themselves. This section collected studies that were centered on aspects of generation or the creation of narratives. The methodological and fundamental mechanisms of narratives and narrative generation were a main theme in the studies contained in *Poetics* by Aristotle (384 BC–332 BC) (1997) and also in contemporary narratology and literary theories. Further, the computational and cognitive modeling of narrative generation and narrative abilities were important topics for AI and cognitive science. For example, the work of Propp (1968), Genette (1972), and the psychoanalysis by Sigmund Freud (1856–1939) (1900) have influenced the fields of AI and cognitive science. The second section of this chapter provided practical examples of computational approaches to narratology through narrative generation researches and described the components of INGS), which corresponded to systematization as the level of the computer program in Ogata's research system.

Chapter 15 by Ogata (2016b) addresses the problem of *kabuki*'s narrative generation-reception and narrative production-consumption modeling from the viewpoint of narrative generation system. The fundamental concept of this modeling was narrative "multiplicity" or a "multiple narrative structures model." Therefore, this chapter's direct goal was to make a first attempt at investigating *kabuki*'s generation consumption models, which consisted of reception and production, based on the above concepts. This system model includes multiple narrative generation and production mechanisms or processes, namely, the multiple narrative structures model. Introductory knowledge, including the history and basic terms, were referenced as the above discussion's background. The next primary goal of this chapter was to describe the phenomenon of multiple narrative structures of *kabuki* using actual examples. Although this chapter simply indicated them, Ogata's ultimate goal was to conduct experiments on the social distribution of INGS, which was the narrative generation synthesizer used in Ogata's research. With the term, social distribution experiment, Ogata refers to a new type of business model for social applications that use the Web, for narrative generation systems themselves, or content systems that include narrative generation systems. Finally, this chapter set the expectation for future researches on the social distribution of INGS through a more advanced understanding of *kabuki* and the modeling of multiple narrative generation.

Chapter 3 was written by Akimoto and Ogata (2016). This chapter revealed the design plan of their socially open narrative generation system, which demonstrated the concept that diverse narrative collections were generated through a collaborative work based on the interaction between the narrative generation program and people. Originally, the multiple narrative structures model by Ogata was related to the social narrative generation, so this chapter was partially related to the narrative model.

The author simply introduces several chapters by the other authors. These chapters by other researchers are also related to the author's narrative generation study in the point that are according to the computational and cognitive approaches to narratology as the common theme of the book. Chapter 2 by Abe (2016) applied the "intertextuality" by Julia Kristeva (1941–) (1980) as a key idea for the computational modeling of literary work generation to explore the possibilities and practices of literary work generation using a computer. Abe and the research group's original idea was *kotoba kōgaku,* or language sense processing engineering relevant to the creativity of the literary computation; this chapter also discussed the possibilities of *waka* and *haiku,* which are Japanese traditional short-form poem. Chapter 4 by Kawamura (2016) showed a narrative generation technology in the advertising field. The chapter introduced concepts and technologies that assisted a commercial image production support system, including image editing techniques and rhetorical knowledge based on the structural analysis of television commercial films. The chapter also used generated advertising images to show the experimental results of way in which advertising stories influenced audiences. Kawamura also demonstrated a strong interest in the social aspect of Ogata's multiple narrative generation structures. In the chapters contained in "Language and Narrative," Chapters 7 (Takeuchi, 2016) and 8 (Seraku, & Ohtani, 2016) approached narratives from the standpoint of natural language processing and linguistics and discussed the relations between narrative and language based on formal and computational studies on linguistic semantics and syntax. Alternatively, Chapters 9 (Seino, Haruna, & Ishizaki, 2016) and 10 (Tsuchiyama, 2016) utilized natural language processing techniques to analyze narratives generated or created by people to understand the concepts of roles such as author or narrator. In particular, Chapter 7 discussed the possibilities and applications of the language resource or semantic dictionary of language developed by Takeuchi in the context of narrative comprehension and generation mechanisms. This chapter introduced the development of a

Japanese language resource system for narrative processing. The resource includes the determination of states, actions, and state changes in narrative texts for narrative processing. Although INGS by Ogata contains conceptual dictionaries, this language resource by Takeuchi was a similar study in its direction and concept. Chapter 9 computationally analyzed clinical knowledge toward effective employment support of handicapped; in their work, many narratives and histories were acquired by questionnaire survey sent through e-mail. Chapter 10 consisted of stylistic research on *Genji Monogatari* (1993, 1994, 1995, 1996, 1997) in the Heian era (794–1185 or 1192). Stylistics was a quantitative analytical method used to survey the writing style of one or more authors, and was particularly useful in identifying a work's author, and determining the creative age and order of a work, among other functions. This chapter addressed the problem of who really wrote *Genji Monogatari* and whether the work was written by an author other than Murasaki Shikibu (She wrote *Genji Monogatari* in the middle of the Heian era. However, her correct dates of birth and death are unknown). The last section mainly focused on perspectives of narrative reception and cultural characteristics. In literature, the narrative aspect of reception was the primary focus and was examined through the theory of reception and the reader-oriented criticisms provided by Wolfgang Iser (1926–2007) (1976), Hans Robert Jauss (1921–1997) (1970), and other researchers. Chapters in this section experientially considered the problem of narrative reception based on cognitive science, neuro science, and information technologies, among others. Moreover, several chapters also discussed rhetorical techniques for computational narrative generation. Alternatively, several chapters employed cultural perspectives, especially narrative genres and systems originating in Japanese culture, including *rakugo*, *kabuki*, and *emoji* communication. In particular, Chapter 15 by Kanai (2016) discussed the nonstory's perspective on narratives. Kanai used his method of cutting techniques from a story to analyze the cognitive effects experienced by the receivers of nostalgic image narratives. Here, the central narratological or literary theory was defamiliarization. Knowledge derived from this analysis, according to the nonstory's narrative concept by Kanai, was applied to the computational model to construct the image based on the narrative rhetoric of nostalgia.

[D] Content Generation Through Narrative Communication and Simulation

[Goals] The next book, *Content Generation Through Narrative Communication and Simulation* (Ogata & Asakawa, 2018a), deals with models, systems, theories, and analyses for the generation of contents, especially narrative contents, in various genres.

[Structure] This book is composed of the following four parts. The introductory part (Part I) includes a chapter by the authors of this book. The next part (Part II), "Systems and Models for Narrative Generation and Simulation" discusses systems and models for generating and simulating narratives. Furthermore, seven chapters in Part III, "Computational and Cognitive Modeling to Narrative Communication and Simulation," provides narrative models using various narrative genres and methods. Three chapters in the last Part IV, "Analyses, Models, and Systems for Social Narratives," focuses on narrative social aspects that frequently overlap with marketing, advertisement, and management.

[Overview] In this book, the term "narrative" was used in a broad sense to include folktales, various types of narration (descriptions of travel experiences, the taste of Japanese *sake*, and free-form answers of people with developmental disabilities), novels, poems, metaphors, *haiku*s, dramas and scenarios, *kabuki* and performing arts, pictures, games, advertisements, social networks, management problem solving, and narratives in the narrow sense. "Narrative content" refers to the informational content conveyed to receivers such as readers and audiences; the concept is associated with reception and consumption through the communication process between the sender and the receiver.

Further, this book also included various narrative levels, namely, the personal and social or collective levels and the automatic, semi-automatic, and interactive levels. This book examined the challenges of automated narrative content generation based on narrative communication and simulation technologies. This research plan was conducted through interdisciplinary methods from such areas as informatics, literary studies, and social sciences. Results of this study contributed to the scientific and technological expansion of information science, engineering and literary studies in the future. Additionally, its results may also be effectively applied to novel digital content ideas as well as creations in more practical fields. This book presented theories, models, systems, and analyses of the generation of narrative contents in various genres based on interdisciplinary methods from informatics, literary studies, and social sciences fields. The chapters in this book discussed the wide variety of narrative forms mentioned above to demonstrate the diversity of narrative genres. At the same time, the book showed that various narrative phenomena could be analyzed, modeled, and systematized based on a methodological framework informed by the method of narrative content generation. This methodological framework integrated previous academic genres into a group. Chapters in this book concerned a diverse range of academic areas, such as informatics (AI, cognitive science, game and entertainment informatics, natural language processing, media technologies, human creative aid, collaborative problem solving, and neuroscience and deep learning), literary (narratology, literary theories, *kabuki* studies, film studies, logics, philosophy, linguistics, discourse theory, and creativity), and social science fields (management science, marketing and advertising studies, communication studies). However, these can all be grouped under the perspective of narrative content generation study. In one of the author's chapters (Ogata, 2018a), the author conducted analyses of a novel by Yukio Mishima (1925–1970) and *kabuki* toward a discussion on integrating the results into narrative generation systems. Although the author could not describe the above aspects as closely connected themes, Mishima and *kabuki* were positioned as two very important narratological symbols that were essentially related to each other.

The following description shows all chapters by Ogata and selected chapters by the other authors.

The introductory chapter (Chapter 1 in Part I) by Ogata and Asakawa (2018b) discussed various problems, including content generation, narrative generation, and narrative simulation, centered on INGS and *geinō* information system by Ogata. In this sense, this chapter demonstrated a synthesized approach and paradigm of the narrative generation studies in this book. These results will be linked to the distributive and social direction of INGS, especially through the *geinō* information system, GIS. This chapter also referred to neural network modeling as a new technique used for narrative generation systems.

Part II includes the following chapters. Chapter 2 by Ogata (2018a) provided a comprehensive description of the structure of his narrative generation study. This chapter had an indirect goal through which the whole of narrative generation study was concisely shown using concrete examples, in addition to the direct goal of considering and discussing related contents. The entire framework of narrative analyses was shown in relation with narrative generation systems, namely, INGS and GIS. Moreover, this chapter conducted analyses of several aspects of a novel by Yukio Mishima and *kabuki* as two examples that were related to either narrative multiplicity or the multiple narrative structures model by Ogata. Additionally, this chapter provided the direction for partially integrating these narrative analysis results into the above narrative generation systems. Thus, this chapter mainly addressed the narrative genres of novel and *kabuki* and their relation with diverse research fields such as AI, cognitive science, narratology, literary theories, and the study of *kabuki*.

The Chapter 3 by Kanai (2018a) showed that non-deterministic elements in the rhetoric created by the techniques of story cutting and nostalgia are essential techniques in narrative simulation; these are used to construct nonstory images by referring to Sergei Eisenstein (1898–1948), Seymour Chatman (1928–2015), and Gilles Deleuze (1925–1995), in addition to Kanai's theories of image rhetoric, story simulation, and narrative simulation. For Kanai, the narrative simulation of image could be investigated based on cognitive effects created through the interaction among story, narrative discourse, and rhetoric. Techniques of nonstory image editing were classified based on categories of rhythm and nostalgia was used for the simulation of narrative image rhetoric. According to Kanai, various types of nostalgia frequently emerge from deterministic places and times and non-deterministic nonstory rhetoric. Non-nostalgic narratives emerging from a non-deterministic place and time could be simulated with nonstory-like editing methods, based on rhythm categories. Based on his theoretical framework, this chapter presented an image narrative simulation mechanism and discussed the results of this approach and its future directions.

Ono and Ogata (2018a), in Chapter 4, presented an approach to automated game systems aimed at narrative generation through the interaction between the system and its users. In this approach, they developed an automatic narrative generation game using the method of table-top (table-talk) role playing games (TRPG). TRPG is a type of analogue game based on the interactive processes of actual humans. A basic idea from the theoretical perspective is that the various gaps between an original story and the changed story contribute to the creation of impressive effects, especially those that constitute a kind of surprise, to result in interesting stories or narratives. This proposed system and its mechanisms were implemented using the INGS by Ogata.

The third part (Part III) was mainly composed of the following chapters. The first chapter by Abe (2018), Chapter 5, discussed the creativity in an "innovative marketplace on data jackets (IMDJ)." The IMDJ was an innovative game in which new production was acquired through various techniques, materials, and the combination of previous products. Abe described the problem from the viewpoint of abduction. Further, he also discussed the role of comfortable communication in useful and creative situations. The goal of Akimoto's (2018) Chapter 7 was to develop a computational model of narrative activity by artificial agents based on the narrative structures provided by hierarchical dimensions and hierarchical narrative graphs. The proposed model includes two elements: the "mental world" as the internal representation of a world formed from many stories and "narration" as the actions taken by the representational parts of the mental world. This chapter represented basic concepts, including narrative ability, mental world, narration, and system framework, based on computational narrative studies such as Roger Schank (1946–), cognitive approaches, and narrative discourse generation. Chapter 8 by Fukushima (2018) presented a comprehensive "multiple faces" representation, or the model describing the taste of Japanese *sake* as the key modes to understanding mental processing of information. This chapter bridged the area of taste with that of language using the method of first-person description drawn from recent cognitive science research. Fukushima achieved this goal by using both language description and image or picture representation. This study tried to extend the field of cognitive science and post-narratological research by showing a new approach to describe and narrate an ambiguous object, which is the taste of *sake*. Nitta (2018) (Chapter 9) formally examined *haiku*, which is, in principle, a very short poem or verse with a fixed form aiming to model future automated *haiku* generation or simulation. The model involved processes of *haiku* interpretation and composition using many actual *haiku* works, an ontology of season words or *haiku* knowledge base, and functional grammar. Utsumi (2018), in Chapter 10, focused on the literary metaphor that appeared in literary works and created rich meanings and aesthetic effects. This unified approach to literary metaphor pursued the development of basic methods for the understanding, generation, and simulation of metaphor. This chapter introduced unified viewpoints on comprehension and the production of literary metaphor based on Utsumi's metaphor theories, namely, the "theory of the interpretive diversity of normative metaphors" and the "theory of the indirect categorization of predicative metaphors."

The fourth part (Part IV) contains the following chapters. First, Seino, Enomoto, and Miyazawa (2018), in Chapter 12, determined the indicators that supported the employment of students with ASD (autism spectrum disorder) by conducting text mining and narrative analysis. This chapter regarded the text data of free-form answering descriptions regarding the employment support of these students as a kind of narrative. In addition to the acquisition of diverse indicators for actual employment support processing, they demonstrated the relations of narrative and story generation from their analytical text studies. The knowledge of narrative and story acquired in this study enabled them to bridge their limited viewpoints to broader, alternative possibilities, namely, stories or narratives. Next, for the advertising and marketing aspects of narrative communication, Chapter 13 by Kawamura (2018a) presented an approach for developing a social and consumptive simulation model. In particular, he concentrated on the reception or consumption aspects of narrative communication and analyzed consumer actions. The differences in many marketing and advertising studies were the points that provided two types of narrative communication strategies, which were sender-driven and receiver-driven, as well as more concrete rhetorical techniques for an automated advertising model or system. In the final chapter, Chapter 15 by Shirozu and Matsushita (2018), explored a system that supports middle-aged travelers by externalizing the information concerning their travel experiences and accumulating this information using narrative forms.

[E] Informational Narratology

[Goals] *Informational Narratology: Artificial Intelligence/Cognition/Social Process and Narrative Generation* (Ogata, Kawamura, & Kanai, 2018a) includes similar problems in this book regarding the development of the "narratology of narrative generation" or post-narratology. However, the book by Ogata, Kawamura, and Kanai limits the narrative genres that are analyzed and considered to movie (image), advertisement (image), and drama (*kabuki*), but provides an in-depth and broad scope of analysis and considerations.

[Structure] This book was composed of the following three parts: "Introduction and Theories" (Part I: Chapter 1 to 6), "Narrative Analyses" (Part II: Chapter 7 to 12), and "Narrative Generation" (Part III: Chapter 13 to 17). First, Part I introduced the attempts and framework for modeling narratives by three authors to lay the foundation for the discussion that followed. Next, Ogata discussed his narrative generation studies from the viewpoint of synthetic modeling and the systematization of diverse narratives phenomena. Following Ogata's analysis of an abstract or general narrative generation process, he presented the methods for synthesizing the results of the analysis. In Kanai's chapters, he referenced narrative techniques and rhetoric and discussed them from the perspective of the positive use of the gaps between a narrative sender and receiver which exist based on differences in their recognition of the story and image. In these chapters, Kawamura's primary topic was the modeling of social narrative generation processes. In Part II, Ogata wrote two continuous chapters about *kabuki* and provided rich, concrete examples to extend his previous *kabuki*-related chapter (Ogata 2016b). Further, in this book, Ogata also extended and revised the description of his Integrated Narrative Generation System (INGS) (Ogata, 2016a, 2018a) and explored the next or future vision for its social/external and personal/internal developments. In these regards, this book is a pioneering work of post-narratology and also provides a partial view of the construction of narratology based on Ogata's description of narrative generation.

[Overview] Chapter 1, which is co-authored by three authors in this book, Kawamura, Ogata, and Kanai (2018) introduced the framework and description of the authors' attempts to attempt to model narratives. In an integrated manner, Ogata progressed the discussion from the viewpoint of modeling and systematizing to that of diverse narratives. In this chapter, although Kanai referred to narrative techniques and rhetoric, he discussed them from a viewpoint that focused on the techniques for the positive use of the gaps in story and image between the sender and receiver. Conversely, Kawamura adopted the viewpoint of modeling the process in which a narrative is socially generated. The "Analyses" part includes chapters that concretely analyze elements, structures, and the generation techniques of narratives such as novels, images, and *geinō*. The final part, "Generation," contains attempts at systems of concrete narrative content, including documentaries, nostalgic narratives, and advertising images. In addition, although narratives and narrative generation discussed were academic objects, the three authors were also drawing from individualized personal and private narrative experiences of their academic careers, which operated in the background. In Chapter 6, each author selected a narrative-related work or a series of works to introduce their personal and private experiences and openly discussed their relations with the chapters they had written. In particular, Ogata, Kawamura, and Kanai (2018b) respectively, used a *kabuki* work, *Kagotsurube Sato no Eizame* (1888, Tokyo: Chitoseza) (1929), advertising films, *Okeihan* series by Keihan Electric Railway (Kyoto and Osaka areas), and *ISAM LIVE* and *Layer3*.

The seven chapters by Ogata are summarized first. In Chapter 2, Ogata (2018b) introduced the multiple narrative structures model, which views narrative phenomena as the holistic representation of a person who is comprised of many levels corresponding to narrative generation and reception systems, and on the social level, corresponds to narrative production and consumption systems. He created a future narrative image of a human-machine symbiotic system in which narrators and receivers were involved as AI. Chapter 3 (Ogata, 2018c) tried to anatomize narrative elements or components based on concepts of "visible narratives" and "invisible narratives." Here, "visible" and "invisible" means "recognized" and "unrecognized." Chapter 3 considered methods for synthesizing separated narrative elements. The idea driving the division and synthesis of narrative elements was systematized in INGS, which was described in Chapter 13 (Ogata, 2018f) of this book.

Chapter 11 (Ogata, 2018d), that discussed *kabuki* as a new narrative genre, mainly showed the motivation of the author toward the analyses of *kabuki*. In particular, this chapter conceptually considered and discussed the process that, at a glance, through plurality and multiplicity, arbitrary editing characteristics in *kabuki* acquire a kind of existential nature that arouses deep narrative feelings and emotions. Continuous Chapter 12 (Ogata, 2018e) considered, in detail, the multiplicity in *kabuki*; namely, it considered the multiple narrative structures model of *kabuki* from fifteen perspectives or topics. These topics used examples of actual *kabuki* and *ningyō jōruri* works, and people, such as characters, actors, authors or writers, historical personas. The final theme was showing the strategic idea for bridging the acquired knowledge with narrative generation systems. Further, the description in this chapter was supported by many *kabuki* researchers and writers, including Tamotsu Watanabe (1936–), who is a contemporary *kabuki* critic.

Regarding Chapter 13, Ogata has been advancing the existing works to integrate previous results, including conceptual considerations of narratives and narrative generation, narrative analyses and modeling, and surveys of narratology and literary theories, into INGS. Here, the entire image of this system was shown. INGS utilized important concepts in narratives, such as story, event, and narrative discourse, as well as several existing models of narratology themselves. Chapter 16 (Ogata, 2018g) considered a direction for using and concretely applying INGS. This was the direction of production and distribution on the social level, which may be realized by introducing the framework of the system that concretizes the level of narrative production and consumption in the multiple narrative structures model, namely, the level of the *Geinō* Information System (GIS). Alternatively, Chapter 17 (Ogata, 2018h) described the internal narrative generation aiming at "*Watakushi Monogatari* (My Story or My Narrative." Although the *Watakushi Monogatari* was different from "*Watakushi Shōsetsu*" ("I-Novel:" a kind of autobiographical story) in modern Japanese literature, the concept of *Watakushi Monogatari* refers to the vision and plan of narratives that have yet to be generated, produced, and distributed by the utilization, editing, and construction of the information related to the author through his narrative generation systems.

Next, Kanai wrote the following four chapters. First, his concept of a "cutting technique" emphasizes all aspects, except for story, in narratives such as novels and movies. Chapter 4 (Kanai, 2018b) utilized the cutting technique for narrative generation using movies as the material. The cutting technique positively uses the story and image gaps between the sender and receiver. In addition, a narrative that includes one or more cuttings has a deep relation with the past and nostalgia as cognitive effects. Next, a receiver has several "viewpoints," when he or she accepts an image. The viewpoint selected by the receiver connects to the change in the narrative that he or she recognizes. Chapter 7 (Kanai, 2018c), using movies by Lars von Trier (1956–) and music videos by Björk Guðmundsdóttir (1965–), discussed the multiplicity of the receiver's narratives and the significance in viewing the differences in perception positively. Chapter 8 (Kanai, 2018d) discussed the significance in seizing an image that is focused on the receiver's cognition using methods from the field of cognitive science through reality, which is described as "a moment in an image." In particular, the chapter described both the reality that was an important characteristic in documentary works and the reality of image that was changed by the application of music based on recognizing a story as familiar and viewing it as strange or foreign. Using a documentary work on the past as its material, Chapter 14 (Kanai, 2018e) considered the method for generating narrative works of images in the past using a computer program; this examination was conducted from the perspective of the various types of generated nostalgias, instead of from the viewpoint of story. The above discussion revealed the methods for dealing with diverse and complex narratives through nostalgia about the past, including different and negative feelings, in addition to the "naive nostalgias."

Kawamura wrote the following four chapters. First, diverse subjects are related through the contents business and social narratives are generated through the interaction among diverse subjects. Chapter 5 (Kawamura, 2018b) treated advertising communication as an example of content business. Kawamura modeled characteristics of an advertising communication process from the viewpoint of narrative generation on the social level and considered the process through which a narrative was generated through social communication. Chapter 9 (Kawamura, 2018c), using the advertising image contents generated by receivers as its material, analyzed the elements, components, and structures of advertising content. In particular, he extracted concrete elements using commercial films of beers and other products (coffee, green tee, personal phones, and personal computers) and analyzed the structures of advertising stories and editing techniques to consider the systems and tendencies of the product categories. Based on advertising contents generated by the sender, Chapter 10 (Kawamura, 2018d) analyzed and examined which images the receivers (consumers) accepted and which responses the receivers returned. First, Kawamura showed the idea of advertising image rhetoric as the combination of advertising image techniques. Next, the relationships between advertising image techniques or rhetoric and reception images or responses were described; the structures of advertising image techniques, their reception, and responses were classified. Based on the analyses in Chapters 9 and 10, Chapter 15 (Kawamura, 2018e) introduced image techniques and rhetoric used by advertising creators into the system and explained the development of an advertising image generation system based on the image techniques or rhetoric and the preferences of consumers. Kawamura presented the plan for advertising image production system based on the collaboration between image techniques, the rhetoric by advertising creators, and the preferences of consumers to describe the system's experiments and results.

[F] Post-Narratology Through Computational and Cognitive Approaches

[Goals] This edited book, *Post-Narratology Through Computational and Cognitive Approaches* (Ogata & Akimoto, 2019), is an enhanced version of *Computational and Cognitive Approaches to Narratology* (Ogata & Akimoto, 2016). As one improvement on the previous version, the concept of a new research system called "post-narratology" is introduced by Ogata (2019b). Post-narratology refers to a synthetic approach to narratives focused on its generative aspect, by inheriting the traditions of previous narratology, literary theories, AI, and cognitive science. As previously described, narratives are ubiquitous in human societies and are studied in various academic fields. Post-narratology is motivated by the requirement of integrating various academic fields in the past into a new research system. More detailed descriptions of the notion of post-narratology are provided in Chapter 2 of this book. Moreover, there are several published books relevant to post-narratology (Ogata & Asakawa, 2018a; Ogata & Kanai, 2010; Ogata, Kawamura, & Kanai, 2018a). Although two of them were written in Japanese, their essences are also introduced in the first chapter. This book contains a variety of studies relevant to narratives based on different research backgrounds, objectives, and methodologies and across the three phases of narrative phenomenon: generation, text, and reception. These studies also examine various narrative genres, including novels, poems, films, comics, commercials, business communications, diaries or personal narratives, and *kabuki*. Moreover, they include various objectives and research approaches: technological studies, including computational modeling and the implementation of narrative processing, conceptual thought, design, and criticism on narrative phenomenon, and qualitative or quantitative analysis of narrative text. This book may seem chaotic. However, throughout its pages runs a common theme: narrative.

[Structure] This book contains thirteen chapters in all. Nine of them (Chapters 1, 3, 4, 5, 6, 7, 8, 9, and 11) are chapters enhanced from the 2016 edition. The other four chapters (Chapters 2, 10, 12, and 13) are newly produced for this book. This book is divided into four parts: Part I "Introduction" (Chapters 1 and 2); Part II "Conceptual Explorations of Narrative Generation" (Chapters 3, 4, and 5); Part III "Computational and Cognitive Systems of Narrative Generation" (Chapter 6, 7, and 8); Part IV "Cognitive and Social Analyses of Human Narratives (Chapter 9, 10, 11, 12, and 13). The theme of each part and outlines of the contained chapters are described below.

[Overview] This book discusses issues of narrative-related information and computational technologies, cognitive mechanisms and analyses, and theories in the context of technology, particularly those from humanities and the social sciences. This book puts special focus on the systems or analysis of narrative- or story-generation processes and its applications.

Chapter 1 by Ogata (2019a) surveyed and discussed interdisciplinary approaches to primarily AI-based computational narrative or story generation systems by introducing cognitive science and narratology and related literary theories. This is also an expansion of ideas, theories, and systems of the "informatics of narratology" presented by Ogata and Kanai (2010). The first part of this chapter provided a general description of narrative or story generation from the perspective of the framework for narrative-related research. The second part presented the research process, theoretical concepts, designs, and implementations of narrative generation. The first part sought to provide a comprehensive overview of narratology and the relevant literary theories, computational and cognitive theories and techniques related to narratology and narrative generation, and narrative generation systems. Readers will be able to understand comprehensively the overview of computational and cognitive approaches to narratology and to aim at the extension of this novel research area in the future. The second part presented an example of the practices of computational approach to narratology through narrative generation studies. It also presented, in relative detail, the components that constitute a systematic study for narrative generation and INGS of all of the previous attempts by the author. This chapter is the enhanced version of

(Ogata, 2016a), including several additional discussions. Chapter 2 by Ogata (2019b) is a new introductory chapter for this book. The concept of a new research system called "post-narratology" was introduced. In particular, this chapter introduced post-narratological studies, including books and papers by the author and his research group, and discusses a narrative generation process from the perspective of post-narratology. Chapter 5 by Ogata (2019c) sought to undertake the problem of a narrative generation-reception and narrative production-consumption modeling of *kabuki* from the viewpoint of an information system or a narrative generation system. A fundamental concept of the modeling is "multiplicity," or "multiple narrative structures" in narrative. Therefore, the direct objective of this chapter is to propose a first approach to a narrative generation-reception and production-consumption model of *kabuki* based on the above concepts. In addition, this chapter associates the model with the concept of the *Geinō* Information System (GIS), representing a system model in which multiple narrative generation and production mechanisms or processes, that is, narrative multiple structures, are included. Introductory knowledge on *kabuki*, including the history and basic terms, is also mentioned as background for the discussion. The next main objective of the chapter is to phenomenally describe the aspects of *kabuki*'s multiple narrative structures with actual examples. Although only indicated in this chapter, the ultimate goal of the author is to conduct social distribution experiments of INGS, which the author has been developing as a synthesis of previous studies. By social distribution, the author means a new type of business model for social applications using the Web, for the narrative generation system itself, or for a content system including narrative generation systems. Finally, the chapter anticipates directions for future research on social distribution of INGS through a greater understanding of *kabuki* and its multiple narrative structures model. This topic will be one of the central issues of the following book. This chapter is the enhanced version of Ogata (2016b).

Chapter 4 by Akimoto and Ogata (2019) discussed the design concept and key issues for developing a co-creative narrative generation system. It provided a vision of a social-level system that generates a collection of diverse narratives from the interaction among two or more agents, including both humans and narrative generation programs. This challenge aims at both a form of narrative generation technology and a computational approach to narratology as a model for the social-level process of narrative production. Partial and experimental implementations are also presented. This chapter is the enhanced version of Akimoto and Ogata (2016). In particular, a discussion about computational principles for narrative co-creation, that is, narrative-based narrative generation, is updated.

The author selects several other authors' chapters that show important approaches to post-narratology. Chapter 3 by Abe (2019) introduced the literary concept of "intertextuality" as a key idea for the computational modeling of literary work generation. After reviewing several related studies in the "*kotoba kōgaku* (language sense processing engineering)," which is an original idea of an AI-based research field by Abe and his research group, and the computational creativity of literature, the chapter discussed a new type of computational method for generating various literary works, including *waka* and *haiku*, contemporary poems, and novels. This chapter is the enhanced version of Abe (2016). This enhanced version explores a method of literary work generation based on a combination of the notion of intertextuality and the reasoning by abduction. Chapter 6 by Kanai (2019) dealt with the aspects of nonstory narrative—in which discourse is considered more important than story, or a clear story does not exist—and the effect of nostalgia in the cognition of film reception, in addition to the story aspect of narrative. In particular, the chapter analyzed cognitive effects of films related to nostalgia through the method of narrative rhetoric of "cutting techniques" (for a flow of a story), which the author has been studying, and "defamiliarization" in literary theories. The knowledge derived from the analysis is applied to a computational modeling for composing films or images of nostalgia-based narrative rhetoric, based on his nonstory narrative concept. This chapter is the enhanced version of Kanai (2016). This enhanced version includes a new discussion on the relationships among memories of a disaster, rhetorical techniques of film, and nostalgia. Chapter 7 by Kawamura (2019) presented the concepts and technologies behind a "Commercial Film Production Support System (CFPSS)," in which image-editing techniques and rhetorical knowledge based on structural analyses of TV commercial films are implemented. The chapter also presented the result of an experiment investigating how an advertising story influences viewers, using commercial films generated by CFPSS. Although this chapter is the enhanced version of the same title in the 2016 book, the enhanced version provided additional analysis and discussion. Chapter 10 by Tadaki and Abe (2019) presented the findings from an experiment to investigate what viewers experience when appreciating artworks (paintings). This study focused on how a story-creation task affects a viewer's experience or impression of an abstract painting. The story-creation task refers to writing a story about the abstract painting. The experimental results were analyzed from multiple perspectives. Moreover, methods of qualitative enrichment of viewers' experiences were discussed. Chapter 11 by Seino, Haruna, and Ishizaki (2019) described clinical knowledge for effective employment support for people with disabilities through a computer-based analysis of the many narratives obtained from mail-survey questionnaires from people with various disabilities who described their work and work life. The chapter presented the mixed analytical method of text mining as a topic in the recent natural language processing field and narratological language analysis for examining effective employment support. This chapter is the enhanced version of Seino, Haruna, and Ishizaki (2016). The enhancements include additional background information and a vision of a story generation for employment support based on the analysis results. Chapter 12 by Fukushima (2019) overviewed three methods for obtaining the mental representation of a gustatory experience of a cup of Japanese *sake* (an alcoholic beverage). In particular, a tool called the "Flavor Wheel of *Sake*" used to aid in a verbal description of the taste of *sake*, an experimental method for observing verbal and visual images evoked from a gustatory experience, and an experimental method for generating pictorial representation of a gustatory experience were presented. Chapter 13 by Nitta (2019) attempted a systematic explanation of *haiku* from communicative and aesthetic perspectives. In a work of *haiku*, various events and situations are narrated in a short and simple sentence composed of three phrases of five, seven, and five syllables, respectively. It can be deemed a method of elegant and efficient communication. Hence, *haiku* is an interesting style from an informatics perspective. General knowledge of *haiku* was also provided in this chapter.

IMPORTANT CONCEPTS AND TERMS FROM A SUMMARY OF THE BOOKS AND PAPERS

The above introduced papers and books include the following important concepts and terms. In the references [E], [F], and [H], the author also shows important and interesting concepts and ideas presented by other authors because they are also related to Ogata's narrative generation study or post-narratology.

[A-1] Master's Dissertation

- **Morphology of the Folktale by Propp:** This literary theory (Propp, 1968) was first introduced in the author's narrative generation study and was used for his macronarrative structures and contents. This theory and a narrative generation mechanism using the theory will be explained in Chapter 1 in this book.

- **Explanation-based Learning:** Explanation-based learning (DeJong, 1993, 2014) is the idea of a learning mechanism for improving the efficiency of intelligent tasks. In this paper, this mechanism holds the results of narrative generation as patterns to use directly from the next-generation cycles to improve the efficiency of narrative generation. Introducing learning mechanisms into the author's research has been considered in the context of the knowledge acquisition of scriptural event sequences in INGS (Ono & Ogata, 2018b).
- **Story Schema or Story Grammar:** Although David Rumelhart (1942–2011) (1975) describes having originally intended the formalization of Propp's theory in his story schema, his actual story grammar models based on AI and cognitive science, including his approach, did not efficiently reflect the event-development logic of Propp's theory. This line's story grammar models were constructed on the basis of the goals and planning processes of human microbehaviors.
- **Story Grammar Based on Propp's Theory:** Different from the above-mentioned story schema, this type of story grammar was originally developed from the author's viewpoint that faithfully reflects Propp's theory as far as possible and will be described in detail in Chapter 1. An advanced idea was that the author positions the story grammar based on AI and cognitive science as the microlevel story-development logic, and he positions the story grammar based on Propp's theory as the macrolevel story to integrate both approaches.
- **Defamiliarization:** Although theoretical framework was presented by Victor Shklovsky (1893–1984) (1990) of Russian formalism and Berthold Brecht (1898–1956) (1964, 1973), a dramatist and poet, it is closely related to rhetorical traditions in the West and Japan. Here, the author's narrative generation study is applied and expanded in a similar mechanism in the context of "norm and deviation" theory (Chapter 4 (Ogata, 2019h) in Ogata (2019d)). The defamiliarization and related topics will appear frequently in this book.
- **Theories of Foreigners, Strangers, Tricksters:** These are common concepts in folklore and anthropology. A folkloric concept of the theory of foreigners ("*Ijin-ron*") presented by Masao Oka (1898–1982) (1994), a Japanese folklorist, was influenced by the *Yamabito-ron* by Kunio Yanagita (1875–1962) (2016), etc. From ancient times in Japan, people of different origins and lifestyles, especially the

"*satobito*" (flatland people) and the "*yamabito*" (mountain people), coexisted in isolation. The *satobito* lived through agriculture and the *yamabito* by hunting, and the *yamabito* were persecuted as the *satobito* constituted the majority in society. To the *satobito*, the *yamabito* were foreigners; their unique characters influenced the Japanese *geinō*s and cultures. Another common tradition in folklore, the trickster, was first presented by Carl Gustav Jung (1875–1961) (1999). The academic theory regarding "tricksters" who rescue organizations and societies in critical situations and management science also exists (Teramoto & Nakanishi, 1993). Although these were topics that previous narrative generation studies by the author did not focus on, in the future, these will be important themes related to the philosophical concepts of narrative generation (discussed in Chapter 4 (Ogata, 2019h) in Ogata (2019d)) and narrative generation as institution (will be referred to in Chapter 4 in this book).

[A-2] Doctoral Dissertation

- **Narrative Techniques:** These are formally defined functions of the Common Lisp programming language (Chapter 1 in this book) that transform and extend the narrative tree structures in the narrative generation system in this doctoral dissertation. As the author will state in detail in Chapter 1, this idea has been inherited by INGS to be used as a core mechanism in the story generation phase. The common architecture of narrative generation systems of the author has a characteristic that enables the gradual extension through adding narrative techniques.
- **Narrative Strategies:** Similarly, these are rules for controlling the actual usage of narrative techniques in the narrative generation system presented in this paper. For example, this paper shows that the narrative generation of a specific genre like advertising is performed though a relatively fixed order of rule application. This idea has also been inherited by INGS and will be described in detail in Chapter 1 in this book.
- **Story:** This is the most important narratological concept based on the temporal development order of narrative events and will be concretely defined in the context of INGS in Chapter 1 in this book. Moreover, Chapters 2 and 4 will discuss this concept from the perspective of the relationship with chronicles and narrative discourse.

- **Plot:** This concept means the detailed event-development order and event sequences developed in a narrative. In this sense, the story is a concept based on actual logic, and plot is a concept based on narrative logic. In subsequent studies, the author uses the more-comprehensive term "narrative discourse" rather than plot. Although this book will explain narrative discourse in detail in Chapter 1, plot is an element included in narrative discourse. Narrative discourse is a larger concept than the concept of plot.
- **Narrative Analysis:** This doctoral dissertation analyzed a detective story and television advertisement works and uses the results to propose a narrative generation system and an advertising-specific narrative generation system. This research style in which various genres' narrative analyses are bridged to design narrative generation systems has continued since the dissertation research. In this book, Chapter 3 will summarize the overview of previous narrative analyses conducted by the author and introduce research regarding brand-introduction rhetoric in advertising from the author's previous narrative analyses.
- **Advertising-specific Narrative Generation System:** Advertising has been one of the most important targets since the first step of the author's narrative generation study. Research into narrative analyses and advertising-specific narrative generation systems has continued since the first design and implementation of the proposed advertising narrative generation system. Currently, the author is conducting a collaborative research project with an advertising production company (Ito, Sasaki, & Ogata, 2018; Ono, Sasaki, & Ogata, 2018, 2019a, 2019b; Ono, Sasaki, Ito, & Ogata, 2018). As mentioned above, Chapter 3 in this book will introduce one of the attempts of this analysis.

[A-3] Two Chapters in an Edited Book

- **System Narratology:** Although currently the author does not use this word, the meaning is similar to the concepts in the later interdisciplinary and expanded narrative generation research, including expanded literary theory and post-narratology.
- **Multiple Narrative Structures Model:** This is one of the central philosophical concepts in the narrative generation study by the author. The concept will be explained in detail in Chapters 2 and 4 in this book and will appear frequently. The multiple narrative structures model regards a narrative phenomenon as an organic collection including

various aspects and levels, such as narrative generation-reception/production-consumption processes and personal (or individual) /social (or collective) levels.

- **Narrative Genre System:** The author presented a comprehensive narrative genre system in a broad sense. In particular, the author broadly surveys the areas of literature, *geinō*, folklore, etc., in Japan to comprehensively arrange the concrete work examples in several categories in the genre system. Chapter 2 in Ogata (2019d) (Ogata, 2019f) described the narrative genre system in detail.
- **Gerald Genette:** A representative researcher of narratology, Gerald Genette published many books on narratological themes, including narrative structural analysis (Genette, 1972), intertextuality (Genette, 1982), and the structure of a book (Genette, 1987). The narrative generation research by the author focuses on Genette's narrative discourse theory to survey, analyze, and redefine it from the narrative generation system's viewpoint. This book will refer to Genette's narrative discourse theory in Chapter 1 in detail. In addition, "narratology" in the narrow sense means Genette's narrative discourse theory.
- **Narrative Discourse:** This is a concept that comprehends narrative methodologies and techniques regarding event development based on narrative logic and narrating methods, including plot. Genette (1972) presented the most systematic narrative discourse theory. However, in the author's narrative generation architecture, the narrative discourse system by Genette was positioned as a part of the whole of narrative discourse. In particular, the narrative discourse in the true meaning should be defined in the area beyond the methods and techniques of the narrative discourse system Genette presented. Chapter 1 in this book will explain concretely the design and implementation based on the Genette narrative discourse theory and the extension aimed at a more comprehensive narrative discourse system for INGS.
- **Hypertext:** Although the author has conducted several narrative analyses using the concept and method of hypertext (Endo & Ogata, 2002, 2003a, 2003b, 2004a, 2004b; Ogata & Endo, 2004), these have not been applied to the narrative generation systems explicitly. Chapter 3 (Ogata, 2019g) in Ogata (2019d) partially discussed the effectiveness and possibilities of introducing the hypertext-like structures into the narrative generation systems.

- ***Geinō* Information System:** In this paper, the *Geinō* Information System (GIS) is described as a mechanism based on a type of knowledge base of *geinō* characters, an idea that has continuously been improving and is currently positioned as a larger narrative generation mechanism that can contain one or more INGS(s). GIS will be mentioned in relation to *kabuki* in Chapter 2 and will be discussed with "external narrative generation" in Chapter 4.

[B] An Introduction to Informatics of Narratology

- **Informatics of Narratology:** The term "informatics of narratology" in this book is basically intended to have a meaning similar to the author's expanded literary theory and post-narratology. Moreover, the term "computational and cognitive approaches to narratology," as the title of the referenced book (Ogata & Akimoto, 2016), is the more detailed and concretized words of the "informatics of narratology."
- **Fluidity and Fixation:** This pair of concepts is one of the philosophical bases of the author's narrative generation study. It means that narrative generation is executed through a type of circulative process of fluid situations and fixed situations at the personal and social levels. In this sense, this concept is closely related to other philosophical bases, including "circulative control" and the "multiple narrative structures model." Although Chapter 4 (Ogata, 2019h) in Ogata (2019d) provided its basic information, Chapters 4 in this book will also include related discussions.
- **The Story-cutting Rhetoric:** Kanai, a co-author of this referenced book, continuously studied this narratological concept (Kanai, 2016, 2018a, 2018b, 2018c, 2018d, 2018e, 2019; Kanai & Ogata, 2004a, 2004b). In a narrative generation process, the level of narrative discourse has been positioned based on a story, in many cases. In contrast, although Kanai has not necessarily denied and ignored the importance of a story, Kanai regards the level of narrative discourse as an element of narrative generation that has equal importance with the story. The story-cutting rhetoric includes narrative techniques that try to destroy positively the smooth progression of a story using the content and semantic elements in a story, the media's elements for language, music, and image. Regarding the author's narrative generation study, this is positioned as a part of the philosophical concepts of norm-deviation introduced in Chapter 4 (Ogata, 2019h) in Ogata (2019d).

- **Kumagusu Minakata:** Although Minakata was a folklorist in Japan who was highly regarded (similar to Kunio Yanagita and Shinobu Orikuchi (1887–1953)), Minakata's research was not limited to folklore. He was also a natural scientist and biologist who researched cells like Myxomycete (slime mold) and a thinker of religion who directed the shrine-consolidation opposition movement. The whole of his prodigious research has not been published completely until now. Minakata also wrote many papers on folktales and discussed, for instance (Minakata, 2017), both the spatial moving of a folktale and the change of its structures and contents in an organic relationship. Considering his narrative-related research and introducing them into the author's narrative generation study is a topic for the future.

[C] Computational and Cognitive Approaches to Narratology

- **Computational and Cognitive Approaches to Narratology:** This term concretely explains previous concepts such as "system narratology," "expanded literary theory," and "informatics of narratology."
- **Integrated Narrative Generation System (INGS):** The author referred to this system in detail and systematically in this book. Although the author continued to develop various individual mechanisms of narrative generation based on two types of systems after his master's and doctoral dissertations, he organically integrated these partial mechanisms into a larger systems framework using a set of conceptual dictionaries as the central part after publishing the above book, [B: *An Introduction to Informatics of Narratology.*] However, as the current version of INGS does not necessarily use all of the mechanisms that the author has previously designed and developed, one of the future directions is to introduce other effective parts in the previous individual research into a part of INGS to expand the entire system. Chapter 1 in this book will correspond to the comprehensive explanation of the current INGS revised and expanded since the referenced book [C].
- ***Kabuki:*** Although the author has conducted narrative analyses of various genres to use the results to design and develop narrative generation systems, since about 2010, the author has also selected *kabuki* as a most primary analysis object and has continued research on the narrative generation of *kabuki* from various viewpoints. Chapter 2

in this book will summarize the previous *kabuki*-related researches and describe the part of experimental systems that is newly proceeding.

- **Intertextuality:** Although this concept was used by Abe (2016) as a core concept in the referenced book, the author has also referred and explained the concept in various papers. For example, Tsuchihashi and Ogata (2009) presented a system that retrieves many fragments of narrative texts based on the narrative structure that is related to Propp's theory to generate consistent sentences from the combinatorial usage to present a type of intertextual narrative generation. Although the current version of INGS (Chapter 1 in this book) does not include the mechanisms to which the intertextuality is explicitly applied, the narrative content knowledge bases, including a story content knowledge base, indicate a method of intertextual narrative generation that collects various types of units from existing narratives and literary works.
- **Japanese Language Resource System:** Takeuchi (2016) discusses this theme in Chapter 8 in this referenced book. He mainly explains the method that represents and describes the semantic elements in each word and proposes the implementation design. This is similar to the conceptual dictionaries that formally describe the concepts of words such as verb concepts and noun concepts in the research of the author (Ogata, 2015). Takeuchi's language resource system and Ogata's conceptual dictionaries can be prototypes of the ontology that gives the semantic descriptions of concepts and their semantic relationships in the world. As the semantic descriptions that support a narrative in the background are given through these language resource systems, conceptual dictionaries, and ontologies, this type of research and development will be a very important basis for the author's narrative generation study. Actually, INGS organically integrates various modules for narrative generation processing through the conceptual and language-related dictionaries. These dictionaries in INGS will be described in detail in Chapter 1 in this book.
- **Handicapped:** This concept was treated by Seino, Haruna, and Ishizaki (2016) in Chapter 9 in this referenced book. The author has recently expanded the applicable possibilities to the direction of the narrative generation model of ASD and for supporting ASD's people (Aoki, Ono, & Ogata, 2018). This type of study requires an explicitly described model of human memory to the corresponding narrative generation system. It is a merit for the development of INGS. Of course, as an interest of the author is to create a narrative generation

model and a system related to a new narratology, post-narratology, the author will directly introduce human memory models into the narrative generation mechanism. The models will be worth considering from the viewpoint of narrative generation control. Chapter 2 (Ogata, 2019f) in Ogata (2019d) included related short descriptions in the context of Freud's dream theory.

- **ASD (Autism Spectrum Disorder):** Regarding the above theme, the author thinks that a new possibility of narrative generation study will be opened through the analysis of linguistic and narratological characteristics of people with ASD, through modeling and system development based on the analysis. One reason is that the unconscious linguistic and narratological acts by such people can become conscious analysis objects, enabling detailed analysis, modeling, and design of various objects and problems. For example, although Sugamura, Kushizaki, Ikemi, & Nabeshima (2018) state that understanding metaphors is difficult for people with ASD, considering the reasons will give effective indications for understanding metaphors and generation research.
- ***Genji Monogatari:*** In Chapter 10, Tsuchiyama (2016) statistically analyzes various texts in *Genji Monogatari* (1993, 1994, 1995, 1996, 1997). This type of study existed before the appearance of computers. On the other hand, narratological analyses of *Genji Monogatari* have also been attempted. For instance, Shimauchi (1989) extracted many motif types of *Genji Monogatari*. This research is also related to motif analyses of folklore (Seki, 1957, 1966, 1977). In INGS, this type of research is closely related to the story content knowledge base that will be mentioned in Chapter 1 in this book. The author currently regards motif units defined and formally described by Common Lisp programming language based on a systematic analysis of Japanese folktales by Seki (1978a, 1978b, 1978c, 1978d, 1978e, 1979a, 1979b, 1979c, 1979d, 1980; Seki, Nomura, & Ōshima, 1979, 1980) and Ozawa (1976, 1997, 1999) (Ito, Ono, & Ogata, 2018a, 2018b, Ono, Ogata, & Ito, 2019).
- **Reception Theory:** Akimoto and Ogata (2013) developed an experimental mechanism using Jauss's (1970) reception theory in relation to the generation control of narrative discourse. This is introduced into INGS as a narrative generation control mechanism

for the narrative discourse mechanism. Chapter 1 will summarize this mechanism.

- **The Nonstory's Perspective of the Narrative:** This is a concept related to the story-cutting rhetoric mentioned in [B: *An Introduction to Informatics of Narratology*]. The "nonstory's perspective" reflects the narrative perspective that emphasizes various other components more than the story in a narrative, including characters, objects, and concepts. The story-cutting rhetoric is used as a narrative technique or a group of narrative techniques according to the narrative nonstory's perspective.

[D] Content Generation Through Narrative Communication and Simulation

- **Narrative Content:** In this referenced book, the "narrative content" means externalized narratives in more broad sense than fixed books. For example, the drafts before an author finally publishes are called narrative content if they are externalized in the form of papers or electric files. This referenced book, [D], also treats the problems of communication and generation in contents more than the narratives in the narrow sense. This book will discuss the problems of narrative contents in Chapter 4.
- **Yukio Mishima:** Ogata (2018a), in the referenced book, [D], shows a narrative analysis using *Gogo no Eikō* by Mishima (1963) and provides his basic information, including his life and main literary works, to the readers. The author's narrative study has been strongly influenced by Mishima. Previously, this has been mainly a psychological influence. However, the author is currently planning his work analyses after the analysis of *Gogo no Eikō*. Although this book does not contain a comprehensive explanation regarding Mishima, his works will be referred to frequently.
- **Film Theory:** The main materials of the study of Kanai (2018a, 2018b, 2018c, 2018d, 2018e) are movies and images. Film theories are his important referential objects. Although film theories partially overlap with narratology (because film is also a type of narrative genre), narratology in film has special characteristics that rely on the image's feature, which transcends the narrative feature in the narrow sense. For instance, Shigehiko Hasumi (1936–) (1979) has constructed his unique film theory for discussing narratives as images that transcend the

narrative of language. However, from a broader perspective, the film genre is also a type of narrative genre. As the part of image processing in INGS that will be referred in Chapter 1 in this book still remains in the primitive stage, introducing the theories of the film and image into the part will be a future topic.

- **Nostalgic Narrative:** This is a recent research topic in Kanai's (2016, 2018a, 2018e, 2019) cognitive and generative study of image. Nostalgic narrative means a narrative phenomenon in which, in films on historical places, a current event has an emerging nostalgic narrative. The non-nostalgic rhetoric for cutting the story's flow of a nostalgic narrative is also a part of this theme. The problem of nostalgic narrative is related to the multiple narrative structures model in the meaning that two or more narratives are overlapped in the place of a nostalgic narrative. Moreover, these techniques are related to the narrative discourse phase in Chapter 1 in the relationship with the INGS architecture.
- **TRPG (Table-talk or Table-top Role Playing Game):** Ono and Ogata (2018a) presented a system called "Automatic Narrative Generation Game," in which the main goal is to automatically generate narratives based on the framework of TRPG. TRPG is called "Table-talk Role Playing Game" in Japanese. Simultaneously, this research also proposes the story processing for giving the cognitive gap and surprising the receivers. This processing is related to the story's cutting by Kanai and Ogata and defamiliarization theories by Shklovsky (1990) and others. In this book, this system and the theories are related to the narrative discourse mechanism in Chapter 1. Chapter 4 (Ogata, 2019h) in Ogata (2019d) also considered the problem of norm-deviation related to defamiliarization in the context of philosophical background of the narrative generation study.
- ***Haiku:*** *Haiku* generation has been attempted using a computer. The form of *haiku* is very simple, and the automatic generation of a short poem with *haiku*'s form is very easy if the quality is not important. However, because of its simple form, *haiku* reveals serious problems and questions, such as "what is creativity?" and "what is the difference or similarity in the generation and creation by humans and machines?" Nitta (2018), who is an actual *haiku* poet though he is a researcher of natural language processing, showed his research based on a deep consideration of *haiku*. On the other hand, the author's *haiku* generation method (Ito & Ogata, 2018a, 2018b) is dependent on the framework of circular generation between *haiku*s and narratives. In this circular

framework, one or more *haiku*(s) made from a narrative correspond to a type of narrative discourse in the framework of INGS.

- **Metaphor:** Metaphor that is a representative figure in language is frequently used in both ordinary language and artistic, literary language. Previous metaphor research on computers has mainly focused on understanding, generation, and identification. However, not much research has been presented on understanding and generating metaphor at the literary and artistic level. In this book, [D], Utsumi (2018) applies his previous metaphor understanding models to the direction of metaphor generation. INGS has not introduced the metaphor-related methods. A direction of the author is to use the method of metaphor at the macro level for generating entire narrative structures instead of metaphor processing in the surface linguistic units. The narratology of histories by Haiden White (1928–2018) (1973), which mentioned in Chapter 3 (Ogata, 2019g) in Ogata (2019d), will give many indications to the direction by showing metaphorical narrative types for narrating a set of historical events.
- **Narrative Communication (Sender and Receiver):** Although narrative generation studies include the problem of communication between narrative senders and receivers as a premise, Kawamura (2018a), in referenced book [D], proposed advertising receivers creating active and positive meanings in their brains and the second-order communication fields regarding produced contents, and he showed narrative generation research of advertising based on that process.

[E] Informational Narratology

- **Movie, Advertisement, and *Kabuki:*** In the book, [E], three authors respectively treat *kabuki* (Ogata), image or movie (Kanai), advertisement (Kawamura) as materials of analysis and modeling. Simultaneously, for Kanai and Kawamura, image and advertisement are objects for design and development of their generation mechanisms, but for Ogata, *kabuki* is not currently a direct generation objective. He treats *kabuki* as a narrative mechanism. Thus, he sees *kabuki* as a new framework for a narrative generation mechanism, one previously used for this purpose. The author's current goal is to acquire, using *kabuki* as a narrative generation mechanism or one of several narrative generation mechanisms, new narrative methods. Furthermore, another goal is to revise his existing narrative generation systems, INGS and *Geinō* Information Systemi (GIS), through the *kabuki*-modeled narrative generation mechanism. From the viewpoint of the multiple narrative structures model, based on *kabuki* as an entire narrative, INGS is a single-narrative generation system within the larger GIS, whereas GIS is a synthetic and social narrative generation system that contains many INGSs. The above content is described in detail in Chapters 3 and 5 in the book, [E].

- **Visible Narrative and Invisible Narrative:** The book, [E], calls the elements represented directly in a narrative, namely explicit narrative elements, "visible narrative," and the potential narrative elements "invisible narrative." The significance is that the story belongs in the invisible narrative. This fact enables diverse narrative discourses (as visible narrative elements) according to a story. Chapter 1 (Ogata, 2019e) in Ogata (2019d) discussed in detail this topic, the relations between invisible narrative and visible narrative.
- ***Ningyō Jōruri*:** The book, [E], refers to *ningyō jōruri*, in addition to *kabuki*, because they are closely related through the historical generation and playing process, especially during the Edo era. If *kabuki* had not introduced the works and narratives in *ningyō jōruri* and had not had its close relationship with *ningyō jōruri*, *kabuki* would have been a poorer art or narrative and may not have had such a long life. Extending the original points in the narratives and works of *ningyō jōruri* has continued to keep *kabuki* alive. Actually, many popular *kabuki* works originated in the works of *ningyō jōruri*.
- **Tamotsu Watanabe:** The author refers to many books by Tamotsu Watanabe, a *kabuki* researcher and critic whose range covers through classical *geinō*, theatrical arts, Japanese dances, and contemporary stage performances. Watanabe is interested in both classical drama and the most contemporary and avant-garde dramas. He thinks that an important mission for contemporary *kabuki* is to hand down diverse types and styles accumulated through the history of *kabuki* and inherit various *kabuki*'s methods, including "*sekai*" ("world") and "*kata*" ("form" or "pattern"). At the same time, he believes that *kabuki* is the most active dramatic art genre in contemporary society with respect to body performance and narrative production. The book, [E], also refers to his books several times.
- ***Watakushi Monogatari*:** "*Watakushi Monogatari*" ("My Story" or "My Narrative." Or "I-Story" or "I-Narrative") is a concept presented by the

author (Ogata) in the book, [E], and means a new narrative genre that should be practiced using narrative generation systems in the future. Although it is a type of story or narrative genre in which narrative worlds and narrations are pursued and represented through "I," this does not necessarily mean that the narrative world to be generated should be narrow and the field related only to the "I"; rather, it means that the first-person world gradually opens to the second-person world and the third-person world to construct an entire narrative dependent on their organic relationships. Chapter 4 in this book will introduce this discussion with a little revision.

[F] Post-Narratology Through Computational and Cognitive Approaches

- **Post-Narratology:** Although this term first appears in the book, [F], its meaning is similar to the previous "expanded literary theory," "informatics of narratology," and "informational narratology," etc. However, previous usage of the terms had a nuance to their expansion based on old narratology and literary theories. In contrast, "post-narratology" more strongly has a nuance different from old narratology and literary theories. Of course, different from original narratology with strong formal and structural tendency, narratology that emphasizes social and political characteristics may be a type of post-narratology. However, post-narratology as used by the author includes the latter meaning and comprehends the narratology dependent on characteristics of the original formal, structural, and technological narratology in the first period. Chapter 3 (Ogata, 2019g) in Ogata (2019d) provided an overview of post-narratology. Further, Ogata (2019d) collected conceptual approaches to post-narratology, whereas this book presented system developing approaches to post-narratology.
- **Narrative Generation-Reception/Narrative Production-Consumption:** These are two related concepts that show the basic construction of the multiple narrative structures model. The former mainly presents the communication structures of narrative generation at the personal level, and the latter shows them at the social level. The multiple narrative structures model is constructed as the unified form of both the above structures. These concepts appear in all of chapters in this book.

- **Language Sense Engineering (*Kotoba Kōgaku*):** This concept presented by Akinori Abe and his research group means natural language processing in the broad sense. In particular, it aims to realize the advanced natural language processing, including the language processing that is necessary artistic and literary, and it is partially related to the author's concept of post-narratology. Moreover, Abe uses the term as the name of a research project in Japanese AI society to establish and extend the concept and theory. The paper by Abe (2019) in this book, [F], is also related to language sense engineering.
- ***Waka:*** *Waka* has the longest history and tradition in the genres of sort poem literature. Whereas a *haiku* follows the syllabic structure of 5-7-5, a *waka* follows the structure of 5-7-5-7-7. *Haiku*s basically describe objects, including natural scenes, through a sequence of words. In contrast, *waka*s use longer words to describe more direct emotional representations. Moreover, a characteristic of *haiku*s is to have a set of "season words" that contain many diverse, previously encountered words. In contrast, a characteristic of *waka*s is to have very complex and literary systems for rhetoric, including typical phrases and collective, associative images. In this sense, in the history of Japanese literature, creating *waka*s is a very rhetorical, technical, and technological work, not just an artistic exercise. Abe (2019) introduces his attempt at computational *waka* generation as an example of text-generation tasks.
- **Story and Nonstory:** Kanai (2019) does not treat a narrative as the object that belongs to a story. Instead, he emphasizes the existence of the nonstory in addition to story. Further, the introduction chapter to Part I in Ogata and Kanai (2010) presents three related concepts: story, nonstory, and anti-story. The story-cutting techniques presented by Kanai (2019) mainly correspond to narratological techniques or rhetoric for creating nonstory narratives and anti-story narratives positively. This idea is reflected throughout this book, and the advertising rhetoric of defamiliarization introduced in Chapter 3 is also associated with this concept.
- **Commercial Film-Production Support System (CFPSS) and Advertising Scenario Generation:** CFPSS is the name of an advertising system developed by Kawamura (2019). As the author will describe in Chapter 3 in this book, the author has also been continuing studies of advertising narrative generation. CFPSS does not automatically generate advertising scenarios or synopses completely. Although several scripts for scenario generation have previously been given by the system developer, small variations in the scripts are automatically

selected. CFPSS displays fragmentary still pictures according to each prepared scenario. In contrast, the most important goal of the author's (Ogata's) advertising generation is to generate automatically diverse advertising scenarios or synopses. Takayoshi Kishimoto (December, 5, 2018), an advertising director in AOI/TSO Holdings, Co. Ltd., stated that planning advertising production requires attention to details about such specific element as talents in a scenario. However, a more general method of advertising production shows that a main task in the planning stage is writing the scenario. Moreover, the general advertising production recognizes that the level of the scenario itself is critical. From such a viewpoint, the author thinks that the automatic generation of advertising scenarios is one of the most important goals.

- **Art Appreciation:** Tadaki and Abe (2019) focus on the problem of the reception of arts in special spaces such as museums, where the reception has different characteristics from the reception of other narrative genres—for instance, novels, movies, and *kabuki*. They highlight the problem of the reception of abstract paintings that do not represent any narrative. Kanai (2019) treats the narrative problem, including nonstories and anti-stories, in the objects that have strong narrativity. In contrast, Tadaki and Abe treated the problem in the objects with, so to speak, weak narrativity. Although this is not discussed directly in this book, [F], as the author does not regard narrative as an object equal to story (as seen in the discussions on story, nonstory, and anti-story), the ideas of Tadaki and Abe do offer various insights into the author's narrative study.

THEMES OF THE BOOK SERIES AND CONFIGURATION OF THE BOOK

The books and relatively long papers in the previous sections show areas and themes of the author's studies that have been finally called post-narratology. In the final part of each study, the author selected important terms that were written in the books and papers. Moreover, Table 1 lists theorists in the field of narratology and literary theories, theoretical frameworks and academic areas, and discussed narrative genres as well as concrete narratives or literary works for reference in various research papers by the author (and partially other authors). Continuing systematic studies of post-narratology in a top-down manner are needed to ensure the inclusion of the above elements in the future.

Table 1. Main objects of the post-narratology seen in the author-related previous books and papers

Theorists	Theoretical Frameworks of Academic Fields	Narrative Genres	Narrative Works
A. Aarne Aristotle M. Bakhtin R. Barthes B. Brecht R. Descartes G. Deleuze S. Eisenstein E. M. Forster S. Freud S. Fujii N. Furuhashi G. Genette M. Gunji S. Hasumi K. Hori W. Iser H. R. Jauss J. Kristeva A. Kurosawa J. McCarthy K. Minakata M. Minsky Y. Mishima K. Miyazawa Murasaki Shikibu S. Natsume K. Noe M. Oka S. Orikuchi Y. Ozu G. Prince V. Propp C. Revi-Strauss D. Rumelhart K. Seki R. Schank V. Shklovsky H. Simon F. Saussure S. Thompson N. Watanabe T. Watanabe H. White K. Yanagita T. Yoshimoto	Advertisement theories Artificial intelligence Clinical psychology Cognitive science Common Lisp Digital story telling Emotional psychology Explanation-based learning Film theories Folklore Folktale theories *Geinō* studies Generative grammar Goal-planning Hypertext theory Intertextuality Japanese dramatic theories *Kabuki* studies Literary theories Literary theory by Natsume Marketing and advertising studies *Manga* studies Media studies Musicology Narrative therapy Narrative discourse theory Narrative generation Narrative grammar Narratology Natural language processing Natural language generation Nostalgia (Psychology) Philosophy Plot theory Polyphonic theory Problem solving Psychoanalysis Reception theory Semiotics Stylistics Story generation Story theory Story schema (grammar) Three illusions theory by Yoshimoto	Advertisement ASD (Autism Spectrum Disorder) Doll (*Ningyō*) Drama Dream Film (Movie) Folktale Game Game book *Geinō* *Haiku* Hypertext novel *Kabuki* *Kyōgen* Jazz *Manga* Metaphor Music video Narration Narrative *Ningyō jōruri* *Nō* Novel Painting Picture-card show Poem Popular music *Rakugo* Short story Sightseeing Television news story TRPG *Waka*	*Genji Monogatari* [novel] *Gogo no Eikō* [novel] *Kagotsurube Sato no Kezairoku* [theory] *Eizame* [*kabuki*] *Kanadehon Chushingura* [*kabuki, ningyō jōruri*] *Kanjincho* [*kabuki*] *Kibazakura Chushingura* [*kabuki*] *Maiko Satsujin Jiken* [short story] *Maison Ikkoku* [*manga*] *Mizukara Waga Namida wo Nuguitau Hi* [novel] *Narukami* [*kabuki*] *Oedipus Tyrannus* [drama] *Okeihan* [advertisement] *Sekai Komoku* [theory] *Shunkinshō* [novel] *Sukeroku* [*kabuki*] *Sugawara Denju Tenarai Kagami* [*kabuki*] *Tokyo monogatari* [film] *Yoshitsune Senbon Zakura* [*kabuki, ningyō-jōruri*] *Sanmai no Ofuda* [Folktale] *Sayogoromo Oshidori no Tsurugiba* [*kabuki*]

Table 2. Big themes treated in each book or paper

Book or Paper	Theme (1)	Theme (2)	Theme (3)	Theme (4)	Theme (5)	Theme (6)
A	○		○			
B	○		○	○		
C	○	○	○	○	○	
D	○	○	○	○	○	○
E	○	○	○	○	○	
F	○	○	○	○	○	
G	○	○	○	○	○	○
H	○	○	○	○	○	

Next, Table 2 shows that the above books and papers respectively dealt with which topics by described below (these topics were explained in detail in the first part in this chapter):

1. Basic standpoints——A unified consideration of narrative phenomena
2. The philosophy and thought of narrative or narrative generation
3. The consideration of narrative generation mechanisms through narrative analyses
4. A fundamental narrative generation system
5. Applications and the social development and distribution of narrative generation
6. The production of literary and artistic works or literary and artistic organizations

In addition, the master's dissertation (A in Table 2) used Japanese folklore and defamiliarization-related theories in relation with Theme (2). The current research structure appeared the doctoral dissertation [B] and two papers in [C]. Although book [D] contained all the above themes, the description was not systematic. However, the book [D] also referred to Theme (6). Three books from IGI Global ([E], [F], and [H]) were collections of papers by ten or more writers, and all the writers, including Ogata, treated important topics in the broad related range. The book [G] concretely described problems related to Theme (6).

Drawing on the above descriptions, the most macro objective of Ogata (2019d) and this book is to describe systematically the author's narrative generation study, including the above themes. In particular, the book series describe more systematically and precisely the six kinds of themes described above, comparing the above previous books and papers. Ogata (2019d), *Toward an Integrated Approach to Narrative Generation: Emerging Research and Opportunities*, was a systematic practice that consistently comprehends from a narrative view of the world and several philosophical concepts. On the other hand, this book, *Internal and External Narrative Generation Based on Post-Narratology: Emerging Research and Opportunities* will systematically deal with topics of the design and development of narrative generation systems, the applications, development and distribution of the systems, and the problems of actual narrative production or creation for external and internal directions.

Finally, the configuration of the entire book is shown.

Chapter 1 [An Integrated Narrative Generation Systems: Synthesis and Expansion] introduces the narrative generation system called Integrated Narrative Generation System (INGS). INGS is one of the central parts of the author's study that aims to develop an integrated approach to narrative generation. Two books, Ogata (2019d) and this book, present interdisciplinary and cross-disciplinary approaches to narrative and narrative generation as well as many related fields and concepts. However, the objectives of these books do not stop with the research of receptive aspects such as consideration, survey, and analysis regarding narratives and narrative generation. In this narrative generation study, all parts are considered components that serve the extreme generative and productive goals; the most important element of the goals corresponds to the design and development of INGS. After introducing the Common Lisp programming language that has been used for the implementation of INGS, this chapter describes the research and development process of INGS. The following parts of this chapter divide INGS into two aspects: mechanisms and contents. The main generation modules in INGS are for story, narrative discourse, and expression including language, music, and picture images. These are performed using the mechanisms of various narrative techniques. On the other hand, INGS has dictionaries, including conceptual and language, and various knowledge, including story content and state-event transformation knowledge bases. These correspond to the aspect of contents. This chapter explains the mechanisms in the order of the entire process, story generation mechanism, narrative discourse mechanism, and narrative representation mechanism. Further, this chapter states in detail the conceptual dictionaries, mainly noun and verb conceptual dictionaries. In addition, this chapter is the revised and expanded version of Ogata (2016a) based on the entire reorganized description.

The next is **Chapter 2 [*Kabuki* as a Synthetic Narrative: Synthesis and Expansion]**. Although the author has previously written several papers on kabuki addressing the narrative generation of kabuki or kabuki as narrative generation, this chapter acts as a kind of synthesis. The chapter first provides a focused summary of what kabuki is, which has already been summarized in the author's previous papers, and presents the synthetic characteristics of kabuki as narrative generation. Next, the chapter describes the fifteen main narrative elements or techniques and several cross-sectional elements or techniques beyond conceptual description. A fundamental concept of the modeling is "multiplicity," or multiple narrative structures. In particular, the author aims to explore narratives of kabuki from the viewpoint of the multiple structures of the narrative generation-reception and narrative production-consumption processes. From the framework of multiple narrative structures, INGS, which is described in Chapter 1, is the mechanism that performs the narrative generation-reception at the simple substance or personal level. By contrast, Geinō Information System (GIS) corresponding to narrative production-consumption does not generate a single narrative work, but is the mechanism that executes narrative generation as the outer system for making sequences of narrative generation using simple substance. Although the author presents an entire image and design of narrative generation mechanisms in kabuki through two types of systems, INGS and GIS, kabuki analyses are conducted for incorporating knowledge mainly into GIS. This chapter focuses on the elements of "person," "story," "tsukushi," and "naimaze" to present ideas related to the concrete design and implementation of kabuki-related experimental narrative generation mechanisms.

Chapter 3 [An Analysis of Advertisements for Narrative Generation] focuses on advertising analysis in the author's previous narrative analyses. Advertising analysis is based on the idea of "advertisement as a narrative." In particular, this chapter first introduces the previous fundamental attempts for an advertising narrative generation system study that was planned as an application of narrative generation study and has been continued in advertising analyses and the design and implementation of experimental systems from two viewpoints of the macro framework and the methods introducing brands into advertising works. In the central part of this chapter, the author presents a detailed rhetorical analysis of television commercial films dependent on three regular or ordinary rhetoric types and nine irregular or extraordinary rhetoric types. This study has been applied to the development of advertising narrative generation techniques in the subsequent study. The last part of this chapter mentions the systematization's topic. This advertising narrative generation is a study regarding destruction and reconstruction in the point that aims to develop combinatorial generation possibilities beyond the simulative generation possibilities based on existing texts, though it uses existing advertising texts. Although the proposed analysis in this chapter is relatively earlier research, the author introduces it as it is an interesting application example of intertextual narrative generation and is bridging to the current more-developed research toward advertising narrative generation through a systematic and precise advertisement analysis (Ito, Sasaki, & Ogata, 2018; Ono, Sasaki, & Ogata, 2018, 2019a, 2019b; Ono, Sasaki, Ito, & Ogata, 2018).

The final main part is **Chapter 4 [Toward External and Internal Practices of Narrative Generation]**. From relationships to the level of multiple narrative structures, the Geinō Information System (GIS) described in Chapter 2 corresponds to a social-level narrative generation mechanism beyond the narrative generation as a simple substance or as something including many narrative generations as a simple substance. From a broader perspective here, this chapter aims to discuss "the future of narrative generation systems or narrative generation study." The author discusses the topic from two directions: "external narrative generation" and "internal narrative generation." The "external" is the outer world beyond the "I" and ordinarily refers to a society or societies. Conversely, the "internal" is equal to the world inside the "I." Therefore, although the practice of narrative generation toward the outer direction ordinarily indicates the social development of narrative generation, the practice of narrative generation toward the internal is equal to the direction of narrative generation study for drawing "I" in any style. In this chapter, the external direction of narrative generation is closely associated with GIS. In particular, GIS is a kind of social narrative production-consumption mechanism that continues to cyclically produce narrative works to create narrative contents at a larger level beyond each narrative work. On the other hand, the internal direction is related to the aspect of narrative content, especially my story's or narrative's level. In contrast to external narrative generation through INGS and GIS that is related to a formal side of narrative generation, internal narrative generation focuses on the contents of narrative works created according to the formal side. Considering this from a broad perspective, the description of internal narrative generation may be a kind of note for a narrative work using narrative generation systems by the author. Therefore, this chapter also deals with individual and personal information, including histories, novels, and writing regarding the author. In addition, this chapter is an extended version of Ogata (2018g, 2018h).

Additionally, **Conclusion** is the final section of the book. It comprehensively summarizes the contents and results of each chapter and shows future directions and visions. In terms of more artistic and literary meanings, the most important issue that needs to be addressed in the future is how to actually create narrative works. Thus, the author's narrative generation study will gradually move in the direction of the practice of narrative generation. In this book, narratives do not necessarily only refer to literary or artistic narratives but include more *geinō*-like and folktale-like narratives. However, in the future, narrative generation studies will aim to examine the creation of an interesting narrative, in a very broad sense. Furthermore, problems and themes of literary and artistic quality will emerge as an important topic at this stage. In the final part, the author simply states the aforementioned future direction.

This introductory section provided an introduction to the book. In the first **TWO BOOKS THAT INTEGRATE A SERIESE OF STUDIES**, the author stated that books are an excellent medium for writing and systematically and consistently showing a series of studies they have undertaken. Further, this introduction section stated main themes in two books, the recent book Ogata (2019d) and this book, which together compose a series. This section simply summarizes the chapters in Ogata (2019d) to show the characteristics of this book. Based on the author's opinion regarding the book medium, the next section, **SYNTHETIC APPROACHES TO NARRATIVE GENERATION THROUGH SYSTEM-ORIENTED BOOKS**, presented an overview of four long papers and four books that the author has published, aiming to create a systematic and consistent research story. This book and a recent book (Ogata, 2019d) correspond to a current synthesis and expansion of these previous papers and books. Finally, the section **THEMES OF THE BOOK SERIES AND PAPERS AND CONFIGURATION OF THE BOOK** reviews themes treated in the previous paper and book series, and introduces the chapters in this book.

In summary, Ogata (2019h) described the theoretical and conceptual aspects of the study of an integrated narrative generation. On the other hand, this book mainly deals with the aspects of concrete systems, including the Integrated Narrative Generation System (INGS) as the core system, the *Geinō* Information System (GIS) as the social-level system with INGS, and the advertisement application system based on the analysis of advertisement films. Although these core system, social-level system, and application system and narrative analyses for system development are the main parts of the integrated approach to narrative generation, or "post-narratology" as used in the title of this book, the final goal of this book is to consider the topics of both external narrative generation and internal narrative generation. This pair in the integrated narrative generation study or post-narratology is related to the area of narrative contents generated, produced, or created through the above systems. Internal narrative generation especially treats the problems that approach the creation of narrative contents as actual works. In particular, though this book does not treat concrete narrative works themselves, an important objective of this book is to bridge formal and mechanical systems of narrative generation to the world to concrete and actual narrative contents or works.

ACKNOWLEDGMENT

This section's research was supported by JSPS KAKENHI Grant Number18K18509, Kayamori Foundation of Information Science Advancement Research Grant, The Telecommunication Advancement Foundation Research Grant, and AOI TYO Holdings Inc.

REFERENCES

Abe, A. (2016). The possibility of the literary work generation by computer. In T. Ogata & T. Akimoto (Eds.), *Computational and cognitive approaches to narratology* (pp. 376–390). Hershey, PA: IGI Global. doi:10.4018/978-1-5225-0432-0.ch002

Abe, A. (2018). Narratology and creativity. In T. Ogata & S. Asakawa (Eds.), *Content generation through narrative communication and simulation* (pp. 187–204). Hershey, PA: IGI Global. doi:10.4018/978-1-5225-4775-4.ch005

Abe, A. (2019). The possibility of the generation of literary works by abduction. In T. Ogata & T. Akimoto (Eds.), *Post-narratology through computational and cognitive approaches* (pp. 144–163). Hershey, PA: IGI Global. doi:10.4018/978-1-5225-7979-3.ch003

Akimoto, T. (2018). Narratives of an artificial agent: Mental world and narrating. In T. Ogata & S. Asakawa (Eds.), *Content generation through narrative communication and simulation* (pp. 241–264). Hershey, PA: IGI Global. doi:10.4018/978-1-5225-4775-4.ch007

Akimoto, T., & Ogata, T. (2013). Towards a discourse mechanism in narrative generation system: Proposal of a system introducing narrative discourse theory and reception theory. *Cognitive Studies*, *20*(4), 396–420.

Akimoto, T., & Ogata, T. (2016). Designing a socially open narrative generation system. In T. Ogata & T. Akimoto (Eds.), *Computational and cognitive approaches to narratology* (pp. 91–117). Hershey, PA: IGI Global. doi:10.4018/978-1-5225-0432-0.ch003

Akimoto, T., & Ogata, T. (2019). Consideration of fundamental methods and principles for human-computer co-creation of narratives. In T. Ogata & T. Akimoto (Eds.), *Post-narratology through computational and cognitive approaches* (pp. 164–191). Hershey, PA: IGI Global. doi:10.4018/978-1-5225-7979-3.ch004

Aoki, S., Ogata, T., & Ono, J. (2018). Narrative generation related to cognitive patterns seen in ASD: From the perspective of "surprise". In *Proceedings of the 35th Annual Conference of the Japan Cognitive Science Society* (pp. 652–660). Tokyo, Japan: Japan Cognitive Science Society.

Aristotle. (1997). Poetics (M. Heath, Trans.). London, UK: Penguin Classics. (Original work published c. 335 BC)

Brecht, B. (1964). *Brecht on theatre: The development of an aesthetics* (J. Willett, Trans. & Ed.). London, UK: Methuen. (Original work published 1950)

Brecht, B. (1973). Engeki no tame no shō shikō genri [A principle for thinking theatre]. In Bertolt Brecht engeki ronshū I [Bertolt Brecht's drama theories collection I] (K. Senda, Trans.). Tokyo, Japan: Kawade Shobō Shinsha. (Original work published 1949)

DeJong, G. (Ed.). (1993). *Investigating explanation-based learning*. Boston, MA: Kluwer Academic Press. doi:10.1007/978-1-4615-3602-4

DeJong, G. (2014). Explanation-based learning. In T. Gonzalez, J. Diaz-Herrera, & A. Tucker (Eds.), *CRC computing handbook: Computer science and software engineering* (Vol. 3, pp. 66.1–66.26). Boca Raton, FL: CRC Press.

Endo, Y., & Ogata, T. (2002). Hyper-comic system as representation field of narrative discourse. In *Proceedings of the 17th Congress of the International Association of Empirical Aesthetics* (pp. 555–558). Rome, Italy: University of Rome Tre.

Endo, Y., & Ogata, T. (2003a). Hyper-comic system as consideration of rhetoric. In *Proceedings of the 4th International Conference on Cognitive Science* (pp. 111–116). Oakbrook Terrace, IL: Cognitive Science Society.

Endo, Y., & Ogata, T. (2003b). Narratology of *manga* and application to hyper-comic. In *Proceedings of the 17th Annual Conference of the Japanese Society for Artificial Intelligence* (2G2-03). Tokyo, Japan: Japanese Society for Artificial Intelligence.

Endo, Y., & Ogata, T. (2004a). A rhetorical analysis of a Japanese comic for hyper-comic system. In *Proceedings of the 18th Congress of the International Association of Empirical Aesthetics* (pp. 502–508). Rome, Italy: University of Rome Tre.

Endo, Y., & Ogata, T. (2004b). Multilayered discourse in hyper-comic. In *Proceedings of the 9th International Symposium on Artificial Life and Robotics*, Vol. 1 (pp. 49–52). Oita, Japan: International Society of Artificial Life and Robotics.

Freud, S. (1900). *Die traumdeutung*. Leipzig und Wien, Germany: Franz Deuticke.

Fujiwara, A., & Ogata, T. (2014). Generalizing and expanding the Propp-based story content grammar using the structural analysis of folktales. *Proceedings of the Forum on Information Technology*, *13*(2), 331–334.

Fujiwara, A., Ono, J., & Ogata, T. (2015). Propp ni motozuku story contents grammar wo riyō shita chishiki tōroku/kakunō kan'i tool ni motozuku kōsatsu [Consideration based on a knowledge registering/storing tool using Propp-based story content grammar]. In *Proceedings of the 48th Special Interest Group on Language Sense Processing Engineering* (pp. 57–66). Tokyo, Japan: The Japanese Society for Artificial Intelligence.

Fukushima, H. (2018). A phenomenological model for generating the tasting description of Japanese sake. In T. Ogata & S. Asakawa (Eds.), *Content generation through narrative communication and simulation* (pp. 265–285). Hershey, PA: IGI Global. doi:10.4018/978-1-5225-4775-4.ch008

Fukushima, H. (2019). Producing a mental representation of a cup of sake: A comparison of experimental methods and a tool for generating cognitive content. In T. Ogata & T. Akimoto (Eds.), *Post-narratology through computational and cognitive approaches* (pp. 430–448). Hershey, PA: IGI Global. doi:10.4018/978-1-5225-7979-3.ch012

Genette, G. (1972). *Discours du récit, Essai de méthode, Figures III.* Paris: Seuil.

Genette, G. (1982). *Palimpsestes: La littérature au second degré.* Paris: Seuil.

Genette, G. (1987). *Seuils.* Paris, Française: Seuil.

Genji Monogatari. (1993). In S. Yanai, S. Murofushi, Y. Ōasa, H. Suzuki, S. Fujii, & Y. Imanishi (Collation Eds.), Shin nihon koten bungaku taikei, 19 [New Japanese classic literature collection, Vol. 19]. Tokyo, Japan: Iwanami Shoten. (Original work published early 11th century, the Heian era)

Genji Monogatari. (1994). In S. Yanai, S. Murofushi, Y. Ōasa, H. Suzuki, S. Fujii, & Y. Imanishi (Collation Eds.), Shin nihon koten bungaku taikei, 20 [New Japanese classic literature collection, Vol. 20]. Tokyo, Japan: Iwanami Shoten. (Original work published early 11th century, the Heian era).

Genji Monogatari. (1995). In S. Yanai, S. Murofushi, Y. Ōasa, H. Suzuki, S. Fujii, & Y. Imanishi (Collation Eds.), Shin nihon koten bungaku taikei, 21 [New Japanese classic literature collection, Vol. 21]. Tokyo, Japan: Iwanami Shoten. (Original work published early 11th century, the Heian era)

Genji Monogatari. (1996). In S. Yanai, S. Murofushi, Y. Ōasa, H. Suzuki, S. Fujii, & Y. Imanishi (Collation Eds.), Shin nihon koten bungaku taikei, 22 [New Japanese classic literature collection, Vol. 22]. Tokyo, Japan: Iwanami Shoten. (Original work published early 11th century, the Heian era)

Genji Monogatari. (1997). In S. Yanai, S. Murofushi, Y. Ōasa, H. Suzuki, S. Fujii, & Y. Imanishi (Collation Eds.), Shin nihon koten bungaku taikei, 23 [New Japanese classic literature collection, Vol. 23]. Tokyo, Japan: Iwanami Shoten. (Original work published early 11th century, the Heian era)

Hasumi, S. (1979). *Eizō no shigaku* [Poetics of the film]. Tokyo, Japan: Chikuma Shobō.

Hori, K. (2007). *Sōzō katsudō shien system no riron to ōyō* [Theories and applications of creative activity support systems]. Tokyo, Japan: Ohmsha.

Iser, W. (1976). *Der akt des lesens*. Munchen, Germany: Wilhelm Fink Verlag.

Ito, T., & Ogata, T. (2018a). A framework for haiku generation from a narrative. *In Proceeding of the 2018 International Conference on Artificial Life and Robotics* (pp. 679–682). Oita, Japan: ALife Robotics Corporation Ltd. 10.5954/ICAROB.2018.OS17-5

Ito, T., & Ogata, T. (2018b). Haiku/monogatari seisei system no tameno haiku-ron no kōsatsu [Consideration of haiku theories for haiku/narrative generation systems]. In *Proceedings of the 57th Special Interest Group on Language Sense Processing Engineering* (pp. 23–24). Tokyo, Japan: The Japanese Society for Artificial Intelligence.

Ito, T., Ono, J., & Ogata, T. (2018a). Monogatari seisei no tame no minwa motif no riyō [Using folktale motifs for narrative generation] In *Proceedings of the 58th Special Interest Group on Language Sense Processing Engineering* (pp. 55–59). Tokyo, Japan: The Japanese Society for Artificial Intelligence.

Ito, T., Ono, J., & Ogata, T. (2018b). Using motifs of folktales for narrative generation. In *Proceedings of IEEE SMC 2018 Workshop on Informational and Cultural Narratology and Cognitive Content Generation (The 59th Special Interest Group on Language Sense Processing Engineering)* (pp. 8–11). Tokyo, Japan: The Japanese Society for Artificial Intelligence.

Ito, T., Sasaki, A., & Ogata, T. (2018). Kōkoku ni kansuru Creative Genome wo mochiita CM concept haiku no seisei [CM concept haiku generation using Creative Genome for advertisements]. In *Proceedings of the 35th Annual Conference of the Japan Cognitive Science Society* (pp. 516–517). Tokyo, Japan: Japan Cognitive Science Society.

Jauss, H. R. (1970). Literaturgeschichte als provokation. Frankfurt am Main: Suhrkamp Verlag.

Jung, C. G. (1999). *Genkei ron* [Psychological types] (M. Hayashi, Trans.). Tokyo, Japan: Kinokuniya Shoten. (Original work published 1921)

Kagotsurube Sato no Eizame. (1929). In S. Atsumi (Ed.), *Nihon gikyoku zenshū, 22* [Japanese theatre collection, Vol. 22] (pp. 2–116). Tokyo, Japan: Shun'yōdō. (Firstly played 1988 at Chitoseza, Tokyo)

Kanai, A. (2016). Non-story, nostalgia, and film cognition: Nostalgia-based narrative rhetoric composition. In T. Ogata & T. Akimoto (Eds.), *Computational and cognitive approaches to narratology* (pp. 376–390). Hershey, PA: IGI Global. doi:10.4018/978-1-5225-0432-0.ch015

Kanai, A. (2018a). Narrative simulation for film rhetoric composition with or without story and nostalgia effects. In T. Ogata & S. Asakawa (Eds.), *Content generation through narrative communication and simulation* (pp. 148–161). Hershey, PA: IGI Global. doi:10.4018/978-1-5225-4775-4.ch003

Kanai, A. (2018b). Setsudan gihō to monogatari [Cutting techniques and narratives]. In T. Ogata, Y. Kawamura, & A. Kanai (Eds.), *Jōhō monogatariron—Jinkōchinō, ninchi, shakai katei to monogatari seisei* [Informational narratology: Artificial intelligence/cognition/social process and narrative generation] (pp. 63–76). Tokyo, Japan: Hakutō Shobō.

Kanai, A. (2018c). Eizō ninchi hōryaku no kanōsei [The possibilities of image cognition methods]. In T. Ogata, Y. Kawamura, & A. Kanai (Eds.), *Jōhō monogatariron—Jinkōchinō, ninchi, shakai katei to monogatari seisei* [Informational narratology: Artificial intelligence/cognition/social process and narrative generation] (pp. 127–140). Tokyo, Japan: Hakutō Shobō.

Kanai, A. (2018d). Eizō kara story wo ninchisurukoto / shinaikoto—Ninchiteki reality no hassei youin [Recognizing/unrecognizing stories from the film: The generating factor of cognitives realities]. In T. Ogata, Y. Kawamura, & A. Kanai (Eds.), *Jōhō monogatariron—Jinkōchinō, ninchi, shakai katei to monogatari seisei* [Informational narratology: Artificial intelligence/cognition/social process and narrative generation] (pp. 141–153). Tokyo, Japan: Hakutō Shobō.

Kanai, A. (2018e). Documentary to nostalgia seisei [Documentaries and nostalgia generation]. In T. Ogata, Y. Kawamura, & A. Kanai (Eds.), *Jōhō monogatariron—Jinkōchinō, ninchi, shakai katei to monogatari seisei* [Informationl narratology: Artificial intelligence/cognition/social process and narrative generation] (pp. 289–302). Tokyo, Japan: Hakutō Shobō.

Kanai, A. (2019). Past, non-story narrative film, and nostalgia. In T. Ogata & T. Akimoto (Eds.), *Post-narratology through computational and cognitive approaches* (pp. 277–291). Hershey, PA: IGI Global. doi:10.4018/978-1-5225-7979-3.ch006

Kanai, A., & Ogata, T. (2004a). Aspect of non-story processing and film rhetoric composition in the narrative generation mechanism. In *Proceedings of the 9th International Symposium on Artificial Life and Robotics*, Vol. 1 (pp. 162–165). Oita, Japan: International Society of Artificial Life and Robotics.

Kanai, A., & Ogata, T. (2004b). Non-story processing on the film rhetoric composition system. In *Proceedings of the 18th Congress of the International Association of Empirical Aesthetics* (pp. 433–436). Rome, Italy: University of Rome Tre.

Kawamura, Y. (2016). An attempt of the commercial film production support system based on the image rhetoric of commercial film. In T. Ogata & T. Akimoto (Eds.), *Computational and cognitive approaches to narratology* (pp. 117–139). Hershey, PA: IGI Global. doi:10.4018/978-1-5225-0432-0.ch004

Kawamura, Y. (2018a). Practice and modeling of advertising communication strategy: Sender-driven and receiver-driven. In T. Ogata & S. Asakawa (Eds.), *Content generation through narrative communication and simulation* (pp. 358–379). Hershey, PA: IGI Global. doi:10.4018/978-1-5225-4775-4.ch013

Kawamura, Y. (2018b). Shakai katei ni yoru monogatari seisei [Narrative generation based on social processes]. In T. Ogata, Y. Kawamura, & A. Kanai (Eds.), *Jōhō monogatariron—Jinkōchinō, ninchi, shakai katei to monogatari seisei* [Informational narratology: Artificial intelligence/cognition/social process and narrative generation] (pp. 77–99). Tokyo, Japan: Hakutō Shobō.

Kawamura, Y. (2018c). Kōkoku eizō no kōsei yōso, kōzō bunseki [The elements and structural analyses of advertisement films]. In T. Ogata, Y. Kawamura, & A. Kanai (Eds.), *Jōhō monogatariron—Jinkōchinō, ninchi, shakai katei to monogatari seisei* [Informational narratology: Artificial intelligence/cognition/social process and narrative generation] (pp. 155–169). Tokyo, Japan: Hakutō Shobō.

Kawamura, Y. (2018d). Kōkoku eizō no image, han'nō bunseki [The images of advertisement films and response analyses]. In T. Ogata, Y. Kawamura, & A. Kanai (Eds.), *Jōhō monogatariron—Jinkōchinō, ninchi, shakai katei to monogatari seisei* [Informational narratology: Artificial intelligence/cognition/social process and narrative generation] (pp. 171–185). Tokyo, Japan: Hakutō Shobō.

Kawamura, Y. (2018e). Kōkoku eizō no seisei system no kaihatsu [The development of an advertisement film generation system]. In T. Ogata, Y. Kawamura, & A. Kanai (Eds.), *Jōhō monogatariron—Jinkōchinō, ninchi, shakai katei to monogatari seisei* [Informational narratology: Artificial intelligence/cognition/social process and narrative generation] (pp. 303–325). Tokyo, Japan: Hakutō Shobō.

Kawamura, Y. (2019). An attempt of the commercial film production support system based on the image rhetoric of commercial film. In T. Ogata & T. Akimoto (Eds.), *Post-narratology through computational and cognitive approaches* (pp. 292–317). Hershey, PA: IGI Global. doi:10.4018/978-1-5225-7979-3.ch007

Kawamura, Y., Ogata, T., & Kanai, A. (2018). Monogatari no wakugumi [The framework of a narrative]. In *Jōhō monogatariron—Jinkōchinō, ninchi, shakai katei to monogatari seisei* [Informational narratology: Artificial intelligence/cognition/social process and narrative generation] (pp. 1–16). Tokyo, Japan: Hakutō Shobō.

Kristeva, J. (1980). *Desire in language: A semiotic approach to literature and art.* Columbia University Press. (Original work published 1969)

Minakata, K. (2017). Neko ippiki no chikara ni yotte daifu to narishi hito no hanashi [A tale of a man who became very rich by the power of a cat]. In N. Watanabe (Ed.), *Nihon hihyō taizen* [Collection of Japanese critiques]. Tokyo, Japan: Kawade Shobō Shinsha. (Original work published 1912)

Mishima, Y. (1963). Gogo no eikō. Tokyo, Japan: Kōdansha.

Nitta, Y. (2018). Haiku-like aesthetic sentence composition through the narrative process. In T. Ogata & S. Asakawa (Eds.), *Content generation through narrative communication and simulation* (pp. 286–309). Hershey, PA: IGI Global. doi:10.4018/978-1-5225-4775-4.ch009

Nitta, Y. (2019). Elegant and efficient communication by haiku-like concise sentences. In T. Ogata & T. Akimoto (Eds.), *Post-narratology through computational and cognitive approaches* (pp. 449–468). Hershey, PA: IGI Global. doi:10.4018/978-1-5225-7979-3.ch013

Ogata, T. (1992). *Setsumei ni motozuku monogatrai seisei system ni kansuru kenkyū* [Study on an explanation-based narrative generation system] (Master's dissertation). Tsukuba University, Tokyo, Japan.

Ogata, T. (1995). *Monogatari seisei—Monogatari no tameno gihō to senryaku ni motozuku approach* [Narrative generation: An approach based on techniques and strategies for narrative] (Doctoral dissertation). The University of Tokyo, Tokyo, Japan.

Ogata, T. (2002). Expanded literary theory: Cognitive/computational expansion of literary theories and narratology. In *Proceedings of the 17th Congress of the International Association of Empirical Aesthetics* (pp. 163–166). Rome, Italy: University of Rome Tre.

Ogata, T. (2003a). Monogatari no tajūsei to kakuchō bungakuriron no gainen—System narratology ni mukete, I [Narrative multiplicity and the concept of expanded literary theory: Toward a system narratology, I]. In M. Yoshida (Ed.), *Fukuzatsu-kei shakai riron no shin chihei* [The new paradigm of complex social system theories] (pp. 127–181). Tokyo, Japan: Senshū Daigaku Shuppankyoku.

Ogata, T. (2003b). Kakuchō bungakuriron no kokoromi—System narratology ni mukete II [Attempts of an expanded literary theory: Toward a system narratology II]. In M. Yoshida (Ed.), *Fukuzatsu-kei shakai riron no shin chihei* [The new paradigm of complex social system theories] (pp. 309–356). Tokyo, Japan: Senshū Daigaku Shuppankyoku.

Ogata, T. (2014). Expanded literary theory for automatic narrative generation. In *Proceedings of the Joint Conference of the 7th International Conference on Soft Computing and Intelligent Systems and the 15th International Symposium on Advanced Intelligent Systems* (pp. 1558–1563). Tokyo, Japan: Japan Society of Fuzzy Theory and Intelligent Informatics.

Ogata, T. (2015). Building conceptual dictionaries for an integrated narrative generation system. *Journal of Robotics. Networking and Artificial Life*, *1*(4), 270–284. doi:10.2991/jrnal.2015.1.4.6

Ogata, T. (2016a). Computational and cognitive approaches to narratology from the perspective of narrative generation. In T. Ogata & T. Akimoto (Eds.), *Computational and cognitive approaches to narratology* (pp. 1–73). Hershey, PA: IGI Global. doi:10.4018/978-1-5225-0432-0.ch001

Ogata, T. (2016b). Kabuki as Multiple Narrative Structures. In T. Ogata & T. Akimoto (Eds.), *Computational and cognitive approaches to narratology* (pp. 391–422). Hershey, PA: IGI Global. doi:10.4018/978-1-5225-0432-0.ch016

Ogata, T. (2018a). An integrated approach to narrative generation: From Mishima and kabuki to narrative generation systems. In T. Ogata & S. Asakawa (Eds.), *Content generation through narrative communication and simulation* (pp. 49–147). Hershey, PA: IGI Global. doi:10.4018/978-1-5225-4775-4.ch002

Ogata, T. (2018b). Monogatari to ningen/shakai/kikai—Tajū monogatari kōzō kara ningen/kikai kyōsei-kei he [Narrative and human/society/machine: To human/machine co-existence systems from multiple narrative structures]. In T. Ogata, Y. Kawamura, & A. Kanai (Eds.), *Jōhō monogatariron—Jinkōchinō, ninchi, shakai katei to monogatari seisei* [Informational narratology: Artificial intelligence/cognition/social process and narrative generation]. Tokyo, Japan: Hakutō Shobō.

Ogata, T. (2018c). Monogatari no bunkai kara gōsei he—Mieru monogatari to mienai monogatari [From the deconstruction of a narrative to its synthesis: Visible narratives and invisible narratives]. In T. Ogata, Y. Kawamura, & A. Kanai (Eds.), *Jōhō monogatariron—Jinkōchinō, ninchi, shakai katei to monogatari seisei* [Informational narratology: Artificial intelligence/cognition/social process and narrative generation] (pp. 45–62). Tokyo, Japan: Hakutō Shobō.

Ogata, T. (2018d). Kabuki ni mukete (1)—Shiisei to henshūsei no monogatari kara tajūsei to jitsuzon no monogatari he [Toward kabuki (1): From narratives of arbitrariness and editing to narratives of multiplicity and real existences]. In T. Ogata, Y. Kawamura, & A. Kanai (Eds.), *Jōhō monogatariron—Jinkōchinō, ninchi, shakai katei to monogatari seisei* [Informational narratology: Artificial intelligence/cognition/social process and narrative generation] (pp. 187–208). Tokyo, Japan: Hakutō Shobō.

Ogata, T. (2018e). Kabuki ni mukete (2)—Tajū monogatari kōzō no shosō [Toward kabuki (2): Aspects of multiple narrative structures]. In T. Ogata, Y. Kawamura, & A. Kanai (Eds.), *Jōhō monogatariron—Jinkōchinō, ninchi, shakai katei to monogatari seisei* [Informational narratology: Artificial intelligence/cognition/social process and narrative generation] (pp. 209–244). Tokyo, Japan: Hakutō Shobō.

Ogata, T. (2018f). Tōgō monogatari seisei system—Mechanism kara contents he [Integrated narrative generation system: From mechanisms to contents]. In T. Ogata, Y. Kawamura, & A. Kanai (Eds.), *Jōhō monogatariron—Jinkōchinō, ninchi, shakai katei to monogatari seisei* [Informational narratology: Artificial intelligence/cognition/social process and narrative generation] (pp. 247–288). Tokyo, Japan: Hakutō Shobō.

Ogata, T. (2018g). Gaibu heno monogatari seisei mataha geinō jōhō system ni mukete [Narrative generation to the outside or toward a geino information system]. In T. Ogata, Y. Kawamura, & A. Kanai (Eds.), *Jōhō monogatariron—Jinkōchinō, ninchi, shakai katei to monogatari seisei* [Informational narratology: Artificial intelligence/cognition/social process and narrative generation] (pp. 327–353). Tokyo, Japan: Hakutō Shobō.

Ogata, T. (2018h). Naibu heno monogatari seisei mataha watakushi monogatari ni mukete [Narrative generation to the inside or toward I narratives]. In T. Ogata, Y. Kawamura, & A. Kanai (Eds.), *Jōhō monogatariron—Jinkōchinō, ninchi, shakai katei to monogatari seisei* [Informational narratology: Artificial intelligence/cognition/social process and narrative generation] (pp. 355–370). Tokyo, Japan: Hakutō Shobō.

Ogata, T. (2019a). A computational, cognitive, and narratological approach to narrative generation. In T. Ogata & T. Akimoto (Eds.), *Post-narratology through computational and cognitive approaches* (pp. 1–84). Hershey, PA: IGI Global. doi:10.4018/978-1-5225-7979-3.ch001

Ogata, T. (2019b). Toward a post-narratology or the narratology of narrative generation. In T. Ogata & T. Akimoto (Eds.), *Post-narratology through computational and cognitive approaches* (pp. 85–142). Hershey, PA: IGI Global. doi:10.4018/978-1-5225-7979-3.ch002

Ogata, T. (2019c). Kabuki as multiple narrative structures and narrative generation. In T. Ogata & T. Akimoto (Eds.), *Post-narratology through computational and cognitive approaches* (pp. 192–275). Hershey, PA: IGI Global. doi:10.4018/978-1-5225-7979-3.ch005

Ogata, T. (2019d). *Toward an integrated approach to narrative generation: Emerging research and opportunities*. Hershey, PA: IGI Global.

Ogata, T. (2019e). What are narrative generation phenomena? In T. Ogata (Ed.), *Toward an integrated approach to narrative generation: Emerging research and opportunities* (pp. 1–58). Hershey, PA: IGI Global.

Ogata, T. (2019f). Areas of narratives or narrative genres. In T. Ogata (Ed.), *Toward an integrated approach to narrative generation: Emerging research and opportunities* (pp. 59–161). Hershey, PA: IGI Global.

Ogata, T. (2019g). Narratology and post-narratology. In T. Ogata (Ed.), *Toward an integrated approach to narrative generation: Emerging research and opportunities* (pp. 162–314). Hershey, PA: IGI Global.

Ogata, T. (2019h). Theoretical or philosophical considerations for an integrated narrative generation approach. In T. Ogata (Ed.), *Toward an integrated approach to narrative generation: Emerging research and opportunities* (pp. 315–403). Hershey, PA: IGI Global.

Ogata, T., & Akimoto, T. (Eds.). (2016). *Computational and cognitive approaches to narratology*. Hershey, PA: IGI Global. doi:10.4018/978-1-5225-0432-0

Ogata, T., & Akimoto, T. (Eds.). (2019). *Post-narratology through computational and cognitive approaches*. Hershey, PA: IGI Global. doi:10.4018/978-1-5225-7979-3

Ogata, T., & Asakawa, S. (Eds.). (2018a). *Content generation through narrative communication and simulation*. Hershey, PA: IGI Global. doi:10.4018/978-1-5225-4775-4

Ogata, T., & Asakawa, S. (2018b). Aspects of content generation through narrative communication and simulation. In T. Ogata & S. Asakawa (Eds.), *Content generation through narrative communication and simulation* (pp. 1–47). Hershey, PA: IGI Global. doi:10.4018/978-1-5225-4775-4.ch001

Ogata, T., & Endo, Y. (2004). Monogatari gensetsu ron to hyper-comic system [Narrative discourse theory and a hyper-comic system]. In *Proceedings of the 18th Annual Conference of the Japanese Society for Artificial Intelligence* (2D2-01). Tokyo, Japan: The Japanese Society for Artificial Intelligence.

Ogata, T., Fujiwara, A., & Imabuchi, S. (2014). A way to relatively freely define the sequence of "functions" for a story: Generalizing a Propp-based story generation system. In *Proceeding of the 28th Annual Conference of the Japanese Society for Artificial Intelligence* (2F5-OS-01b-2in). Tokyo, Japan: The Japanese Society for Artificial Intelligence.

Ogata, T., & Kanai, A. (2010). *Monogatariron no jōhōgaku josetsu—Monogatari seisei no shisō to gijutsu wo megutte* [An introduction to informatics of narratology: Around the thoughts and technologies of narrative generation]. Tokyo, Japan: Gakubunsha.

Ogata, T., Kawamura, Y., & Kanai, A. (2018a). *Jōhō monogatariron—Jinkōchinō, ninchi, shakai katei to monogatari seisei* [Informational narratology: Artificial intelligence/cognition/social process and narrative generation]. Tokyo, Japan: Hakutō Shobō.

Ogata, T., Kawamura, Y., & Kanai, A. (2018b). Taiken toshite no monogatari—Gakumon toshite no monogatari wo sasaeru mono [Narratives as an experience: Things supporting narratives as a study]. In *Jōhō monogatariron—Jinkōchinō, ninchi, shakai katei to monogatari seisei* [Informational narratology: Artificial intelligence/cognition/social process and narrative generation] (pp. 101–123). Tokyo, Japan: Hakutō Shobō.

Ogata, T., Watanabe, K., Hori, K., & Ohsuga, S. (1995). A basic framework of the application of narrative generation system for integrated support of marketing/advertisement. *Journal of the Japan Society for Management Information, 4*(1), 19–42.

Oka, M. (1994). *Ijin sonota* [Foreigner and others]. Tokyo, Japan: Iwanami Shoten. (Original work published 1980)

Ono, J., & Ogata, T. (2018a). Surprise-based narrative generation in an automatic narrative generation game. In T. Ogata & S. Asakawa (Eds.), *Content generation through narrative communication and simulation* (pp. 162–185). Hershey, PA: IGI Global. doi:10.4018/978-1-5225-4775-4.ch004

Ono, J., & Ogata, T. (2018b). Acquiring short scripts and setting a case frame in each acquired script: Toward random story generation. *Journal of Robotics, Networking and Artificial Life, 5*(3), 212–216. doi:10.2991/jrnal.2018.5.3.15

Ono, J., Ogata, T., & Ito, T. (2019). A basic study for using folk tale's motifs to narrative generation. In *Proceedings of the 33rd Annual Conference of the Japanese Society for Artificial Intelligence* (1F2-NFC-1-05). Tokyo, Japan: The Japanese Society for Artificial Intelligence.

Ono, J., Sasaki, A., Ito, T., & Ogata, T. (2018). Creative Genome no parameter ni motozuku kōkoku story seisei ni mukete [Toward advertising story generation based on the parameters of Creative Genome]. In *Proceedings of the 58th Special Interest Group on Language Sense Processing Engineering* (pp. 35–47). Tokyo, Japan: The Japanese Society for Artificial Intelligence.

Ono, J., Sasaki, A., & Ogata, T. (2018). Kōkoku no Creative Genome parameter wo riyō shita shinsō gakusyū kikō no kokoromi—Kōkoku seisei to monogatari seisei no ketsugō ni mukete [Attempt of a deep learning mechanism using the Creative Genome parameters of advertisements: Toward the connection between advertisement generation and narrative generation]. In *Proceedings of the 60th Special Interest Group on Language Sense Processing Engineering* (p. 43). Tokyo, Japan: The Japanese Society for Artificial Intelligence.

Ono, J., Sasaki, A., & Ogata, T. (2019a). CM no bunseki ni motozuku monogatari seisei syuhō no kōsatsu—Tōgō monogatari seisei system to "Creative Genome" no yūki teki ketsugō ni mukete [A consideration of a method for narrative generation based on the analysis of CMs: Toward the organic combination between an integrated narrative generation system and "creative genome"]. In *Proceedings of the 61st Special Interest Group on Language Sense Processing Engineering* (pp. 39–46). Tokyo, Japan: The Japanese Society for Artificial Intelligence.

Ono, J., Sasaki, A., & Ogata, T. (2019b). Tōgō monogatari seisei system to "Creative genome" wo riyō shita kōkoku plot seisei syuhō no ichian [A method of advertisement plot generation using an integrated narrative generation system and "creative genome"]. In *Proceedings of the 33rd Annual Conference of the Japanese Society for Artificial Intelligence* (1F2-NFC-1-04). Tokyo, Japan: The Japanese Society for Artificial Intelligence.

Ozawa, T. (Ed.). (1976). *Nihonjin to minwa* [Japanese and the folktale]. Tokyo, Japan: Gyōsei.

Ozawa, T. (Ed.). (1997). *Mukashibanashi nyūmon* [An introduction to the folktale]. Tokyo, Japan: Gyōsei.

Ozawa, T. (1999). *Mukashibanashi no gohō* [Narrative grammar of the folktale]. Tokyo, Japan: Fukuinkan Shoten.

Propp, V. Y. (1968). *Morphology of the folktale* (L. Scott, Trans.). Austin, TX: University of Texas Press. (Original work published 1928)

Rumelhart, D. E. (1975). Notes on a schema for stories. In D. G. Bobrow & A. Collins (Eds.), *Representation and understanding: Studies in cognitive science*. Academic Press. doi:10.1016/B978-0-12-108550-6.50013-6

Seino, K., Enomoto, Y., & Miyazawa, S. (2018). Narrative analysis of employment support for students with developmental disabilities: Using an objective analysis of free-expression answers. In T. Ogata & S. Asakawa (Eds.), *Content generation through narrative communication and simulation* (pp. 341–357). Hershey, PA: IGI Global. doi:10.4018/978-1-5225-4775-4.ch012

Seino, K., Haruna, Y., & Ishizaki, S. (2016). The employment status and support needs of persons with disabilities in Japan: An analysis of narrative using narratology and text mining on a National Survey. In T. Ogata & T. Akimoto (Eds.), *Computational and cognitive approaches to narratology* (pp. 245–275). Hershey, PA: IGI Global. doi:10.4018/978-1-5225-0432-0.ch009

Seino, K., Haruna, Y., & Ishizaki, S. (2019). The employment status and support needs of persons with disabilities in Japan: An analysis of narrative using narratology and text mining. In T. Ogata & T. Akimoto (Eds.), *Post-narratology through computational and cognitive approaches* (pp. 397–429). Hershey, PA: IGI Global. doi:10.4018/978-1-5225-7979-3.ch011

Seki, K. (1957). *Mukashibanashi to shōwa* [Folktales and funny stories]. Tokyo, Japan: Iwasaki Shoten.

Seki, K. (1966). *Mukashibanashi no rekishi* [The history of the folktale]. Tokyo, Japan: Shibundō.

Seki, K. (1977). *Nihon no mukashibanashi—Hikaku kenkyū josetsu* [Japanese folktales: A introduction to the comparative study]. Tokyo, Japan: NHK Shuppan.

Seki, K. (1978a). *Nihon mukashibanashi taisei, 2—Honkaku mukashibanashi, 1* [Compilation of Japanese folktales, Vol. 2: Ordinary folktales, No. 1]. Tokyo, Japan: Kadokawa Shoten.

Seki, K. (1978b). *Nihon mukashibanashi taisei, 3—Honkaku mukashibanashi, 2* [Compilation of Japanese folktales, Vol. 3: Ordinary folktales, No. 2]. Tokyo, Japan: Kadokawa Shoten.

Seki, K. (1978c). *Nihon mukashibanashi taisei, 4—Honkaku mukashibanashi, 3* [Compilation of Japanese folktales, Vol. 4: Ordinary folktales, No. 3]. Tokyo, Japan: Kadokawa Shoten.

Seki, K. (1978d). *Nihon mukashibanashi taisei, 5—Honkaku mukashibanashi, 4* [Compilation of Japanese folktales, Vol. 5: Ordinary folktales, No. 4]. Tokyo, Japan: Kadokawa Shoten.

Seki, K. (1978e). *Nihon mukashibanashi taisei, 6—Honkaku mukashibanashi, 5* [Compilation of Japanese folktales, Vol. 6: Ordinary folktales, No. 5]. Tokyo, Japan: Kadokawa Shoten.

Seki, K. (1979a). *Nihon mukashibanashi taisei, 1—Dōbutsu mukashibanashi* [Compilation of Japanese folktales, Vol. 1: Animal tales]. Tokyo, Japan: Kadokawa Shoten.

Seki, K. (1979b). *Nihon mukashibanashi taisei, 7—Honkaku mukashibanashi, 6* [Compilation of Japanese folktales, Vol. 7: Ordinary folktales, No. 6]. Tokyo, Japan: Kadokawa Shoten.

Seki, K. (1979c). *Nihon mukashibanashi taisei, 8—Warai-banashi (Shōwa), 1* [Compilation of Japanese folktales, Vol. 8: Jokes and anecdotes, No. 1]. Tokyo, Japan: Kadokawa Shoten.

Seki, K. (1979d). *Nihon mukashibanashi taisei, 9—Warai-banashi (Shōwa), 2* [Compilation of Japanese folktales, Vol. 9: Jokes and anecdotes, No. 2]. Tokyo, Japan: Kadokawa Shoten.

Seki, K. (1980). *Nihon mukashibanashi taisei, 10—Warai-banashi (Shōwa), 3* [Compilation of Japanese folktales, Vol. 10: Jokes and anecdotes, No. 3]. Tokyo, Japan: Kadokawa Shoten.

Seki, K., Nomura, J., & Ōshima, H. (Eds.). (1979). *Nihon mukashibanashi taisei, 12—Kenkyū hen* [Compilation of Japanese folktales, Vol. 12: Researches]. Tokyo, Japan: Kadokawa Shoten.

Seki, K., Nomura, J., & Ōshima, H. (Eds.). (1980). *Nihon mukashibanashi taisei, 11—Shiryō hen* [Compilation of Japanese folktales, Vol. 11: Data]. Tokyo, Japan: Kadokawa Shoten.

Seraku, T., & Ohtani, A. (2016). The word order flexibility in Japanese novels: A dynamic syntax perspective. In T. Ogata & T. Akimoto (Eds.), *Computational and cognitive approaches to narratology* (pp. 213–244). Hershey, PA: IGI Global. doi:10.4018/978-1-5225-0432-0.ch008

Shimauchi, K. (1989). *Genji monogatari no wakei-gaku* [Theory of story's types of The tale of Genji]. Tokyo, Japan: Perikansha.

Shirozu, N., & Matsushita, M. (2018). Collaboration for supporting the externalization of the travel narratives of the elderly: A case study. In T. Ogata & S. Asakawa (Eds.), *Content generation through narrative communication and simulation* (pp. 399–416). Hershey, PA: IGI Global. doi:10.4018/978-1-5225-4775-4.ch015

Shklovsky, V. (1990). *Theory of prose* (B. Sher, Trans.). Dalkey Archive Press. (Original work published 1925)

Sugamura, G., Kushizaki, M., Ikemi, A., & Nabeshima, K. (2018). Shinri-ryōhō to metaphor [Psychotherapy and metaphors]. In *Proceedings of the 60th Special Interest Group on Language Sense Processing Engineering* (pp. 49–50). Tokyo, Japan: The Japanese Society for Artificial Intelligence.

Tadaki, K., & Abe, A. (2019). The potentiality of first-person views: Analyzing narratives during art appreciation. In T. Ogata & T. Akimoto (Eds.), *Post-narratology through computational and cognitive approaches* (pp. 375–396). Hershey, PA: IGI Global. doi:10.4018/978-1-5225-7979-3.ch010

Takeuchi, K. (2016). Thesaurus with predicate-argument structure to provide base framework to determine states, actions, and change-of-states. In T. Ogata & T. Akimoto (Eds.), *Computational and cognitive approaches to narratology* (pp. 192–212). Hershey, PA: IGI Global. doi:10.4018/978-1-5225-0432-0.ch007

Teramoto, Y., & Nakanishi, A. (1993). *Chi no tenkansha-tachi—Business trick-star* [Knowledge's transformers: Business trick-stars]. Tokyo, Japan: NTT Shuppan.

Tsuchihashi, S., & Ogata, T. (2009). Toward the narrative generation by quotations and anagram. In *Proceeding of the 23rd Annual Conference of the Japanese Society for Artificial Intelligence* (1J1-OS2-2). Tokyo, Japan: Japanese Society for Artificial Intelligence.

Tsuchiyama, G. (2016). Quantitative research into narrative: Statistical analysis of "The tale of Genji". In T. Ogata & T. Akimoto (Eds.), *Computational and cognitive approaches to narratology* (pp. 276–302). Hershey, PA: IGI Global. doi:10.4018/978-1-5225-0432-0.ch010

Utsumi, A. (2018). Literary metaphor comprehension and production: A unified view. In T. Ogata & S. Asakawa (Eds.), *Content generation through narrative communication and simulation* (pp. 310–327). Hershey, PA: IGI Global. doi:10.4018/978-1-5225-4775-4.ch010

White, H. (1973). *Metahistory: The historical imagination in nineteenth-century Europe*. Johns Hopkins University Press.

Yanagita, K. (2016). *Tōno monogatari* [Tōno story]. Tokyo, Japan: Shinchōsha. (Original work published 1910)

Yoshida, M. (Ed.). (2003). *Fukuzatsu-kei shakai riron no shin chihei* [New paradigm of complex social system theories]. Tokyo, Japan: Senshū Daigaku Shuppankyoku.

Chapter 1
An Integrated Narrative Generation System:
Synthesis and Expansion

ABSTRACT

This chapter introduces an integrated narrative generation system (INGS), which is one of the central parts in the author's study that aims to define an integrated approach to narrative generation. In the narrative generation study, all parts are considered as components that serve the extreme generative and productive goals, and the most important element of the goals corresponds to the design and development of INGS. This chapter divides INGS into two parts: mechanisms and contents. The main generation modules in INGS are for story, narrative discourse, and expression including language, music, and picture images, which are performed using various narrative techniques. On the other hand, INGS has dictionaries, including conceptual and language, and various knowledge, including story content and state-event transformation knowledge bases, which correspond to the aspect of contents. This chapter explains the mechanisms in the order of the entire process and states in detail the conceptual dictionaries, mainly noun and verb conceptual dictionaries.

INTRODUCTION

The "Integrated Narrative Generation System (INGS)" is one of the central parts in this study that aims at an integrated approach to narrative generation

DOI: 10.4018/978-1-5225-9943-2.ch001

for post-narratology. The integrated approach to narrative generation, or post-narratology as the word in this book's title, deals with interdisciplinary and crossdisciplinary approaches to narrative and narrative generation as well as many related fields and concepts. Moreover, the author introduces philosophical concepts supporting this study and describes narrative analyses of *kabuki* and advertisements. However, the objectives of this study do not stop with the research of the receptive aspects such as consideration, survey, and analysis regarding narratives and narrative generation.

In this narrative generation study, all parts are considered as components that serve the extreme generative and productive goals. The first and the most important element of the goals corresponds to the design and development of INGS. Further, the "*Geinō* Information System (GIS)," discussed in Chapter 2 and again in more detail in Chapter 4, is related to the system's framework that includes an INGS or many INGSs and is executed with an INGS or INGSs. Ogata (2016, 2018a, 2018b, 2019a), which is an enhanced version of Ogata (2016), provides the systematic and comprehensive descriptions of INGS. Although this chapter overlaps with previous papers, with respect to INGS as project, the system has been gradually developed and this new chapter shows the entire system, including newly revised and expanded parts. While the entire structure of the description is based on the above referenced papers, the content is extended and supported with new examples of narrative generation.

The following part describes the chapter's structure. First, **BACKGROUND: COMMON LISP** covers Common Lisp, the main programming language for our system development, including INGS and other systems. Although, in previous papers, the author did not explain in detail such purely technological problems, the discussion here reveals that the Lisp programming language, being a nonprocedural and functional language with flexible scalability, was the most adequate programming language for INGS and related other systems and has been expanded gradually.

Next **GENERATION PROCESSP OF THE INTEGRAYED NARRATIVE GENERATION SYSTEM** describes the research and development process of the author itself used as a kind of narrative generation process, referring to many related papers.

Continuously, **THE ARCHITECTURE OF INTEGRATED NARRATIVE GENERATION SYSTEM** is the central section in this chapter. As in Ogata (2018b), this chapter divides INGS into two parts: "mechanisms" and "contents." Since the literary criticism of Russian formalism (1910s–1930s), narratology has divided a narrative or narrative work into form and content and, especially, emphasized the formal side of a narrative. This had significance or meaning as a kind of antithesis to other literary schools of thought (e.g., "Socialist Realism") that emphasize the contents described in narratives and literary works and consider "what is described" to be the core of values of narratives and literary works. In Japan, many theoretical literary studies, such as Natsume Sōseki (1867–1916) (2007) and Takaaki Yoshimoto (1924–2012) (1965), considered the formal aspect in narratives and literature works to be important. Although the author recognizes an important value in narrative formal aspect, both formal and content aspects are respectively important when we aim at the development of actual narrative generation. Therefore, the author divides a narrative generation process into formal aspects and content aspects. The chief objective of previous narrative generation studies was to present and develop the mechanisms or forms of narrative generation. In contrast, for performing narrative generation, an unavoidable problem is how we richly give content information to a generating narrative. In the current version of INGS, rich content information is mainly given through conceptual dictionaries and narrative content knowledge bases.

Beyond such a research-oriented approach, such as how to produce excellent narrative and literary works using narrative generation systems, arise such important problems as substantial discussions and technological developments regarding the side of narrative contents that emerge in a central position of narrative generation. Accordingly, the author presents a paper's construction in which the formal mechanism and contents in narrative generation are clearly divided.

BACKGROUND: COMMON LISP

This section explains the programming language called Common Lisp, which has been used to develop INGS and narrative generation programs by the author.

Lisp (LISt Processor) (Winston & Horn, 1989; Yuasa & Hagiya, 1986) is a programming language created by John McCarthy (1927–2011), who is an AI researcher and is referred to as the presenter of the "frame problem." Although Lisp has branched off into many dialects, the current main Lisp systems are Common Lisp (Steel, 1990), which is a large and comprehensive programming language and environment. In contrast, Schema is an extremely abstracted programming language based on Lisp. Although Lisp is one of the most classical programming language, it is also the most advanced language.

One of the reasons is that Lisp has been used in the development of many AI programs ranging from several decades ago to the present, especially in 1980s, to describe programming resources such as editors, operating systems, and compilers. Moreover, Lisp has influenced the development of subsequent programming technololgies. For example, dynamic linking, interactive programming execution environments, and object-oriented programming are examples of programming environments and flexible developmental styles derived from Lisp.

Lisp has many technological features. Theoretically, Lisp is a language executed using only functional processing based on mathematical lambda calculus and is a computational system modeled and abstracted via the evaluation of one or more argument(s) to a function and the application of computational execution. Different from many other programming languages, Lisp is a formula-oriented programming language in which formulas and data sentences are not differentiated and all of programming codes and data are described as formulas. When a formula is evaluated, it generates value(s) as one or more symbol(s) or one or more list(s). A formula can be embedded in another formula. Further, Lisp has the characteristic of the programming language that facilitated the programming. It is related to a feature of the descriptive form, namely, data and program have the same form. In particular, when a description of data is written in a form that can be executed as a description of the program, the data description can be treated as a program description based on the evaluation. Macro functions are also used for the role. Although each macro function is used as an ordinary function, it represents the transformation of the programming source codes instead of representing a formula to be evaluated. Macro functions accept that Lisp programmers create new syntactic forms in the Lisp language. Further, Lisp has various data types including integer type, character type, and array, similar to other programming languages. Function is also a data type and can define a function and changes to the definition of a function rather in the execution process of a program. Moreover, Lisp sometimes uses the recursive method or inductive definition that a function contains itself in the function. To avoid infinite regress, "itself" in the function needs to be carefully defined. Lisp is a very good language and environment for the prototyping approach to program the development of simple structures, descriptive simplicity, and high modularity. These are the reasons that Lisp has been used in many AI programming projects. Given that Lisp was originally an interpretive programming language, it was easy to repeat the programming cycle between description and execution. This characteristic facilitates rapid repetition of the development and test in the small-scale program stage. As such, the highly independent definition of functions in Lisp facilitates high-quality system development at the level of partial components. However, currently, given that Lisp compilers are available, many Lisp systems have processing speeds that are comparable to other high-speed programming languages.

The author was involved in a project that transformed Golden Common Lisp (GCL) on personal computers, developed at Massachusettes Institute of Technology (MIT), into the personal computers of Fujitsu. The software system and developed AI systems (a Common Lisp tutorial system called *San Marco Lisp Explorer* written by P. Winston) were sold with an expert system development tool written in GCL. In particular, the author developed the Japanese-language version based on the translation. Thereafter, all of the narrative generation systems by the author have been described using Common Lisp. Although the present INGS is a system that integrates previous narrative generation systems into an integrated framework, the integration has been conducted without large design changes because partial program systems have been written. Moreover, although INGS required various extensions and changes throughout the development process, they were comparatively easy for the characterization of Common Lisp that used a small-scale function as a basic unit.

Furthermore, another merit in Lisp programming is the point that can represent the program theories and methods developed by artificial intelligence (AI) and cognitive science (Anzai, Saeki, & Muto, 1981; Anzai, Saeki, & Namba, 1982; Tanaka, Motoyoshi, & Yamanashi, 1983). INGS also utilizes the descriptions that transform the frame representation by Marvin Minsky (1927–201619) (1975) and the semantic networks of Quillian (1968).

GENERATION PROCESS OF THE INTEGRATED NARRATIVE GENERATION SYSTEM

Currently, one of the main objectives is to synthesize all the work of previous studies of the author into an organic system called Integrated Narrative Generation System (INGS). INGS is intended as a foundation for the production and practice of narrative personal and social levels. The design and development are a necessary process to the next step.

INGS can comprehend all steps from the narrative deep, semantic, and structural level to the level of surface representation using several representation media to process consistently them under the mechanism of transformation. Moreover, at the same time, it can synthesize diverse types of narrative rhetorical techniques. INGS consists of both a generation mechanism for a story (narrative content), narrative discourse, and narrative representation, and knowledge elements including several types of conceptual dictionaries. Moreover, although INGS is currently being developed, it is already in operation.

INGS has various characteristics. In this system, we can conduct individual research and development of partial modules. Moreover, we can also investigate the part researched and developed individually by incorporation into the entire system of INGS. Thus, INGS facilitates the conducting of partial research and integrated research in parallel. Moreover, the "versatile" feature means that INGS aims to acquire the ability to be used and applied to achieve a variety of various goals related to narrative generation without the limitation and constraint of themes and worlds. By contrast, many existing narrative generation systems deal with limited themes, knowledge, and narrative worlds. However, the author also intends to make intentionally specific narrative genres, such as *kabuki*, which is described in Chapter 2 in addition to personal and individual narratives that will be discussed in Chapter 4. These aspects indicate that INGS can be applied to achieve a variety of goals related to narrative generation without the limitation of narrative genres. Furthermore, narrative "diversity" is realized through which diverse narrative techniques for narrative structures and representations are executed. In addition, the corresponding knowledge contents are selected and used in each generation point in a narrative generation process under the consistent management of several control methods.

This chapter divides the INGS architecture into two aspects of form or mechanism and content. The side of mechanism corresponds to a narrative formal generation process and the content is the side of objects used by the mechanism. Although it is possible to adopt knowledge-oriented research and development, a formal method for storing and executing knowledge, for the mechanism used in this case in particular, should be the preposition of the truly large-scaled organizational collection of knowledge contents. For example, enlarging the content requires developed mechanisms. Based on the research and development of INGS, narrative generation mechanisms have been established. Therefore, a central part of the future development will be changed to the aspect of contents. Although, in this chapter, the author introduces the current version of INGS based on the division between mechanisms and contents, this section provides the author's research history of INGS as a generation process or a story generation, as a preparation for the description of INGS architecture in the next section.

As in previous papers related to INGS, papers such as Ogata (2010b,2013), Ogata, Akimoto, and Ono (2014), Ogata and Kanai (2010a), and Oishi, Akimoto, Ono, Onodera, Zhan, and Ogata (2011) showed the entire plan and design of INGS. Akimoto and Ogata (2012a, 2012c, 2014a), Ogata (2013, 2014a, 2014b, 2015), and Ogata and Akimoto (2010a, 2010b, 2011) explained the system architectures for each instance and further, Ogata (2016) introduced the version at the time point depending on the background and theoretical consideration of INGS.

In the first stage of the author's narrative generation research, the two systems that exhibited two sides of narrative generation, i.e., the interdisciplinary characteristics with narratology and literary theories and the integration of modules were used. These ideas have been used in several studies and development investigation. A proposed idea is a story generation system using a kind of story grammar based on the classical folktale analysis theory by Vradimir Propp (1895–1970) (1968) and a planning method by Roger Schank (1946–) and Abelson (1977) was used in the generation part for the characteristics of the action sequences. Moreover, a learning theory, explanation-based learning, using AI (Dejong, 1993, 2014) was also used for generation control. This system was also the first concretization of the "expanded literary theory" (Chapter 4 (Ogata, 2019d) in Ogata (2019c)) by the author (Ogata, 1992, 2007, 2010b; Ogata & Terano, 1991a, 1991b, 1991c, 1992a, 1992b, 1992c). Furthermore, the author aimed to perform narrative generation using an organizational collection of a variety of methods instead of using only one method to develop a system based on narrative techniques, narrative strategies, and other knowledge bases (Ogata, 1995; Ogata, Hori, & Ohsuga, 1994, 1995, 1996a, 1996b).

These basic narrative generation systems are connected to the partial design, development, and experiments of the respective parts (INGS corresponds to a current integration of the results). For instance, the author developed conceptual dictionaries for event generation (Ogata, 2015; Ogata, Ono, Herai, Igarashi, & Shirai, 2015; Oishi & Ogata, 2012; Oishi, Ogata, & Onodera, 2011; Oishi, Kurisawa, Kamada, Fukuda, Akimoto, & Ogata, 2012) and a story content knowledge base used in the story techniques (Ogata, 2010c, 2014a, 2014b; Akimoto & Ogata, 2014a). The story grammar in the Propp-based story generation system was extended to story content grammar (Imabuchi & Ogata, 2012b, 2013, 2014). Moreover, a state-event transformation knowledge base (Onodera, Akimoto, & Ogata, 2012a, 2012b; Onodera & Ogata, 2012a,

2012b; Fukuda & Ogata, 2014) was developed to infer and generate the former state and later state of generated events in a story. Other main current topics include the survey of the sequential and structural knowledge for stories, their storage into the story content knowledge base, and the acquisition of automatic or semi-automatic event sequential knowledge (Arai, Ono, & Ogata, 2016; Imabuchi & Ogata, 2012a; Ono & Ogata, 2013). A new idea for generation control of the narrative discourse mechanism (Akimoto & Ogata, 2009, 2010a, 2013b, 2014b) was also investigated as a mechanism based on a virtual narrator and narratee, using the reception theory by Hans Robert Jauss (1921–1997) (1970).

There are studies related to the expanded literary theory. Regarding the narrative discourse techniques (Ogata, 1999a, 1999b, 2003c, 2004a; Ogata, Mukouyama, & Kanai, 2000; Ogata & Yamakage, 2003, 2004; Ogata & Yazawa, 2000) based on the narrative discourse theory by Gerald Genette (1930–2018) (1972), the following studies were conducted and experimental programs were developed: temporal order transformation (Akimoto & Ogata, 2013a, 2013b; Mukouyama & Ogata, 2002; Mukouyama, Kanai, & Ogata, 1999); viewpoint or perspective (Akimoto & Ogata, 2014b; Ueda & Ogata, 2004a, 2004b); duration, speed, or tempo (Umehara & Ogata, 2003); distance (Yamakage & Ogata, 2003; Ogata & Yamakage, 2003, 2004); voice (Ueda & Ogata, 2004a, 2004b). Moreover, recently, the author has begun the systematization of motifs using Keigo Seki's (1899–1990) (1957, 1966, 1977) and Seki, Nomura, & Ōshima, 1979, 1980) folktale study (Ito, Ono, & Ogata, 2018a, 2018b; Ono, Ogata, & Ito, 2019). Although this research is related to the story generation mechanism and story content knowledge base in INGS, it is also positioned as a practice of expanded literary theory in addition to the Propp-based story generation mechanism and story content grammar.

Although a direction or objective of INGS is one that does not use a fixed generation order among the modules, this idea was chiefly investigated using the music generation mechanism in INGS (Akimoto & Ogata, 2013c, 2014c, 2014d; Kobayashi & Ogata, 2004a, 2004b; Ogata & Akimoto, 2007; Ogata & Kobayashi, 2004). This system also used a musical theory, "Generative Theory of Tonal Music (GTTM)" (Lerdahl & Jackendoff, 1983).The language generation mechanism produces sequential sentences from narrative conceptual structures owing to the story generation process and the narrative discourse mechanism arrangement of the generated word notations (Kumagai, Funakoshi, Akimoto, & Ogata, 2012; Sato, Kobayashi, Yasuda, & Ogata, 2008) using a language notation dictionary (Kurisawa, Michimata, Kamada, & Ogata, 2012; Ogata, Ono, Herai, Igarashi, & Shirai, 2015; Ogata & Ono,

2015). The frequency and co-occurrence information of verbs were also used to adjust the readability and understandability of vocabularies (Kamada & Ogata, 2013a, 2013b, 2013c). Although the author has initiated a study of automatic CG representation of a generated conceptual representation of a story, several preliminary studies have already been conducted on this topic. For example, a visual image transformation processing from narrative conceptual representation was attempted using a simple mapping method based on an image database and image scripts (Tomite, Ogata, & Hanada, 2009). Ogata (2008) and Ogata, Matsuda, Naito, Manage, Takahashi, Nakashima, Yoshio, and Numata (2007) showed a basic consideration of the visual imaging of narratives. Other image-relating studies include rhetorical analyses of image techniques (Kanai, Mukouyama, Ogata, & Kawamura, 2000; Kanai & Ogata, 2004a; Kanai, Ogata, & Shinohara, 2002, 2003; Mukouyama, Kanai, & Ogata, 2001); a camera-work analysis of a movie film (Ogata, Tachibana, & Tomite, 2009; Tachibana & Ogata, 2009a, 2009b).

In addition to these concrete system design and development, the author has presented the following theoretical concepts to direct the system development from the early stage of the narrative generation study.

Firstly, the proposition of a *Geinō* Information System (GIS) has been based on the survey and analysis of modern and contemporary *geinōjins* (entertainers and performers) and *geinō* production companies (entertainment agencies), and Japanese folklore, *kabuki*, and *ningyō jōruri* (Kawamura & Ogata, 1999, 2000a, 2000b, 2000c, 2000d; Ogata, 1998, 2002a; Ogata & Kawamura, 1998a, 1998b, 1999a, 1999b, 1999c). The multiple narrative structures model produces a value of narrative generation in multiplicity. Although the basic structure was in the multiplicity according to the development and complication of an original communication structure seen in folktale-like narratives, the author expanded this concept into the multiplicity in various elements in narrative generation, such as stories and characters (Ogata, 1997, 2000a, 2000b, 2004b). Furthermore, the expanded literary theory meant the organic and organizational introduction of narratology and literary theories into narrative generation study as artificial intelligence and the introduction of design and operational features in narratology and literary theories. However, it was not a method for incorporating narratology and literary theories into AI and cognitive science, but instead was a practical vision for the development of a new narratology or literary theory (Ogata, 2001, 2002b, 2002c, 2003a, 2003b, 2014d; Akimoto & Ogata, 2010b, 2012b; Ogata, Kanai, & Aoyagi, 2003; Ogata & Morita, 2002; Ogata, Tokaji, & Morita, 2002).

Additionally, the author has explored several other narrative generation concepts for INGS. Firstly, in a theoretical concept called "fluidity-fixation" (Ogata, 2010a, 2014c), the author regarded a narrative generation process itself and a produced work (content) as a fluid process and the fixation respectively. In previous narrative contents, narrative fluid processes were conducted through the narrators' brains and human groups but was never implemented. By contrast, narrative generation systems enable the representation of a fluid process and facilitate a narrative work. Through the concept of fluidity and fixation, narrative representation forms and the creation processes will be diverse and expanded to enable different narrative productions and experiments from the previous forms. The aim of the author is not only the visualization and explanation of the fluid process of a narrative work, but also the determination of the mechanisms of a narrative generation process or the fluid process itself. In this approach, in addition to the previous form through which a novel written by a writer becomes a work (fixation), a new possibility is revealed in which the novel writing process itself by the writer is also a narrative work. Concerning the next concept, "norm-deviation," the conceptual dictionaries and other knowledge components in INGS are dependent on "realistic standard," namely the standard (norm) of ordinary possibility in the real-world, and the normal narrative generation generates narratives according to the realistic standard. By contrast, the author attempted to apply various deviation techniques, especially using the defamiliarization theory by Viktor Shklovsky (1893–1984) (1990), Friedrich Brecht (1898–1956) (1964) to the generated normal narratives to produce narratives that deviated organizationally from the norm. Originally, the norm-deviation was an idea for synthesizing various previous studies by the author, including consideration of the relationships between melodies and variations in music, thereby systematizing the story-cutting techniques with Kanai using a movie by Sergei Parajanof (1924–1990) (1968) and so on, in addition to the analyses of advertising rhetorical techniques based on defamiliarization (Abe, Ogata, & Hanada, 2008; Abe, Ogata, & Onodera, 2009; Ogata & Kanai, 2010b; Ogata, Watanabe, Hori, & Ohsuga, 1995a, 1995b; Ono, Zhang, Oishi, Ogata, & Onodera, 2011; Zhang, Ono, & Ogata, 2011, 2012). Further, the relation of a story with the narrative discourse (Ogata, 2000c) is also a kind of norm and deviation. A simulation experiment based on the acquisition of defamiliarization techniques (Ono, Zhang, & Ogata, 2012) was partially incorporated into INGS (Akimoto & Ogata, 2015a; Kurisawa & Ogata, 2013a, 2013b, 2014; Ogata, 2014d). Finally, a control concept of INGS does not preliminarily determine the generation order and also accepts repetitive generation. The author also aimed to develop a type of processing based on circular narrative generation and the selection and synthesis of specific results. This kind of processing is called "circular narrative control" as third philosophical concept for the author's narrative generation or INGS.

The preceding philosophical concepts for the narrative generation will frequently be referenced, mentioned, and explained in various places in this book. Moreover, the author designed and developed several application systems according to the partial utilization and development of INGS and other already developed narrative generation systems.

KOSERUBE version 1 is a kind of automated picture-story show that generates a story and its simple representation, including animation, still picture, music, and natural language (Akimoto, Imabuchi, Endo, Ono, Kurisawa, Kamada, & Ogata, 2013; Imabuchi, Akimoto, Ono, & Ogata, 2012; Ono & Ogata, 2016). A hyper-comic system is an electrical *manga* (comic) that has semi-automatic narrative generation functions using hypertext structures. The author presented the concept and implementation of "multi-layered narrative discourse" (Endo & Ogata, 2002, 2003a, 2003b, 2004a, 2004b; Mori & Ogata, 2005; Ogata & Endo, 2004) which is the technique for representing several narrative discourse techniques in a *koma* (frame) simultaneously using the *manga* work by Rumiko Takahashi (1957–) (2003a, 2003b, 2004a, 2004b, 2004c, 2004d, 2004e, 2004f, 2005a, 2005b, 2005c, 2005d, 2005e, 2005f, 2006). A musical variation system transforms a narrative to a piece of music by applying the narrative discourse techniques by Genette (1972). In particular, the system transforms a narrative structure to the corresponding musical form using the narrative discourse techniques for generating narrative discourse structures. In addition, the system could transform a musical structure to other narrative structures (Ogata & Kobayashi, 2004; Kobayashi & Ogata, 2004a, 2004b; Ogata & Akimoto, 2007). "Narrative Forest (NF)" is composed of a narrative generation part that automatically generates narrative conceptual structures and the surface representations of natural language and music, and the user interface part that corresponds to the visual image of a tree to the generation process of a narrative tree structure. The user can see a narrative generation process based on the user interface with a figurative image (Akimoto, Ono, & Ogata, 2012a, 2012b). The author generally referred to the user interfaces, such as NF, that can visualize both the fluid process and fixed process of narrative interfaces (Ogata & Ono, 2013). Although the author developed applicable advertising generation systems that bridge the analysis results of television advertisement films and the development of narrative generation mechanisms, the focus of the analysis was the consideration of the event sequences of advertising scenarios in addition to the defamiliarization in the semantic elements that construct each event. The latter defamiliarization was regarded as a generalized and expanded norm-deviation mechanism and it was partially used as an adjustment mechanism for deviating a generated narrative from the realistic standard or ordinary actions. The author also developed an experimental story-cutting system (Kanai, Ogata, & Shinohara, 2002, 2003) that applied the story-cutting theory by Kanai (Kanai & Niwa, 2008) to the author's narrative generation system. Moreover, fragmental application systems or experimental systems based on the intertextual literary theory by Julia Kristeva (1941–) (1980) were also developed (Oishi, Ogata, Nakashima, & Akimoto, 2007; Ogata & Nakashima, 2006; Ogata & Ono, 2013, 2014).

INGS corresponds to the narrative generation and reception system at the personal level in relationships with multiple narrative structure models. This multiple narrative generation research by the author includes both the personal level of narrative generation-reception system and the level of a narrative production-consumption system that multiply comprehends personal narrative generation-reception systems. GIS is corresponded to the concrete design for the narrative production and consumption system.

THE ARCHITECTURE OF INTEGRATED NARRATIVE GENERATION SYSTEM

This section explains the architecture of INGS in the order of the side of mechanism (**The Generation Mechanism of INGS**) and the side of content (**Content Knowledge in INGS**). Main generation mechanisms contain entire process, story generation mechanism, narrative discourse mechanism, and narrative representation mechanism. The generation controls are also included in the explanations. Main components in content knowledge are conceptual dictionaries and narrative content knowledge bases. Two large sections, **The Generation Mechanism of INGS** and **Content Knowledge in INGS**, describe the above topics.

Figure 1 shows the overall INGS architecture. The main generation modules are for story, narrative discourse, and expression including language, music, and picture images. The story and narrative discourse are represented by respective conceptual representation forms. All modules are controlled by each local control mechanism under the global control. Story and discourse

Figure 1. INGS architecture
Source: Ogata (2016)

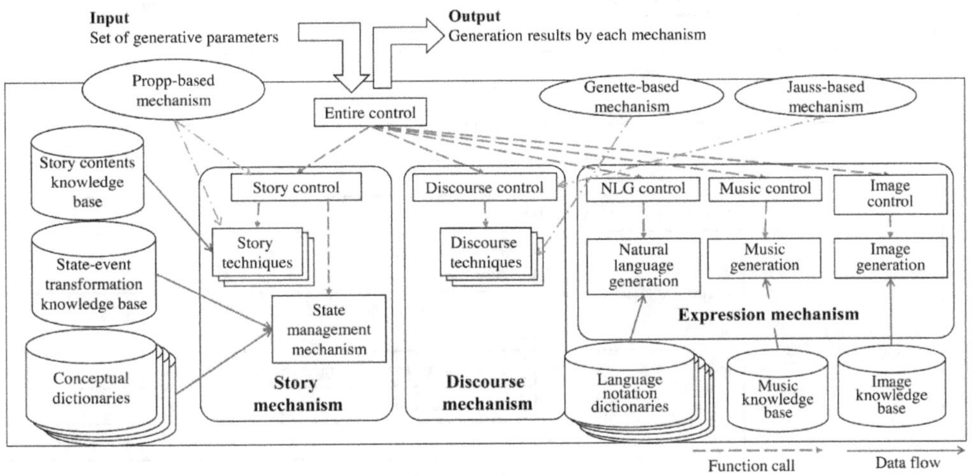

generation are performed using story and discourse techniques, respectively. Furthermore, INGS has dictionaries, including conceptual and language, and various knowledge, including story content and state-event transformation knowledge bases. In addition, as stated in the previous section, several literary theories are incorporated into INGS for the practice of expanded literary theory.

All the program components of the implementation in the current version are listed in Table 1. The main development programming language is Common Lisp as mentioned in BACKGROUND, but several types of development tools are also used partially for particular purposes, such as a part of music processing.

The Generation Mechanism of INGS

As mentioned previously, the aspects of INGS architecture in the division of mechanism or form and content is considered. Firstly, this section describes the formal aspect of mechanism.

Entire Generation Process

INGS proposes both mechanisms for the control of the level of the entire generation process and control in several points The main function of the control mechanism for each generation stage is the operation of the

Table 1. Components of the current version of INGS

Program Structure/Data				Function	Data
Main functions				12	-
Common library				64	-
Conceptual structures generation mechanism					
	Story generation mechanism	Main functions		11	-
		Control functions		20	-
		Story techniques selection functions		19	-
		Event concepts generation	Event generation functions based on verb concept	1	-
			Event generation functions based on instance	1	-
		Story techniques	Story techniques application functions	18	-
			Story content grammar	38	-
			Pair relationships	18	-
			Defamiliarization rhetoric	4	-
		States generation rules	States generation rules based on event	18	-
			Event complement functions based on states	9	-
	Narrative discourse generation mechanism	Main functions		13	-
		Control functions		16	-
		Narrative discourse techniques selection functions		9	-
		Narrative discourse techniques application functions		10	-
		Narrative discourse techniques		21	-
		Evaluation functions		19	-
		Discourse generation functions based on variation music		17	-
		Common library		9	-
Surface expression generation mechanisms					
	Control functions			3	-
	Sentence generation mechanism	Main functions		23	-
		Single sentence generation mechanism	Basic form generation functions	7	-
			Name generation functions for instances	7	-
			Location addition functions	1	-
			Conjugation decision functions	9	-
			Word order transformation function	1	-
			Language notation decision functions	3	-
			Basic form transformation functions	2	-
		Complex sentence gesneration mechanism	Application rules	37	-
			Connection rules	142	-
		Structural transformation functions from narrative relationships to language relationships		3	-

continued on following page

Table 1. Continued

Program Structure/Data				Function	Data
	Music generation mechanism	Original music generation mechanism	Main functions	1	-
			Transformation functions from story to original music	14	-
			Transformation functions from original music to original music	13	-
			Transformation functions from variation music to original music	4	-
		Variation music generation mechanism	Main functions	2	-
			Transformation functions from narrative discourse to variation music	15	-
			Transformation functions from narrative discourse to variation music	6	-
			Transformation functions from original music to variation music	17	-
		Midi generation functions (including "Text music SAKURA") 5 mm		15	-
		Common library		14	-
	Image generation mechanism	Main functions		2	-
		Animation generation functions		8	-
		Image expression interface (by HSP: Procedural programming language)		31	-
Dictionaries					
	Conceptual dictionaries	Noun conceptual dictionary		-	115765
		Verb conceptual dictionary		-	23978
		Adjective conceptual dictionary		-	1428
		Adjective verb conceptual dictionary		-	2382
	Language notation dictionaries	Noun notation dictionary		-	70079
		Verb notation dictionary		-	4920
		Adjective notation dictionary		-	12
		Adjective verb notation dictionary		-	20
		Particle notation dictionary		-	40
	Paraphrase dictionaries	Dialect transformation dictionaries (Iwate)	Noun dialect transformation dictionary	-	11
			Verb dialect transformation dictionary	-	31
			Adjective dialect transformation dictionary	-	22
			Adjective verb dialect transformation dictionary	-	3
			Particle dialect transformation dictionary	-	1
		Verb paraphrase dictionary		-	45
	Name dictionary			-	45093
Knowledge bases					

continued on following page

Table 1. Continued

Program Structure/Data				Function	Data
State-event transformation knowledge base				-	5241
Story content knowledge bases	Story content grammar	Propp-based story content grammar	Functions	-	32
			Sub-functions	-	271
		Story content grammar based on "Introduction- development-turn-conclusion"	Functions	-	6
			Sub-functions	-	13
		Story content grammar based on "The Story of the White Crane"	Functions	-	7
			Sub-functions	-	7
	Cause-effect knowledge base			-	17
	Villainy-liquidation knowledge base			-	3
	Interdiction-violation knowledge base			-	3
	Lack-liquidation knowledge base			-	1
	Order-obey knowledge base			-	2
	Pursuit-rescue knowledge base			-	1
	Reconnaissance-delivery knowledge base			-	3
	Struggle-victory knowledge base			-	5
	Transfer-return knowledge base			-	1
	Trickery-complicity knowledge base			-	4
	Theme knowledge base			-	10
	Script knowledge base		Bi-gram script	-	23751142
			Three or more than	-	60
	Noun-verb relationship knowledge base			-	6
Musical knowledge base	Motif knowledge base			-	2
	Scale knowledge base			-	12
	Instruments knowledge base			-	3
	Narrative structure-music structure transformation knowledge base			-	8
Animation knowledge base				-	170
Picture knowledge base				-	53520
Total				727	24074380

Source: Ogata (2019a)

characteristics of a narrative to be generated to direct the use of narrative techniques. Although many such mechanisms are simple, the Jauss-based mechanism that is comparatively precisely designed is used only in the narrative discourse mechanism.

One of the characteristics of narrative generation control in INGS is circulative control based on the use of one of the basic concepts of the narrative generation by the author. INGS aims to implement a form of the concept of circular narrative control. The INGS mechanisms are partially developed through the diversity of generation routes. Ideally, this control mechanism means that all of the generation mechanisms can receive any type of input information. The current version of INGS partially implements the generation routes shown in Figure 2. The user can select the generation route at the starting time in a session in any of the following ways:

1. The user selects a possible route: The user can select a free combination of "story, narrative discourse, sentence, visual image, original music, and variation music."
2. The system automatically determines a generation route.
3. The system automatically determines the next module at the ending time of a generation module.
4. If the user does not select any of the previous ways, the system determines the order, "story -> narrative discourse -> sentence -> visual image -> variation music."

Table 2 lists the input information of each generation module and the corresponding output information. In general, a generation module uses the previous output data as the input data. However, for example, in the case of the order "story -> sentence -> discourse," the discourse module takes the output data from the story module as the input data. Such details are explained

Figure 2. Possible generation routes in INGS
Source: Ogata (2016)

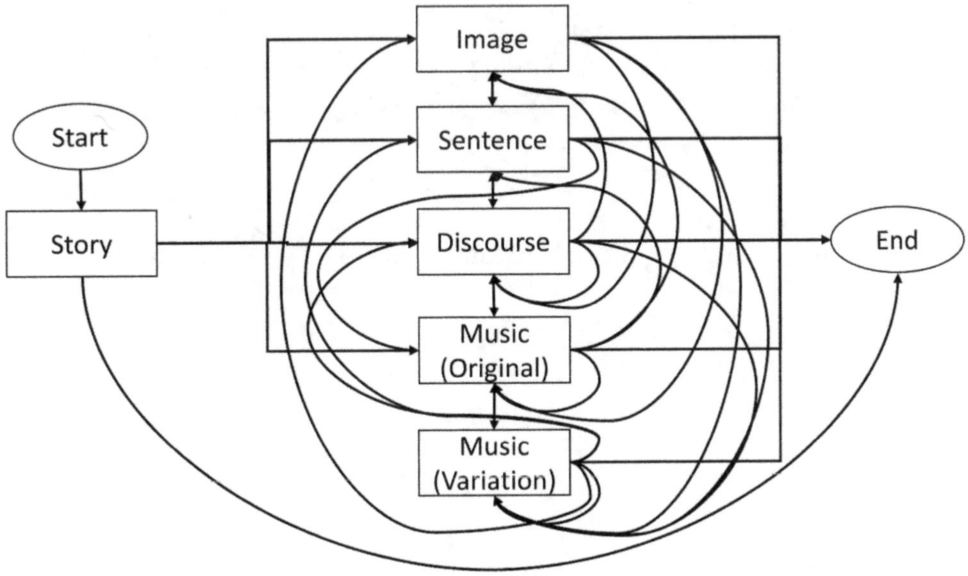

Table 2. Input and output information in each module

Generation Mechanism (Module)	*Input*	*Output*
Story generation	Generation parameters	Story structure
Narrative discourse generation	1) Generation parameters 2) Story structure or variation music structure	Discourse structures
Sentence generation (expression)	1) Generation parameters 2) Story structure or narrative generation structure	Sentences (text file)
Original music generation (expression)	1) Basic setting data 2) Story structure, narrative discourse structure, original music structure, or variation music structure	Original music structure (MIDI files)
Variation music generation (expression)	1) Basic setting data 2) Discourse structure, original music structure, or variation music structure	Variation music structure (MIDI file)
Image generation (expression)	Story structure or narrative discourse structure	Command file

Source: Ogata (2016)

Figure 3. The hierarchical conceptual structure of a story
Source: Ogata (2016)

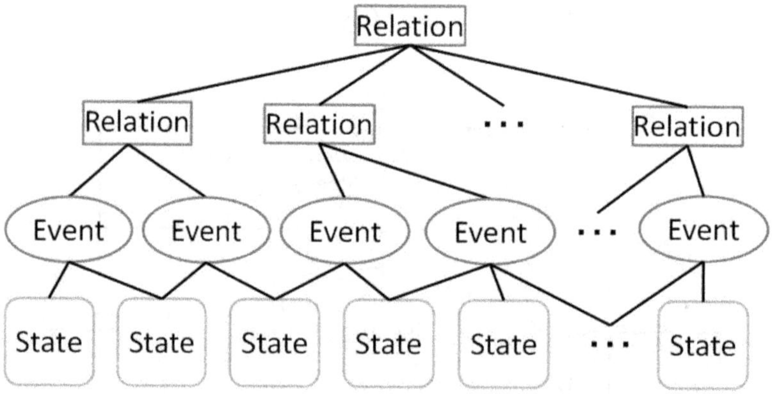

Figure 4. Transformation from a story tree structure to the expanded story tree structure using a story technique
Source: Ogata (2019a)

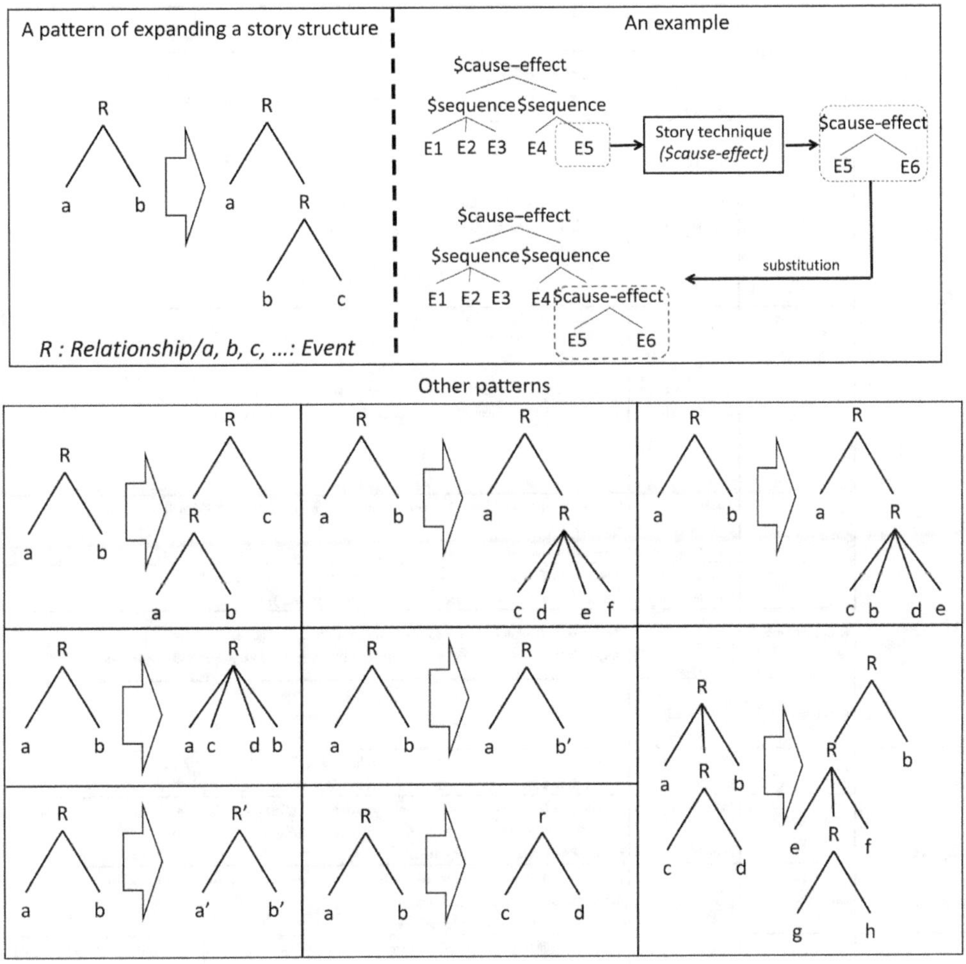

Figure 5. Description form of a story technique and an example
Source: Ogata (2019a)

Description form	Example
((relationship {story-technique-name}) (eve-pos {pair; 1|2|3| pattern; 'root | 'leaf}); type of story content (relationship-type {'rel | 'pat}); pair type or pattern type)	((relationship '$原因-結果); cause-effect (eve-pos 2); pair 2 content type (type 'rel)); pair relationship type

Table 3. Names of story techniques and story content knowledge bases

Type of Story Technique	Name of Story Technique	Story Knowledge Base	Example
Pair type	Effect	Cause-effect knowledge base	((event 飲む2[drink] (agent (&v age1)) (object (&v obj1) (&sc 酒[liquor]))) (event 酔う2[get-drunk] (agent (&v age1)) (object (&v obj1))))
	Cause		
	Villainy-liquidation	Villainy-liquidation knowledge base	((event 略奪する2[plunder] (agent (&v age1)) (counter-agent (&v age2)) (object (&v obj1))) (event 奪還する1[recapture] (agent (&v age2)) (object (&v obj1)) (from (&v age1))))
	Lack-liquidation	Lack-liquidation knowledge base	((event 欠如する1[be-lacking] (agent (&v age1)) (counter-agent (&v age2))) (event 出会う1[meet] (agent (&v age2)) (counter-agent (&v age1))))
	Struggle-victory	Struggle-victory knowledge base	((event 戦う1[fight] (agent (&v age1)) (counter-agent (&v age2))) (event 敗れる2[be-defeated] (agent (&v age2)) (counter-agent (&v age1))))
	Trickery-complicity	Trickery-complicity knowledge base	((event 使う3[spend] (agent (&v age1)) (object (&sc 錠剤[tablet]@薬品類{医用}[medicine])) (to (&v age2))) (event 眠る1[sleep] (agent (&v age2))))
	Transfer-return	Transfer-return knowledge base	((event 移動する1[move] (agent (&v age1)) (from (&v loc1)) (to (&v loc2))) (event 脱出する2[escape] (agent (&v age1)) (from (&v loc2))))
	Reconnaissance-delivery	Reconnaissance-delivery knowledge base	((event 捜す2[look-for] (agent (&v age1)) (counter-agent (&v age2))) (event 示す2[indicate] (agent (&sc 人間[human])) (object (&sc 噂[gossip]@噂[gossip]))))
	Repetition	<No use a knowledge base>	<None>
Pattern type (sequence)	Pursuit-rescue	Pursuit-rescue knowledge base	((event 妨げる2[disturb] (agent (&v age1)) (object (&sc 道[road]@道路[street]))) (event 退ける3[refuse] (agent (&v age2)) (object (&sc 誘惑[temptation]@勧誘[invitation]))))
	Interdiction-violation	Interdiction-violation knowledge base	((event 禁じる1[forbid] (agent (&v age1)) (counter-agent (&v age2)) (object (&v-eve 2))) (event 食べる2[eat] (agent (&v age2)) (object (&sc 果樹[fruit-tree]))) (event 罹る1[suffer] (agent (&v age2))))
	Order-obey	Order-obey knowledge base	((event 命令する1[order] (agent (&v age1)) (counter-agent (&v age2)) (object (&v-eve 2))) (event 食べる2[eat] (agent (&v age2))) (event 回復する3[recaover] (agent (&v age2))))
Pattern type (inclusion)	Script (inclusion)	Script knowledge base	(script0001 (働く1[work] ((or 焼く3[burn] 茹でる1[boil]) 盛る2[serve] 渡す2[give])) ((event 働く1[work] (agent (&v age1)) (location (&v loc1) (&sc 飲食店[restrain]))) ((or (1a (event 焼く3[burn] (agent (&v age1)) (object (&v obj1) (&sc 食料[food])) (location (&v loc1)))) (1b (event 茹でる1[boil] (agent (&v age1)) (object (&v obj1) (&sc 食料[food])) (location (&v loc1))))) (2 (event 盛る2[serve] (agent (&v age1)) (object (&v obj1) (&sc 食料[food])) (location (&v loc1)))) (3 (event 渡す2[give] (agent (&v age1)) (object (&v obj1) (&sc 食料[food])) (location (&v loc1)))))))
	Theme (inclusion)	Theme knowledge base	(theme0002 (刑事[police] (殺す1[kill] 発見する2[discover] 捜査する1[investigate] 発見する1[find] 逮捕する1[arrest])) ((theme 刑事[police]) ((1 (event 殺す1[kill] (agent (&v age1)) (counter-agent (&v age2)) (location (&v loc1)))) (2 (event 発見する2[discover] (agent (&v age3)) (counter-agent (&v age2)) (location (&v loc1)))) (3 (event 捜査する1[investigate] (agent (&v age4) (&sc 刑事[police]@警官[police-officer])) (counter-agent (&v age1)))) (4 (event 発見する1[find] (agent (&v age4)) (object (&sc 証拠[evidence]@証拠[evidence])))) (5 (event 逮捕する1[arrest] (agent (&v age4)) (counter-agent (&v age1)))))))

continued on following page

Table 3. Continued

Type of Story Technique	Name of Story Technique	Story Knowledge Base	Example
Pattern type (expansion)	Script (expansion)	Same as "Script (inclusion)"	
	Theme (expansion)	Same as "Theme (inclusion)"	
<None>	<None>	Noun-verb relationship knowledge base	((食料[food]) (食べる2[eat] 試食する2[taste] 料理する2[cook] 調理する1[cook]))
Defamiliarization type	Character defamiliarization	<These techniques are not used in the current version of INGS>	
	Object defamiliarization		
	Location defamiliarization		
	Defamiliarization scenario		
Macro type	Propp-based story techniques (for story content grammar)	Propp-based story content grammar	See the sub-section of "Propp-based Mechanism"

Source: Ogata (2019a)

in the respective subsections. The detailed information in this table will be shown in the sections for each module.

Story Generation Process

INGS provides mechanisms that control both the entire narrative generation process and each point of narrative generation. During the story generation phase, the system generates a story using a tree structure with three basic elements: event, relation, and state, organized by time. Each event is constructed using an event concept and semantically related noun concepts. The conceptual representation form is used as a frame structure to semantically categorize each noun concept, using eight basic cases ("agent," "counter-agent," "object," "location," "time," "instrument," "from," and "to") and nine special cases ("adverb," "possessive," "situation," "purpose," "experiencer," "source," "idiom," "information," and "as"). Each event accompanies the preposition state and the resulting state at the beginning and end of the event. At the same time, each relationship links an event or aspect of the tree structure (which includes events) to another event or aspect of the tree to create a new structure, where events and states have a top-level relationship. Many relationships, such as "cause-effect," "sequence," and "script," are created. Figure 3 shows an example of a story structure that includes relations, events, and states.

States and events correspond to static and dynamic informational elements, respectively. The main functions of a state are to manage knowledge about the story world and to manage the flow, from event to event. Although a story is a temporary organization of events, its semantic consistency is managed by relationships and states.

Story Techniques and Story Content Knowledge

The tree structure of a story including events and relations is expanded by applying a story technique to any objects in the structure. When the story technique is applied to a point, the knowledge contents from conceptual dictionaries and the story content knowledge base are used. At this point, the mechanism of narrative generation is related to the aspect of narrative contents. Moreover, when an event is generated in a story generation process, the states relating to the event are generated. In addition, in a narrative generation system (Ogata, 1995) in which this basic structuring method was presented, any nodes in a structure could recursively be the object of expansion and transformation. Although only terminal nodes in a story tree structure could be operated in the previous version of INGS (Ogata, 2016, 2019a), the current version adopts the original method.

A story structure that consists of events and relations can be generated using story techniques as part of the applicable narrative techniques, and any states associated with the created events are developed according to another mechanism. Figure 4 (the upper part) shows the expansion process using a story technique. The concrete content knowledge for expanding “E5” to add “E6” is kept in the story content knowledge base. Thus, although this example extends an original story structure and transforms it, another type of technique is also prepared to transform only the elements. The Propp-based mechanism, which will be mentioned in the next sub-section, is also a type of both the story technique and story content knowledge. In other words, stories are represented by a tree structure with events bound by various relations, and it is gradually expanded using rules composed of differing functions of structural transformation; each technique uses the information stored in a story content knowledge base. “Other patterns” in Figure 4 shows various patterns of the transformation of story trees by applying story techniques. Figure 5 is the description form of a story technique and example. Table 3 lists story techniques and the corresponding story content knowledge. The names in the corresponding story content knowledge bases are also described in this table for use in the next section. Additionally, as a set of story techniques and the corresponding story content knowledge, the author also uses the techniques and knowledge based on Propp’s narrative morphologic theory.

In summary, a story generated by such mechanism is defined by either the content of a narrative or temporal sequence of events. It is represented by a tree structure that consists of the three types of basic units: event, relation, and state. An event is described in conceptual representation form as a frame or case structure that includes a verb concept and related noun concepts. It is the principal combination of preceding and subsequent states. A relation semantically combines events or sub-structures (such as "cause-effect," "continuation," and "script"). A state is the static information within a story, while "event" corresponds to the dynamic information. The chief functions of the state are to maintain the knowledge about a story's world and manage the coherency of the flow of events.

The input data to the story generation module is listed in Table 4. These are set as global parameters through each generation session. Using values between one and three, a user can input parameters to provide the direction for a story. The mechanism manipulates the different entered techniques in order to generate a story structure through the process flow. In addition, the mechanism necessarily uses a story technique from previously prepared candidates at the starting point to generate a first story structure. For each parameter, the user can input a value from 1 to 3 to direct the story generation. The story generation mechanism applies different story techniques according

Table 4. Input data to story generation module

Parameter	Overview	Procedure
Length	The number of events contained in a story to be generated is controlled by the parameter of "length."	The system accounts for the number of events.
Macrostructure	"Macrostructure" adjusts macro level story relations through an entire story controls of the number of macro relations.	The system accounts for the number of the following events to be contained in the generated story: $Theme, $Villainy-liquidation, $Order-obey, $Interdiction-violation, $Russian_forktail, $Reserve_portion, $Beginning, $Reserve_trial, $Strugle_and_victory, $Task_and_solution, $Solution_of_problem, $Arrival_and_trial, and $End."
Non-reality	"Non-reality" controls the number of the relation, $defamiliarization_scenario contained in a story to be generated.	The system accounts for the number of $Defamiliarization_scenario events.
Repetition	"Repetition" means repeated similar events and this parameter controls the number of $repetition and "$defamiliarization_scenario in a story.	The system accounts for the number of $Repetition and $Defamiliarization-scenario events.

Source: Ogata (2019a)

to the input value. In addition, when the system generates the first story structure, the generation mechanism necessarily selects a story technique from preliminarily prepared candidates.

In addition, INGS creates states for an event in the story generation phase using a state-event transformation knowledge base. Each of the 5241 state-event transformation rules corresponds to a verb concept. This knowledge base is described in the sub-section of *State-Event Transformation Knowledge Base* in the section of Content Knowledge in INGS.

Figure 6. PBSG hierarchical structure
Source: Ogata (2019a)

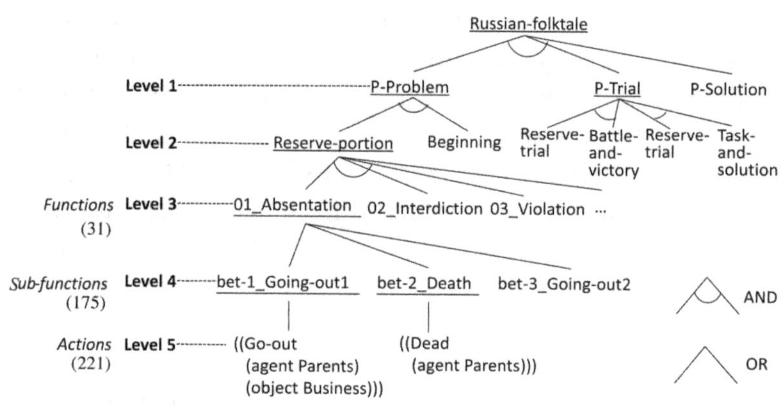

Table 5. PBSG hierarchical structure

Level	Definition
Level 1	**Definition of a macro or entire story structure (1).**
Level 2	**Definition of a macro or entire story structure (2).**
Level 3	**Division of 31 "functions" into several parts.**
Level 4	**Generation of "sub-functions" from "functions": A "function" means the abstract and comprehensive definition of an actor's action, while a sub-function corresponds a concrete way to actualize a "function". PBSG also implements "pairs of functions" to connect two distant functions (pairs between sub-functions are actually used).**
Level 5	**Actualization of each sub-function to an action or a sequence of actions: Each action is described as a case structure including a verb concept and necessary noun concepts.**

Source: Ogata (2019a)

Propp-Based Mechanism

The central argument of Propp's theory is a structural method based upon certain "functions," but the overall accomplishment includes other knowledge related to various aspects of folktales and general narratives. Most of the generation systems that use Propp's theory focus only on these functions, but the theory includes productive and theoretical ideas that could contribute to story creations. In particular, Propp presented an array of examples and techniques in the lower level for materializing the functions as abstract definitions by analyzing collected folktales. The two methods he used are "interdiction" and "order or a suggestion," and correspond to the description for materializing the former abstract "functions"; these levels of analysis are known as "sub-functions."

In his studies, Propp noted that specific functions have a strong tendency to be combined with other specific functions in order to form different pairs. For example, "inhibition and the violation" are strongly connected in many narratives. Such parts can be abstracted simply to advance smaller story functions. The aurhor utilized this idea to define a type of story content knowledge dependent on the function pairs. For example, "Pair type" and "Pattern type (sequence)" in Table 3 contain story content units based on the function pairs. Likewise, each function is connected to one or more role(s) of characters including "hero," "villain," "victim," "helper," "dispatcher," and "princesse." These are formed as a specific role that becomes a major agent for driving a group of functions.

(1) Propp-Based Story Grammar (PBSG)

As shown in Figure 6, "Propp-Based Story Grammar (PBSG)" is rooted in the reorganization of Propp's theory and is divided into four hierarchical levels. This figure shows a portion of Common Lisp's actual description of PBSG, which has five levels in the hierarchical structure, as listed in Table 5.

PBSG opens to different types of operational methods: top-down, bottom-up, and hybrid. The top-down method is the standard and simply expands PBSG from higher to lower levels by the automatic choice of subsequent options using a random function. When the process reaches the lowest level in the grammar's description, the operation generates an event concept using conceptual dictionaries in INGS and develops a PBSG tree structure using depth-first search and constructs the overall structure of a story. In this process, if a sub-function is selected as an element in a pair, the corresponding sub-function is selected automatically. The pairs contribute to maintain a story's consistency and its comparatively natural flow. Although Propp prepared both pairs of functions and sub-functions, this method uses only the latter.

The bottom-up method ascends the levels of a hierarchy within a story from the lowest level of its PBSG case structure as determined by the user. The generated result is often a collection of fragmented parts. The hybrid method is a blending of the two methods and completes a story structure by expanding the results of a bottom-up operation using sub-function pairs and PBSG.

One of the basic standards used for evaluating the Propp-based system is the coherence of a story or resulting sequence of events. One system used for breakdown and interpretation is KOSERUBE (Imabuchi, Akimoto, Ono, & Ogata, 2012; Akimoto, Imabuchi, Endo, Ono, Kurisawa, Kamada, & Ogata, 2013), an application system that uses the PBSG as its main mechanism for story generation, and can clarify problems relevant to a story's cohesion. By finding most of the results that overlap with the problems of the Propp-based system, the author prepared natural language sentences generated by the mechanism within KOSERUBE. The experimenter asked the subjects (including eight university students) to identify and describe coherence problems according to categorized questions. The object of the evaluation is not the aspect of natural language description, but the aspect of s story's logical development. The results showed problems such as "hard to understand; the meaning of an event; the reason of a character's action; the relationship between an event and the entire story." These indicate future studies. Althrough, as described in the following KOSERUBE: Parallel Generation Processing sub-section, KOSERUBE is an experimental system for synthesizing multimedia narrative representations, as it uses a relatively simple story generation mechanism based on PBSG, it can function as a platform system for considering the qualitative problems of generated sentences.

Figure 7. Transformation from story structure to discourse structure
Source: Ogata (2019a)

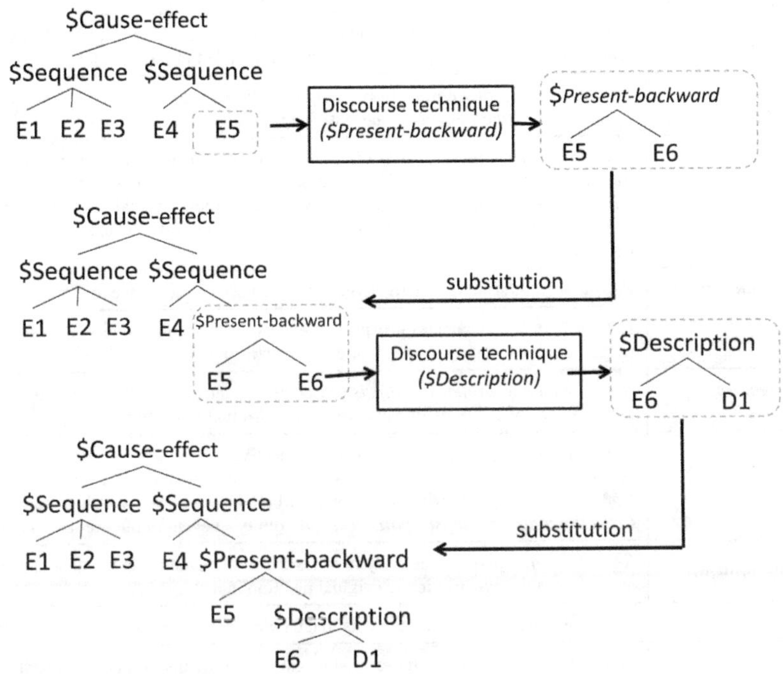

(3) Incorporating Into INGS

Although the Propp-based mechanism can function as an independent system, INGS can also use it as a part. PBSG and the structure generation function correspond to a story technique and its content knowledge base. Many story techniques exist to define micro-story structures (such as the causal relationship) and promote the continuing relationships between events to form a partial story structure. Adversely, the Propp-based mechanism constructs macro-story structures where the names of the "functions" and other elements (with the exception of events in the PBSG hierarchy) are equivalent to the names of the relationships. Three types of development method that concern PBSG correspond to a group of story techniques capable of generating a tree structure from the input information. A structure built by the Propp-based mechanism can be the portion of an overall story structure that uses other approaches. Story techniques based on Propp include the top-down process, bottom-up process, and hybrid processing methods. PBSG corresponds to the knowledge to be contained in the story content knowledge base.

Table 6. Used narrative discourse techniques

Name	Overview
(1) External_analepsis	*External_analepsis* retroactively narrates the past event(s) located outside a story's temporal range (i.e., the past event(s) are not included in the story).
(2) Complementary_analepsis_ellipsis	*Complementary_analepsis_ellipsis* retroactively narrates the past event(s) that was missing or omitted in the original position in a story.
(3) Complementary_analepsis_paralipsis	*Complementary_analepsis_paralipsis* retroactively narrates a past event that was missing or omitted in the original position in a story. In *Complementary_analepsis_ellipsis*, all of the inserted events are missing from their original place. By contrast, in *Complementary_analepsis_paralipsis*, only an inserted event is missing in the original position.
(4) Repetitive_analepsis	*Repetitive_analepsis* retroactively narrates the past event(s) once again.
(5) External_prolepsis	*External_prolepsis* narrates prospective events positioned outside a story's temporal range (i.e., the events are not included in the story).
(6) Complementary_prolepsis_ellipsis	*Complementary_prolepsis_ellipsis* prospectively narrates the future event(s) in a generated story and omits the event(s) in the original position in a story.
(7) Complementary_prolepsis_paralipsis	*Complementary_prolepsis_paralipsis* prospectively narrates the future event(s) in the generated story and omits the event(s) in the original position of a story. Although *Complementary_prolepsis_ellipsis* omits all the inserted events in the original position, *Complementary_prolepsis_paralipsis* only omits an event in the events to be inserted.
(8) Repetitive_prolepsis	*Repetitive_prolepsis* retroactively narrates the future event(s) in the generated story and narrates the event(s) in the original position in a story.
(9) Achronie	*Achronie* narrates an event or events that cannot decide the temporal position in a story.
(10) Pause	*Pause* stops the temporal progression in a story via the insertion of the event(s) of "description" and "explanation."
(11) Implicit_ellipsis	*Implicit_ellipsis* omits a sequence of events in a story and does not show the omission.
(12) Repeating	*Repeating* narrates an event or events that appeared only one time in a story.
(13) Paralipsis	*Paralipsis* omits the information that should be narrated in the current range of narrative discourse, namely, they show less information than the information required in principle. Although *Implicit_ellipsis* deletes of the events, *Paralipsis* deletes only an event in the sequence of events.

The Propp-based mechanism is designed to be combined simultaneously with INGS through the conceptual dictionaries. Stories in the system essentially amount to sequential events linked hierarchically by story relationships. Each event is described with a frame representation, including a verb concept and case elements with a variety of noun concepts for "agent," "object," and "location," and is associated with conceptual hierarchical dictionaries for both verbs and nouns. Using this mechanism, the lowest level of PBSG addresses the same event structures.

Narrative Discourse Mechanism

Similar to the story generation mechanism, the narrative discourse mechanism is also composed of narrative discourse techniques applied to its developing structure. The similarity between the story generation and narrative discourse mechanisms are based on the following characteristic in INGS. Namely, both of the mechanisms play the aspect of conceptual processing in narrative generation in INGS. The conceptual processing means the deep and semantic narrative generation processing in INGS in contrast to the surface narrative processing using various representation media in the narrative expression mechanisms in INGS. Although the concrete processing of the narrative discourse mechanism is different from the story generation processing, they performe similar procedures in the formal aspect.

The module has implemented generation and circular control mechanisms that use the narrative discourse theory by Genette (1972) and reception theory by Jauss (1970), respectively. These represent a concretization in the narrative discourse mechanism of the expanded literary theory along with the introduction of Propp's theory into the story generation mechanism. The latter mechanism based on Jauss's reception theory shows a possible implementation of narrative generation control in various possibilities. A narrative discourse generated or changed by the Genette-based mechanism corresponds to the structural aspect of narrating a story tree structure including a sequence or series of events. The discourse structure to be narrated and its required surface media, such as language and image, are points distinguished within the system, and are described accordingly in conceptual representation form as stated above; the narrative discourse should correspond to a transformed story structure. From the viewpoint of expanded literary theory that simply uses narratological and literary theories, in the main portion of the process, Genette's theory develops a set of discourse techniques as the root for the narrative techniques; the Genette-based mechanism and Jauss-based control mechanisms are related to this phase. However, INGS is a comprehensive system that incorporates existing narratological and literary theories and methods into the entire system framework. Therefore, the generalization of theory-based mechanisms has been conducted through the design and development of INGS.

Regarding the concrete process of narrative discourse, a generated story structure is transformed into a narrative discourse as the narration's structure according to a temporal sequence of events and states. A narrative discourse structure can reorganize, further ignore, the temporal order of events in the story. Additionally, in INGS, the narrative discourse structure to be narrated is differentiated from the narrative representation by the surface representation media, such as language, image, and music. Narrative discourse structures are described using corresponding conceptual representation forms such as in story structures. Thus, a narrative discourse corresponds to a transformed story structure. Although it is a tree structure in which events are the main elements, the different relations from story structures are used. A narrative discourse structure is described by the tree form transformed from a story tree and the events and states in the narrative discourse structure are same as the story structure.

Currently, the narrative discourse mechanism uses the following seven types of narrative discourse relations:

1. In $Recall, one or more past events associated with the "recall" of a character are introduced or inserted into the front of an event at a point. In the generated discourse tree structure, the left node is the event of "recall" and the right node corresponds to the content.
2. $Present-past inserts the past event or events into the left node and more than one or more events into the right node.
3. $Prediction introduces or inserted the future event or events by the "prediction" of a character into the front of an event. The left node is the event of "predict" and the right node is the content.
4. $Present-future sets the future event or events in the right node more than one or more events in the left node.
5. $Episode inserts one or more events with an unclear relation to the story into the right node as an episode.
6. $Description inserts the event or events that describe the event or events in the left node into the right node.
7. $Repetitin_discourse connects two same event or events.

Furthermore, the narrative discourse mechanism prepares five kinds of primitive procedures, including "deletion," "copy," "connection," "exchange," and "generation," for a narrative discourse structure. For example, in the operational process of a narrative discourse technique called "Complementary_prolepsis_paralipsis," these primitive procedures can move one or more past events in a narrative discourse structure to a different point from the original point in the input story.

Genette-Based Mechanism

In his work on narrative discourse theory, Genette (1972) never commented on narrative generation. This study changed the theory's context using a computational perspective to create a technological method for designing or formalizing the narrative discourse generation process. This allows each of the classified techniques applied by Genette to be used as a rule for associating an input structure with a transformed output discourse structure. An "input structure" is equal to a complete story, the partial structure of a story or part of the discourse structure. Figure 7 shows an example where a story structure is transformed into a discourse structure; a relation specified to discourse, $Present-backward, is added to the structure. Both structures for story and discourse have similar tree forms. Each leaf node in a story's tree corresponds to an event described with conceptual representation, which is further described using a case structure that consists of one verb concept and eight types of cases, such as "agent" and "object". Each internal node in the tree is equivalent to a relation such as $Cause-effect and $Serial combined with the child nodes.

A series of investigations by the author considered various narrative discourse techniques, including temporal order, viewpoint (, perspective, or focalization), distance, speed or tempo, respectively, and a plan for incorporating the narrative discourse mechanism into an integrated INGS. Because currently developed narrative techniques in INGS (Table 6) are related mainly to tense, the results of the narrative discourse techniques are not fully introduced. However, this research is developing a new possible direction of narratological approach to our narrative generation through connection to a Jauss-based control mechanism. However, theoretically, the Genette-based narrative discourse mechanism is placed as only a part in the entire narrative discourse system by the author as well as the Propp-based mechanism in the story generation mechanism in INGS. Although designing and implementing the Genette-based narrative discourse mechanism itself is an important topic, a more comprehensive approach to narrative discourse including it is required in the future (Ogata, 2003c, 2004a, 2004c).

The seven types of relations used for discourse trees are:

1. $Recall
2. $Present-backward
3. $Prophecy

4. $Present-feature
5. $Episode
6. $Description
7. $Repetition_discourse

The following nine types of input parameters are prepared:

1. satisfaction point
2. generation cycle
3. complexity
4. suspense
5. length
6. hiding
7. description
8. repetition
9. diffuseness
10. implication

As described in the preceding, although Genette presented a comprehensively categorized system of narrative discourse techniques that includes various techniques regarding time, narration, and so on, they mainly correspond to the level of "what" instead of the level of "how." Originally, narrative discourse deals with the side of "how" in a narrative in the point of how to narrate a story as the aspect of "what" in a narrative. However, the Genette's narrative discourse theory treats the aspect of relatively "what" in the narrative discourse as "how." For instance, Genette showed a detailed classification of narrative discourse techniques as shown in Ogata (1999b). His explanation of the problem with respect to the actual use in a narrative states the description using examples. However, in actual narrative discourses, the structural discourse techniques include various more concrete or lower level techniques, such as "recall," "prediction." Such concrete or lower level techniques can be systematically considered based on a clear operational perspective for narrative generation. Further, the description by Genette was limited to the non-systematized introduction of necessary techniques for the concretization of narrative discourse, such as "description" and "explanation." Solving these problems requires the adoption of a hierarchical viewpoint on narrative discourse techniques in addition to the operational viewpoint. Moreover, although Genette referred to knowledge as the extended version of the aforementioned "how," in particular, the knowledge for control and

Figure 8. Overview of Jauss-based narrative discourse control mechanism
Source: Akimoto and Ogata (2013b)

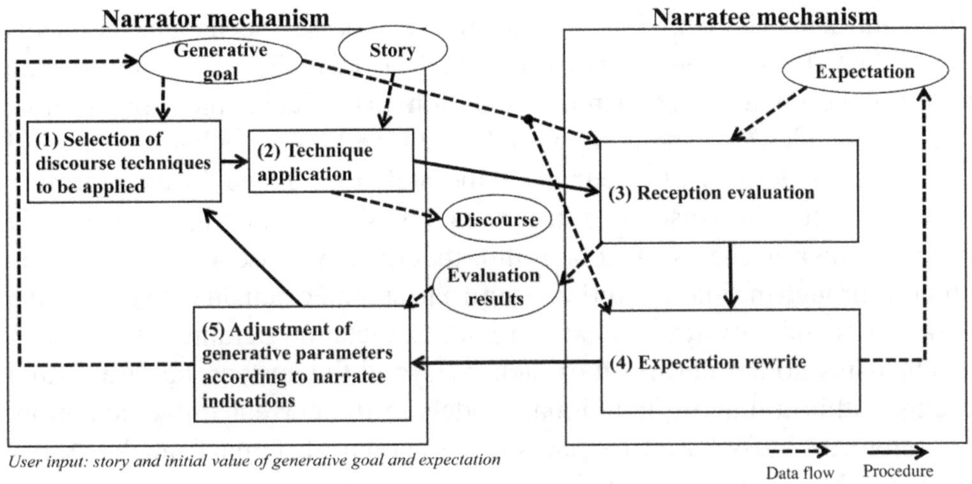

strategy of narrative discourse techniques, such as "for increase of the effect of X, use the technique Y" and "the usage of the technique Y increases the effect of X," these have not been systematically considered in later studies. This is related to the problem of "voice" in Genette's theory. Given that this problem is very difficult to approach using a rule-based method, more data-driven experiential approaches may be required in the future. In a broader perspective, the basic direction in this study regarding these problems is to develop a more comprehensive, systematic, and hierarchical mechanism for narrative discourse techniques based on Genette's theory as a basic reference framework, from the perspective of the operation and processing of narrative generation. In a typical processing operation of narrative discourse techniques, a fragmental story is inputted to the system and a transformed representation structure as the result to which a narrative discourse technique is applied is outputted. Furthermore, a required mechanism is the system in which executed higher level techniques influence the execution of lower level techniques according to the organized method.

The Genette-based mechanism plays the part of discourse structure generation, and the Jauss-based mechanism, in the next sub-section, controls the processing. Input parameters for the narrative discourse generation are used to make one discourse structure, whereas the Jauss-based part performs circulative generation based on another control mechanism. It also controls the parameters' changes.

Jauss-Based Mechanism

The control mechanism for the narrative discourse mechanism connects Jauss's (1970) reception theory to Genette's narrative discourse theory in the current experimental system implementation. In particular, the application of Jauss is intended to work along with the Genette-based mechanism. Several discourse categories by Genette are elaborately formalized for transforming a story structure into discourse structures. Jauss's reception theory is simply interpreted as a mechanism for controlling the circulative discourse construction process through the interaction between a narrator mechanism with generative parameters and narratee mechanism with expectation parameters. The two mechanisms do not define an existence beyond the system, but are virtual agents within it and are individual models in the current implementation. Therefore, the Jauss-based mechanism is essentially different from the Propp-based mechanism because the Genette portion uses structural processing and the Jauss section controls it. The Jauss-based control mechanism is currently applied only to the narrative discourse phase in INGS, but a future goal is the expansion of a similar mechanism to the entire range.

The reception theory by Jauss is a narratological literary theory that aims to understand narrative and literary phenomena from the perspective of readers. His central concept is called the "horizon of expectation" and his reception theory based on the concept of the horizon of expectation states that the collection of expectations of the readers in an age (context) to a literary work determines the evaluation of the work. The reader in this theory does not mean individual readers but the collective readers in an age and a context. If a literary work satisfies the expectations of the readers, they are satisfied. Contrastingly, if it does not satisfy the readers' expectations, they do not experience satisfaction. Although these responses occur via the continuation of reception acts by the readers, their likeliness gradually changes based on repetition. In particular, the readers are disinterested in literary works as they were satisfied previously and begin to experience satisfaction with other types of literary works. Given that the readers, in this case, are collective readers, the satisfaction connects to social trend and tendency. A chain or sequence of literary works that results in greater satisfaction of the readers forms a dominant literary history. Thus, one of Jauss's aims was to present a new idea concerning the history of literature. He criticized the literary theory of evolution as being dependent upon the dynamism of change in narrative techniques and descriptions. He sought to center his literary design on the

Figure 9. Examples of an intermediate sentence and the transformed forms

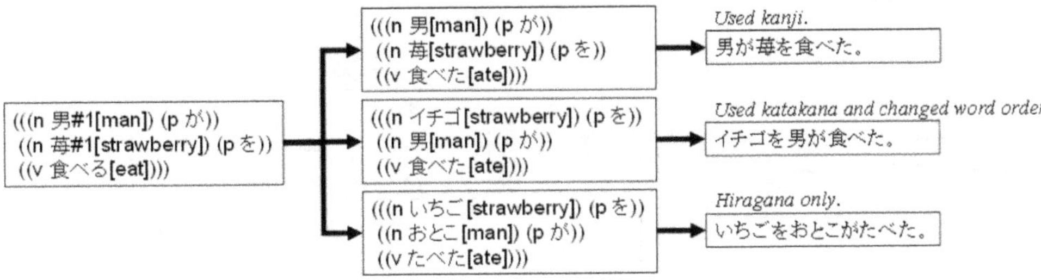

dynamism of the mutual relationship between a literary work/author and the readers. Jauss developed seven types of *these* (thesis) for the essence of his reception theory. His idea involved designing the literary history of various relationships between a reader's "horizon of expectation" and literary works, including those estrangements, and the change of expectation through a continual reception process.

The Jauss-based mechanism focuses on *these II* and *these III* for describing the horizon of expectation and changing catalyst to design a Jauss-based mechanism. His *these II* describes the "horizon of expectation" as being the reader's preliminary knowledge for ranking a literary work in the context of reading experience, and it can be acquired by known genres and exposure to famous literary works. In *these III,* Jauss proposed that readers understand literary artistic features as the distance between the work and their horizon of expectation. If the distance is small, the work has strong recreational purpose and a familiar beauty. Those with a large distance have various and distinguished influences on the readers: to some the work might be highly regarded creative work, but might be a waste of time for other readers. The horizon of expectation can change so that the impression of a work becomes ordinary for a particular reader.

For system design, the author created a model of Jauss's reception theory where literary works are repeatedly produced according to a reader's expectations and the corresponding responses from an author; the interaction can continuously or historically change a literary work. In particular, texts of narrative discourse are repeatedly produced via a process according to parameters for a reader (or narratee) as the expectation and using the same parameters for an author (or narrator) as the generative goals. These parameters are the same as those in the Genette-based mechanism (*satisfaction point*, *generation cycle*, *complexity*, *suspense*, *length*, *hiding*, *description*, *repetition*,

Figure 10. The transformation of a conceptual representation to sentences using property information

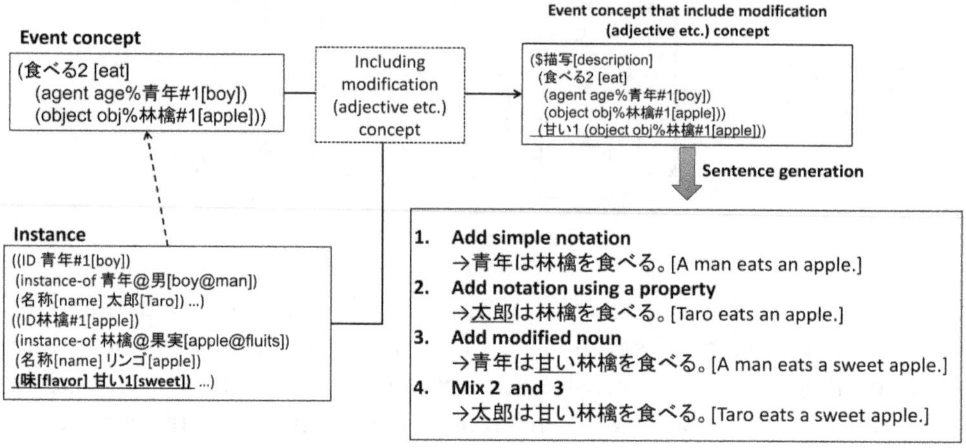

diffuseness, and *implication*). In addition, the narrator and narratee exist inside the system as virtual agents, and as personal and individual models.

The cycle continues according to the interaction between a narrator and narratee. A narrator mechanism has a set of rules for controlling the application of discourse techniques based on the generative parameters to transform a discourse structure. The direct transformation of a discourse structure is conducted using techniques mainly defined by Genette's theory. A narratee mechanism evaluates the results by comparing the narratee's expectation parameters with the narrator's generative parameters. In the next cycle, the narrator applies corrections according to the narratee's evaluation in order to achieve reader satisfaction. The process, however, eventually reaches a point where the narratee becomes tired or his/her satisfaction begins to fall. When this happens, "deviation" occurs and the narrator abandons a portion of the old generative parameters and moves to a new cycle of discourse grounded on the newly found strategy. The narratee's expectations change according to the reconstruction.

Figure 8 is a detailed process flow of the narrative discourse control based on the interaction between the narrator and narratee mechanisms (see Akimoto and Ogata (2011, 2012a, 2012b, 2012c, 2014c) in detail).

1. The system selects the corresponding techniques according to the parameter values.

Figure 11. The descriptive form of a word in the language notation dictionary and an example

Description form	*Example*
(<終端概念名> ; Concept name	(男[man]
(<漢字を含んだ言語表記>) ; Kanji	(男)
(<ひらがなのみ言語表記>) ; Hiragana	(おとこ お おのこ)
(<カタカナのみ言語表記>) ; Katakana	(オトコ オ オノコ)
(<アルファベット（ローマ字）のみ言語表記>)) ; Roma-ji	(otoko o onoko))

2. The Genette-based techniques fulfill the functions in the context of the Jauss-based control.
3. Higher satisfaction can be obtained when a more strongly desired parameter is satisfied. The narratee indicates one parameter with the lowest satisfaction.
4. The narratee rewrites an expectation parameter through two different procedures. The first rewrites the sufficiency value and degree of desire. The sufficient value is increased each time the narratee receives the sufficed discourse.
5. This processing includes "deviation" from the expectation.

In addition, this model is simplified by comparison with the description by Jauss. It indicates individual and psychological characteristic. By contrast, the horizon of expectation by Jauss has a collective and social characteristic and is related to the problem of normative literary rhetoric. In this sense, Jauss's model is not individualistic. The current application by the author is different from the original theory in the comparatively individual characterization. If we can make a model in which collective readers are formed from many agents of the reader, it may be more similar to his original theory. Ogata (2016, 2018b, 2019a) showed the concrete method and process.

Narrative Representation Mechanism

This mechanism for generating surface representations contains the following sub-mechanisms:

1. **Natural language generation mechanism**: This generate natural language representation or expression.
2. **Image generation mechanism**: This generates image preresentation or expressin, namely visual images of moving and still pictures.

3. **Music generation mechanism**: This generates musical representations or expressions.

Previous INGS has focused on generating conceptual representations, while the above mechanisms for producing surface representations, especially language and images, remain at the invention or framework stage. The following sub-section discusses natural language generation (1) and music generation (3).

In addition, an experimental function in the mechanism for generating narrative images (2) creates pictures that correspond to elements in the conceptual dictionaries in INGS, using image management rules to synthesize them. In response to a story or narrative discourse structure, the mechanism generates a text command file, which delivers a series of processing commands to the image representation interface. In the future, the author aims to construct an image generation mechanism in INGS based on the similar concept of the music generation mechanism. In particular, the plan treats the generated tree structure of narrative conceptual representation as the tree structure of image representation; it inserts various rhetorical elements of imagery, corresponding to rhythm and tempo in the case of music. The basic grammar of image representation, which corresponds to language-based case structures and sentence patterns, may effectively control the concrete image representation through a camera work function or other means. Characters can be used in the generated narrative visual representation.

Natural Language Generation Mechanism

The natural language generation mechanism transforms all events in the structure of a generated narrative conceptual representation (a story structure or narrative discourse structure) into a sequence of sentences. Originally, although the generated states need to be transformed, the current system addresses only event transformations. This mechanism first generates sentences using basic sentence patterns described in each verb conceptual item in the verb conceptual dictionary. Then, entire sentences are generated based on both the processing inside a sentence, such as tense, and the processing of connecting several sentences. Although, as the author will explain in detail in **Content Knowledge in INGS**, the mainly used conceptual dictionaries include verb and noun conceptual dictionaries, adjective and adjective-verb concepts are also used partially. Further, each concept item in the conceptual dictionaries is directly transformed into the corresponding word using a developed language notation dictionary. This processing became possible because items in the conceptual dictionaries are not extremely abstract and are represented using concrete and surface words. For example, when the conceptual description, "食べる[taberu: eat]#2," is used in a story conceptual representation, the natural language generation mechanism initially transforms it to the word representation, "食べる[taberu: eat]." Moreover, the language notation dictionary can transform the basic word representation to various notations, including "たべる[taberu: eat]" and "タベル[taberu: eat]."

The overall processing flow of this mechanism is as follows:

1. **Generating basic forms of intermediate structural representation**: Using the sentence patterns in the verb conceptual dictionary, the case structures of a story or narrative discourse are transformed into the basic forms of an intermediate structure. The recursive form of a sentence is also treated here.
2. **Transforming the basic forms of intermediate representation into the changed forms**: This mechanism uses various forms based on the order change from the sentence patterns, random generation, and the addition or deletion of specific cases.
3. **Transforming the intermediate forms into the surface sentence forms**: The developed surface sentence generation program transforms the aforementioned forms into the surface sentences.

Other selection elements are as follows:

1. **Selection of names for all of the instances**: 1) concept name (the representation in the conceptual dictionaries), 2) name, 3) concept name and name, 4) upper concept name, and 5) concept name and name in the character that appears first, name in other cases.
2. **Selection of word notations**: 1) concept name (same as the representation in the conceptual dictionaries), 2) *kanji* (Chinese character), 3) *katakana* (the square from of *hiragana*), and 4) *hiragana* (the Japanese cursive syllabary)
3. **Selection of end of each word**: 1) original form, 2) past, 3) polite expression, past, and 4) random mixing of original form and past.
4. **Selection of postpositional particle ("が (*ga*)" and "は (*ha*)")**: The basic representation is "*ga*" and "*ha*" in descriptive sentences.
5. **Selection of connective representation**: 1) no connective representation, 2) transformation using the complex sentence generation mechanism,

Figure 12. Roles of content knowledge in narrative generation

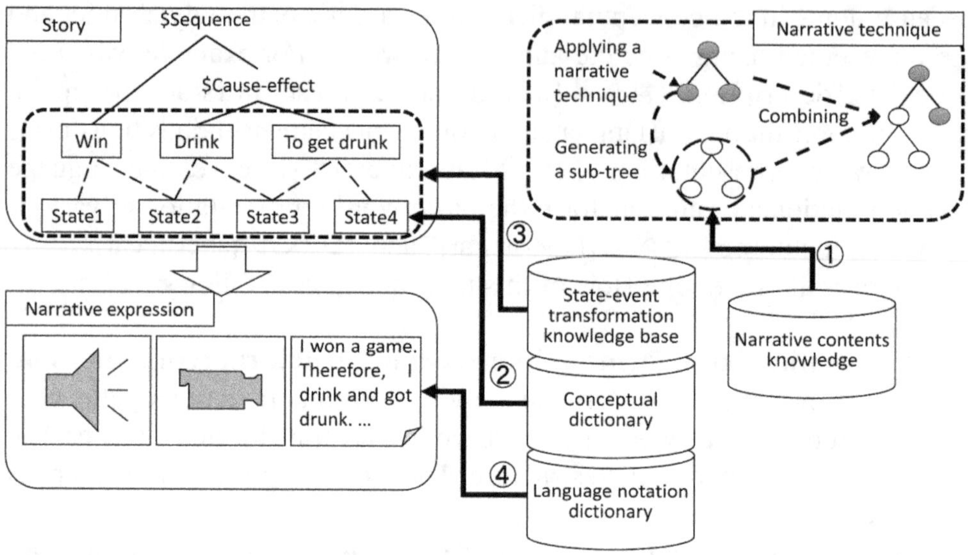

3) using "そして ("*soshite*")," "それから ("*sorekara*")," etc. for the conceptual relation of "$継起 ($Sequence)", and 4) insertion of temporal representations in a timing.

As optional processing, the generated simple sentences are partially or wholly transformed into one or more complex sentences by selecting the names of instances, deciding the ending of words, selecting postpositional particles, connecting methods of sentences, and so on.

Figure 9 shows examples of an intermediate sentence and its transformed forms.

An important processing that is not included in the aforementioned list is the extension of sentences using the following property information of concepts. In particular, noun concepts hold each property using the elements in the conceptual dictionaries. This mechanism generates sentences using the property information to expand previously generated simple sentence forms to more figurative representation with descriptive and explanatory representations. By contrast, event sentences represent the progress of a character's action. Figure 10 shows the result of the experimental implementation. Based on this mechanism, a simple conceptual representation can be transformed into several sentences using descriptive and explanatory sentences. For instance, a basic sentence such as "A young man eats an apple." results in different sentences, such as "Taro ate an apple. It was produced in Aomori. It was sweet." or "Taro ate a sweet apple produced in Aomori. He liked flutes."

In addition, the author has conducted investigations on the diversity and adjustment of Japanese language notation (Kamada & Ogata, 2013a, 2013b, 2013c). Our analyses included Jun'ichiro Tanizaki's (1886–1965) novel (1951) that contained numerous *hiragana* and a Kenzaburo Ōe's (1935–) novel (1972) with a mixture of various character types. In particular, the natural language mechanism defines the syntax in each of the sentences, selected vocabularies, and their letters of notation. This module uses language notation (or representation for the Japanese language) dictionaries. Figure 11 shows the description method in each unit of the language notation dictionary for noun concepts. The structures of the language notation dictionaries correspond to the conceptual dictionaries. For a terminal concept in the conceptual dictionaries, the language notation dictionary can provide the following four types of word representations: (1) that which includes one or more *kanji*, (2) that by only *hiragana*, (3) that by only *katakana*, and (4) that by only *Romaji* (Roman character). In addition, each concept in the conceptual dictionaries is described using the most popular representation. Naturally, the concept representation is also included in the corresponding language notation dictionaries.

Ogata (2016, 2019a) showed various generation examples from INGS, conceptual representations, and their natural language sentences. The following descriptions are the main focus of the revision and expansion:

1. **The description and explanation of the elements in an event**: After an event first appears, the system generates the explanation and description of character(s), place(s), and object(s). These are inserted using the property information stored in each noun unit in the noun conceptual dictionaries. Currently, for the automatic knowledge acquisition and utilization, Ono, Akimoto, and Ogata (2014) and Ono and Ogata (2014c) utilized Wikipedia articles that corresponded to the general noun concepts and proper noun concepts in the conceptual dictionaries in INGS to include the property information.

In addition, it is also possible to add an explanation of general information regarding each concept:

2. **Concretization of each event**: This is a method that utilizes typical event sequence patterns or scriptural knowledge, in particular, in the insertion

Figure 13. Example of an event description in a story

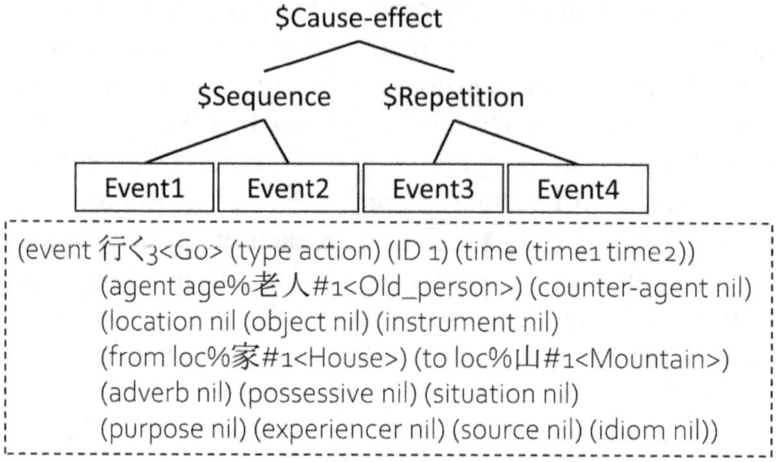

of a script as the detailed development of an event into a part of a story. To improve this procedure, the author has developed a semi-automatic setting tool of scripts (Arai, Ono, and Ogata, 2016), that multiply scripts via the synthesis of existing scripts and so on (Ogata, Arai, & Ono, 2016), and other methods for script acquisition and utilization.

3. **Elaboration of the sentence representation of nested sentences**: The author considers the methods for a more natural representation of the nested sentence.

Music Generation Mechanism or Music Representation Mechanism

This mechanism transforms a conceptual narrative structure to a musical tree structure to determine the musical instruments and their tones, chord progressions, melody patterns, etc. The music in INGS is not necessarily the accompaniment of the narrative representation and has been understood to be an independent representative element that is equal to other representation media, such as language and image. The author has specifically aimed to treat musical elements as linguistic elements and introduce penetrate musical logics into linguistic representations. The transformation of the input information into the music representation mechanism is as follows:

Table 7. Optional cases

Meaning	Sentence Pattern	Okada (1991)	Takeuchi (2011)		Case Name in INGS
Adverbial concept	N1が N2を N3 に **N4** 縮める (N1 shorten N2 by **N4** to N3)	*OC: complement of attribute*	*numeral (corresponds to adverb)*		*adverb*
Possessive case	N1が **N2**の N3 に/へ 潜る (N1 go under **N2**)				*possessive*
Event	N1が N2を **N3** で 破る (N1 beat N2 at/ in **N3**)		*situation*		*situation*
Purpose of action	N1が N2から/ より N3へ/に/ まで **N4**に 出かける (N1 go to N3 from N2 for **N4**)		*purpose*	®	*purpose*
Experiencer	**N1**が N2に 疲れる (**N1** get tired of N2)		*experiencer*		*experiencer*
Criterion of comparison	N1が N2を **N3** の 倍にする (N1 make N2 twice as much as **N3**)	*OS: base*			*criterion of comparison*
Idiom	N1は **N2(足)** が N3から 遠のく (N1 go to N3 less frequently)		*usage*		*idiom*
Information	N1 が **N2**を N3 に 書き取る (N1 write down **N2** in/on N3)				*information*
As	N1は **N2**としての 腕が 鈍る (N1 become les capable as **N2**)				*as*

Source: Ogata (2015)

Table 8. The number of elements in a general noun conceptual dictionary

Hierarchy	Terminal Concepts	Intermediate Concepts	Average Number of Terminal Concepts	
1	42	-	-	
2	15	2	7.50	
3	11	6	1.83	
4	121	23	5.26	
5	5,043	113	44.63	
6	13,957	337	41.42	
7	23,786	796	29.88	
8	31,239	1,367	22.85	
9	26,168	1,650	15.86	
10	11,480	1,068	10.75	
11	2,803	334	8.39	
12	855	83	10.30	
13	245	28	8.75	
Total	115,765	5,808	**Average**	19.93

Table 9 Number of elements in a verb conceptual dictionary

Hierarchy	Terminal Concepts	Intermediate Concepts	Average Number of Terminal Concepts	
1	0	-	-	
2	410	6	68.33	
3	159	5	31.80	
4	11,382	24	638.54	
Total	11,951	36	**Average**	331.97

1. **Length of motif**: This determines the length or patterns of musical basic units and has the following types and pattern: *quarter_rest - quarter_rest - quater_note - quaver_note - quarter_note - quaver_note*.
2. **Key**: This represents the musical keys and C is used in this case.
3. **Scale**: This represents the scales in the music to be generated and C Major is used in this case.

Figure 14. Story structure and related conceptual dictionaries

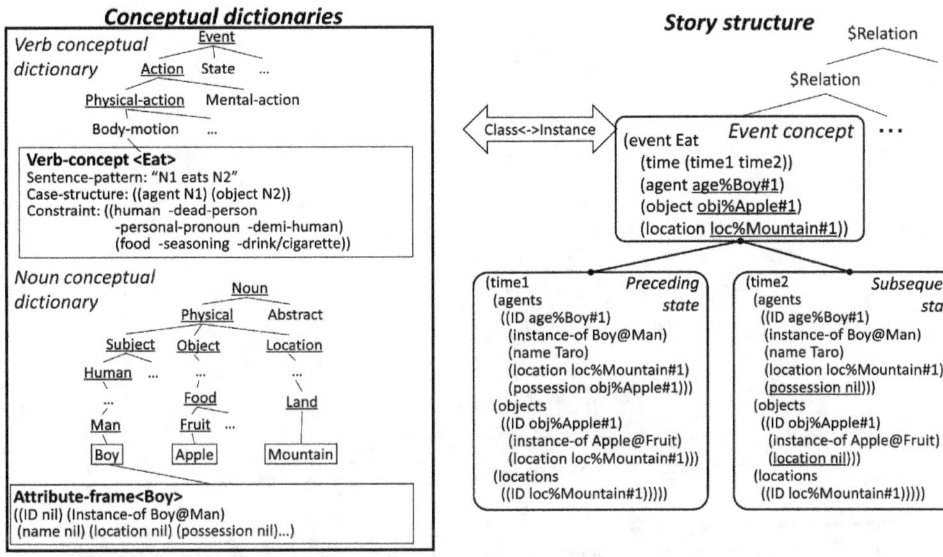

4. **Instruments**: This current system uses thc following seven types instrumental tones: flute, oboe, English horn, clarinet, bassoon, piccolo, Japanese drum, and piano for seven trucks.
5. **Basic tempo**: This is the number of beats per minute and 100 is used.

In addition, these values are applied to only the cases in which a story or narrative discourse is the input information. The detailed description of the musical representation processes for several inputs was presented in Ogata (2018b).

1. **Original music generation (1) [The input is a story]** The process is as follows: 1) Setting the basic information of music (currently, fixed values), 2) Generating motifs according to instances in the input story, 3) Transforming each relation in the input story into the medieval node in the musical structure to generate a "medieval music structure 1," 4) Deciding the harmonic progression based on the medieval music structure 1 to generating the "harmonic progression information," 5) Transforming each event in the input story into a musical event to generate "the musical structure of the original music," and 6) Outputting the "musical structure of the original music" to finish the process.

Figure 15. Hierarchical structure of noun conceptual dictionary
Source: Ogata (2015)

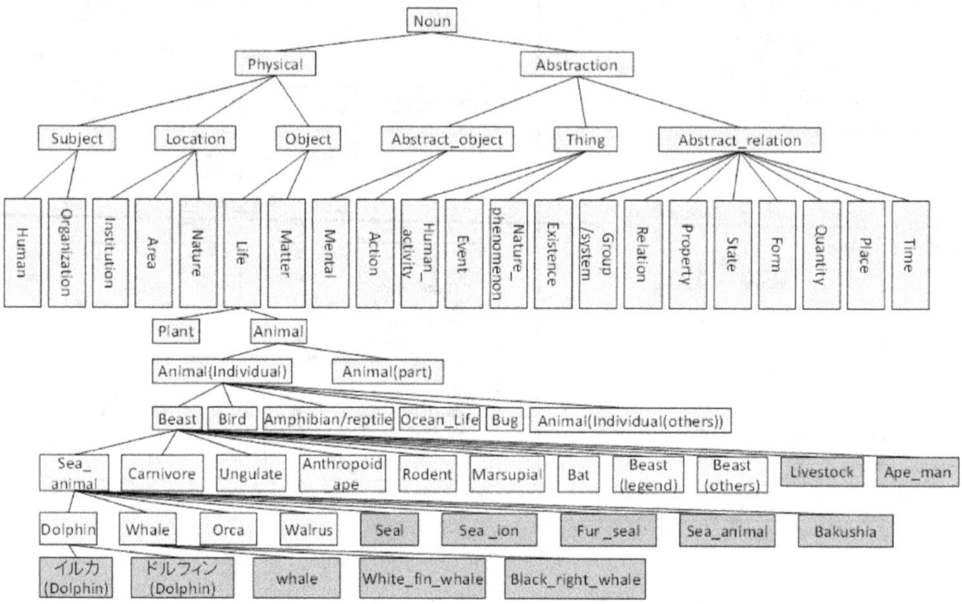

2. **Original music generation (2) [The input is original music]** The process is as follows: 1) Selecting a musical technique, 2) Applying the musical technique, 3) Is the musical technique applicable? Yes - Moving to 4), No - Returning to 1), 4) Deciding the harmony, and 5) Setting the harmony to each musical event.
3. **Original music generation (3) [The input is a variation musical piece]** 1) Transforming the terminal elements in the musical structure in the variation music into musical events in the original musical piece and 2) Changing the relations of variation techniques in the musical structure into the "$sec-pri" in the original piece.
4. **Variation music generation (1) [The input is narrative discourses]** 1) Transforming the narrative discourse structures as the input information into a tree structure by deleting the relations of "recall" and "prediction," 2) Changing the relations in the narrative discourse structure into the relations of the musical structure, and 3) Transforming the terminal events in the musical structure into musical events.
5. **Variation music generation (2) [The input is an original musical piece]** 1) Transforming the original music into the data form of a variation on the piece, and 2) Applying variation techniques to the musical structure

Figure 16. Description format of a noun concept and an example of the hierarchy
Source: Ogata (2015)

```
(set [noun-concept]
 '([noun-concept]
(hierarchy
     (depth [number])
     (hype [superordinate-concept])
     (hypo [list of subordinate-concepts])
     (terminal [list of subordinate-terminal-concepts]))
    (frame [nil])))
-----------------------------------------------------------------
(set '獣[海獣]<animal[sea_animal]>
   '(獣[海獣]<animal[sea_animal]>
    (hierarchy
     (depth 8)
     (hype 獣<animal>)
     (hypo 獣[イルカ]<animal[dolphin]> 獣[鯨]<animal[whale]> 獣[シャチ
獣[セイウチ]<animal[walrus]>)
     (terminal  海豹<seal>  海驢<sea_lion>  膃肭臍<fur_seal>  海獣<se
<dugong> バークシア<bakushia>))
    (frame nil)))
(set '獣[イルカ]<animal[dolphin]>
   '(獣[イルカ]<animal[dolphin]>
    (hierarchy
     (depth 9)
     (hype 獣[海獣]<animal[sea_animal]>)
     (hypo nil)
     (terminal イルカ<dolphin> ドルフィン<dolphin>))
    (frame nil)))
(set '獣[鯨]<animal[whale]>
   '(獣[鯨]<animal[whale]>
    (hierarchy
     (depth 9)
     (hype 獣[海獣]<animal[sea_animal]>)
     (hypo nil)
     (terminal 鯨<whale> 白長須鯨<white_fin_whale> せみ鯨<black_rig
     (frame nil)))
(set '獣[シャチ]<animal[orca]>
   '(獣[シャチ]<animal[orca]>
    (hierarchy
     (depth 9)
     (hype 獣[海獣]<animal[sea_animal]>)
     (hypo nil)
     (terminal 逆叉<orca> 鯱<orca>))
    (frame nil)))
```

Figure 17. Hierarchical structure of the example in Figure 16
Source: Ogata (2015)

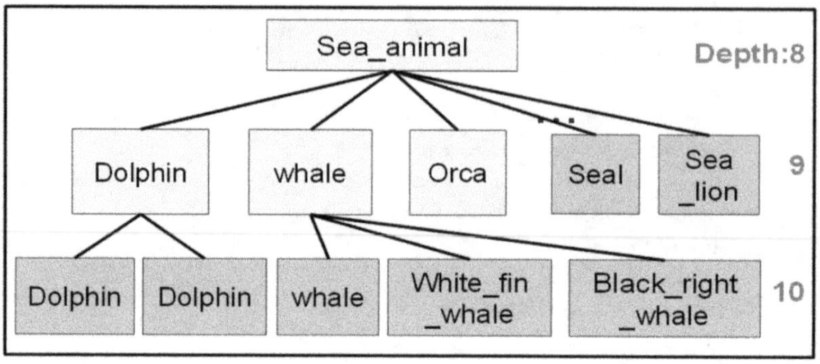

according to the following order: *analepsis*, *prolepsis*, *summary*, *pause*, ellipsis, *scene*, *repeating*, *distance (far)*, *distance (near)*.

Content Knowledge in INGS

The content knowledge in INGS contains conceptual and language dictionaries, narrative content knowledge bases, and a state-event transformation knowledge base. Although narrative content knowledge is also related to the generation and processing of conceptual stories, narrative discourses, and surface representations, this section focuses mainly on conceptual dictionaires.

As Figure 12 illustrates, story and narrative discourse are commonly represented by hierarchically structured conceptual forms, using states related to narrative elements, such as characters or agents, objects, and places in each timeframe, as well as the events described by verb concepts, the corresponding case structures, and relations connecting events or narrative sub-trees. As Figure 12 shows, content knowledge has four roles and functions:

1. To provide narrative content knowledge for narrative techniques (especially story techniques)
2. To add concrete information to the elements of events
3. To associate events with states
4. To add elements of surface representation to the conceptual narrative structure

Figure 18. Hierarchical Structure of verb conceptual dictionary
Source: Ogata (2015)

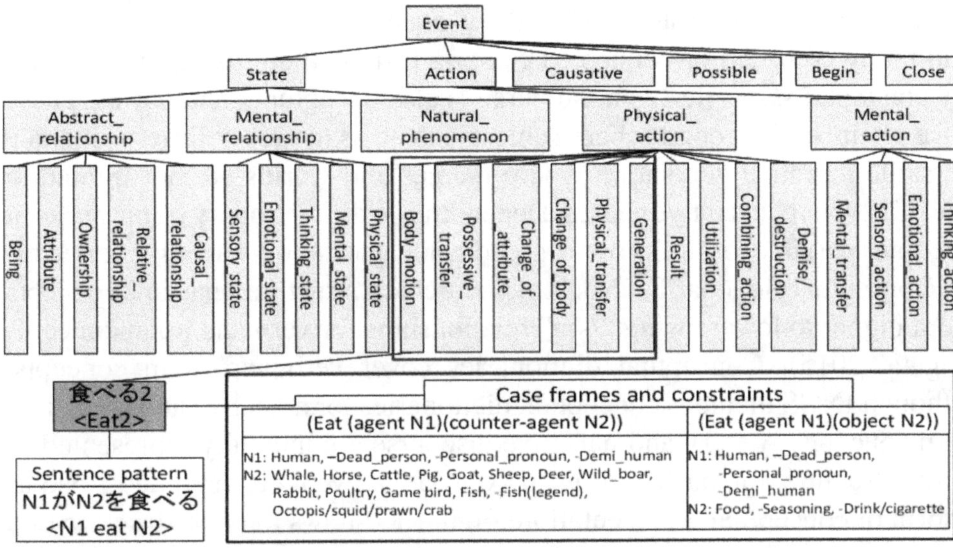

Conceptual Narrative Representations Using Conceptual Dictionaries

The current version of INGS offers both conceptual and language dictionaries. Japanese has several methods of writing or representing characters to create a surface representation of words; the same word can be written using many different types of notation. This is a unique characteristic of Japanese. In INGS, information about word notation is provided by language notation dictionaries that correspond to conceptual dictionaries. The conceptual dictionaries are used mainly to provide semantic information for event elements in the conceptual narrative structure. Although conceptual dictionaries were originally written by hand, the author has developed knowledge acquisition and editing methods for noun and verb conceptual dictionaries; these include methods for automatically editing conceptual knowledge, using information about how often, and in what context, concepts occur (Ono & Ogata, 2014a, 2014b, 2015a, 2015b, 2015c; Ogata & Ono, 2016). They also include a mechanism for automatically acquiring knowledge using property frames to describe the characteristics and state of each noun concept (Ono, Akimoto, & Ogata, 2014; Ono & Ogata, 2014c).

The content knowledge stored in conceptual dictionaries corresponds to the general and objective knowledge associated with each concept. Many previous narrative generation systems offered few conceptual and linguistic elements and rarely collected or connected existing texts without using dictionaries. By contrast, INGS offers general, large-scale conceptual dictionaries. These have a number of strengths, being able to extend usage possibilities to include versatile narrative generation where themes and worlds are not limited in specific types of narrative genres. Depending on the dictionary frameworks, it becomes possible to gain knowledge, edit, and learn about narrative concepts.

Conceptual dictionaries consist of hierarchical structures that rank concepts from higher to lower, using "is-a" relationships for verb and noun concepts (Ogata, 2015). Conceptual dictionaries cover verb and noun concepts, although INGS includes conceptual dictionaries for adjective and adjective-verb concepts. A story and narrative discourse are mutually represented as a tree structure; the basic unit is an event or an event concept, described as a form of conceptual representation. Figure 13 shows the tree structure of a story and provides an example of an event: "An old person leaves home to go to the mountain." In the story structure, each terminal node corresponds to an event, while each intermediate node is a relationship that binds the lower nodes. An event is described as a case frame; it includes a verb concept and several noun concepts. The case frame includes the following eight basic cases:

1. *agent*: The subject of an action.
2. *counter-agent*: The object of an action and a living thing.
3. *object*: The object of an action and a non-living thing.
4. *instrument*: The tool used for an action.
5. *location*: The place in which an action takes a place.
6. *time*: When the action starts and ends.
7. *from* Where the action starts from.
8. *to*: Where the action ends.

In addition to these basic cases, nine optional cases are used to address concepts not by the basic cases (Table 7). According to Okada (1991) (and Okada and Endo (1992)) and Takeuchi (2011), some optional cases overlap with special cases. The first function of INGS conceptual dictionaries is thus to define the constituents of an event.

Currently, the verb conceptual dictionary includes 36 intermediate concepts and 11,951 terminal concepts. The noun conceptual dictionary includes 5808 intermediate concepts and 115,765 terminal concepts. The noun concepts

Figure 19. The descriptive format of verb concepts and an example
Source: Ogata (2015)

Description Format (Verb Concept)	(set [verb-concept] '((name [verb-concept]) (sentence-pattern [sentence-pattern]) (case-cons-set ((case-frame ((agent [*]) (counter-agent [*]) (location [*]) (object [*]) (instrument [*]) (from [*]) (to [*]) (adverb [*]) (possessive [*]) (situation [*]) (purpose [*]) (experiencer [*]) (source [*]) (idiom [*]) (information [*]) (as [*]))) ;; *Each required case takes a "case number" described as 'N' + number like "N1" and "N2". Unrequired cases are "nil". (constraint (([N1's constraint]) ([N2's constraint]) ... [Nn's constraint]))) ;; A "constraint" is a list of noun concepts. ... ;; One or more pairs of "case-frame" and "constraint" are described.) (is-a ([list of superordinate-concept])))
Example	(set '食べる2<eat2> '((name '食べる2<eat2>) (sentence-pattern "N1が N2を 食べる"<N1 eat N2>) (case-cons-set ((case-frame ((agent N1) (counter-agent N2) (location N3) (object nil) (instrument nil) (from nil) (to nil) (adverb nil) (possessive nil) (situation nil) (purpose nil) (experiencer nil) (source nil) (idiom nil) (information nil) (as nil))) (constraint ((人<human> -死人<dead_person> -人間〈人称<personal_pronoun>〉 -準人間<semi_human>) (獣[鯨]<animal[whale]>] 獣[馬]<animal[horse]>] 獣[牛]<animal[cattle]>] 獣[豚]<animal[pig]>] 獣[山羊]<animal[goat]>] 獣[羊]<animal[sheep]>] 獣[鹿]<animal[deer]>] 獣[猪]<animal[wild_boar]>] 獣[兎]<animal[rabbit]>] 鳥[家禽]<bird[poultry]>] 鳥[猟鳥]<bird[game_bird]>] 魚<fish> -魚[伝説]<fish[legendary]>] たこ・いか・えび・かに<octopus・squid・prawn・crab>) (場所<location> -交通路<traffic_route> -公共施設<public_facility>{その他<others>} -地域<region> -崖<cliff> 山<mountain>{部分<part>} 平地<plains> 岸<coast>)))) ((case-frame ((agent N1) (counter-agent nil) (location N3) (object N2) (instrument nil) (from nil) (to nil) (adverb nil) (possessive nil) (situation nil) (purpose nil) (experiencer nil) (source nil) (idiom nil) (information nil) (as nil))) (constraint ((人<human> -死人<dead_person> -人間〈人称<<personal_pronoun>>〉 -準人間<semi_human >) (食料<food> -調味料<seasoning> -飲物・たばこ<drink・cigarette>) (場所<location> -交通路<traffic_route> -公共施設<public_facility>{その他<others>} -地域<region> -崖<cliff> 山<mountain>{部分<part>} 平地<plains> 岸<coast>))))) (is-a (v身体動作<body_motion>))))

Figure 20. Attribute frame and roles in a noun concept

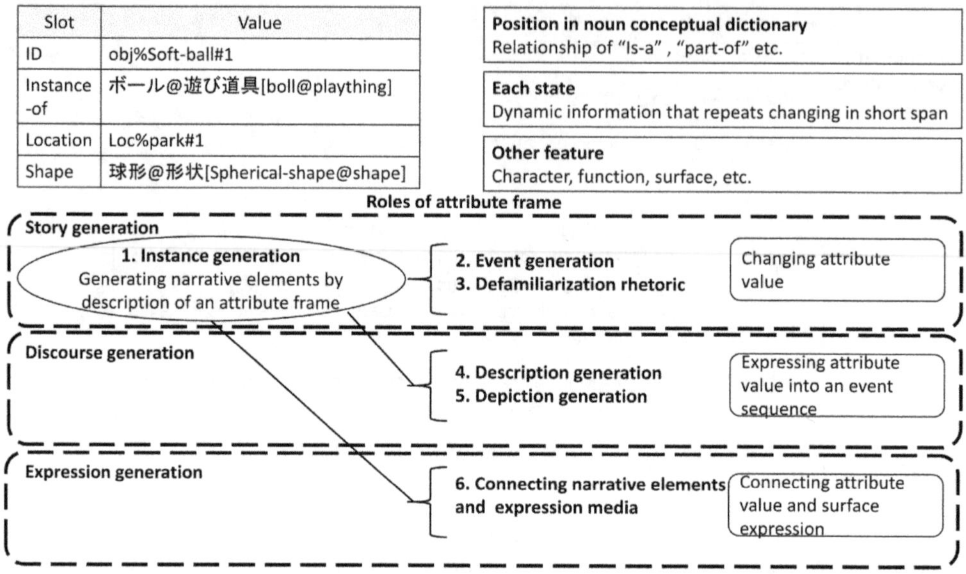

are equal to "general" noun concepts; the author has also developed a proper noun conceptual dictionary (Terada, Akimoto, Ono, & Ogata, 2014), which includes 226,026 concepts, defined under the 987 terminal concepts of the general noun conceptual dictionary. The author has also described adjective, adjective-verb, and adverb conceptual dictionaries. Tables 8 and 9 reveal the number of terminal and intermediate concepts in the noun and verb conceptual dictionaries, respectively. In Table 8, the general noun conceptual dictionary is hierarchically divided into 13 levels. This table presents the numbers of terminal and intermediate concepts at each level and the average number of the terminal concepts positioned directly below intermediate concepts. In Table 9, the verb conceptual dictionary is hierarchically divided into 4 levels; and this table presents the number of terminal and intermediate concepts at each level and the average number of terminal concepts directly below intermediate concepts.

Figure 14 illustrates a relationship between verb and noun conceptual dictionaries and the narrative conceptual representation structures of a story in INGS. The form of a story's conceptual representation is equivalent to a tree structure that hierarchically describes a story. Two or more events are linked within the tree structure. Through recursive processing, the structure to be linked through a relationship sometimes includes two or more partial structures with other relationships. At the start and finish of a story, a precedent and subsequent state are added to store background information, including characters, places, and objects. In addition, when an event is generated, INGS automatically creates corresponding "instance(s)." An instance is concrete, substantial, or actual information about a noun concept. One function of a noun conceptual dictionary is to create an instance from each general noun concept described in the case frame, reflecting the description of the noun concept's attribute frame (Ono, Akimoto, & Ogata, 2014). In the current INGS, the instance-generation mechanism is tentative because the attribute frames are incomplete.

Noun conceptual dictionary

Figure 15 depicts the hierarchical structure of the noun conceptual dictionary; the elements are described in Japanese. In the following description, which relates to actual conceptual dictionaries, the author will show each concept as a pair in the original Japanese description, providing an English translation where necessary. The terminal nodes account for most concrete concepts; the intermediate nodes correspond to more abstract concepts, which bind the lower parts. Figure 16 presents the description format and an actual example; Figure 17 illustrates the hierarchical structure of the example. Each "depth" value represents a number from the highest hierarchical level.

As a chief principle developing a noun conceptual dictionary, the author uses terminal concepts close to language-level words as much as possible. A natural language generation system in INGS directly uses the concepts to a generated conceptual story structure or narrative discourse structure to transform the concepts into words. As mentined in ***Natural Language Generation Mechanism***, in Japanese, words usually have several different notations. For example, the word "*tanuki*," which means "raccoon dog" can be written as "狸" ("*tanuki*" in *kanji*), "タヌキ" ("*tanuki*" in *katakana*), and "たぬき" ("*tanuki*" in *hiragana*). The INGS process has two aims, to generate concepts and surface representations such as language. A noun concept generated using the noun conceptual dictionary is represented by an actual word or a notation that uses a word (or letter) notation dictionary for surface language representation. The author has developed a dictionary, "lanauge notation dictionary," in which various word notations are provided for each noun concept.

Two important knowledge-based mechanisms relate to the noun conceptual dictionary. Since the present version of INGS is not fully developed, the present study refers to its basic concept and design for future development. Each noun concept in a generated event is defined as an instance in INGS. An instance has a proper existence; it can often be defined using a proper noun. For example, "富士山<*Fujisan* (mount_Fuji)>" is a proper noun. A person's name, such as "花子 <*Hanako*>" is the name of an actual, concrete person. Although "富士山 <*Fujisan* (mount_*Fuji*)>" and "花子<*Hanako*>" are words, each instance requires a set of information for the content. Proper nouns, such as "富士山 <*Fujisan* (mount_*Fuji*)>, must be archived in the noun conceptual dictionary. In particular, each proper noun is linked to a corresponding general noun concept, as a proper-noun concept. A "concept" includes a "(noun) attribute frame," which defines the content information. The author is currently adding proper-noun concepts to the noun conceptual dictionary.

A "conceptual" noun dictionary is needed to define attribute frames, which can represent the content of each instance of a noun concept. The author have analyzed novels to prepare a tentative list of attributes of narrative characters. The list includes social, external, internal, and psychological characteristics. Some aspects of the INGS elements are used. The current attribute information is not sufficient, and the author are beginning to study the automatic acquisition of various types of texts (including novels) from the web.

The other main roles of attribute frames are as follows. The first is to generate explanations and descriptions of a noun concept and group. The mechanism is able to generate texts that explain abstract concepts related to an instance, such as social characteristics of a character; it also produces texts that describe external concepts, such as a character's face or body. These functions relate to both the construction of an explanation within the conceptual structure and natural language generation. The second role is especially relevant to natural language generation because it uses a noun phrase to create an attribute frame for an instance. Figuratively speaking, if narrative events and states are equivalent to a narrative tree trunk or branch, the explanations and descriptions that use attribute frames are nuts on the tree. One of the most significant future goals of INGS must be to systematically build attribute frames based on the noun conceptual dictionary.

Verb Conceptual Dictionary

Figure 18 illustrates the overall hierarchical structure of the verb conceptual dictionary. The lower part of this figure uses the example "食べる2<Taberu2 (eat2)>"—a terminal verb concept that describes a sentence pattern, one or more case frame(s), and constraint(s).

Figure 19 presents the description format and example of a verb-concept. The example includes the following values:

1. **A sentence pattern**: "N1" and "N2" correspond to the values of *agent* and *counter-agent* or *agent* and *object* in the case frames. "N3" is an optional value, not included in the sentence pattern.
2. **Two case frames**: "(eat2 (agent N1) (counter-agent N2))" and "(eat2 (agent N1) (object N2))" are included.
3. **Constraints for each case frame**: For example, the constraint for "N1" in the first case frame is "(人<human> -死人<dead_person> -人間〈人称<personal_pronoun>〉 -準人間<semi_human>)." The first case frame has three lists of constraints for "N1," "N2," and "N3."
4. **The list of super-ordinate concepts using "is-a"**: The list expresss the concept relation. In this example, the super-ordinate concept is "(v身体動作<body_motion>)."

A story event is described as a case structure; the value in each case is an instance of the corresponding noun concept in the noun conceptual dictionary. For example, "obj%Apple#1" indicates an instance or concrete object of the noun concept, "Apple." INGS automatically creates the corresponding instance when an event is produced during the story generation phase. The verb concept in an event ("eat" in this example) must provide constraint conditions for the case values. The constraint information for "eat" in Figure 19 is "*human*," *with* the exception of "*dead-person*," "*personal-pronoun*," and "*semi-human*" for "agent," and *"food,"* with the exception of *"seasoning"* and *"drink/cigarette"* for "object."

A terminal verb concept uses a basic sentence pattern for natural language generation and one or more case structures. The case structure is comprised of several noun concepts required by the case of a verb concept. Each instance in a case structure corresponds to a specific element in the noun conceptual dictionary; each noun concept has a set of values in the attribute frame. Figure 20 shows an attribute frame and these roles. Finally, a basic sentence pattern is the structure of a basic sentence used when the natural language

Figure 21. Mutual relationship between a state and an event

Description form	*Example*
((Changing state condition <Case frame to be changed>) (Precondition (Condition1 <Case frame before changing>)) (Case frame <Case frame of corresponding verb concept>) (Name of verb concept (<Verb concept name>))	((変化内容 [Changing state condition] ((actor agent) (slot 所持[Possession]) (op nil) (order (del object)))) (前提条件 [Precondition] (条件1 [Condition1] (条件名[Name] 所持[Possession]) (op nil) (actor agent) (slot 所持 [Possession]) (val (contain object)))) (格フレーム [Case frame] ((agent N1) (object N2))) (動詞概念名 [Verb concept name] (食べる2 [eat])))

generation mechanism transforms each event into a sentence. The natural language generation mechanism adds figurative processing to the basic sentence structure, as described in ***Natural Language Generation Mechanism***.

A Summary of Conceptual Dictionaries in INGS

The author describes the validity of the constraints. Adequate case frames for each of the noun concepts were defined based on a sentence pattern and constraints. An example of the definition based on a sentence pattern, "N1 eat N2," are shown. We simply hypothesize that the constraints of "N1" and "N2" can respectively be "(human, animal)" and "(food, animal)." This means that "N1" takes the subordinate concepts of "human" or "animal" and "N2" takes the subordinate concepts of "food" or "animal." In "N2," "a_living_thing(animal)" is distinguished from "a_nonliving_thing(food)" to define the following two types of case frames, "(eat2 (agent N1) (object N2))" and "(eat2 (agent N1) (counter-agent N2))." The constraints of "N1" in the two case frames are commonly "human" and "animal," while the constraints of "N2" are "food" and "animal" respectively. As previously indicated, a verb concept sometimes holds some different case frames. (There are many concepts in "eat" and the "eat2" is the one type.)

In the present implementation, case frames of 266 verb concepts are defined using optional nine types of cases (Table 7) instead of basic cases. This is explained using the case of "purpose." For a verb concept of "go3," the sentence pattern is "N1 go to N3 from N2 for N4" and the constraint of "N4" is "(abstraction)" which illustrates the purpose of "go." For instance, "a man goes to a mountain from a house for mowing." If we interpret it as "to," we are not able to distinguish "N4" from "N3" ("(place, field)").

As shown the parts in Table 7, as a related study, Okada and Takeuchi, respectively, proposed ten types of cases and more detailed 71 types to represent argument structures of verbs that define the relationship between a verb and a noun in a sentence. The author's study tentatively uses only nine types of cases with an "optional" sign because the description may become too long and complex if all types are used.

Additionally, there are various semantic knowledge resources that are relevant to our research. FrameNet (Fillmore & Baker, 2010) and Japanese FrameNet (Ohara, 2008) define the meaning of a word using a semantic frame, which represents the structured knowledge about a typical scene. VerbNet (Kipper-Schuler, Dang, & Palmer, 2000; Kipper-Schuler, 2005) is a hierarchical verb lexicon in which each class is described using semantic predicates, thematic roles, and basic syntactic frames. These resources are advanced semantic knowledge for addressing complex narrative knowledge units. In the modular approach of INGS, each common and relatively large-scaled conceptual dictionary treats the semantic knowledge, which has a simple structure, and another mechanism handles more complex semantic processing. In particular, the state-event transformation mechanism generates the preceding and subsequent states for an event using the state-event transformation knowledge base, which defines transformation rules between an event and the states. These semantic resources will be more closely connected to the component in INGS.

Next, the conceptual dictionaries enable to edit constraints. A previous evaluation of constraints described in the verb conceptual dictionary revealed a difficult and ambiguous result based on the standard of "actual naturalness." Therefore, the author re-defined the constraints according to the standard of "physical possibility" (common sense in our daily life) as a basic policy. Although the method has some challenges, it is relatively straightforward to define. For example, the purpose and role of the constraints described in "A Japanese Lexicon" (Ikehara, Miyazaki, Shirai, Yokoo, Nakaiwa, Ogura,

Figure 22. Temporal structure of generated discourse

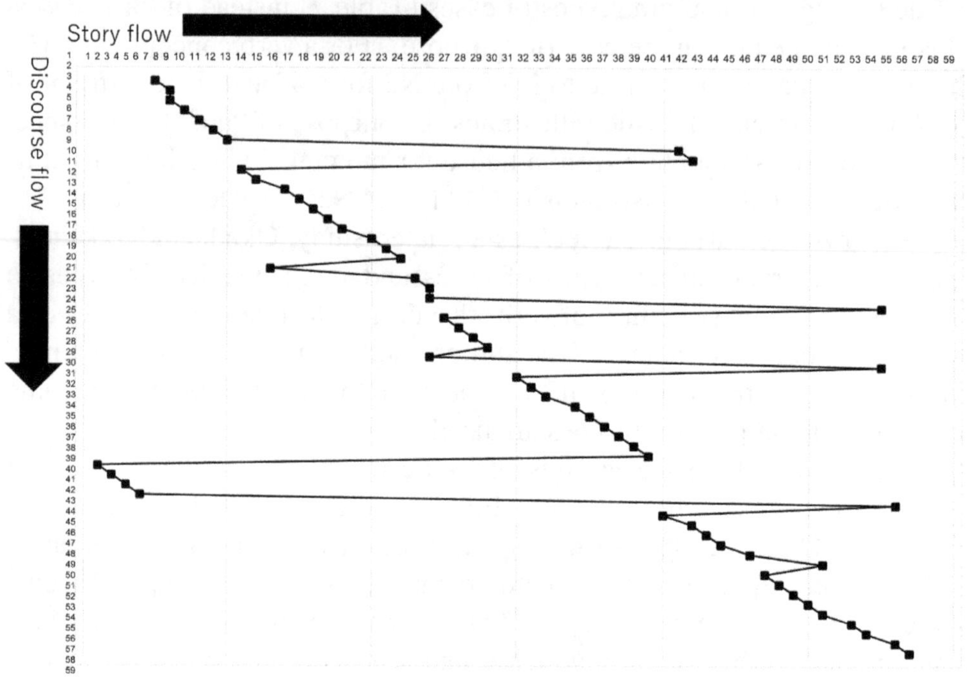

Figure 23. Trends in the levels of listener satisfaction during 100 cycles of narrative discourse

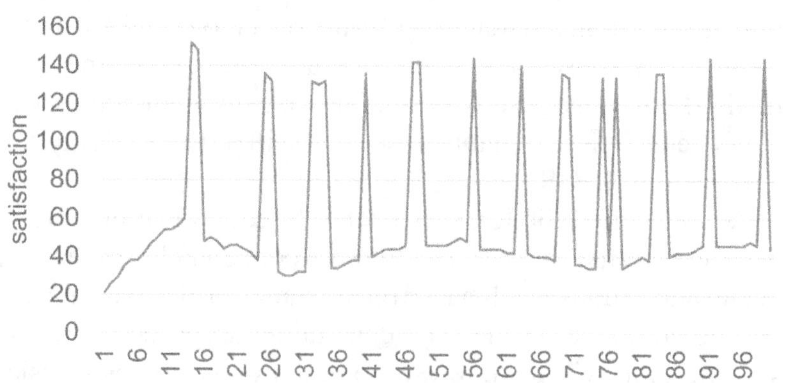

Figure 24. Trends in the parameter levels during the application of 100 cycles of discourse techniques

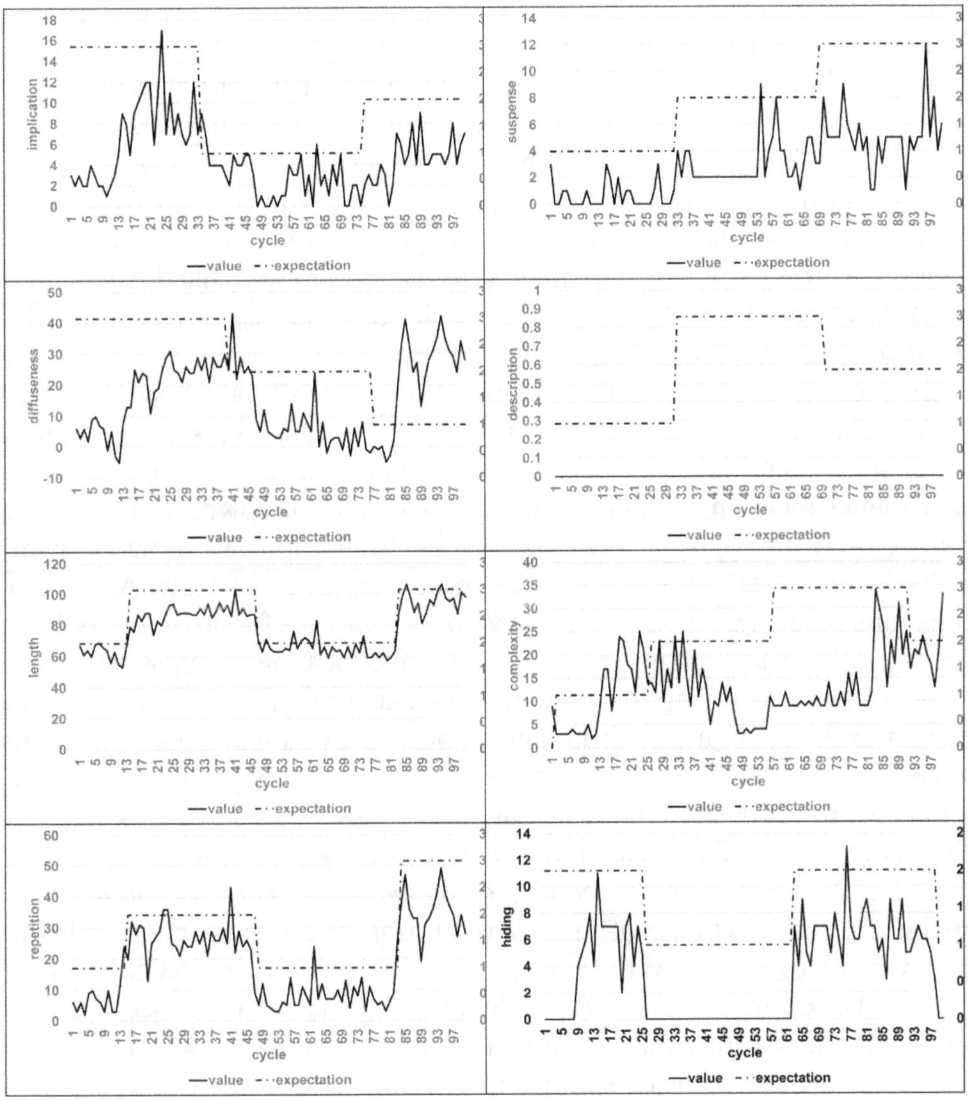

Ooyama, & Hayashi, 1999) is originally conflict resolution for Japanese-English translation. The definition is too wide to limit adequately the semantic range of a noun concept to be generated in the context of narrative generation. In particular, in the case of "eat2," the subordinate concepts of a constraint "food" corresponding to the "N2" include inadequate concepts such as "drink." One of the primary policies of developing constraints was to avoid such ambiguity and inaccuracy.

Constraints have been defined according to the following two ways: the first is based on hyponymy concepts, for example, "furniture" is substituted with "chair" and "bed," while the second is based on the partial exclusion of hyponymy concepts. An instance of the latter is "dead-person" is excluded from "human" using a minus sign such as "-dead -person." As an actual example, for the constraints of "N3" in "N1 return from N2 to N3", we have defined "(lodging housing area -area(scope) -area(human_activity) -land -world foot_of_mountain mountain_pass valley ground island/cape shore -bank farm site point_of_compass edge distance house(body)[housing])." However, this procedure is time-consuming.

To provide a more effective solution, many "sets" have been prepared for constraint description. A "set" is a group of constraints to be applied commonly to similar case frames. If a common "set" can be applied to the case frames including "a place where a person can sit down," the "set" can also be used for opposite concepts such as "stand up." We utilize a name for showing a "set" of noun concepts to be used as constraints. A "set" of constraints with a name can be applied to various case frames commonly as a whole. For instance, the "set" of "inhabitable location" is equal to a "set" of constraints, "(lodging housing area -area(scope) -area (human_activity) -land -world foot_of_mountain mountain_pass valley ground island/cape shore -bank farm site point_of_compass edge distance house(body)[housing])." The set is commonly used for the constraints in various events, such as "N2" in "N1 move to/into N2." As another example, for "N1 take out N2 from N3", the set of "*a person of such age who can take care of herself or himself*" is applied to "N1" and the set of "transportable object" is applied to "N2." In the first development stage of the verb conceptual dictionary in INGS, we have defined the constraints for only agent, counter-agent, and object because they are the minimal information required to make a sentence. Moreover, on this basis, we have added the constraints for location cases. Finally, 38 sets have been defined and used in the present version of the verb conceptual dictionary.

Narrative Content Knowledge Bases

These knowledge bases, which are currently divided into a story content knowledge base and narrative representation knowledge base, store fragmentary, partial, or macro narrative content knowledge. The story

Figure 25. Overall architecture of KOSERUBE
Source: Ogata (2019a)

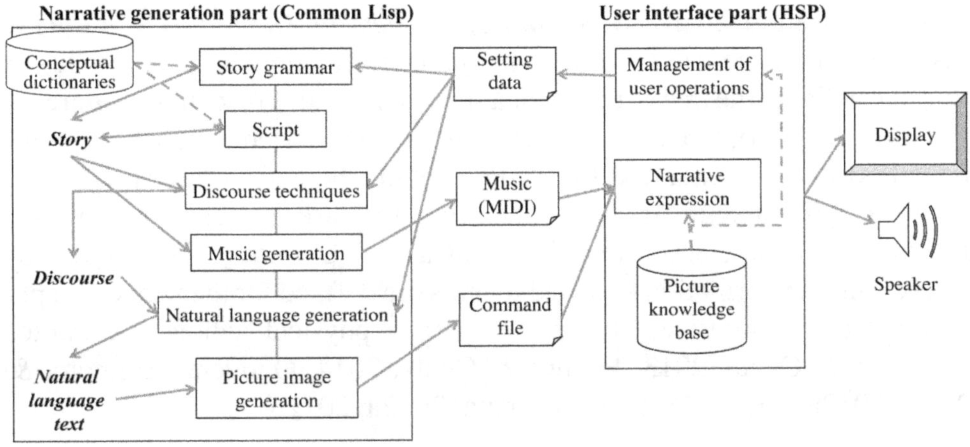

content knowledge base provides concrete story information, particularly fragmentary information, such as event sequences corresponding to a story's part. However, this content knowledge base can also store macro event synopses representing an entire event structure. On the other hand, the narrative representation knowledge base describes sound information as a medium of narrative representation; it includes motifs such as fragmental melodies and image information, including still images and animations. The story content knowledge base contains concrete story content knowledge, reflecting the story techniques that extend the conceptual representation of a story. Table 3 includes story techniques and the corresponding story content knowledge; however, some story techniques have no corresponding story content knowledge. As Table 3 illustrates, the current descriptive form is divided into a pair of elements and a longer pattern with three or more elements. The latter is divided into simple event-sequence patterns and the method of extending a single event into event-sequence patterns with many elements. In addition, the story content knowledge bases contain knowledge needed to generate larger-scale story structures, including story content grammar based on Propp's narrative theory (PBSG) (Imabuchi & Ogata, 2012a, 2012b, 2013, 2014).

State-Event Transformation Knowledge Base

As already mentioned in ***A Summary of Conceptual Dictionaries in INGS***, this knowledge base is used to generate static information that corresponds to terminal nodes in the tree structure of a story or narrative discourse. Figure 21 provides the descriptive form of each state-event knowledge transformation, as well as a concrete example. In this description, the state in "time1" changes to the changing state in "time2." In addition, the case elements in a state are described as concrete and individual instances. Currently, 5,241 types of state-event transformation knowledge units are defined for the verb concepts included in the intermediate verb concept, "physical action" (Akimoto, Kurisawa, & Ogata, 2013; Fukuda & Ogata, 2014; Onodera, Akimoto, & Ogata, 2012a, 2012b; Onodera & Ogata, 2012a, 2012b).

LANGUAGE NOTATION DICTIONARIES

As Ian Hideo Levy (2014) has argued, one feature that differentiates Japanese from other languages is the fact that each word has various different notations; different types of notation result in different effects. Especially in artistic language, such as novels and poems, selecting the effective notation is an important strategy for writers. In INGS, each terminal element in the conceptual dictionaries is associated with the most typical word notation. Although the mechanism for generating natural language can use the previously described notation, it can also change the notation. Various types of word notation, including Chinese characters, *hiragana*, *katakana*, and Roman characters (*Romaji*), are made ready, as shown in Figure 11. Although mechanical transformation is easy, more strategic transformations that consider the effects and timing lead to more difficult processing. Currently, ordinary methods of processing use the notations described in conceptual dictionaries as a default; the author has been exploring various possibilities for more advanced notation processing (Kamada & Ogata, 2013a, 2013b, 2013c). Although the author has described the notation method sometimes in this chapter, the author showed its positioning in INGS as a language notation dictionary here.

Generation Examples

Propp-Based Generation

APPENDIX 1 shows a generated story according to the top-down method using PBSG stated in the section of Propp-based Mechanism. The upper conceptual structure, (1), is the actual output description of the generated story, and the lower text, (2), is an English translation of the Japanese text generated from the conceptual structure using a natural language generation mechanism in INGS. In the generated story, the name of each instance is concretized using property information. Additionally, the author shows another generated story using new mechanisms in this enhanced version (APPENDIXs 1 (3) and (4)).

Generation Example in INGS

Incorporation of the Propp-based mechanism into INGS allows the creation of stories where the macro-development (using Propp-based story generation) and micro-development (based on other story techniques) are mixed. APPENDICIES 2 (1) and (2) again show the generated example in the previous chapter (Ogata, 2016). As indicated in APPENDIX 2 (1), such macro-story can be refined by other story techniques as the higher level of the hierarchy. APPENDIX 2 (2) shows a Propp-based macro-story inserted into the portion of a micro-sequence that does not belong to a larger structure.

In addition, APPENDICIES 2 (3) and (4) show another generated story using new mechanisms in this enhanced version. In the generated story, the possibility of story generation is extended based on enhanced functions: the event generation from intermediate nodes in the story tree and the use of scriptural knowledge for detailing the event development.

Jauss-Based Narrative Discourse Generation

APPENDIX 3 show execution examples with input data. All 13 types of discourse techniques are applied. These examples commonly used a generated story in **APPENDIXs 1 (1)** and **(2). APPENDIX 3** is a narrative discourse that was automatically generated completely. Figure 22 illustrates the temporal structures with the horizontal axis corresponding to the temporal progress events ordered from 1 to 59. The number description in **APPENDIX 3** includes the 59 events and inserted events (such as "14-1") in the timing of state generation using the state-event transformation knowledge base in the story generation phase. However, in figure 22, the inserted events are not used for simplification. Connecting nodes allow the order of the discourse to be narrated. Six events not included in the input story are inserted by the discourse techniques of "external analepsis," "external prolepsis," and "achronie.

Figure 23 shows trends in the levels of listener satisfaction during the application of 100 cycles of narrative discourse techniques. It can be seen from this figure that listener satisfaction moves up and down between approximately equal levels. Figure 24, on the other hand, shows movements to the various parameters used in the calculation of Figure 23 (1: implication; 2: diffuseness; 3: complexity; 4: suspense; 5: length; 6: hiding; 7: description; 8: repetition). If the dotted line (expectation values) and the solid line (parameter values) follow a similar trajectory, this indicates that the generation was in keeping with expectations. (An expectation value is taken for each of 1 (small), 2 (medium), and 3 (large) parameter cycle, and the higher the expectation value, the higher the corresponding parameter value is expected to be)

From Figure 24, it can be seen that results for five of the eight parameters (1: implication; 4: suspense; 5: length; 6: hiding; 8: repetition) were in line with expectations. Trends for the remaining three parameters (2: diffuseness; 3: complexity; 7: description) saw fluctuations that varied from expectation. For example, the "implication" parameter changed in line with listener expectation. However, with "diffuseness," changes in the parameter that diverge from expectation are seen from loop 82. There are two reasons for the occurrence of changes to parameters that differ from expectations—

(1) Parameter overlap affecting narrative discourse techniques

The mechanism involves the selection of narrative discourse techniques used for generation on the basis of listener expectation. With multiple parameters, however, the overlap of narrative discourse techniques corresponding to various parameters causes interference in parameter processing. Additionally, for parameters other than "length" and "diffuseness," the application of narrative discourse techniques only increases parameter values.

To give a specific example, the same techniques are used to increase "complexity" and to increase "implication." In Figure 24 "implication" was processed after "complexity;" the result of this is that trends in the "complexity" parameter diverged from expectation due to "implication" processing.

(2) Formula dependence

The parameter formulae are not independent, and it is possible that contradictions may arise in generation targets. For example, the "hiding" value is derived from the "diffuseness" and "repetition" values. It is therefore extremely difficult to create a mechanism in which "hiding" and "diffuseness" are low, but "repetition" is high.

KOSERUBE: PARALLEL GENERATION PROCESSING

The standard processing in INGS is sequential; namely, a generation or expression module is executed at one generation timing. However, the author seeks to explore parallel narrative generation processing. An experimental system called KOSERUBE (Akimoto, Imaguchi, Endo, Ono, Kurisawa, Kamada, & Ogata, 2013) intended to be an application system for INGS is an example or the first experimental framework of parallel narrative generation. Figure 25 shows the overall architecture of KOSERUBE. The system is mainly divided into narrative generation and interface parts. The system has an "Iwate region database" related to Iwate prefecture and the near area in Japan. The database contains conceptual data that includes 34 characters, 21 objects, 85 places, explanation sentences, and other data. First, the user selects the characters, story type, narrator, etc. Seven types of narrators are prepared in order to change the narration style. Although the generation flow of each narrative mechanism is as follows, the system represents the generated discourse structure, music, and image synchronously:

1. **Generating a story structure**: (a) Expanding the story content grammar by top-down method, (b) Generating states, (c) Generating scripts of the particular events using a script technique, and (d) Complimenting the states.
2. **Generating a narrative discourse structure**: (a) Generating a first narrative discourse structure, (b) Inserting descriptions, and (c) Using discourse techniques according to a selected narrator;
3. **Generating a musical structure**
4. **Generating visual image expression**

This synchronized narrative representation architecture will be able to applied to INGS. However, in KOSERUBE, after all of narrative structures and representations are generated, the system syncronoously displays them. This is a kind of offline processing. By contrast, INGS will require an online processing that enales narrative representation synchronizing with narrative generation.

FUTURE RESEARCH DIRECTIONS

As described in the previous section, the author has developed an Integrated Narrative Generation System (INGS), including the framework of the system, philosophical concepts, and mechanisms such as generation and content mechanisms and application systems. INGS is currently being executed at the experimental level. INGS developed the aspects of "expanded literary theory" through integrating the literary theories of Propp (1968), Genette (1972), and Jauss (1970) into the system and covered the implementation of story, narrative discourse, and narrative representation. On the other hand, in the developmental process of INGS, the parts and mechanisms were newly added, including conceptual dictionaries, state-event transformation knowledge base, and story generation mechanisms using them. The aforementioned implementation of INGS enabled effective constructive research on the complex mechanisms of narrative generation in which the entire performance becomes possible through the interaction of diverse techniques and knowledge. An important merit is that INGS architecture is implemented as a modularized and flexible program that it is easy to extend, change, and revise. This characteristic is linked to the possibility of development that can gradually incorporate various ideas and modules into the system. In the future, the author plans to conduct development based on the circulative and repetitive process through organically integrating partial revisions into the entire architecture using its plastic feature. In the future, in addition to continuing the gradual development of INGS, the author aims to bridge the system to the social and personal "practices" of narrative generation. The following topics will be covered in the continual system development of INGS:

1. **Basic generation methods:** The system admits various types of input information for each generation module and allows the flexible flow of generation order in connection to the concept of "circulative control."

2. **Content knowledge for generation:** A central work in the next step is related to the aspect of content, which is the knowledge to be used, especially outside of the dictionaries for concepts and words. The main parts contain narrative patterned knowledge, such as scripts, and story grammar for story generation and property knowledge for the noun conceptual dictionary. After Ogata and Akimoto (2016) was published, the author has been analyzing and introducing motif content knowledge used in folktales. This knowledge acquisition is related to the next automatic functions. These have been partially achieved. In addition, this study is planning to use Japanese narratological theories, including the structural narrative theories of *nō* by Zeami (c. 1363–c. 1443) (Zeami Zenchiku, 1974), for the expansion and reorganization of the story content knowledge base in INGS.
3. **Expanding the content knowledge**: Furthermore, the author has plans to extend the range of narrative content knowledge bases. A possibility is a "narrative discourse knowledge base." The knowledge base may be used for storing the structured and abstracted examples related to narrative discourse. For instance, it defines the structural transformation of the temporal order of a story showing the correspondence relations with the original story. Narrative content knowledge is also related to the aspect of narrative representation. The fragments and motifs of music and novels can be utilized as a function of narrative generation through processing, including partial changes and synthesis. A musical motif knowledge base, the knowledge base for visual representation, including still image and motion pictures and the fragments of natural language novels are also included in the same group knowledge base containing the story content knowledge base. The musical content knowledge base will store various elements composed of musical structures, such as motifs, as examples. Similarly, the image content knowledge base holds various elements or components composed of image works as examples. The sentence content knowledge base stores actual sentences from novels and stories as structural data. Although the natural language generation mechanism in INGS generates the tree structures of narrative surface sentences using generation techniques and one or more knowledge base(s), a tree structure of a narrative sentences is connected to the generated story and narrative discourse structures. In this plan, all the structures described by the conceptual generation processing and natural language knowledge processing are formally identical and are unified into an organically connected structural description.

4. **Knowledge acquisition and learning mechanisms:** As content knowledge increase, it is particularly necessary to automatically acquire knowledge using formal and semantic methods. This theme also includes automatic data acquisition using statistical analysis methods to control or adjust the use of conceptual knowledge. Further, after Ogata and Akimoto (2016) was published, the author has been attempting to develop the automatic and semi-automatic acquisition of scriptural knowledge.
5. **Natural language generation:** This includes formal processing and semantic processing. The latter requires previous various content knowledge. In INGS, property knowledge regarding the noun conceptual dictionary provides the main content knowledge regarding natural language generation. Hence, this topic is associated with points (2) and (3).
6. **GIS (*Geinō* Information System) or social narrative production and distribution system:** GIS is not only an application system of INGS, it is also an extension of INGS that allows it to be connected to a social or business framework for the production and distribution of generated narratives; GIS is designed to contain one or more INGS.
7. **Expanded literary theory:** This section should discuss the introduction of previous narratology and literary theories into the INGS system. First, previously analyzed literary theories, such as Mikhail Bakhtin (1895–1975), Sigmund Freud (1856–1939), Julia Kristeva, or intertextuality, will be applied to various parts of narrative generation. In addition, other important literary theories, such as Rorand Barthes (1915–1980), Claude Lévi-Strauss (1908–2009), and especially Japanese theories and narratology, will be incorporated as a cultural approach into INGS.
8. **Japanese literary theories and narratology**: An especially important direction of the expanded literary theory is to survey and analyze traditional or modern and contemporary narratological research and literary theories in Japan and to incorporate them into INGS (and GIS). Regarding detailed discussions and considerations, refer to several descriptions in Ogata (2019c) that is a series of this book.
9. **Philosophical concepts related to expanded theory**: The author will extend and elaborate "circulative narrative control" and more concretely apply it to the INGS system. A perfect free processing order among the modules will also be related to the concept of "fluidity and fixation." In "norm-deviation," in the relative sense, each cycle of narrative generation corresponds to a system of the deviation from the current state (the norm). In particular, preliminarily constructed scripts and dictionaries

correspond to the norm, whereas various rhetorical techniques deviate. The author seeks to consider the application of this concept for the entire INGS system. Regarding fluidity-fixation, for instance, the Jauss-based mechanism indicates one of the future directions. Namely, the circulative process that corresponds to the fluidity and user-viewing of one or more generated results is equivalent to fixation. As stated above, three of these concepts are associated with each other.

10. **Multiple narrative structures:** The current INGS does not reflect the concept of the "multiple narrative structures model," which is one of the essential characteristics in the author's integrated approach to narrative generation study, especially in the aspect of expanded literary theory or post-narratology. Although the model has various aspects, the author has considered model-making through a novel by Yukio Mishima (1925–1970) (1963) (Ogata, 2018a). In order to introduce INGS, the system requires each character agent to be actualized as a system, and the entirety of the narrative generation system should be actualized as the synthesis of each narrative generation system. In addition, multiple narrative structures are closely related the analysis of *kabuki* from the narrative generation perspective. The implementation of the multiple narrative structures of INGS in relationships with GIS and *kabuki* analyses is a very important theme in the future.
11. **Social narrative distribution (in plans for social and individual development):** This topic implies that narratives produced by a narrative generation system, which include narratives created in collaboration with humans, and the narrative generation system itself are connected to society—the outside of the system. Although one of the typical forms is business, there are other possibilities. "Virtual publishing" and "virtual entertainment production" (by GIS) also show images for introducing the aspect of business into narrative generation systems. This is at the most outside field of a multiple narrative generation framework by the author. This area is directly related to the contents in Chapters 2 and 4 in this book.
12. **"My (the author's)" narrative creation (in plans for social and individual development)**: This concept refers to narratives based on one's private themes and materials. The INGS system introduces private content into the knowledge content. In particular, INGS introduces private knowledge content into the story content knowledge base and other content knowledge bases. This theme is also associated with Chapters 2 and 4 directly.

13. **Evaluating narrative generation systems**: Evaluating narrative generation systems or generated results is a very difficult problem. This problem is originally related to the extreme difficulty of finding and determining a universal and versatile standard for evaluating narratives, literary works, and arts. The author's conclusion is that the evaluation standards related to narrative generation systems must be plural or diverse. Moreover, when we evaluate narrative generation systems and generate narratives with specific evaluation standards limited from the viewpoint of a specialized goal, we will acquire effective results and implications. In INGS, the author evaluated the constraint of information of noun concepts described in the verb conceptual dictionaries using the standard of "physically possible" to change the constraint descriptions based on the results (Ogata, 2015). Furthermore, an extreme evaluation for INGS will be related to plans for social or external and personal or internal directions in Chapters 2 and 4 in this book.

CONCLUSION

In this chapter, first, **BACKGROUND: COMMON LISP** covered Common Lisp, the main programming language for our system development, including INGS and other systems. The present INGS organically integrates previously developed narrative generation systems into an integrated framework; the integration has been conducted without large design changes because partial program systems have been written. Moreover, although INGS required various extensions and changes throughout the development process, they were comparatively easy for the characterization of Common Lisp that used a small-scale and simple function as a basic unit. **GENERATION PROCESS OF THE INTEGRATED NARRATIVE GENERATION SYSTEM** described the research and development process the author used as a kind of narrative generation process, referring to many related papers.

Next, **THE ARCHITECTURE OF INTEGRATED NARRATIVE GENERATION SYSTEM** is the core section in this chapter that introduced INGS in detail. Although Ogata (2016, 2018a) also described the INGS architecture, this chapter divided it into two parts, mechanisms and contents, to aim at a systematic description of the construction. Moreover, several expansions were conducted after the aforementioned papers. The first sub-section, **The Generation Mechanism of INGS**, explained INGS's mechanisms. In particular, the author describes the entire generation process from the viewpoints of the module division and generation control mechanisms. Although it is not completely implemented, an important characteristic of INGS generation control is that the generation order is not fixed according to the philosophical concept of circular narrative generation mentioned by Ogata (2019d). On the other hand, the primary modules in aspects of the mechanisms are the three modules of story generation, narrative discourse, and narrative representation. In each sub-section, the mechanism is explained in detail. The following **Content Knowledge in INGS** mainly contains conceptual and language dictionaries, content knowledge bases, and a state-event transformation knowledge base. This section focused mainly on conceptual dictionaries, including the noun conceptual dictionary and verb conceptual dictionary. These conceptual dictionaries are used to generate and edit event concepts in the story and discourse generation mechanisms. On the other hand, content knowledge bases are related to the generation and processing of conceptual stories and narrative discourses, and surface representations for providing concrete, fragmentary, or more completed narrative knowledge to the modules to generate, expand, and edit generating narratives.

As stated above, in this chapter, "contents" referred to the content knowledge used in INGS to generate the various levels of a narrative. However, in the entire integrated approach to narrative generation or post-narratology by the author (Ogata, 2019b), narrative contents have another meaning at a different level. In particular, narrative contents at a different level refer to narrative works generated by narrative generation mechanisms. If the content knowledge described in this chapter is at a level that is more similar to system mechanisms, the level of narrative contents or narrative works is further different from the system mechanisms. The description related to the problem of "internal narrative generation" corresponds to the contents in the latter meaning. Narrative contents in this sense will be generated or created by INGS, the *Geinō* Information System (GIS) in Chapter 2, and application systems such as the advertising narrative generation system described in Chapter 3. Thus, from a broader perspective, INGS, GIS, and other application systems correspond with the side of narrative mechanisms and, by contrast, the narrative contents through the mechanical systems are equal to the side of contents. At the same time, creating narrative contents in this sense is the ultimate goal of the integrated approach to narrative generation or post-narratology posited by the author.

ACKNOWLEDGMENT

This chapter's research was supported by JSPS KAKENHI Grant Number 18K18509, Kayamori Foundation of Information Science Advancement Research Grant, The Telecommunication Advancement Foundation Research Grant, and AOI TYO Holdings Inc.

REFERENCES

Abe, H., Ogata, T., & Hanada, K. (2008). Televi kōkoku no shot ni okeru brand dōnyū rhetoric no bunseki: Kōkoku no monogatari seisei system no ichi kikō no tameni [An analysis of brand introduction rhetoric in a shot in television commercial films: For a mechanism of an advertising narrative generation system]. In *Proceedings of the 16th Meeting of Literature, Cognition, and Computer Research Group II* (16G-05). Tokyo, Japan: Japanese Cognitive Science Society.

Abe, H., Ogata, T., & Onodera, K. (2009). An analysis of products introducing rhetoric in advertising and the prototype system implementation. In *Proceedings of the 23rd Annual Conference of the Japanese Society for Artificial Intelligence* (1J1-OS2-4). Tokyo, Japan: Japanese Society for Artificial Intelligence.

Akimoto, T., Imabuchi, S., Endo, J., Ono, J., Kurisawa, Y., Kamada, M., & Ogata, T. (2013). Development of a generation/expression system of narratives in the style of a folk tale, KOSERUBE Version 1. *Transactions of the Japanese Society for Artificial Intelligence*, *28*(5), 442–456. doi:10.1527/tjsai.28.442

Akimoto, T., Kurisawa, Y., & Ogata, T. (2013). A mechanism for managing the progression of events by states in integrated narrative generation system. In *Proceedings of the 2nd International Conference on Engineering and Applied Science* (pp. 1605–1614). Copenhagen, Denmark: ICEAS Secretariat.

Akimoto, T., & Ogata, T. (2009). Monogatari gensetsuron no system-ka ni mukete—Genette to Jauss no kakuchō bungaku riron [Toward the systematization of narrative discourse theories: The expanded literary theory of Genette and Jauss]. In *Proceedings of the Japanese Cognitive Science Society 18th Workshop of Literature in Cognition and Computer II* (18G-02). Tokyo, Japan: Japanese Cognitive Science Society.

Akimoto, T., & Ogata, T. (2010a). Monogatari gensetsuron to juyō riron no shisō wo toriireta monogatari gensetsu seisei system no shisaku ni okeru seisei nōryoku [The generation ability in the experimental system of a narrative discourse generation system introducing the thoughts of a narrative discourse theory and a reception theory]. In *Proceedings of Japanese Cognitive Science Society 20th Workshop of Literature in Cognition and Computer II* (20W-03). Tokyo, Japan: Japanese Cognitive Science Society.

Akimoto, T., & Ogata, T. (2010b). Implementing a narrative discourse system as a mechanism in narrative generation: A hybrid approach of AI and narratology. In *Proceedings of the 7th International Conference on Cognitive Science* (pp. 359–360). Oakbrook Terrace, IL: Cognitive Science Society.

Akimoto, T., & Ogata, T. (2011). Computational model of narrative discourse theory and reception theory in narratology and its implementation. In *Conference Handbook of the 13th Annual International Conference of the Japanese Society for Language Sciences* (pp. 155–156). Tokyo, Japan: The Japanese Society for Language Sciences.

Akimoto, T., & Ogata, T. (2012a). A prototypc of the integrated narrative generation system consisting of narrative structural operation techniques and the control mechanism. In *Proceedings of the 26th Annual Conference of the Japanese Society for Artificial Intelligence* (1N2-OS-1b-13, 1N2-OS-1b-13in). Tokyo, Japan: Japanese Society for Artificial Intelligence.

Akimoto, T., & Ogata, T. (2012b). A narratological approach for narrative discourse: Implementation and evaluation of the system based on Genette and Jauss. In *Proceedings of the 34th Annual Conference of the Cognitive Science Society* (pp. 1272–1277). Tokyo, Japan: Japanese Cognitive Science Society.

Akimoto, T., & Ogata, T. (2012c). Macro structure and basic methods in the integrated narrative generation system by introducing narratological knowledge, in *Proceedings of the 11th IEEE International Conference on Cognitive Informatics and Cognitive Computing* (pp. 253–262). The Institute of Electrical and Electronics Engineers. 10.1109/ICCI-CC.2012.6311156

Akimoto, T., & Ogata, T. (2013a). Temporal order transformation techniques in narrative discourse: The current state and future work in an integrated narrative generation system. *Proceedings of Forum on Information Technology*, *2*, 207–210.

Akimoto, T., & Ogata, T. (2013b). Towards a discourse mechanism in narrative generation system: Proposal of a system introducing narrative discourse theory and reception theory. *Cognitive Studies*, *20*(4), 396–420.

Akimoto, T., & Ogata, T. (2013c). The expansion of paths in the mutual transformation mechanism of music and narrative. *International Journal of Cognitive Informatics and Natural Intelligence*, *7*(4), 44–63. doi:10.4018/ijcini.2013100103

Akimoto, T., & Ogata, T. (2014a). Tōgō monogatari seisei system ni okeru monogatari naiyō gihō to monogatari naiyō contents chishiki base [Story techniques and the story content knowledge base in the integrated narrativegeneration system]. In *Proceedings of the 20th Annual Meeting of the Association for Natural Language Processing* (pp. 224–227). Tokyo, Japan: The Association for Natural Language Processing.

Akimoto, T., & Ogata, T. (2014b). Towards the focalization mechanism in an integrated narrative generation system. In *Proceedings of the 31st Annual Meeting of the Japanese Cognitive Science Society* (pp. 344–353). Tokyo, Japan: Japanese Cognitive Science Society.

Akimoto, T., & Ogata, T. (2014c). An information design of narratology: The use of three literary theories in a narrative generation system. *The International Journal of Visual Design*, *7*(3), 31–61. doi:10.18848/2325-1581/CGP/v07i03/38747

Akimoto, T., & Ogata, T. (2014d). Circulative narrative generation based on the mutual transformation between narrative conceptual structures and music in the integrated narrative generation system. *Journal of Robotics. Networking and Artificial Life*, *1*(3), 198–202. doi:10.2991/jrnal.2014.1.3.6

Akimoto, T., & Ogata, T. (2015a). Evaluation of a narrative discourse generation system based on the concept of "norm and deviation". *Journal of Robotics, Networking and Artificial Life*, *2*(1), 50–53. doi:10.2991/jrnal.2015.2.1.12

Akimoto, T., Ono, J., & Ogata, T. (2012a). Narrative Forest: A trial of integrated application of narrative generation system. In *Proceedings of the 29th Annual Meeting of the Japanese Cognitive Science Society* (pp. 342–352). Tokyo, Japan: Japanese Cognitive Science Society.

Akimoto, T., Ono, J., & Ogata, T. (2012b). Narrative Forest: An automatic narrative generation system with a visual narrative operation mechanism. In *Proceedings of the 6th International Conference on Soft Computing and Intelligent Systems & the 13th International Symposium on Advanced Intelligent Systems* (pp. 2164–2167). Tokyo, Japan: Japan Society of Fuzzy Theory and Intelligent Informatics. 10.1109/SCIS-ISIS.2012.6505319

Anzai, Y., Saeki, Y., & Muto, T. (1981). *Lisp de manabu ninchi sinrigaku, 1—Gakusyū* [Cognitive psychology for learning by Lisp, 1: Learning]. Tokyo, Japan: Tokyo Daigaku Shuppankai.

Anzai, Y., Saeki, Y., & Namba, K. (1982). *Lisp de manabu ninchi sinrigaku, 2—Mondai kaiketsu* [Cognitive psychology for learning by Lisp, 2: Problem solving]. Tokyo, Japan: Tokyo Daigaku Shuppankai.

Arai, T., Ono, J., & Ogata, T. (2016). Semi-automatic generation of events sequence knowledge for narrative generation: The use in an integrated narrative generation system. In *Proceedings of the 30th Annual Conference of the Japanese Society for Artificial Intelligence* (3P1-1in2). Tokyo, Japan: Japanese Society for Artificial Intelligence.

Brecht, B. (1964). *Brecht on theatre: The development of an aesthetics* (J. Willett, Trans. & Ed.). London, UK: Methuen. (Original work published 1950)

DeJong, G. (Ed.). (1993). *Investigating explanation-based learning*. Boston, MA: Kluwer Academic Press. doi:10.1007/978-1-4615-3602-4

DeJong, G. (2014). Explanation-based learning. In T. Gonzalez, J. Diaz-Herrera, & A. Tucker (Eds.), *CRC computing handbook: Computer science and software engineering* (Vol. 3, pp. 66.1–66.26). Boca Raton, FL: CRC Press.

Endo, Y., & Ogata, T. (2002). Hyper-comic system as representation field of narrative discourse. In *Proceedings of the 17th Congress of the International Association of Empirical Aesthetics* (pp. 555–558). Rome, Italy: University of Rome Tre.

Endo, Y., & Ogata, T. (2003a). Hyper-comic system as consideration of rhetoric. In *Proceedings of the 4th International Conference on Cognitive Science* (pp. 111–116). Oakbrook Terrace, IL: Cognitive Science Society.

Endo, Y., & Ogata, T. (2003b). Narratology of manga and application to hyper-comic. In *Proceedings of the 17th Annual Conference of the Japanese Society for Artificial Intelligence* (2G2-03). Tokyo, Japan: Japanese Society for Artificial Intelligence.

Endo, Y., & Ogata, T. (2004a). A rhetorical analysis of a Japanese comic for hyper-comic system. In *Proceedings of the 18th Congress of the International Association of Empirical Aesthetics* (pp. 502–508). Rome, Italy: University of Rome Tre.

Endo, Y., & Ogata, T. (2004b). Multilayered discourse in hyper-comic. In *Proceedings of the 9th International Symposium on Artificial Life and Robotics*, Vol. 1 (pp. 49–52). Oita, Japan: International Society of Artificial Life and Robotics.

Fillmore, C. J., & Baker, C. (2010). A frames approach to semantic analysis. In B. Heine & H. Narrog (Eds.), *The Oxford handbook of linguistic analysis* (pp. 313–339). Oxford, UK: Oxford University Press.

Fukuda, I., & Ogata, T. (2014). The current status and issues of the state/event transformation knowledge base in an integrated narrative generation system. In *Proceedings of the 28th Annual Conference of the Japanese Society for Artificial Intelligence* (2F4-OS-01a-8in). Tokyo, Japan: The Japanese Society for Artificial Intelligence.

Genette, G. (1972). *Discours du récit, Essai de méthode, Figures III.* Paris: Seuil.

Ikehara, S., Miyazaki, M., Shirai, S., Yokoo, A., Nakaiwa, H., Ogura, K., ... Hayashi, Y. (Eds.). (1999). *Nihongo goi taikei (CD-ROM)* [A Japanese lexicon (CD-ROM)]. Tokyo, Japan: Iwanami Shoten.

Imabuchi, S., Akimoto, T., Ono, J., & Ogata, T. (2012). KOSERUBE: An application system with a Propp-based story grammar and other narrative generation techniques. In *Proceedings of the 6th International Conference on Soft Computing and Intelligent Systems & the 13th International Symposium on Advanced Intelligent Systems* (pp. 248–253). Tokyo, Japan: Japan Society of Fuzzy Theory and Intelligent Informatics. 10.1109/SCIS-ISIS.2012.6505320

Imabuchi, S., & Ogata, T. (2012a). Propp-based story grammar ni okeru rule no jidō kakutoku no kōan [Considering the automatic acquisition of rules in the Propp-based story grammar]. In *Proceedings of Japanese Cognitive Science Society 29th Workshop of Literature in Cognition and Computer II* (29G-03). Tokyo, Japan: Japanese Cognitive Science Society.

Imabuchi, S., & Ogata, T. (2012b). A story generation system based on Propp theory: As a mechanism in an integrated narrative generation system. In *Lecture Notes in Artificial Intelligence* (Vol. 7614, pp. 312–321). Berlin, Germany: Springer.

Imabuchi, S., & Ogata, T. (2013). Methods for generalizing the Propp-based story generation mechanism. In Lecture Notes in Computer Science (Vol. 8210, pp. 333–344). Berlin, Germany: Springer. doi:10.1007/978-3-319-02750-0_36

Imabuchi, S., & Ogata, T. (2014). Integrating the event generation mechanism in the Propp-based story generation mechanism into the integrated narrative generation system. *Journal of Robotics, Networking and Artificial Life, 1*(2), 164–168. doi:10.2991/jrnal.2014.1.2.14

Ito, T., Ono, J., & Ogata, T. (2018a). Monogatari seisei no tame no minwa motif no riyō [Using folktale motifs for narrative generation] In *Proceedings of the 58th Special Interest Group on Language Sense Processing Engineering* (pp. 55–59). Tokyo, Japan: The Japanese Society for Artificial Intelligence.

Ito, T., Ono, J., & Ogata, T. (2018b). Using motifs of folktales for narrative generation. In *Proceedings of the IEEE SMC 2018 Workshop on Informational and Cultural Narratology and Cognitive Content Generation (The 59th Special Interest Group on Language Sense Processing Engineering)* (pp. 8–11). Tokyo, Japan: The Japanese Society for Artificial Intelligence.

Jauss, H. R. (1970). Literaturgeschichte als provokation. Frankfurt am Main: Suhrkamp Verlag.

Kamada, M., & Ogata, T. (2013a). Analysis and imitation of the Japanese letter notation in narrative texts. In *Proceeding of the 27th Annual Conference of the Japanese Society for Artificial Intelligence* (2I4-5in). Tokyo, Japan: The Japanese Society for Artificial Intelligence.

Kamada, M., & Ogata, T. (2013b). Monogatari bunshō ni okeru moji hyōki—Monogatai seisei system ni okeru moji hyōki chōsetsu kikō ni mukete [Character notation in narrative sentences: Toward a character notation adjustment mechanism in narrative generation systems]. In *Proceedings of the 26th Special Interest Group on Language Sense Engineering Research Group in Japanese Artificial Intelligence Society* (p. 29). Tokyo, Japan: The Japanese Society for Artificial Intelligence.

Kamada, M., & Ogata, T. (2013c). Monogatari seisei system ni okeru bun seisei kikō to moji hyōki kikō [The sentence generation and character representation mechanisms in a narrative generation system]. In *Proceedings of the 19th Annual Conference of the Association of Natural Language Processing* (pp. 698–701). Tokyo, Japan: Association of Natural Language Processing.

Kanai, A., Mukouyama, K., Ogata, T., & Kawamura, Y. (2000). Film generation based on the rhetoric. In *IPSJ SIG Notes 2000, 67(2000-CH-047)* (pp. 9–14). Tokyo, Japan: Information Processing Society of Japan.

Kanai, A., & Niwa, Y. (2008). *Eizō henshū no riron to jissen* [Theories and practices of image editing]. Tokyo, Japan: Hosei Daigaku Shuppankyoku.

Kanai, A., & Ogata, T. (2004a). Cutting techniques on film rhetoric cognition and composition. In *Proceedings of the 30th Annual Conference of the Japanese Society for Artificial Intelligence* (2D2-07). Tokyo, Japan: The Japanese Society for Artificial Intelligence.

Kanai, A., Ogata, T., & Shinohara, K. (2002). Film rhetoric generation system for intensive cognitive system. In *Proceedings of IEEE International Conference on Systems, Man and Cybernetics (CD-ROM)*. The Institute of Electrical and Electronics Engineers.

Kanai, A., Ogata, T., & Shinohara, K. (2003). Identical and different elements in film rhetoric generation. *Transactions of the Japanese Society for Artificial Intelligence, 18*(2G), 114–121. doi:10.1527/tjsai.18.114

Kawamura, Y., & Ogata, T. (1999). An introduction to the information system of performing arts. In *IPSJ SIG Technical Report, 5(99-CH-41)* (pp. 61–68). Tokyo, Japan: Information Processing Society of Japan.

Kawamura, Y., & Ogata, T. (2000a). A conceptual framework on geinō organization model for artificial geinō performers with lives. In *Proceedings of the 5th International Symposium on Artificial Life and Robotics* (pp. 793–796). Oita, Japan: International Society of Artificial Life and Robotics.

Kawamura, Y., & Ogata, T. (2000b). Geinōjin image marketing system no tameno geinōjin homepage no bunseki [An analysis of geinōjins' homepages for a geinōjin image marketing system]. *Journal of Advertising Science, 40*, 181–192.

Kawamura, Y., & Ogata, T. (2000c). Kasō geinōjin character no jinsei [Lives of virtual geinō characters]. In Y. Kawaramu, S. Hamada, & T. Ogata (Eds.), Japanese Cognitive Science Society Technical Report "Language and Literature," 00-No. 32 (pp. 52–60). Tokyo, Japan: Japanese Cognitive Science Society.

Kawamura, Y., & Ogata, T. (2000d). Geinō soshiki model to image senryaku [A geinō organization model and the image strategies]. *Journal of Osaka University of Economics and Law, 23*(2), 130–164.

Kipper-Schuler, K. (2005). *VerbNet: A broad-coverage, comprehensive verb lexicon* (Doctoral dissertation). Computer and Information Science Department, University of Pennsylvania.

Kipper-Schuler, K., Dang, G. T., & Palmer, M. (2000). Class-based construction of a verb lexicon. *Proceedings of AAAI, 2000*, 691–696.

Kobayashi, F., & Ogata, T. (2004a). Narrative and music as variation: Transformation of musical structure based on narrative discourse theory. In *Proceedings of the 9th International Symposium on Artificial Life and Robotics*, Vol. 1 (pp. 170–173). Oita, Japan: International Society of Artificial Life and Robotics.

Kobayashi, F., & Ogata, T. (2004b). A musical variation system by the structural correspondence between music and narrative. In *Proceedings of 18th Congress of the International Association of Empirical Aesthetics* (pp. 555–559). Rome, Italy: University of Rome Tre.

Kristeva, J. (1980). *Desire in language: A semiotic approach to literature and art*. Columbia University Press. (Original work published 1969)

Kumagai, S., Funakoshi, S., Akimoto, T., & Ogata, T. (2012). Development of a language dictionary and a simple narrative sentence generation mechanism. In *Proceedings of the 26th Annual Conference of the Japanese Society for Artificial Intelligence* (1N1-OS-1a-3). Tokyo, Japan: The Japanese Society for Artificial Intelligence.

Kurisawa, Y., Michimata, R., Kamada, M., & Ogata, T. (2012). From general conceptual/language dictionaries to the use as narrative generation. In *Proceedings of the 29th Annual Conference of the Cognitive Science Society* (pp. 363–369). Tokyo, Japan: Japanese Cognitive Science Society.

Kurisawa, Y., & Ogata, T. (2013a). A consideration of a norm-deviation mechanism in the integrated narrative generation system. In *Proceedings of the 43th Language Sense Engineering Research Group in Japanese Artificial Intelligence Society* (pp. 35–45). Tokyo, Japan: The Japanese Socirty for Artificial Intelligence.

Kurisawa, Y., & Ogata, T. (2013b). A consideration of a norm-deviation mechanism in the integrated narrative generation system. In *Proceedings of the 44th Language Sense Engineering Research Group in Japanese Artificial Intelligence Society* (pp. 25–35). Tokyo, Japan: The Japanese Society for Artificial Intelligence.

Kurisawa, Y., & Ogata, T. (2014). Incorporating a norm-deviation mechanism into the integration narrative generation system. In *Proceedings of the 28th Annual Conference of the Japanese Society for Artificial Intelligence* (2F4-OS-01a-5). Tokyo, Japan: The Japanese Society for Artificial Intelligence.

Lerdahl, F., & Jackendoff, R. (1983). *A generative theory of tonal music*. Cambridge, MA, USA: MIT Press.

Levy, I. H. (2014). Renzoku-teki na identity no bōken [Adventure of continuous identity]. In Watakushi shōsetsu handbook [I novels handbook] (pp. 234–244). Tokyo, Japan: Bensei Shuppan.

Minsky, M. (1975). A framework for representing knowledge. In P. H. Winston (Ed.), *The psychology of computer vision*. McGraw-Hill.

Mishima, Y. (1963). Gogo no eikō. Tokyo, Japan: Kōdansha.

Mori, Y., & Ogata, T. (2005). Hyper-comic no kōsō to hyper-link oyobi koma gōsei no jidōka [A hyper-comic and the automatization of its hyper-link and frame synthesis]. In *Proceedings of the 19th Special Interest Group on Language Sense Processing Engineering* (pp. 1–11). Tokyo, Japan: The Japanese Society for Artificial Intelligence.

Mukouyama, K., Kanai, A., & Ogata, T. (1999). Monogatari ni okeru jikan junjo no henkan—Eiga ni motozuku kōsatsu. [Temporal order transformation in narrative: Considering based on films]. In *Proceedings of the 13th Annual Conference of the Japanese Society for Artificial Intelligence* (pp. 492–493). Tokyo, Japan: The Japanese Society for Artificial Intelligence.

Mukouyama, K., Kanai, A., & Ogata, T. (2001). An analysis of the rhetoric of the film and an attempt to an editing system. In *Proceeding of the 15th Annual Conference of the Japanese Society for Artificial Intelligence* (1F1-11). Tokyo, Japan: The Japanese Society of Artificial Intelligence.

Mukouyama, K., & Ogata, T. (2002). Expansion of narrative discourse theory based on story: A system for time order transformation. In *Proceedings of the 16th Annual Conference of the Japanese Society for Artificial Intelligence* (3f2-09). Tokyo, Japan: The Japanese Society for Artificial Intelligence.

Natsume, S. (2007). *Bungaku ron, 1, 2* [Literary theory, Vols. 1, 2]. Tokyo, Japan: Iwanami Shoten. (Original work published 1907)

Ōe, K. (1972). Mizukara waga namida wo nuguitamau hi [The day he himself shall wipe my tears away]. Mizukara waga namida wo nuguitamau hi [The day he himself shall wipe my tears away], 9–137. Tokyo, Japan: Kōdansya. (The day he himself shall wipe my tears away (J. Nathan, Trans., 1994). In Teach us to outgrow our madness: Four short novels. NY, USA: Grove Press.)

Ogata, T. (1992). *Setsumei ni motozuku monogatrai seisei system ni kansuru kenkyū* [Study on an explanation-based narrative generation system] (Master's dissertation). Tsukuba University, Tokyo, Japan.

Ogata, T. (1995). *Monogatari seisei—Monogatari no tameno gihō to senryaku ni motozuku approach* [Narrative generation: An approach based on techniques and strategies for narratives] (Doctoral dissertation). The University of Tokyo, Tokyo, Japan.

Ogata, T. (1997). Tajū monogatari kōzō no model [The model of multiple narrative structures]. In *Proceeding of the 9th Japan Association of Simulation and Gaming* (pp. 107–110). Tokyo, Japan: Japan Association of Simulation and Gaming.

Ogata, T. (1998). Geinō character heno monogatari-sei no huyo [Giving narrativity to geinō characters]. *Journal of Advertising Science*, *37*, 69–75.

Ogata, T. (1999a). An attempt for systematization of narrative discourse theory from the viewpoint of narrative generation system. *Proceedings of IPSJ SIG Computers and the Humanities*, *99*(85), 31–38.

Ogata, T. (1999b). Bungaku riron no sekkeiteki saikōsei no kokoromi (zoku)—Monogatari gensetsuron no kaisōteki kōsei [An attempt of reconstruction based on the design of literary theories (II): A hierarchical structure of a narrative discourse theory]. In *Proceedings of Japanese Cognitive Science Society 4th Workshop of Literature in Cognition and Computer*. Tokyo, Japan: Japanese Cognitive Science Society.

Ogata, T. (2000a). Tajū monogatari kōzō no macro model—Simulation toshite no monogatari josetsu [A macro model of multiple narrative structures: Introduction to the narrative as simulation]. *Studies in Simulation and Gaming*, *10*(1), 35–46.

Ogata, T. (2000b). Tajū monogatari kōzō model to sono tenkai he mukete—Simulation toshite no monogatari josetsu [Toward multiple narrative structures model and its development: Introduction to the narrative as simulation]. In Y. Kawamura, S. Hamada, & T. Ogata (Eds.), Japanese Cognitive Science Society Technical Report "Literature and cognition, computer 6: Language and literature," 00-No. 32 (pp. 87–119). Tokyo, Japan: Japanese Cognitive Science Society.

Ogata, T. (2000c). On the narrative content and the narrative discourse. In IPSJ SIG Technical Report (67(2000-CH-047) (pp. 1–8.). Tokyo, Japan: Information Processing Society of Japan.

Ogata, T. (2001). Expanded literary theory: Concept, methodology, and trials. *Cognitive Studies*, *8*(4), 405–416.

Ogata, T. (2002a). Geinō jōhō system [Geinō information system]. In T. Kitagawa, O. Sudo, T. Nishigaki, J. Hamada, S. Yoshimi, & S. Yonemoto (Eds.), *Jōhōgaku jiten* [Dictionary of informatics] (p. 258). Tokyo, Japan: Kōbundō.

Ogata, T. (2002b). Expanded literary theory: Cognitive/computational expansion of literary theories and narratology. In *Proceedings of the 17th Congress of the International Association of Empirical Aesthetics* (pp. 163–166). Rome, Italy: University of Rome Tre.

Ogata, T. (2002c). The concept of system narratology: From the viewpoint of expanded literary theory. In *Proceedings of the 3rd international workshop of literature in cognition and computer (iw LCC) held in PRICAI 2002* (W7-1, pp. 1–10). Tokyo, Japan: Japanese Cognitive Science Society.

Ogata, T. (2003a). Monogatari no tajūsei to kakuchō bungakuriron no gainen—System narratology ni mukete I [Narrative multiplicity and the concept of expanded literary theory: Toward a system narratology]. In M. Yoshida (Ed.), *Fukuzatsu-kei shakai riron no shin chihei* [New paradigm of complex social system theories] (pp. 127–181). Tokyo, Japan: Senshū Daigaku Shuppankyoku.

Ogata, T. (2003b). Kakuchō busngakuriron no kokoromi—System narratology ni mukete II [Attempts of expanded literary theory: Toward a system narratology]. In M. Yoshida (Ed.), *Fukuzatsu-kei shakai riron no shin chihei* [New paradigm of complex social system theories] (pp. 309–356). Tokyo, Japan: Senshū Daigaku Shuppankyoku.

Ogata, T. (2003c). The framework of "expanded" narrative discourse theory. In *Proceedings of the 17th Annual Conference of the Japanese Society for Artificial Intelligence* (3C1-01). Tokyo, Japan: The Japanese Society for Artificial Intelligence.

Ogata, T. (2004a). On the integration of narrative discourse techniques. In *Proceedings of the 18th Annual Conference of Japanese Society for Artificial Intelligence* (2D1-03). Tokyo, Japan: The Japanese Society for Artificial Intelligence.

Ogata, T. (2004b). Monogatari no <ninchi/text/shakai>—Tajū monogatari kōzō wo meguru kōsatsu [Narrative <cognition/text/society>: Consideration on multiple narrative structures]. In T. Ogata (Ed.), Japanese Cognitive Science Society Technical Report "<cognition, text, society>: Records of intensive lecture in Yamanashi universtiy at 2002" (No. 52, pp. 88–113). Tokyo, Japan: Japanese Cognitive Science Society.

Ogata, T. (2004c). A computational approach to literary and narrative production: Toward computational narratology. In *Proceedings of the 18th Congress of the International Association of Empirical Aesthetics* (pp. 509–516). Rome, Italy: University of Rome Tre.

Ogata, T. (2007). To the rhetoric of story from Propp: Centering around the rhetoric of de-composition and re-composition. *Cognitive Studies*, *14*(4), 532–558.

Ogata, T. (2008). Monogatari seisei system ni okeru eizō kōsei he mukete [Toward the image editing in a narrative generation system]. In A. Kanai & Y. Niwa (Eds.), *Eizō henshū no riron to jissen* [Theories and practices of image editing] (pp. 165–235). Tokyo, Japan: Hosei Daigaku Shuppankyoku.

Ogata, T. (2010a). Shōsetsu—Ryūdō to kotei, sakuhin no hō he [Novels: Fluidity and fixation, toward works]. In T. Ogata & A. Kanai (Eds.), *Monogatariron no jōhōgaku josetsu—Monogatari seisei no shisō to gijutsu wo megutte* [An introduction to informatics of narratology: Toward the thoughts and technologies of narrative generation] (pp. 130–169). Tokyo, Japan: Gakubunsha.

Ogata, T. (2010b). "Monogatari seisei" no haikei oyobi monogatari to bungaku no aida [Backgrounds of "narrative generation" and between narrative and literature]. In T. Ogata & A. Kanai (Eds.), *Monogatariron no jōhogaku josetsu—Monogatari seisei no shisō to gijutsu wo megutte* [An introduction to the informatics of narratology: Around the thoughts and technologies of narrative generation] (pp. 186–258). Tokyo, Japan: Gakubunsha.

Ogata, T. (2010c). Monogatari seisei system no taikyoku-teki kekkō—Monogatari no gijutsu to keiei heno jo [The macro architecture of a narrative generation system: An introduction to narrative technologies and managements]. In T. Ogata & A. Kanai (Eds.), *Monogatariron no jōhogaku josetsu—Monogatari seisei no shisō to gijutsu wo megutte* [An introduction to the informatics of narratology: Around the thoughts and technologies of narrative generation] (pp. 259–340). Tokyo, Japan: Gakubunsha.

Ogata, T. (2013). Jiko to sono saihen-ka/soshiki-ka no junkan—Tōgō monogatari seisei system no saki no contents he no kōsō: Ryūdō to kotei (20) [Self and the circulation between its fragmentaiot and structuring: The plan to the concepts beyond the integrated narrative generation system: Fluidity and fixation (20)]. In *Proceedings of Japanese Cognitive Science Society 30th Workshop of Literature in Cognition and Computer II* (30G-01). Tokyo, Japan: Japanese Cognitive Science Society.

Ogata, T. (2014a). Form and content in narrative generation: From the view point of the implementation of an integrated narrative generation system. In IPSJ SIG Technical Report (4(2014-NL-217), pp. 1–11). Tokyo, Japan: Information Processing Society of Japan.

Ogata, T. (2014b). Developing an integrated narrative generation system: Its current status and problems [Tokyo, Japan: Information Processing Society of Japan.]. *Proceedings of Forum on Information Technology*, *2*, 323–330.

Ogata, T. (2014c). Monogatari seisei no concept—Ryūdō to kotei no gainen wo chūshin ni [Concepts of narrative generation: Focusing on the concept of fluidity and fixation]. In *Proceedings of the 46th Special Interest Group on Language Sense Processing Engineering* (pp. 29–38). Tokyo, Japan: The Japanese Society for Artificial Intelligence.

Ogata, T. (2014d). Expanded literary theory for automatic narrative generation. In *Proceedings of the Joint Conference of the 7th International Conference on Soft Computing and Intelligent Systems and the 15th International Symposium on Advanced Intelligent Systems* (pp. 1558–1563). Tokyo, Japan: Japan Society of Fuzzy Theory and Intelligent Informatics.

Ogata, T. (2015). Building conceptual dictionaries for an integrated narrative generation system. *Journal of Robotics. Networking and Artificial Life*, *1*(4), 270–284. doi:10.2991/jrnal.2015.1.4.6

Ogata, T. (2016). Computational and cognitive approaches to narratology from the perspective of narrative generation. In T. Ogata & T. Akimoto (Eds.), *Computational and cognitive approaches to narratology* (pp. 1–73). Hershey, PA: IGI Global. doi:10.4018/978-1-5225-0432-0.ch001

Ogata, T. (2018a). An integrated approach to narrative generation: From Mishima and kabuki to narrative generation systems. In T. Ogata & S. Asakawa (Eds.), *Content generation through narrative communication and simulation* (pp. 49–147). Hershey, PA: IGI Global. doi:10.4018/978-1-5225-4775-4.ch002

Ogata, T. (2018b). Tōgō monogatari seisei system [Integrated narrative generation system]. In T. Ogata, Y. Kawamura, & A. Kanai (Eds.), *Jōhō monogatariron—Jinkōchinō, ninchi, shakai katei to monogatari seisei* [Informational narratology: Artificial intelligence/cognition/social process and narrative generation] (pp. 247–288). Tokyo, Japan: Hakutō Shobō.

Ogata, T. (2019a). A computational, cognitive, and narratological approach to narrative generation. In T. Ogata & T. Akimoto (Eds.), *Post-narratology through computational and cognitive approaches* (pp. 1–84). Hershey, PA: IGI Global. doi:10.4018/978-1-5225-7979-3.ch001

Ogata, T. (2019b). Toward a post-narratology or the narratology of narrative generation. In T. Ogata & T. Akimoto (Eds.), *Post-narratology through computational and cognitive approaches* (pp. 85–142). Hershey, PA: IGI Global. doi:10.4018/978-1-5225-7979-3.ch002

Ogata, T. (2019c). *Toward an integrated approach to narrative genertion: Emerging research and opportunities*. Hershey, PA: IGI Global.

Ogata, T. (2019d). Theoretical or philosophical considerations for an integrated narrative generation approach. In T. Ogata (Ed.), *Toward an integrated approach to narrative genertion: Emerging research and opportunities* (pp. 315–403). Hershey, PA: IGI Global.

Ogata, T., & Akimoto, T. (2007). Towards the circular narrative generation based on the correspondences between language-based narrative and music: Systems development based on narrative rhetoric and fundamental considerations about the systems and the circular narrative generation. *Cognitive Studies*, *14*(3), 355–379.

Ogata, T., & Akimoto, T. (2010a). Experimental merging of modules for the integrated narrative generation system. In *Proceedings of the 24th Annual Conference of Japanese Society for Artificial Intelligence* (1I2-OS1b-8). Tokyo, Japan: The Japanese Society for Artificial Intelligence.

Ogata, T., & Akimoto, T. (2010b). For constructive narratology: Through an implementation of the pilot version of integrated narrative generation system. In *Proceedings of the 27th Annual Conference of the Cognitive Science Society* (pp. 767–776). Tokyo, Japan: Japanese Cognitive Science Society.

Ogata, T., & Akimoto, T. (2011). Towards integrated narrative generation system: Development of version 0.4. In *Proceedings of the 25th Annual Conference of Japanese Society for Artificial Intelligence* (1H2-OS1-6in). Tokyo, Japan: The Japanese Society for Artificial Intelligence.

Ogata, T., & Akimoto, T. (Eds.). (2016). *Computational and cognitive approaches to narratology*. Hershey, PA: IGI Global. doi:10.4018/978-1-5225-0432-0

Ogata, T., Akimoto, T., & Ono, J. (2014). Towards narrative generation based on the concept of "fluidity-fixation": Basic concepts for the expansion of an integrated narrative generation system (1). In *Proceedings of the 31st Annual Conference of the Cognitive Science Society* (pp. 354–363). Tokyo, Japan: Japanese Cognitive Science Society.

Ogata, T., Arai, T., & Ono, J. (2016). Random narrative generation in an integrated narrative generation system. *Proceedings of Forum on Information Technology*, *2*, 133–134.

Ogata, T., & Endo, Y. (2004). Monogatari gensetsuron to hyper-comic system [Narrative discourse theory and hyper-comic system]. In *Proceedings of the 18th Annual Conference of Japanese Society for Artificial Intelligence* (2D2-01). Tokyo, Japan: The Japanese Society for Artificial Intelligence.

Ogata, T., Hori, K., & Ohsuga, S. (1994). Towards narrative text generation based on narrative techniques and strategies. In *Proceedings of International Federation for Information and Documentation* (pp. 296–300). Brussels, Belgium: International Federation for Information and Documentation.

Ogata, T., Hori, K., & Ohsuga, S. (1995). A basic framework of narrative generation system as creative interface. In Y. Anzai, K. Ogawa, & H. Mori (Eds.), *Symbiosis of human and artifact (20A)* (pp. 679–684). Amsterdam, The Netherlands: Elsevier. doi:10.1016/S0921-2647(06)80107-3

Ogata, T., Hori, K., & Ohsuga, S. (1996a). A basic framework for narrative conceptual structure generation based on narrative techniques and strategies. *Jinkō Chinō Gakkaishi*, *11*(1), 148–159.

Ogata, T., Hori, K., & Ohsuga, S. (1996b). An analysis of narrative structure and generation process for narrative generation system. *Cognitive Studies*, *3*(1), 72–109.

Ogata, T., & Kanai, A. (2010a). *Monogatariron no jōhōgaku josetsu—Monogatari seisei no shisō to gijutsu wo megutte* [An introduction to informatics of narratology: Around the thoughts and technologies of narrative generation]. Tokyo, Japan: Gakubunsha.

Ogata, T., & Kanai, A. (2010b). Story to shōhin = brand—Seisei ni muketa kōkoku no bunseki [Stories and products = brands: An analysis of advertisements toward narrative generation]. In T. Ogata & A. Kanai (Eds.), *Monogatariron no jōhōgaku josetsu—Monogatari seisei no shisō to gijutsu wo megutte* [An introduction to informatics of narratology: Toward the thoughts and technologies of narrative generation] (pp. 42–115). Tokyo, Japan: Gakubunsha.

Ogata, T., Kanai, A., & Aoyagi, E. (2003). Literary theories and cognitive science: Regarding expanded literary theory. In *Proceedings of the 20th Annual Conference of the Cognitive Science Society* (pp. 402–405). Tokyo, Japan: Japanese Cognitive Science Society.

Ogata, T., & Kawamura, Y. (1998a). Geinōjin no monogatari simulation no kihon shisō [The basic thought of the narrative simulations of geinōjin]. *Nikkei Kōkoku Ken'kyūjo Hō, 180*, 26–31.

Ogata, T., & Kawamura, Y. (1998b). Geinō produce no gihō taikei no bunseki—Geinō simulation no tame no kisoteki kosatsu [Analyzing the system of techniques of geinō producing: A basic consideration for geinō simulation 1]. In *Proceeding of the 10th Japan Association of Simulation and Gaming* (pp. 107–111). Tokyo, Japan: Japan Association of Simulation and Gaming.

Ogata, T., & Kawamura, Y. (1999a). Monogatari sanshutsu soshikiron to geinō soshiki model [Theory of narrative production organizations and a geinō organization model]. In *Transactions of the Academic Association for Organizational Science* (pp. 219–222). Tokyo, Japan: Academic Association for Organizational Science.

Ogata, T., & Kawamura, Y. (1999b). Service to monogatari seisei no kanten kara no geinō system no kōsatsu to model-ka [Consideration and modelization of a geinō organization system from the viewpoint of services and narrative generation]. In *Abstracts of Annual Conference of Japan Society for Management Information 1999 Spring* (pp. 307–310). Tokyo, Japan: Japan Society for Management Information.

Ogata, T., & Kawamura, Y. (1999c). Monogatari sanshutsu soshikiron to geinō soshiki model [Theory of narrative production organizations and a geinō organization model]. In *Proceedings of Japanese Cognitive Science Society 2nd Workshop of Literature in Cognition and Computer in Tokyo*, 1999, Winter (pp. 55–70). Tokyo, Japan: Japanese Cognitive Science Society.

Ogata, T., & Kobayashi, F. (2004). Hensō kara no monogatari seisei he no sekkin—Monogatari to ongaku no henkan oyobi ongaku hensō system no shisaku ni motozuku sho-kōsatsu [Approaching to narrative generation from variation: Considerations based on the transformation between narrative and music and an experimental music variation system]. In *Proceedings of the 17th Special Interest Group on Language Sense Processing Engineering* (pp. 1–33). Tokyo, Japan: The Japanese Society for Artificial Intelligence.

Ogata, T., Matsuda, A., Naito, Y., Manabe, Y., Takahahi, N., Nakajima, M., ... Numata, M. (2007). Monogatari seisei system ni okeru eizō hyōgen [Image representation in narrative generation systems]. In *Proceedings of the 25th Special Interest Group on Language Sense Processing Engineering* (pp. 19–59). Tokyo, Japan: The Japanese Society for Artificial Intelligence.

Ogata, T., & Morita, H. (2002). Kakuchō bungaku riron no kihon concept—Simulation tosite no Monogatari [Basic concepts of an expanded literary theory: Narratives as simulation]. *Studies in Simulation and Gaming*, *12*(1), 13–23.

Ogata, T., Mukouyama, K., & Kanai, A. (2000). Toward a formal system of narrative discourse theory and its expansion: An approach by language and movies. In IPSJ SIG Technical Report (100(2000-CH-048), pp. 9–16). Tokyo, Japan: Information Processing Society of Japan.

Ogata, T., & Nakashima, M. (2006). Monogatari seisei system to kan-text-sei—Taikeiteki kenkyū ni muketa kokoromi [Narrative generation systems and inter-textuality: An attempt to a systematic study]. In *Proceedings of the 23rd Annual Conference of the Cognitive Science Society* (pp. 270–271). Tokyo, Japan: Japanese Cognitive Science Society.

Ogata, T., & Ono, J. (2013). Designing narrative interface with a function of narrative generation. In *Proceeding of the 2013 International Conference on Cyberworlds* (pp. 214–221). Berlin, Germany: Springer. 10.1109/CW.2013.74

Ogata, T., & Ono, J. (2014). The integrated narrative generation system, inter-textulality, text mining. In *IEICE Technical Report (NLC 2013-53), 113(429)* (pp. 33–38). Tokyo, Japan: The Institute of Electoronics, Information and Communication Engineers.

Ogata, T., & Ono, J. (2015). Language notation dictionaries and the use in an integrated narrative generation system. In *IEICE Technical Report (NLC 2015-5), 115(70)* (pp. 25–30). Tokyo, Japan: The Institute of Electoronics, Information and Communication Engineers.

Ogata, T., & Ono, J. (2016). A way for using the verb conceptual dictionary in an integrated narrative generation system: The use of co-occurrence information on verb concepts. *Journal of Robotics. Networking and Artificial Life*, *3*(3), 209–212. doi:10.2991/jrnal.2016.3.3.15

Ogata, T., Ono, J., Herai, Y., Igarashi, W., & Shirai, S. (2015). Tōgō monogatari seisei system no tame no gainen jisho oyobi gengo hyōki jisho no kōsei/riyō no kakuchō kōsō [A plan for expanding and utilizing the conceptual dictionaries and language notation dictionaries for an integrated narrative generation system]. In *Proceedings of the 48th Special Interest Group on Language Sense Processing Engineering* (pp. 51–56). Tokyo, Japan: The Japanese Society for Artificial Intelligence.

Ogata, T., Tachibana, S., & Tomite, S. (2009). Animated movie generation from narrative conceptual representation and the automatic camerawork: Analysis and simulation of "Tokyo Story." In *Proceedings of the 26th Annual Meeting of the Japanese Cognitive Science Society* (P2–32). Tokyo, Japan: Japanese Cognitive Science Society.

Ogata, T., & Terano, T. (1991a). Katarite no ito wo kōryoshita monogatari seisei system no kōsō [The plan of a narrative generation system considering the narrator's intention]. In *Proceedings of the 5th Annual Conference of Japanese Society for Artificial Intelligence*, Vol. 2 (pp. 561–564). Tokyo, Japan: The Japanese Society for Artificial Intelligence.

Ogata, T., & Terano, T. (1991b). Katarite no ito ni motozuku monogatari seisei system [A narrative generation system based on the narrator's intention]. In *Proceeding of the 14th Knowledge System Symposium* (pp. 317–324). Tokyo, Japan: The Society of Instrument and Control Engineers.

Ogata, T., & Terano, T. (1991c). Explanation-based narrative generation using semiotic theory. In *Proceedings of Natural Language Processing Pacific Rim Symposium* (pp. 321–328). Tokyo, Japan: Special Interest Group of Natural Language Processing of Information Processing Society of Japan.

Ogata, T., & Terano, T. (1992a). EBL wo mochiita monogatari seisei system ni okeru arasuji no shōsaika [Detailed description of synopses in the narrative generation system using explanation-based learning]. In Proceeding of Joint Workshiop of Japanese Society for Artificial Intelligence and the Institute of Electronics, Information and Communication Engineers (SIG-KBS-9104, pp. 117–124). Tokyo, Japan: The Japanese Society for Artificial Intelligence.

Ogata, T., & Terano, T. (1992b). Setsumei ni motozuku monogatairi seisei system ni okeru plot no tenkai [Plot development in an explanation-based narrative generation system]. In *Proceedings of the 6th Annual Conference of Japanese Society for Artificial Intelligence*, Vol. 2 (pp. 521–524). Tokyo, Japan: The Japanese Society for Artificial Intelligence.

Ogata, T., & Terano, T. (1992c). Plot generation and expansion in explanation-based narrative generator. In *Proceedings of the 1st Singapore International Conference on Intelligent Systems* (pp. 549–554). Singapore: Singapore International Conference on Intelligent Systems (SPICIS).

Ogata, T., Tokaji, A., & Morita, H. (2002). Bungaku-teki contents no sanshutsu to shōhi—"Bungaku to ninchi/computer kenkyū" ni okeru business to bungaku-teki jikken he no approach [The production and consumption of literary contents: Approaches to business and literary experiments in a "literature and cognition/computer study"]. In *Proceeding of Annual Convention, Japanese Association of Administrative Science*. Tokyo, Japan: Japanese Association of Administrative Science.

Ogata, T., Watanabe, K., Hori, K., & Ohsuga, S. (1995a). A basic framework of the application of narrative generation system for integrated support of marketing/advertisement. *Journal of the Japan Society for Management Information*, *4*(1), 19–42.

Ogata, T., Watanabe, K., Hori, K., & Ohsuga, S. (1995b). Monogatari seisei system ni yoru kōkoku sōsaku shien wo mokuteki to shita televi-comercial no kōzō bunseki [Structural analyses of television commercial films for advertising creation support by a narrative generation system]. *Journal of Advertising Science*, *30*, 1–22.

Ogata, T., & Yamakage, S. (2003). The framework of "expanded" narrative discourse theory. In *Proceedings of the 17th Annual Conference of Japanese Society for Artificial Intelligence* (3C1-02). Tokyo, Japan: The Japanese Society for Artificial Intelligence.

Ogata, T., & Yamakage, S. (2004). A computational mechanism of the "distance" in narrative: A trial in the expansion of a literary theory. In *Proceedings of the 8th World Multiconference on Systemics, Cybernetics and Informatics* (pp. 179–184). Winter Garden, FL: World Multiconference on Systemics, Cybernetics and Informatics.

Ogata, T., & Yazawa, K. (2000). On the rhetoric for narrating the world. In N. Yoshimine, H. Akama, & A. Tokosumi (Eds.), *Japanese Cognitive Science Society Technical Report "Literature, Cognition and Computer 2: The Expansion of Literature" (00-No. 40)* (pp. 10–13). Tokyo, Japan: Japanese Cognitive Science Society.

Ohara, K. H. (2008). Lexicon, grammar, and multilinguality in the Japanese FrameNet. In *Proceedings of the 6th International Conference on Language Resources and Evaluation* (pp. 3264–3268). Academic Press.

Oishi, K., Akimoto, T., Onodera, K., Zhag, Y., Ono, J., & Ogata, T. (2011). Conceptual dictionaries in a narrative generation system: Toward an integrated narrative generation system. In *IEICE Technical Report (NLC 2011-29), 111(228)* (pp. 39–44). Tokyo, Japan: The Institute of Electoronics, Information and Communication Engineers.

Oishi, K., Kurisawa, Y., Kamada, M., Fukuda, I., Akimoto, T., & Ogata, T. (2012). Building conceptual dictionary for providing common knowledge in the integrated narrative generation system. In *Proceedings of the 34th Annual Conference of the Cognitive Science Society* (pp. 2126–2131). Tokyo, Japan: Japanese Cognitive Science Society.

Oishi, K., & Ogata, T. (2012). Development and application of noun/verb conceptual dictionaries for a narrative generation system. In *IEICE Technical Report (NLC 2011-60), 111(427)* (pp. 25–30). Tokyo, Japan: The Institute of Electoronics, Information and Communication Engineers.

Oishi, K., Ogata, T., Nakashima, M., & Akimoto, T. (2007). Monogatair seisei to kan-text-sei no kōsatsu [Considering narrative generation and intertextuality]. In *Proceedings of the 26th Special Interest Group on Language Sense Processing Engineering* (pp. 55–62). Tokyo, Japan: The Japanese Society for Artificial Intelligence.

Oishi, K., Ogata, T., & Onodera, K. (2011). Development and use of verb concept case frames and the noun conceptual system for a narrative generation system. In *Proceedings of the 25th Annual Conference of Japanese Society for Artificial Intelligence* (1H2-OS1-10). Tokyo, Japan: The Japanese Society for Artificial Intelligence.

Okada, N. (1991). *Go no gainen no hyōgen to chikuseki* [Representation and storage of word concepts]. Tokyo, Japan: Denshi Jōhō Tsūshin Gakkai.

Okada, N., & Endo, T. (1992). Story generation based on dynamics of the mind. *Computational Intelligence*, *8*(1), 123–160. doi:10.1111/j.1467-8640.1992.tb00341.x

Ono, J., Akimoto, T., & Ogata, T. (2014). A trial for the automatic acquisition of attribute frames in an integrated narrative generation system. In *IEICE Technical Report (NLC 2014-9), 114(81)* (pp. 47–52). Tokyo, Japan: The Institute of Electronics, Information and Communication Engineers.

Ono, J., & Ogata, T. (2013). A framework of the inter-textual knowledge acquisition and creation mechanism for an integrated narrative generation system. *Proceedings of Forum on Information Technology*, *2*, 201–204.

Ono, J., & Ogata, T. (2014a). Concept/word selection based on novel data in an integrated narrative generation system. In *Proceedings of the 47th Special Interest Group on Language Sense Processing Engineering* (pp. 47–53). Tokyo, Japan: The Japanese Society for Artificial Intelligence.

Ono, J., & Ogata, T. (2014b). Selecting noun concepts based on accounting data: As a mechanism in the integrated narrative generation system. In *IEICE Technical Report (NLC 2014-38), 114(366)* (pp. 49–54). Tokyo, Japan: The Institute of Electronics, Information and Communication Engineers.

Ono, J., & Ogata, T. (2014c). Automatic acquisition of attribute frames of noun concepts using the classification of noun phrases: As a mechanism of an integrated narrative generation system. *Proceedings of Forum on Information Technology*, *2*, 201–204.

Ono, J., & Ogata, T. (2015a). Noun concepts selection in events based on the co-occurrence relation in verb concepts and noun concepts: Mechanisms for the story generation in an integrated narrative generation system. *Proceedings of Forum on Information Technology*, *2*, 239–242.

Ono, J., & Ogata, T. (2015b). Concept selection/language representation selection and the control in an integrated narrative generation system. In *Proceedings of the 29th Annual Conference of Japanese Society for Artificial Intelligence* (3G4-OS-05a-3). Tokyo, Japan: The Japanese Society for Artificial Intelligence.

Ono, J., & Ogata, T. (2015c). Selecting words and notation using literary data in the integrated narrative generation system. *Journal of Robotics, Networking and Artificial Life*, *2*(3), 194–199. doi:10.2991/jrnal.2015.2.3.13

Ono, J., & Ogata, T. (2016). An attempt for expansion of the "KOSERUBE: Folk tales of Iwate." *Proceedings of the Information Processing Society of Japan Interaction, 2016*, 837–840.

Ono, J., Ogata, T., & Ito, T. (2019). A basic study for using folk tale's motifs to narrative generation. In *Proceeding of the 33rd Annual Conference of the Japanese Society for Artificial Intelligence* (1F2-NFC-1-05). Tokyo, Japan: The Japanese Society for Artificial Intelligence.

Ono, J., Zhang, Y., & Ogata, T. (2012). Defamiliarization rhetoric and scenario generation for events using the constraints in conceptual. In *Proceedings of the 26th Annual Conference of Japanese Society for Artificial Intelligence* (1N1-OS-1a-1). Tokyo, Japan: The Japanese Society for Artificial Intelligence.

Ono, J., Zhang, Y., Oishi, K., Onodera, K., & Ogata, T. (2011). An advertising rhetorical mechanism for a single event combined with conceptual systems. In *Proceedings of the 28th Annual Conference of the Cognitive Science Society* (pp. 580–584). Tokyo, Japan: Japanese Cognitive Science Society.

Onodera, K., Akimoto, T., & Ogata, T. (2012a). Development of state-event transformation knowledge base and circular generation for story world and story line. In *Proceedings of the 26th Annual Conference of Japanese Society for Artificial Intelligence* (1N2-OS-1b-9). Tokyo, Japan: The Japanese Society for Artificial Intelligence.

Onodera, K., Akimoto, T., & Ogata, T. (2012b). A state-event transformation mechanism for generating micro structures of story in an integrated narrative generation system. In *Proceedings of the 34th Annual Conference of the Cognitive Science Society* (pp. 2150–2155). Tokyo, Japan: Japanese Cognitive Science Society.

Onodera, K., & Ogata, T. (2012a). Story generation mechanism based on state-event transformation. In *IEICE Techical Report (NLC2011-60), 111(427)* (pp. 77–82). Tokyo, Japan: The Institute of Electronics, Information and Communication Engineers.

Onodera, K., & Ogata, T. (2012b). Sequence generation based on mutual relationship between state and action: As a mechanism in narrative generation system. In *Proceedings of the 4th IEEE International Conference on Digital Game and Intelligent Toy Enhanced Learning (DIGITEL 2012)* (pp. 159–161). The Institute of Electrical and Electronics Engineers. 10.1109/DIGITEL.2012.45

Parajanof, S. (Director) (1968). *The color of pomegranates* [Motion picture]. Armenia, Soviet Union.

Propp, V. Y. (1968). *Morphology of the folktale* (L. Scott, Trans.). Austin, TX, USA: University of Texas Press. (Original work published 1928)

Quillian, M. (1968). Semantic memory. In M. Minsky (Ed.), *Semantic information processing* (pp. 227–270). Cambridge, MA: MIT Press.

Sato, H., Kobayashi, K., Yasuda, T., & Ogata, T. (2008). Monogatari seisei system ni okeru kan'i na tanbun/fukubun seisei hōshiki [Simple generation methods of simple and complex sentences in a narrative generation system]. In *Proceedings of the 70th National Convention of Information Processing Society of Japan*, Vol. 4 (pp. 803–804). Tokyo, Japan: Information Processing Society of Japan.

Schank, R. C., & Abelson, R. P. (1977). *Scripts, plans, goals, and understanding: An inquiry into human knowledge structures*. Lawrence Erlbaum.

Seki, K. (1957). *Mukashibanashi to shōwa* [Folktales and funny stories]. Tokyo, Japan: Iwasaki Shoten.

Seki, K. (1966). *Mukashibanashi no rekishi* [The history of the folktale]. Tokyo, Japan: Shibundō.

Seki, K. (1977). *Nihon no mukashibanashi—Hikaku kenkyū josetsu* [Japanese folktales: An introduction to the comparative study]. Tokyo, Japan: NHK Shuppan.

Seki, K., Nomura, J., & Ōshima, H. (Eds.). (1979). *Nihon mukashibanashi taisei, 12—Kenkyū hen* [Compilation of Japanese folktales, Vol. 12: Researches]. Tokyo, Japan: Kadokawa Shoten.

Seki, K., Nomura, J., & Ōshima, H. (Eds.). (1980). *Nihon mukashibanashi taisei, 11—Shiryō hen* [Compilation of Japanese folktales, Vol. 11: Data]. Tokyo, Japan: Kadokawa Shoten.

Shklovsky, V. (1990). *Theory of prose* (B. Sher, Trans.). Dalkey Archive Press.

Steele, G. L. Jr. (1990). Common lisp. Digital Equipment.

Tachibana, S., & Ogata, T. (2009a). Eizō satsuei ni okeru kisoku no bunseki—"Tokyo monogatari" wo sozai to shite [Extracting rules of cameraworks: Using "Tokyo Story" as a material]. In *Proceedings of the 71st National Convention of Information Processing Society of Japan*, Vol. 4 (pp. 759–760). Tokyo, Japan: Information Processing Society of Japan.

Tachibana, S., & Ogata, T. (2009b). A camerawork setting system based on rules and simulation of "Ozu rules." In *Proceedings of the 23rd Annual Conference of Japanese Society for Artificial Intelligence* (1J1-OS2-11). Tokyo, Japan: The Japanese Society for Artificial Intelligence.

Takahashi, R. (2003a). Maison Ikkoku, Vol. 1. VIZ Media LLC. (Original work published 1980–1987)

Takahashi, R. (2003b). Maison Ikkoku, Vol. 2. VIZ Media LLC. (Original work published 1980–1987)

Takahashi, R. (2004a). Maison Ikkoku, Vol. 3. VIZ Media LLC. (Original work published 1980–1987)

Takahashi, R. (2004b). Maison Ikkoku, Vol. 4. VIZ Media LLC. (Original work published 1980–1987).

Takahashi, R. (2004c). Maison Ikkoku, Vol. 5. VIZ Media LLC. (Original work published 1980–1987)

Takahashi, R. (2004d). Maison Ikkoku, Vol. 6. VIZ Media LLC. (Original work published 1980–1987)

Takahashi, R. (2004e). Maison Ikkoku, Vol. 7. VIZ Media LLC. (Original work published 1980–1987)

Takahashi, R. (2004f). Maison Ikkoku, Vol. 8. VIZ Media LLC. (Original work published 1980–1987)

Takahashi, R. (2005a). Maison Ikkoku, Vol. 9. VIZ Media LLC. (Original work published 1980–1987)

Takahashi, R. (2005b). Maison Ikkoku, Vol. 10. VIZ Media LLC. (Original work published 1980–1987)

Takahashi, R. (2005c). Maison Ikkoku, Vol. 11. VIZ Media LLC. (Original work published 1980–1987)

Takahashi, R. (2005d). Maison Ikkoku, Vol. 12. VIZ Media LLC. (Original work published 1980–1987)

Takahashi, R. (2005e). Maison Ikkoku, Vol. 13. VIZ Media LLC. (Original work published 1980–1987)

Takahashi, R. (2005f). Maison Ikkoku, Vol. 14. VIZ Media LLC. (Original work published 1980–1987)

Takahashi, R. (2006). Maison Ikkoku, Vol. 15. VIZ Media LLC. (Original work published 1980–1987)

Takeuchi, K. (2011). Construction of thesaurus of predicate-argument structure for Japanese verbs. In *Proceedings of the 25th Japanese Society for Artificial Intelligence* (3H2-OS3-5). Tokyo, Japan: The Japanese Society for Artificial Intelligence.

Tanaka, H., Motoyoshi, F., & Yamanashi, M. (1983). *Lisp de manabu ninchi shinrigaku, 3—Gengo rikai* [Cognitive psychology for learning by Lisp, 3: Language uderstanding]. Tokyo, Japan: Tokyo Daigaku Shuppankai.

Tanizaki, J. (1951). *Shunkinshō*. Tokyo, Japan: Shinchōsha. (Original work published 1933)

Terada, T., Akimoto, T., Ono, J., & Ogata, T. (2014). A systematic description of proper noun concepts in the integrated narrative generation system. In *Proceeding of the 20th conference of Natural Language Processing* (pp. 217–220). Tokyo, Japan: The Association for Natural Language Processing.

Tomite, S., Ogata, T., & Hanada, K. (2009). A system for constructing animated movie from conceptual representation and a consideration on the structure of actions. In *Proceedings of the 23th Annual Conference of Japanese Society for Artificial Intelligence* (1J1-OS2-10). Tokyo, Japan: The Japanese Society for Artificial Intelligence.

Ueda, K., & Ogata, T. (2004a). The diversity of narrative discourse by perspective and voice. In *Proceedings of the 18th Annual Conference of Japanese Society for Artificial Intelligence* (2D1-05). Tokyo, Japan: The Japanese Society for Artificial Intelligence.

Ueda, K., & Ogata, T. (2004b). A computational modeling of perspective and voice in the narrative rhetoric. In *Proceedings of the 18th Congress of the International Association of Empirical Aesthetics* (pp. 480–486). Rome, Italy: University of Rome Tre.

Umehara, S., & Ogata, T. (2003). An analysis of the duration in narrative and its application. In *Proceedings of the 17th Annual Conference of Japanese Society for Artificial Intelligence* (2G2-02). Tokyo, Japan: The Japanese Society for Artificial Intelligence.

Winston, P. H., & Horn, B. K. P. (1989). *LISP* (3rd ed.). Addison Wesley.

Yamakage, S., & Ogata, T. (2003). A computation mechanism of the "distance" in narratives. In *Proceedings of the 17th Annual Conference of Japanese Society for Artificial Intelligence* (2G2-05). Tokyo, Japan: The Japanese Society for Artificial Intelligence.

Yoshimoto, T. (1965). *Gengo nitotte bi toha nanika* [What is the beauty for language?]. Tokyo, Japan: Keisō Shobō.

Yuasa, T., & Hagiya, M. (1986). *Common Lisp nyūmon* [Introduction to Common Lisp]. Tokyo, Japan: Iwanami Shoten.

Zeami Zenchiku. (1974). In A. Omote & S. Katō, (Eds.), *Nihon shisō taikei, 24* [Japanese philosophical thoughts collection, Vol. 24]. Tokyo, Japan: Iwanami Shoten.

Zhang, Y., Ono, J., & Ogata, T. (2011). An advertising rhetorical mechanism for single event combined with conceptual dictionaries in narrative generation system. In *Proceedings of the 7th International Conference on Natural Language Processing and Knowledge Engineering* (pp. 340–343). The Institute of Electrical and Electronics Engineers. 10.1109/NLPKE.2011.6138221

Zhang, Y., Ono, J., & Ogata, T. (2012). Single event and scenario generation based on advertising rhetorical techniques using the conceptual dictionary in narrative generation system. In *Proceedings of the 4th IEEE International Conference on Digital Game and Intelligent Toy Enhanced Learning* (pp. 162–164). The Institute of Electrical and Electronics Engineers. 10.1109/DIGITEL.2012.46

KEY TERMS AND DEFINITIONS

Conceptual Dictionary: It formally describes the concepts of words, such as verb concepts and noun concepts, as a component of the author's "Integrated Narrative Generation System: INGS"; further, it is a kind of ontology that gives the semantic descriptions of concepts and their semantic relationships in the world. INGS organically integrates various modules for narrative generation processing through the conceptual and language-related dictionaries.

Gerald Genette (1930–2018): As a representative researcher of narratology, Genette published many books on narratological themes, including narrative structural analysis, intertextuality, and the structure of a book. Narratology, in the narrow sense, means Genette's "narrative discourse theory." The author's narrative generation research focuses on this theory to survey, analyze, and redefine it from the narrative generation system's viewpoint.

Integrated Narrative Generation System (INGS): INGS is a synthetic narrative generation system architecture integrating the previous studies by the author. From the broadest perspective, INGS is divided into two types of part, knowledge and procedure. INGS aims at a kind of narrative synthesizer that integrates a variety of narrative techniques, methods, rhetoric, and knowledge into an organic and dynamic generation framework.

Literary Theories and: Narratology: Various sub-areas in literary theories commonly aim at general literary approaches to diverse literary phenomena beyond the implementation and criticism of each literary work. Narratology is an area of literary theories and specially focuses on narrative phenomena, such as the structure of a narrative text, aspects of a narrative sender and receiver, and narrative media.

Narrative Discourse and Narrative Representation: Narrative discourse, in the broad sense, is divided into structural level (narrative discourse in the narrow sense) and representative level (narrative representation). The former (simply narrative discourse) corresponds to a kind of story adaptation and refers to narrative components based on the composition and order of a story, the organization of events, narrative viewpoints, and tempo. Narrative representation, on the other hand, includes language representation, picture or image representation, music representation, etc.

Story: The story in a narrative is the equivalent of a temporal sequence of events, and an event is a unit that changes a particular state in a story. This most important narratological concept is clearly different from the concept of narrative discourse. Although a narrative discourse structure is a narrative organization based on the elements directly appearing in a represented narrative itself, a story is a narrative organization that implicitly exists in a represented narrative. For example, a story is constructed through a receiver's inferential and imaginative abilities through a narrative reception process.

Story Grammar Based on Propp's Theory: Such story grammar has originally been developed from the author's viewpoint that faithfully reflects Propp's theory as far as possible. The author positions the story grammar based on AI and cognitive science as the micro-level story-development logic, and that based on Propp's theory as the macro-level story to integrate both approaches in the story generation module in "Integrated Narrative Generation System: INGS."

APPENDIX 1

Table 10. The conceptual representation of a generated story (partially omitted)

```
Input:
(st-propp ‘((武部[Takebe] 勇士[warrior]@勇者[warrior]) (スネ吉[Sunekichi] 蛇[snake]@爬虫類{蛇}[reptiles{snake}]) (フローラ[Flora] 皇女[princess]@女[woman]) (杖蔵[Tsuezo] 杖[wand]@棒{杖}[rod{wand}]) (ツァーリ[Tsar] 王様@君主{王}[monarch{king}]) (バーバ・ ヤガー[Baba-Yaga] 老婆[older-woman]@女[woman]) (メロス[Melos] 男[man]@男[man])) ‘PROPP ‘topdownC 0 ‘(ロシア民話[Russian-folktale])))
Output:
($ロシア魔法昔話
 ($問題
 ($予備部分
 ($01_留守
 (event 死ぬ1 (type action) (ID 1) (time (time1 time2)) (agent age%親#1) (counter-agent nil) (location loc%瀑布#1) (object nil) (instrument obj%病#1) (from nil) (to nil) (adverb nil) (possessive nil) (situation nil) (purpose nil) (experiencer nil) (source nil) (idiom nil) (information nil) (as nil)))
 ($04_探り出し
 (event 考える2 (type action) (ID 2) (time (time2 time3)) (agent age%蛇#1) (counter-agent age%皇女#1) (location loc%着駅#1) (object nil) (instrument nil) (from nil) (to nil) (adverb nil) (possessive nil) (situation nil) (purpose nil) (experiencer nil) (source nil) (idiom nil) (information nil) (as nil))
 (event 派遣する1 (type action) (ID 3) (time (time3 time4)) (agent age%蛇#1) (counter-agent age%諸姉#1) (location loc%着駅#1) (object nil) (instrument nil) (from nil) (to nil) (adverb nil) (possessive nil) (situation nil) (purpose nil) (experiencer nil) (source nil) (idiom nil) (information nil) (as nil))
 (event 捜す2 (type action) (ID 4) (time (time4 time5)) (agent age%諸姉#1) (counter-agent age%皇女#1) (location loc%着駅#1) (object nil) (instrument nil) (from nil) (to nil) (adverb nil) (possessive nil) (situation nil) (purpose nil) (experiencer nil) (source nil) (idiom nil) (information nil) (as nil)))
 ($05_情報漏洩
 (event 発見する2 (type action) (ID 5) (time (time5 time6)) (agent age%諸姉#1) (counter-agent age%皇女#1) (location loc%着駅#1) (object nil) (instrument nil) (from nil) (to nil) (adverb nil) (possessive nil) (situation nil) (purpose nil) (experiencer nil) (source nil) (idiom nil) (information nil) (as nil))))
 ($発端
 ($08_加害
 ($-
 (event 出産する2 (ID 2) (time (time6 time6-1)) (agent age%蛇#1) (counter-agent age%皇女#1) (location loc%着駅#1) (object nil) (instrument nil) (from nil) (to nil) (adverb nil) (possessive nil) (situation nil) (purpose nil) (experiencer nil) (source nil) (idiom nil) (information nil) (as nil))
 (event 潜る2 (ID 3) (time (time6-1 time6-2)) (agent age%蛇#1) (counter-agent nil) (location loc%着駅#1) (object nil) (instrument nil) (from nil) (to loc%流#1) (adverb nil) (possessive nil) (situation nil) (purpose nil) (experiencer nil) (source nil) (idiom nil) (information nil) (as nil))
 (event 変える2 (type action) (ID 6) (time (time6-2 time7)) (agent age%蛇#1) (counter-agent nil) (location loc%流#1) (object age%皇女#1) (instrument nil) (from nil) (to age%山猫#1) (adverb nil) (possessive nil) (situation nil) (purpose nil) (experiencer nil) (source nil) (idiom nil) (information nil) (as nil)))
 ($-
 (event 赴任する1 (ID 4) (time (time7 time7-1)) (agent age%皇女#1) (counter-agent nil) (location loc%流#1) (object age%国#1) (instrument nil) (from nil) (to nil) (adverb nil) (possessive nil) (situation nil) (purpose nil) (experiencer nil) (source nil) (idiom nil) (information nil) (as nil))
 (event 追い出す1 (type action) (ID 7) (time (time7-1 time8)) (agent age%蛇#1) (counter-agent age%皇女#1) (location loc%流#1) (object nil) (instrument nil) (from age%国#1) (to nil) (adverb nil) (possessive nil) (situation nil) (purpose nil) (experiencer nil) (source nil) (idiom nil) (information nil) (as nil))))
 ($09_仲介
 (event 連れる1 (type action) (ID 8) (time (time8 time9)) (agent age%王様#1) (counter-agent age%勇士#1) (location loc%縦坑#1) (object nil) (instrument nil) (from nil) (to obj%旅#1) (adverb nil) (possessive nil) (situation nil) (purpose nil) (experiencer nil) (source nil) (idiom nil) (information nil) (as nil)))
 ($10_対抗開始
 (event 決意する2 (type action) (ID 9) (time (time9 time10)) (agent age%勇士#1) (counter-agent nil) (location loc%縦坑#1) (object obj%冒険#1) (instrument nil) (from nil) (to nil) (adverb nil) (possessive nil) (situation nil) (purpose nil) (experiencer nil) (source nil) (idiom nil) (information nil) (as nil)))
 ($11_出立
 (event 出国する1 (type action) (ID 10) (time (time10 time11)) (agent age%勇士#1) (counter-agent nil) (location loc%縦坑#1) (object nil) (instrument nil) (from loc%縦坑#1) (to loc%ケーブル#1) (adverb nil) (possessive nil) (situation nil) (purpose nil) (experiencer nil) (source nil) (idiom nil) (information nil) (as nil)))))
<omitted>...
```

(English sentences were translated from Japanese sentences using Google Translate)

Table 11. The sentences by the conceptual representation in table 10

親が瀑布で病で死ぬ。蛇のスネ吉が着駅で皇女のフローラを考える。スネ吉が諸姉を着駅に派遣する。諸姉がフローラを捜す。諸姉がフローラを発見する。スネ吉がフローラを出産する。スネ吉が門を潜る。スネ吉がフローラを山猫へ変える。フローラが邦に赴任する。スネ吉がフローラを邦から追い出す。王様のツァーリが勇士の武部を旅へ連れる。武部が冒険を決意する。武部は縦坑からケーブルに出国する。それから一週間たちました。スネ吉がケーブルに埋もれる。老婆のバーバ・ヤガーがケーブルでスネ吉と肉を取り合う。バーバ・ヤガーが「武部が肉を複数で分配する」ことを武部に頼む。武部が頼みを聞く。武部が肉を複数で分配する。バーバ・ヤガーが杖の杖蔵を落書で作る。バーバ・ヤガーが杖蔵を偽造する。バーバ・ヤガーから杖蔵を武部に手渡す。武部が耕地で杖蔵に乗る。杖蔵が武部をミュージアムに連れる。武部が耕地からミュージアムへ移動する。そして、スネ吉がケーブルでミュージアムへ埋まる。武部がミュージアムでスネ吉と戦う。スネ吉が指輪を懐から足下に落とす。武部が指輪を拾う。スネ吉が武部に敗れる。それから三日たちました。スネ吉が呪文を武部に教える。武部が呪文を唱える。魔法が解ける。フローラがミュージアムの草取りをスル。フローラが「魔法が解ける」ことを喜ぶ。武部がフローラを帰路に連れる。武部がミュージアムから脱出する。そして、武部がミュージアムより縦坑に到着する。男が言い張る。男が報酬をツァーリに要求する。そして、ツァーリが武部に会う。フローラが武部に会う。武部が指輪を持参する。フローラが武部で縦坑を荒らす。武部が縦坑で指輪をフローラに返す。フローラが真実を知る。ツァーリが縦坑へ倒れる。フローラが縦坑で真実を肉に伝える。ツァーリが真実を知る。ツァーリが「男がツァーリに「孫がカトリック教徒を助ける」ことと嘘を言う」ことに気付く。男のフローラが嘘偽りが露見する。ツァーリが岳麓に帰化する。ツァーリが岳麓で宮殿を岳麓へ建てる。武部が宮殿で宮殿に住む。ツァーリが岳麓から縦坑に転落する。ツァーリが男を縦坑から追放する。フローラが縦坑から岳麓に戻る。武部が岳麓でフローラと婚約する。[The parents die of the disease with a waterfall. The snake's sister thinks of the royal flora at the arrival station. Sisters look for Flora. My sisters discover Flora. Sune is to make Flora. Sunekichi dives in the gate. Suneichi turns Flora into a wildcat. The floor will be assigned to Japan. Sunekichi kicks Flora out of the country. The king 's tour will take the warrior' s Takebe for his journey. Takebe's determination to adventure. Takebe departs from the mine to the cable. Then it was a week. Sune is buried in the cable. An old woman's bar and yaga keeps meat with sisters on cable. Barba Yagar listens to "Takebe's allocating more than one meat" This is what Mr. Takebe begins to ask. Takebe will allocate more meat. To counterfeit. From Barbaya - hand the cane to Takebe. Takebe's riding on arable land. Kane Kura takes me to the museum in Takebe. Takebe's move from arable land to music. It's sisters are wrapped in music in cherub. Takebe fights against Snoji at a certain meam. Snaji drops the ring on a nostalgic feet. Takebe picks up the ring. Sunekichi loses to Takebe. Then I did it on the third day. Sune is teaching spells to Takebe. Takebe casts a spell. The magic melts. The frame makes Muam weeding out. Flurr is pleased with this "magic melts". Takebe will bring Flor to his way home. Takebe's escape from Muamu. That, Takebe arrives at the shaft from Miam. A man insists. A man requests reward from a tour. Tsari also meets Takebe. Flurler will meet Takebe. Fuller who brings a ring with Takebe devides the body. The weapon returns the ring to the flower with a shaft. Flower knows the truth. The toy collapses on the shaft. Flora tells the truth to the flesh on the shaft. Tsari knows the truth. Tsari says, "A man tells Tari" The grandchildren will help the Catholics "and lie" This flora of a man who notices that a lie / fake is revealed. Tsari naturalized to a mountain. I will build a toure to greet the palace with the mountain tower. Takebe lives in the palace at the palace. The toilet falls from the mountains to the shaft. Tsari expels the man from the shaft. Flora returns from the shaft to the foot of Mountain. Takebe is engaged to Flora at Bucheon.]

Table 12. The conceptual representation of a generated story (partially omitted)

```
Input:
(st-propp '((イワン 勇士@勇者) (スネーク 蛇@爬虫類[蛇]) (マリア 皇女@女) (魔法の杖 杖@棒[杖]) (王様 王様@君主[王]) (バーバ・ヤガー 老婆@女) (メロス 男@男)) 'PROPP 'topdownC 0 '(ロシア民話))
Output:
($ロシア魔法昔話
 ($問題
 ($予備部分
 ($01_留守
 (event 出かける1 (type action) (ID 1) (time (time1 time2)) (agent age%親#1) (counter-agent nil) (location loc%茶屋#1) (object obj%仕事#1) (instrument nil) (from nil) (to nil) (adverb nil) (possessive nil) (situation nil) (purpose nil) (experiencer nil) (source nil) (idiom nil) (information nil) (as nil)))
 ($02_禁止
 (event 説く4 (type action) (ID 2) (time (time2 time3)) (agent age%侯#1) (counter-agent nil) (location loc%茶屋#1) (object obj%あれこれ#1) (instrument nil) (from nil) (to age%勇士#1) (adverb nil) (possessive nil) (situation nil) (purpose nil) (experiencer nil) (source nil) (idiom nil) (information nil) (as nil))
 (event 禁止する1 (type action) (ID 3) (time (time3 time4)) (agent age%侯#2) (counter-agent nil) (location loc%茶屋#1) (object (event 出る18 (type action) (ID 4) (time nil) (agent age%勇士#1) (counter-agent nil) (location loc%碧落#1) (object obj%仕事#1) (instrument nil) (from loc%楼閣#1) (to loc%宮城#1) (adverb nil) (possessive nil) (situation nil) (purpose nil) (experiencer nil) (source nil) (idiom nil) (information nil) (as nil))) (instrument nil) (from nil) (to age%勇士#1) (adverb nil) (possessive nil) (situation nil) (purpose nil) (experiencer nil) (source nil) (idiom nil) (information nil) (as nil)))
 ($03_違反
 ($-
 (event 歩行する1 (ID 2) (time (time4 time4-1)) (agent age%勇士#1) (counter-agent nil) (location loc%楼閣#2) (object nil) (instrument nil) (from nil) (to nil) (adverb nil) (possessive nil) (situation nil) (purpose nil) (experiencer nil) (source nil) (idiom nil) (information nil) (as nil))
 (event 出る18 (type action) (ID 5) (time (time4-1 time5)) (agent age%勇士#1) (counter-agent nil) (location loc%茶屋#1) (object obj%仕事#1) (instrument nil) (from loc%楼閣#2) (to loc%碧落#1) (adverb nil) (possessive nil) (situation nil) (purpose nil) (experiencer nil) (source nil) (idiom nil) (information nil) (as nil)))
 ($-
 (event 独走する1 (ID 3) (time (time5 time5-1)) (agent age%勇士#1) (counter-agent nil) (location loc%茶屋#1) (object nil) (instrument nil) (from nil) (to nil) (adverb nil) (possessive nil) (situation nil) (purpose nil) (experiencer nil) (source nil) (idiom nil) (information nil) (as nil))
 (event 破る1 (type action) (ID 6) (time (time5-1 time6)) (agent age%勇士#1) (counter-agent nil) (location loc%茶屋#1) (object obj%禁物#1) (instrument nil) (from nil) (to nil) (adverb nil) (possessive nil) (situation nil) (purpose nil) (experiencer nil) (source nil) (idiom nil) (information nil) (as nil)))))
 ($発端
 ($08a_欠如
 ($-
 (event 通う2 (ID 4) (time (time6 time6-1)) (agent age%勇士#1) (counter-agent nil) (location loc%茶屋#1) (object nil) (instrument nil) (from loc%茶屋#1) (to loc%地下鉄#1) (adverb nil) (possessive nil) (situation nil) (purpose nil) (experiencer nil) (source nil) (idiom nil) (information nil) (as nil))
 (event 欠如する1 (type action) (ID 7) (time (time6-1 time7)) (agent obj%金銭#1) (counter-agent age%勇士#1) (location loc%地下鉄#1) (object nil) (instrument nil) (from nil) (to nil) (adverb nil) (possessive nil) (situation nil) (purpose nil) (experiencer nil) (source nil) (idiom nil) (information nil) (as nil)))
 (event 考える3 (type action) (ID 8) (time (time7 time8)) (agent age%勇士#1) (counter-agent nil) (location loc%地下鉄#1) (object (event 発見する2 (type action) (ID 9) (time nil) (agent age%勇士#1) (counter-agent obj%金銭#2) (location loc%グローバルビレッジ#1) (object nil) (instrument nil) (from nil) (to nil) (adverb nil) (possessive nil) (situation nil) (purpose nil) (experiencer nil) (source nil) (idiom nil) (information nil) (as nil))) (instrument nil) (from nil) (to nil) (adverb nil) (possessive nil) (situation nil) (purpose nil) (experiencer nil) (source nil) (idiom nil) (information nil) (as nil)))
 ($09_仲介
 (event 悲しむ2 (type action) (ID 10) (time (time8 time9)) (agent age%遅進児#1) (counter-agent nil) (location loc%碧落#1) (object obj%霜枯れ#1) (instrument nil) (from nil) (to nil) (adverb nil) (possessive nil) (situation nil) (purpose nil) (experiencer nil) (source nil) (idiom nil) (information nil) (as nil))
 (event 歌う1 (type action) (ID 11) (time (time9 time10)) (agent age%遅進児#1) (counter-agent nil) (location loc%碧落#1) (object obj%呪文#1) (instrument nil) (from nil) (to nil) (adverb nil) (possessive nil) (situation nil) (purpose nil) (experiencer nil) (source nil) (idiom nil) (information nil) (as nil))
 ($-
 (event 航海する1 (ID 5) (time (time10 time10-1)) (agent age%勇士#1) (counter-agent nil) (location loc%地下鉄#1) (object nil) (instrument nil) (from nil) (to loc%碧落#1) (adverb nil) (possessive nil) (situation nil) (purpose nil) (experiencer nil) (source nil) (idiom nil) (information nil) (as nil))
 (event 聞く4 (type action) (ID 12) (time (time10-1 time11)) (agent age%勇士#1) (counter-agent nil) (location loc%碧落#1) (object obj%呪文#2) (instrument nil) (from nil) (to nil) (adverb nil) (possessive nil) (situation nil) (purpose nil) (experiencer nil) (source nil) (idiom nil) (information nil) (as nil)))
 (event 知る1 (type action) (ID 13) (time (time11 time12)) (agent age%勇士#1) (counter-agent nil) (location loc%碧落#1) (object obj%被害#1) (instrument nil) (from nil) (to nil) (adverb nil) (possessive nil) (situation nil) (purpose nil) (experiencer nil) (source nil) (idiom nil) (information nil) (as nil)))
 ($10_対抗開始
 (event 備える1 (type action) (ID 14) (time (time12 time13)) (agent age%勇士#1) (counter-agent nil) (location loc%碧落#1) (object obj%冒険#1) (instrument nil) (from nil) (to nil) (adverb nil) (possessive nil) (situation nil) (purpose nil) (experiencer nil) (source nil) (idiom nil) (information nil) (as nil)))
 ($11_出立
 (event 出国する1 (type action) (ID 15) (time (time13 time14)) (agent age%勇士#1) (counter-agent nil) (location loc%緑地帯#1) (object nil) (instrument nil) (from loc%碧落#1) (to loc%三角州#1) (adverb nil) (possessive nil) (situation nil) (purpose nil) (experiencer nil) (source nil) (idiom nil) (information nil) (as nil)))))
<omitted>...
```

Table 13. The sentences by the conceptual representation in table 12

親が茶屋で仕事に出かけた。侯がアレコレとイワンに説いた。侯がイワンに「イワンが楼閣から宮城へ仕事に出る」ことを禁止した。イワンが楼閣で楼閣を歩行した。イワンが楼閣から碧落に仕事に出た。イワンが茶屋を独走した。イワンが禁物を破った。イワンが茶屋から地下鉄へ通った。金銭が地下鉄でイワンに欠如した。イワンが「イワンが金銭を発見する」ことを考えた。遅進児が碧落で霜枯を悲した。遅進児が呪文を歌った。イワンが碧落へ航海した。イワンが碧落で呪文を聞いた。イワンが被害を知った。イワンが冒険に備えた。イワンは碧落から三角洲へ出国した。バーバ・ ヤガーが魔法の杖をイワンに見せた。イワンがバストをバーバ・ ヤガーに見せた。バーバ・ ヤガーがバストを刈った。バーバ・ ヤガーがバストを魔法の杖と交換した。イワンが魔法の杖をバーバ・ ヤガーに適用した。イワンがバーバ・ ヤガーを苦しめた。バーバ・ ヤガーが呪文をイワンに教えた。イワンが呪文を唱えた。魔法の杖が其処らから目の前へ出現した。魔法の杖が「魔法の杖がイワンを助ける」ことと申し出た。イワンが宮城へ来襲した。魔法の杖が宮城で路をイワンに教えた。イワンが魔法の杖に従った。イワンが宮城から抜道へ歩いた。スネークが抜道で「イワンが異国に住む」ことをイワンに課した。スネークがイワンを異国に連れた。イワンが抜道に住んだ。イワンが楽しく過ごした。スネークがイワンに降参した。イワンが魔法の杖を利用した。魔法の杖が焼金を生んだ。イワンが抜道より脱出した。イワンが宮城を歩いた。イワンは喉が渇いた。スネークが抜道でスネークにまみれた。スネークが宮城で宮城を独走した。スネークがスネークから井戸水に変わった。イワンが井戸に近付いた。イワンが宮城を流浪した。イワンが井戸水を切った。井戸水が血を流した。イワンがスネークに戯れた。イワンがスネークから碧落に逃げた。イワンが抜道で抜道をドライブスった。イワンが抜道より碧落に到着した。メロスが言い張った。メロスが報酬を王様に要求した。マリアが真実を知った。マリアが真実を名子に伝えた。王様が真実を知った。王様が「メロスが王様に「侯がコンキューを助ける」ことと嘘を言う」ことに気付いた。メロスの金銭が嘘偽りが露見した。イワンが峠で峠を独走した。イワンがピジャマを着た。イワンが美しさで輝いた。スネークが宮城で碧落を独走した。王様が碧落でスネークを容赦した。王様がメロスを容赦した。親が峠で峠を回転した。王様が峠を参観した。王様が報酬をイワンに与えた。[The parents went to work at a tea house. Hou preached Alekole and Ivan. Hou prohibited Ivan from "Ivan will go to work from the castle to Miyagi." Ivan walked over the castle in a castle tower. Ivan got off to work from the castle tower. Ivan traveled alone in the tea room. Ivan broke the prohibited thing. Ivan passed from the tea house to the subway. Money was lacking in Ivan on the subway. Iwan thought "Ivan will find money". The late boy fell frost dying with a falling fall. A lazy boy sang a spell. Ivan sailed to the falls. Ivan listened to the spell with a falling fall. Ivan knew the damage. Ivan prepares for the adventure. Ivan departed from the falling to the triangle. Barba Yaga showed a magic wand to Ivan. Ivan showed the bust to Barba Yaga. Barba Yaga mowed the bust. Barba Yaga replaced bust with a magical wand. Ivan applied the magic wand to Barba Yaga. Ivan afflicted Barba Yaga. Barba Yaga taught the spell to Ivan. Ivan cast a spell. A magic wand appeared from front of us in front of our eyes. The magic wand has offered that "Magic wand helps Ivan". Ivan came to Miyagi. A magic wand taught Ivan the way in Miyagi. Ivan followed the magic wand. Ivan walked down from Miyagi. I imposed Ivan that Snake was unlucky and "Ivan lives in a foreign country". Snake took Ivan to a foreign country. Ivan lived in the unexpected. I have had a good time. Snake abandoned Ivan. Ivan used a magic wand. A magic wand has given burned money. Ivan escaped from the excavation. Ivan walked through Miyagi. Ivan was thirsty. The snake covered with a snake in the unexpected. Snake moved alone in Miyagi in Miyagi. The snake changed from snake to well water. Ivan approached the well. Ivan wandered through Miyagi. Ivan cut the well water. Well water shed blood. Ivan played Snake. Ivan ran away from snake to falling. Ivan drifted out of the way with an unexpected break. Ivan arrived at a falling fall from the unexpected. Meros insisted. Melos demanded the king a reward. Mary knew the truth. Mary conveyed the truth to his baby. The king got to know the truth. The king noticed that Meros "told the king," Hou helps the conduit "and tells a lie. The money of Meros revealed a false lie. Ivan took a run over the pass at the pass. Ivan wore a pajama. Ivan shined with beauty. Snake made a runaway in Miyagi. The king pardoned the snake with fallen fall. The king forgave Meros. My parents turned the pass at the pass. The king visited the pass. The king gave his reward to Ivan.]

(English sentences were translated from Japanese sentences using Google Translate)

APPENDIX 2

Table 14. Underlined parts are refined by other story techniques

親は病で死にました。蛇のスネ吉は皇女のフローラを考えました。スネ吉は諸姉を着駅に派遣しました。スネ吉は杖の杖蔵を撮りました。嘘偽りが指環よりファンまで飛びました。スネ吉は孫に冒険に対して弔文を言いました。諸姉はフローラを捜しました。諸姉は落書きを間違いました。諸姉はフローラを発見しました。スネ吉はフローラを出産しました。スネ吉は門を潜りました。スネ吉はフローラを山猫へ変えました。フローラは国に赴任しました。スネ吉はフローラを国より追い出しました。スネ吉は真実に立腹しました。呪文が呪文より来ました。指環が腫れました。指環に起伏がありました。国に粘りがありました。王様のツァーリは勇士の武部を旅へ連れました。武部は冒険を決意しました。武部は縦坑からケーブルに出国しました。スネ吉はケーブルに埋もれました。老婆のバーバ・ヤガーはスネ吉と肉を取り合いました。バーバ・ヤガーは「武部は肉を複数に分配する」ことを武部に頼みました。バーバ・ヤガーは守り(まもり)に同意しました。バーバ・ヤガーはツァーリに落書きの責任を問いました。バーバ・ヤガーはツァーリに気持ちを打ち明けました。武部は頼みを聞きました。武部は肉を複数に分配しました。バーバ・ヤガーは杖蔵を落書きで作りました。バーバ・ヤガーは杖蔵を偽造しました。武部は杖蔵をバーバ・ヤガーに注文しました。バーバ・ヤガーは杖蔵を武部へ手渡しました。バーバ・ヤガーは杖蔵を武部まで手渡しました。バーバ・ヤガーは順境にありました。バーバ・ヤガーはトーストに余裕がありました。バーバ・ヤガーは杖蔵を武部に手渡しました。武部は杖蔵に乗りました。杖蔵は武部をミュージアムへ連れました。武部は耕地からミュージアムへ移動しました。武部はnilに当たりました。武部はトーストをドゥミタスに分けました。武部は肉を明るみに出しました。スネ吉はミュージアムで埋まりました。武部はスネ吉と戦いました。武部にアリバイがありました。スネ吉は指環を懐より足下へ落としました。武部は指環を拾いました。武部は指環を身に付けました。武部は指環を指環と当てました。武部は宮殿を売り出しました。スネ吉は武部に敗れました。難事が済みました。禍事はバーバ・ヤガーへ軍配が上がりました。スネ吉は呪文を武部に教えました。武部は呪文を唱えました。魔法が解けました。フローラはミュージアムの草取りをしました。フローラは「魔法が解ける」ことを喜びました。銅が事故止を起こしました。武部はフローラを帰路に連れました。武部はミュージアムから脱出しました。武部はミュージアムから縦坑に到着しました。武部は耕地を縦坑から縦坑に南下しました。男は「男はツァーリを助ける」ことと言い張りました。男は真実の状態にありました。男は報酬をツァーリに要求しました。頼みに勢いがありました。男は魔法を収めました。男は武装を解除しました。ツァーリは武部と会いました。フローラは武部と会いました。武部は指環を持参しました。フローラは縦坑を荒らしました。フローラはカトリック教徒に縦坑と知恵を出しました。縦坑に男から事故止が掛かりました。武部は指環をフローラへ返しました。武部は夢枕より覚めました。フローラは雑人を時代劇に撮りました。フローラは真実を知りました。フローラは報酬に応答しました。ツァーリは縦坑へ倒れました。フローラは真実を肉に伝えました。フローラは真実を積みました。ツァーリは真実を知りました。ツァーリはファンを他郷に撒きました。ツァーリは「男はツァーリに「孫はカトリック教徒を助ける」ことと嘘を言う」ことに気付きました。男のフローラが嘘偽りが露見しました。ツァーリは岳麓に帰化しました。ツァーリは宮殿を岳麓へ建てました。真実が雑人の胸を潰しました。武部は宮殿に住みました。武部は生き馬に乗りました。武部は宮殿をバーバ・ヤガーへ願いました。ツァーリで武部は光りました。武部は国を潰しました。肉にがたが来ました。ツァーリは岳麓より縦坑に転落しました。ツァーリは男を縦坑から追放しました。フローラは縦坑から岳麓に戻りました。フローラは銅を離しました。武部はフローラと婚約しました。[Parents died of illness. Snake 's snake thought of the Princess' Flora. Senichi dispatched his sisters to the station. Suneko took a cane warehouse. A false lie flew from the ring to the fans. Suneichi gave his grandson a memorial service for adventure. My sisters searched for Flora. My sisters mistook graffiti. My sisters have found Flora. Senichi gave birth to Flora. Sunekich went under the gate. Suneich changed Flora into a wildcat. Flora moved to the country. Suneich drove flora out of the country. Suneichi was upset about the truth. The spell came from a spell. The finger ring swelled. There was a relief in the finger ring. The country was sticky. The Tsarri of the King took the warrior 's Takebe for his journey. Takebe 's decision to adventure. Takebe left the cable from the shaft. Sune was buried in the cable. Old woman Barba Yaga caught sneakichi with meat. Barba Yaga asked Takebe that "Takebe distributes meat more than once". Barba Yaga agreed to defend (marriage). Barba Yaga asked Tsari about the responsibility of graffiti. Barba Yagar confirmed his feelings to Tsari. Takebe asked for a favor. Takebe distributed the meat to more than one. Barba Yaga made a wood store with graffiti. Barba Yaga forged the wrestler. Takebe ordered a wood store from Barba Yaga. Barba Yaga handed the wizard store to Takebe. Barba Yaga handed out the war store to Takebe. Bhuba Yaga was in a quarrel. Barba Yaga had enough toast to spare. Barba Yaga handed the wizard store to Takebe. Takebe riding the storehouse. Kanezo took Takebe to the museum. Takebe moved from the arable land to the museum. Takebe has hit nil. Takebe divided the toast into Dumitas. Takebe put out the meat to light. Sune is buried in the museum. Takebe fought against Sunekichi. There was an alibi in Takebe. Sunekichi dropped the finger ring to his feet from the bosom. Takebe picked up the ring. Takebe has worn a finger ring. Takebe applied the finger ring to the finger ring. Takebe developed the palace. Sunekichi was defeated by Takebe. Difficulties have been done. The trouble went up to Barba Yaga. Suneichi taught spells to Takebe. Takebe cast a spell. The magic has been solved. Flora weeded the museum. Flora was pleased that "magic is melted". Copper caused the accident to break. Takebe has taken Flora home. Takebe has escaped from the museum. Takebe arrived at the mine from the museum. Takebe led the arable soil from the shaft to the shaft. The man insisted that "a man will help the Tsari." The man was in the state of truth. The man demanded compensation from Tsari. There was a momentum for reliance. The man enchanted magic. The man canceled the armed. Tsarri met with Takebe. Flora met with Takebe. Takebe brought a finger ring. Flora devastated the shaft. Flora issued a shaft and wisdom to the Catholics. An accident was taken from a man on the shaft. Takebe returned the finger ring to Flora. Takebe woke up more than Yume Pillow. Flora took a miscellaneous in a historical drama. Flora knew the truth. Flora responded to the reward. Tsari fell to the shaft. Flora conveyed the truth to the flesh. Flora accumulated the truth. Tsar learned the truth. Zarri sprinkled his fans elsewhere. Tsari noticed that "a man tells Tsari" that grandchildren will help Catholics "and lie. A male Flora revealed a false lie. Tsari was naturalized at the foot of the mountain. Tsar built a palace at the foot of the mountain. The truth crushed the chest of a miscellaneous person. Takebe lived in the palace. Takebe riding the live horse. Takebe wished the palace to Barba Yaga. Takebe shined in Tsar. Takebe crushed the country. I have come to the meat. Tsari fell down the shaft from the foot of the mountain. Tsari banished the man from the shaft. Flora returned from the shaft to the foot of Mountain. Flora separated copper. Takebe engaged to Flora.]

Table 15. Underlined portion is Propp based macro-story

鍵っ子はお腹が空きました。鍵っ子はお腹が空きました。悪知恵が落書きの前兆と/に成りました。鍵っ子は肉を食べました。鍵っ子はお腹が空きました。鍵っ子は粥を食べました。鍵っ子は老婆のバーバ・ヤガーに悪知恵を付けました。親は病で死にました。蛇のスネ吉は皇女のフローラを考えました。スネ吉は諸姉を着駅に派遣しました。スネ吉は杖の杖蔵を撮りました。嘘偽りが指環よりファンまで飛びました。スネ吉は孫に冒険に対して弔文を言いました。諸姉はフローラを捜しました。諸姉は落書きを間違いました。諸姉はフローラを発見しました。スネ吉はフローラを出産しました。スネ吉は門を潜りました。スネ吉はフローラを山猫へ変えました。フローラは国に赴任しました。スネ吉はフローラを国より追い出しました。スネ吉は真実に立腹しました。呪文が呪文より来ました。指環が腫れました。指環に起伏がありました。国に粘りがありました。王様のツァーリは勇士の武部を旅へ連れました。武部は冒険を決意しました。武部は縦坑からケーブルに出国しました。スネ吉はケーブルに埋もれました。老婆のバーバ・ヤガーはスネ吉と肉を取り合いました。バーバ・ヤガーは「武部は肉を複数に分配する」ことを武部に頼みました。バーバ・ヤガーは守り(まもり)に同意しました。バーバ・ヤガーはツァーリに落書きの責任を問いました。バーバ・ヤガーはツァーリに気持ちを打ち明けました。武部は頼みを聞きました。武部は肉を複数に分配しました。バーバ・ヤガーは杖蔵を落書きで作りました。バーバ・ヤガーは杖蔵を偽造しました。武部は杖蔵をバーバ・ヤガーに注文しました。バーバ・ヤガーは杖蔵を武部へ手渡しました。バーバ・ヤガーは杖蔵を武部まで手渡しました。バーバ・ヤガーは順境にありました。バーバ・ヤガーはトーストに余裕がありました。バーバ・ヤガーは杖蔵を武部に手渡しました。武部は杖蔵に乗りました。杖蔵は武部をミュージアムへ連れました。武部は耕地からミュージアムへ移動しました。武部はnil に当たりました。武部はトーストをドゥミタスに分けました。武部は肉を明るみに出しました。スネ吉はミュージアムで埋まりました。武部はスネ吉と戦いました。武部にアリバイがありました。スネ吉は指環を懐より足下へ落としました。武部は指環を拾いました。武部は指環を身に付けました。武部は指環を指環と当てました。武部は宮殿を売り出しました。スネ吉は武部に敗れました。難事が済みました。禍事はバーバ・ヤガーへ軍配が上がりました。スネ吉は呪文を武部に教えました。武部は呪文を唱えました。魔法が解けました。フローラはミュージアムの草取りをしました。フローラは「魔法が解ける」ことを喜びました。銅が事故止を起こしました。武部はフローラを帰路に連れました。武部はミュージアムから脱出しました。武部はミュージアムから縦坑に到着しました。武部は耕地を縦坑から縦坑に南下しました。男は「男はツァーリを助ける」ことと言い張りました。男は真実の状態にありました。男は報酬をツァーリに要求しました。頼みに勢いがありました。男は魔法を収めました。男は武装を解除しました。ツァーリは武部と会いました。フローラは武部と会いました。武部は指環を持参しました。フローラは縦坑を荒らしました。フローラはカトリック教徒に縦坑と知恵を出しました。縦坑に男から事故止が掛かりました。武部は指環をフローラへ返しました。武部は夢枕より覚めました。フローラは雑人を時代劇に撮りました。フローラは真実を知りました。フローラは報酬に応答しました。ツァーリは縦坑へ倒れました。フローラは真実を肉に伝えました。フローラは真実を積みました。ツァーリは真実を知りました。ツァーリはファンを他郷に撒きました。ツァーリは「男はツァーリに「孫はカトリック教徒を助ける」ことと嘘を言う」ことに気付きました。男のフローラが嘘偽りが露見しました。ツァーリは岳麓に帰化しました。ツァーリは宮殿を岳麓へ建てました。真実が雑人の胸を潰しました。武部は宮殿に住みました。武部は生き馬に乗りました。武部は宮殿をバーバ・ヤガーへ願いました。ツァーリで武部は光りました。武部は国を潰しました。肉にがたが来ました。ツァーリは岳麓より縦坑に転落しました。ツァーリは男を縦坑から追放しました。フローラは縦坑から岳麓に戻りました。フローラは銅を離しました。武部はフローラと婚約しました。鍵っ子が皇女のフローラを夢枕に誘いました。鍵っ子は当てに成りました。[The key child was hungry. The key child was hungry. The wickedness became a sign of graffiti. The key player ate meat. The key child was hungry. The key player ate porridge. The keeper gave birth to an old woman, Barba Yaga. Parents died of illness. Snake 's snake thought of the Princess' Flora. Senichi dispatched his sisters to the station. Suneko took a cane warehouse. A false lie flew from the ring to the fans. Suneichi gave his grandson a memorial service for adventure. My sisters searched for Flora. My sisters mistook graffiti. My sisters have found Flora. Senichi gave birth to Flora. Sunekich went under the gate. Suneich changed Flora into a wildcat. Flora moved to the country. Suneich drove flora out of the country. Suneichi was upset about the truth. The spell came from a spell. The finger ring swelled. There was a relief in the finger ring. The country was sticky. The Tsarri of the King took the warrior 's Takebe for his journey. Takebe 's decision to adventure. Takebe left the cable from the shaft. Sune was buried in the cable. Old woman Barba Yaga caught sneakichi with meat. Barba Yaga asked Takebe that "Takebe distributes meat more than once". Barba Yaga agreed to defend (marriage). Barba Yaga asked Tsari about the responsibility of graffiti. Barba Yagar confirmed his feelings to Tsari. Takebe asked for a favor. Takebe distributed the meat to more than one. Barba Yaga made a wood store with graffiti. Barba Yaga forged the wrestler. Takebe ordered a wood store from Barba Yaga. Barba Yaga handed the wizard store to Takebe. Barba Yaga handed out the war store to Takebe. Bhuba Yaga was in a quarrel. Barba Yaga had enough toast to spare. Barba Yaga handed the wizard store to Takebe. Takebe riding the storehouse. Kanezo took Takebe to the museum. Takebe moved from the arable land to the museum. Takebe has hit nil. Takebe divided the toast into Dumitas. Takebe put out the meat to light. Sune is buried in the museum. Takebe fought against Sunekichi. There was an alibi in Takebe. Sunekichi dropped the finger ring to his feet from the bosom. Takebe picked up the ring. Takebe has worn a finger ring. Takebe applied the finger ring to the finger ring. Takebe developed the palace. Sunekichi was defeated by Takebe. Difficulties have been done. The trouble went up to Barba Yaga. Suneichi taught spells to Takebe. Takebe cast a spell. The magic has been solved. Flora weeded the museum. Flora was pleased that "magic is melted". Copper caused the accident to break. Takebe has taken Flora home. Takebe has escaped from the museum. Takebe arrived at the mine from the museum. Takebe led the arable soil from the shaft to the shaft. The man insisted that "a man will help the Tsari." The man was in the state of truth. The man demanded compensation from Tsari. There was a momentum for reliance. The man enchanted magic. The man canceled the armed. Tsarri met with Takebe. Flora met with Takebe. Takebe brought a finger ring. Flora devastated the shaft. Flora issued a shaft and wisdom to the Catholics. An accident was taken from a man on the shaft. Takebe returned the finger ring to Flora. Takebe woke up more than Yume Pillow. Flora took a miscellaneous in a historical drama. Flora knew the truth. Flora responded to the reward. Tsari fell to the shaft. Flora conveyed the truth to the flesh. Flora accumulated the truth. Tsar learned the truth. Zarri sprinkled his fans elsewhere. Tsari noticed that "a man tells Tsari" that grandchildren will help Catholics "and lie. A male Flora revealed a false lie. Tsari was naturalized at the foot of the mountain. Tsar built a palace at the foot of the mountain. The truth crushed the chest of a miscellaneous person. Takebe lived in the palace. Takebe riding the live horse. Takebe wished the palace to Barba Yaga. Takebe shined in Tsar. Takebe crushed the country. I have come to the meat. Tsari fell down the shaft from the foot of the mountain. Tsari banished the man from the shaft. Flora returned from the shaft to the foot of Mountain. Flora separated copper. Takebe engaged to Flora. A key player invited the Princess' Flora to a dream crane. The key player became reliable.]

Table 16. Underlined parts are refined by other story techniques

親が茶屋で仕事に出かけた。親が書き残した。侯がアレコレとイワンに説いた。侯が熱中した。侯がイワンに「イワンが楼閣から宮城へ仕事に出る」ことを禁止した。イワンが禁を犯す。イワンが楼閣で楼閣を歩行した。イワンが楼閣で休息する。イワンが楼閣から碧落に仕事に出た。イワンが悟る。イワンが茶屋を独走した。イワンが禁物を破った。イワンが茶屋から地下鉄へ通った。金銭が地下鉄でイワンに欠如した。イワンが「イワンが金銭を発見する」ことを考えた。遅進児が碧落で霜枯を悲した。遅進児が呪文を歌った。イワンが碧落へ航海した。イワンが碧落で呪文を聞いた。イワンが被害を知った。イワンが冒険に備えた。イワンは碧落から三角洲へ出国した。バーバ・ヤガーが魔法の杖をイワンに見せた。イワンがバストをバーバ・ヤガーに見せた。バーバ・ヤガーがバストを刈った。バーバ・ヤガーがバストを魔法の杖と交換した。イワンが魔法の杖をバーバ・ヤガーに適用した。イワンがバーバ・ヤガーを苦しめた。バーバ・ヤガーが呪文をイワンに教えた。イワンが呪文を唱えた。魔法の杖が其処らから目の前へ出現した。魔法の杖が「魔法の杖がイワンを助ける」ことと申し出た。イワンが宮城へ来襲した。魔法の杖が宮城で路をイワンに教えた。イワンが魔法の杖に従った。イワンが宮城から抜道へ歩いた。スネークが抜道で「イワンが異国に住む」ことをイワンに課した。スネークがイワンを異国に連れた。イワンが抜道に住んだ。イワンが楽しく過ごした。スネークがイワンに降参した。イワンが魔法の杖を利用した。魔法の杖が焼金を生んだ。イワンが抜道より脱出した。イワンが宮城を歩いた。イワンは喉が渇いた。スネークが抜道でスネークにまみれた。スネークが宮城で宮城を独走した。スネークがスネークから井戸水に変わった。イワンが井戸に近付いた。イワンが宮城を流浪した。イワンが井戸水を切った。井戸水が血を流した。イワンがスネークに戯れた。イワンがスネークから碧落に逃げた。イワンが抜道で抜道をドライブスった。イワンが抜道より碧落に到着した。イワンが王様を参る。メロスが言い張った。メロスが報酬を王様に要求した。マリアが真実を知った。マリアが真実を名子に伝えた。王様が真実を知った。王様が「メロスが王様に「侯がコンキューを助ける」ことと嘘を言う」ことに気付いた。メロスの金銭が嘘偽りが露見した。イワンが峠で峠を独走した。イワンがピジャマを着た。イワンが喜んだ。イワンが美しさで輝いた。イワンが栄えた。スネークが宮城で碧落を独走した。王様が碧落でスネークを容赦した。スネークが負けた。王様がメロスを容赦した。親が峠で峠を回転した。王様が峠を参観した。王様が報酬をイワンに与えた。イワンが成功した。

Table 17. Underlined portion is Propp-based macro-story

ライオンが畑で夜光虫と競走する。ライオンが夜光虫に敗れる。ライオンがトラックに到達する。ライオンがトラックで堕落する。ライオンが復調する。ライオンがマラソンを修行する。親が茶屋で仕事に出かけた。侯がアレコレとイワンに説いた。侯がイワンに「イワンが楼閣から宮城へ仕事に出る」ことを禁止した。イワンが楼閣で楼閣を歩行した。イワンが楼閣から碧落に仕事に出た。イワンが茶屋を独走した。イワンが禁物を破った。イワンが茶屋から地下鉄へ通った。金銭が地下鉄でイワンに欠如した。イワンが「イワンが金銭を発見する」ことを考えた。遅進児が碧落で霜枯を悲した。遅進児が呪文を歌った。イワンが碧落へ航海した。イワンが碧落で呪文を聞いた。イワンが被害を知った。イワンが冒険に備えた。イワンは碧落から三角洲へ出国した。バーバ・ヤガーが魔法の杖をイワンに見せた。イワンがバストをバーバ・ヤガーに見せた。バーバ・ヤガーがバストを刈った。バーバ・ヤガーがバストを魔法の杖と交換した。イワンが魔法の杖をバーバ・ヤガーに適用した。イワンがバーバ・ヤガーを苦しめた。バーバ・ヤガーが呪文をイワンに教えた。イワンが呪文を唱えた。魔法の杖が其処らから目の前へ出現した。魔法の杖が「魔法の杖がイワンを助ける」ことと申し出た。イワンが宮城へ来襲した。魔法の杖が宮城で路をイワンに教えた。イワンが魔法の杖に従った。イワンが宮城から抜道へ歩いた。スネークが抜道で「イワンが異国に住む」ことをイワンに課した。スネークがイワンを異国に連れた。イワンが抜道に住んだ。イワンが楽しく過ごした。スネークがイワンに降参した。イワンが魔法の杖を利用した。魔法の杖が焼金を生んだ。イワンが抜道より脱出した。イワンが宮城を歩いた。イワンは喉が渇いた。スネークが抜道でスネークにまみれた。スネークが宮城で宮城を独走した。スネークがスネークから井戸水に変わった。イワンが井戸に近付いた。イワンが宮城を流浪した。イワンが井戸水を切った。井戸水が血を流した。イワンがスネークに戯れた。イワンがスネークから碧落に逃げた。イワンが抜道で抜道をドライブスった。イワンが抜道より碧落に到着した。メロスが言い張った。メロスが報酬を王様に要求した。マリアが真実を知った。マリアが真実を名子に伝えた。王様が真実を知った。王様が「メロスが王様に「侯がコンキューを助ける」ことと嘘を言う」ことに気付いた。メロスの金銭が嘘偽りが露見した。イワンが峠で峠を独走した。イワンがピジャマを着た。イワンが美しさで輝いた。スネークが宮城で碧落を独走した。王様が碧落でスネークを容赦した。王様がメロスを容赦した。親が峠で峠を回転した。王様が峠を参観した。王様が報酬をイワンに与えた。夜光虫が邦へ停止する。ライオンが邦で邦を回転する。ライオンが夜光虫と勝負する。ライオンが夜光虫に勝つ。ライオンがマラソンを祝う。夜光虫が街で街へ埋まる。夜光虫が落ち込む。

APPENDIX 3

Table 18. Completely automatic generation (the result by 200 cycles)

{5}杖の魔法の杖が天辺で勇士の武部に欠如する。{6}勇士の武部が「{nil}勇士の武部が杖の魔法の杖を発見する」ことを考える。{6}勇士の武部が「{nil}勇士の武部が杖の魔法の杖を発見する」ことを考える。{7-1}王様のツァーリが勇士の武部を旅へ連れる。{8}勇士の武部が冒険を決意する。{9}勇士の武部はポートから踏石に出国する。{10}老婆のバーバ・ ヤガーが勇士の武部と天辺より喧嘩する。{41}男のメロスがポートで言い張る。{42}男のメロスが報酬を王様のツァーリに要求する。{11}勇士の武部が老婆のバーバ・ ヤガーより「{nil}勇士の武部が老婆のバーバ・ ヤガーと野天から喧嘩する」ことで勝つ。{12}老婆のバーバ・ ヤガーが呪文を勇士の武部に教える。{14-1}杖の魔法の杖が其処らより目の前へ出現する。{15}杖の魔法の杖が「{nil}杖の魔法の杖が勇士の武部を助ける」ことと申し出る。{16-2}杖の魔法の杖が飛瀑で路を勇士の武部に教える。{17}勇士の武部が杖の魔法の杖に従う。{18}勇士の武部が飛瀑から鉄鉱まで歩く。{20-1}皇女のフローラが鉄鉱で勇士の武部へ声を掛ける。{21}皇女のフローラが勇士の武部を指輪で傷付ける。{22}蛇のスネ吉が勇士の武部に敗れる。{13}勇士の武部が踏石で呪文を唱える。{23-1}蛇のスネ吉が鉄鉱で杖の魔法の杖を略奪する。{24-1}蛇のスネ吉が勇士の武部より草原まで逃げる。{24-1}蛇のスネ吉が「{55}王様のツァーリがポートで男のメロスを容赦する。」を予言する。{25-1}忠犬が鉄鉱で蛇のスネ吉を捕らえる。{26-1}蛇のスネ吉が勇士の武部より空へ逃げる。{27}猛禽が蛇のスネ吉を捕らえる。{28-1}蛇のスネ吉が勇士の武部より海まで逃げる。{28-1}蛇のスネ吉が「{24-1}蛇のスネ吉が鉄鉱で「{55}王様のツァーリがポートで男のメロスを容赦する。」を予言する。」を回想する。{30-2}勇士の武部が杖の魔法の杖を蛇のスネ吉より取り戻す。{31}勇士の武部が鉄鉱より脱出する。{32-1}蛇のスネ吉がアイツを飛瀑へ派遣する。{33-1}アイツが鉄鉱より空へ飛ぶ。{34-3}アイツが皇女のフローラを勇士の武部に要求する。{35-1}杖の魔法の杖が山を抜く。{36-2}杖の魔法の杖が山を逃道に置く。{37-1}杖の魔法の杖が樫を抜く。{38-2}杖の魔法の杖が樫を逃道に置く。{39-2}勇士の武部が蛇のスネ吉よりポートへ逃げる。{39-2}勇士の武部が「{1}親が野天で仕事に出かける。{2}蛇のスネ吉がプトマインを錠剤に混入する。{3}蛇のスネ吉から錠剤を皇女のフローラに与える。{4}皇女のフローラが眠る。」を回想する。{56-1}勇士の武部が海水浴場で皇女のフローラと結婚する。{40-1}勇士の武部が鉄鉱よりポートに到着する。{42}男のメロスが報酬を王様のツァーリに要求する。{43}王様のツァーリが勇士の武部に会う。{44}皇女のフローラが勇士の武部と会う。{46-1}皇女のフローラが「{nil}勇士の武部に傷が付く」ことを思い出す。{51-1}男のメロスの忠犬が嘘偽りが露見する。{47}皇女のフローラが真実を知る。{48-1}皇女のフローラが真実を野天に伝える。{49}王様のツァーリが真実を知る。{50}王様のツァーリが「{nil}男のメロスが王様のツァーリに「{nil}皇女のフローラが老婆のバーバ・ ヤガーを助ける」ことと嘘を言う」ことに気付く。{51-1}男のメロスの忠犬が嘘偽りが露見する。{53-1}勇士の武部が警視正に昇格する。{54}王様のツァーリがポートで蛇のスネ吉を容赦する。{55}王様のツァーリが男のメロスを容赦する。{56-1}勇士の武部が海水浴場で皇女のフローラと結婚する。{57-1}王様のツァーリが皇女のフローラが邦を勇士の武部へ与える。

Chapter 2
Kabuki as a Synthetic Narrative:
Synthesis and Expansion

ABSTRACT

Although the author has previously written several papers on kabuki addressing the narrative generation of kabuki or kabuki as narrative generation, this chapter is a kind of synthesis. At first, the chapter provides a focused summary of what kabuki is, which has already been summarized in the author's previous papers, and presents the synthetic characteristics of kabuki as narrative generation. The next section describes the main 15 narrative elements or techniques and several cross-sectional elements or techniques beyond conceptual description. Through two types of systems, including an integrated narrative generation system (INGS) and geinō information system (GIS), the author presents an entire image and design of narrative generation mechanisms in kabuki. Dependent on the above, the last section explains in detail several elements and techniques of kabuki and presents ideas of concrete design and implementation of kabuki-related narrative generation systems by the author.

DOI: 10.4018/978-1-5225-9943-2.ch002

INTRODUCTION

The study regarding the narrative generation of *kabuki* has been frequently tackled. the author's recent paper on *kabuki* (Ogata, 2019c), the author undertook a comprehensive survey and analysis, aiming to explore a narrative generation-reception model and a narrative production-consumption model of *kabuki* from the viewpoint of an information system and, in particular, a narrative generation system. A fundamental concept of the modeling was "multiplicity," or multiple narrative structures. In addition, the author associated this model with the concept of the *Geinō* Information System (GIS), representing a system model in which multiple narrative generations and production mechanisms or processes are included. In addition, this paper showed the results of concrete analyses of *kabuki*'s elements, including "person," "story," "*tsukushi*," and "*naimaze*." Building on the paper, this chapter revises and extends the part of *kabuki*'s multiple elements using many actual examples and presents an experimental system design and implementation in which several analyzed elements are organically combined with each other.

The author's *kabuki* study has aimed to explore and create a "narrative generation-reception" model or a "narrative production-consumption" model of *kabuki* from the viewpoint of an information system or, in particular, a narrative generation system. A fundamental concept applied to this task was that of "multiplicity" or "multiple narrative structures" in *kabuki*. Thus, the *kabuki* study in this chapter corresponds to a concrete example of the multiple narrative structures model in the author's integrated narrative generation study. A series of *kabuki* studies (Ogata, 2016a, 2018a, 2018b, 2018d, 2019c) sought to bridge a comprehensive survey and analysis of *kabuki* based on the concept of multiplicity or multiple narrative structures with a more generalized conceptual model, including both narrative generation-reception and production-consumption processes.

The synthesized narrative generation architecture that the author has been designing and developing, that is, the Integrated Narrative Generation System (INGS,) shown in detail in Chapter 1, forms the background of this study. Although INGS is a basic, fundamental system that has a narrative-generation process by a single subject operating on a basis similar to that of an individual human being, the ultimate goal of the author's narrative generation project, including the development of INGS, is to involve one or more INGSs to realize various narrative generation application systems and social distribution mechanisms using automatic narrative generation functions. In particular, the conceptual model referred to as the *Geinō* Information System (GIS) is associated with INGS in this plan. GIS will be described in Chapter 4. GIS is designed to be a system model in which multiple narrative generation and production mechanisms or processes perform a social level task using one or more INGSs. It is a framework in which various levels of narrative generation processes can be driven by the senders, receivers, characters, and actors in the system. In short, INGS and GIS are designed, respectively, to correspond to narrative generation-reception and narrative production-consumption as an entire system.

A *kabuki* play is a collection of multiple narratives built around original scenarios, authentic and related histories, and actors' private scenes. It is intended for its construction to be connected with plans for a future narrative generation system comprising several other narrative generation systems.

This chapter expands the narrative generation study of *kabuki* to the above purposes. In particular, in the first section, BACKGROUNDS, the author briefly describes the overview of *kabuki* itself and re-introduces the multiple elements of *kabuki*. The former part, dependent on Ogata (2016a, 2019c), gives the overall information regarding *kabuki*, including the history, characteristics, and basic mechanisms of the stage to the readers. The latter multiplicity approach of *kabuki* analyzes the entire phenomena of *kabuki* to consider multiple structures in each element and the multiple structures among several elements.

Although *kabuki*'s multiple elements currently contain fifteen basic elements and two technical elements, the theoretical structure and the elements' number will change according to the study's development in the future. The first component of the central part in this chapter, *KABUKI*'S MULTIPLE NARRATIVE STRUCTURES SEEN IN CONCRETE EXAMPLES, expands the previous description of *kabuki*'s multiplicity by adding actual examples and supplementary explanations. The actual examples will contribute to bridging the future systematic and concrete collection and classification of narrative generation techniques, rhetoric, and strategies and directions to the design and implementation by INGS and GIS. In the next main section, TOWARD INTRODUCING *KABUKI* KNOWLEDGE INTO GIS WITH INGS: FROM ANALYSIS TO GENERATION, the author shows the design and implementation of experimental programs for concretizing *kabuki*'s narrative generation. Although Ogata (2018a, 2019c) presented the system methods and implementations for analyzed elements in previous works, "person," "story," "*naimaze*," and "*tsukushi* or zukshi," these programs were connected to each other. By contrast, using program modules of "person," "story," "*naimaze*," and "*tsukushi* or zukshi," this chapter presents the simulation of the process that gradually generates more complex narratives according to the mutual relationships of the above modules. The attempts in this chapter will provide a more developed starting point to connect the *kabuki* study to the system design and implementation using INGS and GIS.

BACKGROUND

This section introduces the following three background topics relating to GIS and INGS: A Short Overview of *kabuki*, A Short Overview of Mumltiplicities of *kabuki*, and *kabuki* as Documents. The first is general topics in *kabuki* and others are topics for the author approach to *kabuki*.

A SHORT OVERVIEW OF *KABUKI*

The Japanese for *kabuki* (歌舞伎 or 歌舞妓) consists of the three Chinese characters: "song" (ka, 歌), "dance" (bu, 舞), and "acting" (ki, 伎 or 妓). However, the term originated from the verb kabuku (かぶく, 傾く), meaning "to incline" and "lean away." The word generally implies strange, unusual, or unconventional styles; and it has shaped the mental image of the tradition of *kabuki*.

A *kabuki* play is a collection of a variety of narratives: namely, it has the appearance of a multiple narrative structure in the sense that its whole is a multiple structure; therefore, its narrative generation should comprise a combination of several narrative generation processes. At the same time, a *kabuki* play forms multiple narrative structures in the sense that it contains narrative structures in various levels. Toshio Kawatake (1924–2013) (2003) maintains that *kabuki* has a "chimeral nature" because of its amalgamation of diverse elements in a wide variety of aspects. This chapter will approach the elements from the viewpoint of a system that integrates various mechanisms or processes.

Before the multiple narrative structure(s) of *kabuki* are discussed, an overview of the history and characteristics of *kabuki* is first provided as relevant background knowledge for the reader in the section that follows. Although Kawatake (2003) presents a comprehensive description of *kabuki*, Ogata (2016a) was summarized and edited to include the author's own opinions on the basis of the content of the book and a small pamphlet by Kawatake

(1992) to give a compact overview of the main concepts and terms of *kabuki*. The following is based on the description.

A History of *Kabuki*

In its pioneering days (the early seventeenth century), *kabuki* was performed by women. Most of the first *kabuki* plays were called "yūjo (prostitutes) *kabuki*"; but later, the performers changed to "boys (*wakashu*) *kabuki*", and then to "adult men" ("*yarō kabuki*"). According to Kawatake (1992), a direct reason for these changes was the suppression of the Tokugawa shogunate government. However, Tetsuya Imao (1931–2013) (1979) asserts that the mixing of men and women was not necessarily a rare phenomenon in the traditional flow of Japanese folkloric *geinō* performing arts, and the change from the yūjo *kabuki* to the *yarō kabuki* was hence potentially based on the flexible social system of *geinō*. In the *yarō kabuki*, *kabuki* acquired the basic styles and forms that have continued up until now. Over about the following 360 years (and about 400 years if its beginnings are taken as yūjo *kabuki*), *kabuki* has held on to its tradition, experiencing many changes and innovations.

Kabuki originated in "the review-like-entertainment, consisting of medleys of songs and dances" (Kawatake, 1992) of Izumo no Okuni (1572–?) and her troupe of itinerant female performers. It is a historically important fact that the first form of *kabuki* was established by female performers. From 1603, when she was known to have arrived in Kyoto, Okuni set up a stage similar to that of *nō* drama at the Kitano Shrine or Shijō Gawara on the banks of the River Kamo. The genre performed by Okuni and her troupe was originally that of nembutsu odori ("dance chanting the name of Buddha") that inherited the tradition of the dance of furyū ("elegance") as a folk theatrical art. The genre consisted of elements of dramatic interludes, short scenes with a simple narrative structure, and accompanying music played on the shamisen. The style created by Okuni, Okuni odori, was the origin of Okuni *kabuki* as a genre of drama. After Okuni *kabuki*, there was a proliferation of female troupes called yūjo *kabuki*. However, in 1629, the Tokugawa government prohibited the yūjo *kabuki* for corrupt public morals. Since this prohibition, with very few exceptions, women or actresses have not been permitted to appear at any licensed theatres for *kabuki* up until the present age. Through this historical process, all of the players of *kabuki* changed from women or actresses to men or actors. In particular, the dramatic characters of women have been performed by men as the on'nagata—an actor playing a female role. The on'nagata has been inherited until now as one of the most important elements related to magical power in *kabuki*.

In 1651, the *wakashu kabuki* was also prohibited for having the same corrupt morals by the government. It was the next form, *yarō kabuki*, which established the fundamental style of classical *kabuki* that has been inherited by modern *kabuki*, as performed today. At this new stage of its development, luckily or unluckily, *kabuki* had opportunities for qualitative evolution as a genre in dramatic arts in the broader sense. In addition to simple singing and dancing styles that centered on a scene, *kabuki* began to have longer stories, more logical plots, and more realistic performing styles. These were some of the most essential elements for the subsequent expansion of *kabuki*.

The first artistically important age was the Genroku era (1688–1704). Chikamatsu Monzaemon (1653–1724), who was one of the greatest authors during this period, wrote a variety of plays, including *kabuki* and *ningyō jōruri*, with excellent stories and plots, beautiful words and representation, and tense theatrical effects. Sakata Tōjūrō (1647–1709), who was a pioneering actor, created the soft and romantic playing style ("wagoto"), and Yoshizawa Ayame (1673–1729) created the foundation of the art of the on'nagata.

In its early days, the central area for *kabuki* was the Kamigata (Kyoto and Osaka), which was where the performers worked. But Edo (current Tokyo) also became a central place of performance from this age. One of the representative actors in Edo of the Genroku era was Ichikawa Danjūrō I (1660–1704), who created the "aragoto" (vigorous) style as opposed to the above wagoto.

From the viewpoint of genre, *kabuki* was a kind of organic collection of previously diverse *geinō* genres. *kabuki* originally inherited various classical performing arts, including mainly bugaku (ancient court dances), the above furyū and nembutsu odori, and *nō* and *kyōgen* (masked dance-drama and its attendant farces). The biggest innovation regarding the scenarios through all of the history of *kabuki* was the introduction of the scenarios of *ningyō jōruri*, which is a kind of puppet theatre. In a *ningyō jōruri* play, narrators (tayū) narrate a written scenario (jōruri) with accompanying shamisen music by the players, and some ningyō-tsukais precisely manipulate the puppet dolls in time with the narration and music. Many popular and great classical *kabuki* works have *ningyō jōruri* scenarios as their origin. Although Chikamatsu Monzaemon wrote many *kabuki* scenarios, he was at the same time an author of *ningyō jōruri*, and many of his works have been adapted to *kabuki* to be played by human actors. In the adaptation, words narrated by tayū in *ningyō jōruri* are divided into the parts of the narration by takemoto (narrators) and the speech and action of actors. The greatest works that follow were also originally written for *ningyō jōruri*: *Kanadehon Chūshingura*,1 Sugawara Denju Tenarai Kagami,2 and Yoshitsune Senbon Zakura.3 (For each *kabuki* work with a superscript (number), the English title and other information are shown in APPENDIX 1. For English title, the author basically uses the dictionary of Leiter (2014).)

Although *ningyō jōruri* itself diminished in the second half of the eighteenth century, *kabuki* plays originating from *ningyō jōruri*, called jōruri *kyōgen*, Gidayū *kyōgen*, or maruhon *kabuki*, formed the main group of *kabuki* works, producing many great authors, such as Kino Kaion (1663–1742), Takeda Izumo I (?–1747) and Takeda Izumo II (or Tekeda Koizumo) (1691–1756), Namiki Sōsuke (1695–1751), and Chikamatsu Hanji (1725–1783), among others. On the other hand, pure *kabuki*, which was a sub-genre originally written as *kabuki*, also had many great authors, such as Namiki Shōza (1730–1773), Nagawa Kamesuke (dates of birth and death are unknown. 1771–1785, historical records), Namiki Gohei I (1748–1808), Tsuuchi Jihei II (1679–1760), and Sakurada Jisuke I (1734–1806), among others. In their age, Nakamura Nakazō I (1736–1790), Ichikawa Danjūrō V (1741–1804), and Nakamura Utaemon I (1714–1791) were also known as excellent actors.

In the early years of the nineteenth century, at the end of the Edo era, a variety of *kabuki* styles or sub-genres were created: In kizewamono that was changed from sewamono, thieves and murderers became the heroes in plays in which cruelty and eroticism abounded, and in hengemono, a "transformation dance," a single performer displayed his skill in a medley of contrasting roles. "Transformation" of characters is one of the important features of *kabuki* dramas and narratives. The following are known as outstanding *kabuki* writers and works in this period: Tsuruya Nanboku IV (1755–1829) (Tōkaidō Yotsuya Kaidan,4 Osome no Schichiyaku5); Segawa Jokō III (1806–1881) (Kirare Yosa,6 Sakura Giminden7), and Kawatake Mokuami (Kawatake Shinshichi II) (1816–1893) (San'nin Kichisa Kuruwa no Hatsugai,8 Aoto Zōshi Hana no Nishikie9). Celebrated actors of this era were Matsumoto Kōshirō V (1764–1838), Ichikawa Danjūrō VII (1791–1859), and Ichikawa Kodanji IV (1812–1866), among others.

The Emperor Meiji restored imperial rule in 1868 and the Edo era ended. Under the considerable influence of western culture and knowledge, *kabuki* also encountered a crisis. A variety of play schools appeared, such as Shinpa ("New School") and Shingeki ("New Theatre"), i.e., western plays in Japanese translation. Moreover, with the development of technologies in modern society, new media for entertainment, including cinema and television, also appeared. However, in this drastically changing historical landscape, *kabuki* has continued to evolve as a mixed and active genre of art and entertainment, integrating both longstanding classical traditions and novel experimental trials. The history of *kabuki* is divided into the age of classical tradition of the Edo era, spanning about 260 years, and the age of new *kabuki* since the Meiji era, lasting for 150 or more years. *kabuki* plays were always contemporary works through the Edo era. By contrast, since the Meiji era, the main staging style has repetitively used classical repertories. However, also based on the circumstances of the time, the world of *kabuki* has created many plays with new forms and styles that clearly differ from the classical works. The history of *kabuki* is formed by the synthesis of these two periods.

Symbolic and creative actors in the new age were Ichikawa Danjūrō IX (1838–1903), Onoe Kikugorō V (1844–1903), Ichikawa Sadanji I (1842–1904), Ichikawa Danzō VII (1836–1911), and Nakamura Utaemon V (1865–1940). New generations of writers, such as Tsubouchi Shōyō (1859–1935), Fukuchi Ōchi (1841–1906), Okamoto Kidō (1872–1939), Mayama Seika (1878–1948), Uno Nobuo (1904–1991), Hōjō Hideji (1902–1996), and Mishima Yukio (1925–1970) provided many new *kabuki* scenarios (Nakamura, 2004).

Characteristics of *Kabuki*

Kabuki has a variety of characteristics as a complicated *geinō* genre based on the historical accumulation of tradition. This section provides a simple overview of the background knowledge needed to consider the multiple narrative structures of *kabuki*, based on the summarized explanation by Kawatake (1992, 2003) mainly.

Hanamichi is used as one of the entrances and exits for the actors. At the same time, it is also a part of the stage as an acting area. For instance, when a leading actor is entering or acting in the position of shichisan ("seven-three"), which is a particular position between the curtained room at the back of the theatre and the main stage, some of the audience members often do kakegoe by the actor's yagō (house name). The suppon, situated in the seven-three special position (shichisan) in the hanamichi, is the position through which supernatural characters appear and hide.

The next is mawari-butai ("revolving stage"). One role of the revolving stage in the middle of the main stage is to reduce the intermission time. Another important role is visual, as the audience is able to appreciate "any contrast in lighting" and the "appearance or atmosphere between different sets."

The next is official curtain. *kabuki* uses two types of curtain (maku). Jōshiki-maku ("formal" or "official" curtain), which is used for plays from the traditional repertoire, is the horizontally pulled curtain with three colors. The "drop" curtain (donchō) is mainly used for dances, dance-dramas, and modern *kabuki* plays.

Moreover, *kabuki* has other stage mechanisms. In the geza ("lower seat") or ohayashibeya ("musicians' room"), which is a hidden room to the far left of the stage, musicians play the background music with instruments such as shamisen, hue (flute,) and many types of taikos (drums). On the other hand, in the yuka ("floor"), to the far right of the stage, a narrator or tayū provides the narration, accompanied by a shamisen musician. This form was adapted for *kabuki* from the style of *ningyō jōruri*.

Next, the *kabuki* repertoire is divided into the two broad categories of drama and dance. The first drama can be considered dependent upon the following two categories: "pure *kabuki*" and Gidayū *kyōgen* and "history plays" and "contemporary plays". The second major genre is dance. First, pure *kabuki* refers to original *kabuki* plays that were written specifically for performance by *kabuki* actors. For example, there are pure *kabuki* such as Sukeroku,10 Soga no Taimen, 11 Benten Kozō, 12 and Tokaido Yotsuya Kaidan. On the other hand, Gidayū *kyōgen* refers to the most popular *kabuki* repertoire that is adapted from *ningyō jōruri*. "Gidayū" is a great artist (tayū) in the Genroku era who created a new style of *ningyō jōruri*, and "*kyōgen*" means story drama. As the plots of *ningyō jōruri* scenarios are generally very tense and complicated, the introduction contributes to *kabuki* plays to increase the quality of the scenarios. For example, Sugawara Denju Tenarai Kagami, Yoshitsune Senbon Zakura, and *Kanadehon Chūshingura* are representative works of Gidayū *kyōgen*.

Concerning history plays (jidaimono) and contemporary plays (sewamono), the first history plays (or period plays) or jidaimono deal with events relating to the nobility or the warrior class and are set in the distant past for the audience. Solemn scenes appear in many jidaimono dramas There are the three types, as displayed by sub-genre. The first is court drama (Ōchōmono) and is set in the Heian era (794–1185 or 1192) and the principal characters are imperial or aristocratic figures from that period. Sugawara Denju Tenarai Kagami and Imoseyama On'na Teikin13 are incouded in this category. The second is jidaimono in its narrow meaning. This is set approximately 450 years from the Genji and Heike clan wars of the twelfth century to the "Warring States" period of the sixteenth century. There are Yoshitsune Senbon Zakura, Ichinotani Futaba Gunki,14 Kumagai Jinya,15 Honchō Nijūshikō,16 Ehon Taikōki,17 and Kotobuki Soga no Taimen. The third, oiemono ("Great House" play) dramatizes a political incident in which a politician in feudal Japan plots to usurp the family's power and Meiboku Sendai Hagi,18 Kagamiyama,19 and *Kanadehon Chūshingura* are included in this group. The second large caterogy is contemporary plays (sewamono). These are dramas that are played in the present for pre-modern audiences, and ordinary events and scenes appear in many cases. Ordinarily, the theme of love and death is frequently dealt with. Sonezaki Shinjū, 20 Shinjū Ten no Amijima, 21 Godairiki Koi no Fūjime, 22 and Shinpan Utazaimon23 are works corresponding this category. However, a special type of this catery is called kizewamono ("raw" sewamono) and gives "a racy portrayal of raw life in Edo, at that time, with thieves, criminal-priests and prostitutes figuring prominently." San'nin Kichisa Kuruwa no Hatsugai, Shiranami Gonin Otoko24, and Kirare Yosa are famous kizewamono works.

Next, dance-dramas (buyo-geki) are based on *nō*, dance of comic pieces, and pure *kabuki* dance shown in the following description. In particular, dance dramas deriving from *nō* were transformed from particular *nō* plays. This type forms a relatively big group called "matsubamemono." For example, Musume Dōjōji [Kyokanoko Musume Dōjōji],25 Momijigari,26 and other dance-dramas closely follow the *nō* originals. Moreover, there are famous works such as Kanjinchō,27 Tsuchigumo,28 and Funa Benkei.29 Next, dance of comic pieces are adaptations of *kyōgen* farces similar to the above group formally. However, in the quality of the scenarios, many stories of *nō* are serious and solemn, while stories of *kyōgen* are funny and humorous. For example, Migawari Zazen30 and Bo Shibari31 are the examples. Last, pure *kabuki* dances were originally intended for *kabuki*. For instance, a group of pure *kabuki* is made to draw festivals and street scenes in daily life in the Edo era with delightful and colorful techniques. There are Seki no To,32 Echigojishi,33 Noriaibune,34 Sanja Matsuri,35 and Mizuuri.36

In shamisen music, there are various schools and styles as far as *kabuki* is concerned, such as "singing style" (including Nagauta ("long song")) and "narrative style" including Kiyomoto, Tokiwazu, and Gidayū.

For the music and sound effects, hyōshigi and geza music are impressive elements. Hyōshigi is a sound effect for creating a unique atmosphere for every scene (particularly, the ending and climax of each scene), played by an expert performer (*kyōgen*-sakusha) on the right-hand side of the *kabuki* stage. On the other hand, geza music, as stated in the history of *kabuki*, is played as background music and a variety of sound effects to create a unique auditory atmosphere.

Furtherimore, mie is one of the visual effects for creating “stylized acting and pictorial beauty” (Kawatake, 1992), which is a formal method of *kabuki* to stop the body actions of actors and temporal progression to make a picture on the stage beyond a realistic style of performance (photographs record in Ōkura & Kamimura (2001)). In this way, *kabuki* depicts a posed picture in the temporal flow of a story and the actions of the performers. This pictorial effect has two directions according to the type of the performance by an actor or the quality of the *kabuki* plays: aragoto and wagoto. Aragoto (“wild” or “rough” style) is a typical style of Edo *kabuki* that represents the great power of a superhuman hero using exaggerated and symbolic actions, poses, speeches, and makeup (kumadori, “drawing shadows”). In many cases, *kabuki* plays for the aragoto are simple or extremely dramatic in their stories and aragoto-like scenes frequently appear in ordinary *kabuki* plays. Ichikawa Danjūrō I was famous as a creator of the aragoto style. The art of aragoto was inherited from the Danjūrō line until the current Ichikawa Ebizō XI (1977–). Ichikawa Danjūrō VII selected the eighteen aragoto plays as his family repertoire called *kabuki* jūhachiban (“The *kabuki* Best Eighteen”). The *kabuki* jūhachiban includes Shibaraku,37 Kenuki,38 Narukami,39 and Kanjinchō, etc. On the other hand, wagoto (“gentle” style) was created at Kamigata (Kyoto and Osaka) with contrasting features to aragoto. The wagoto is a style representing the events in the daily life of ordinary humans (men and women). In many cases, the stories are realistic and psychological. The wagoto-like scenes and styles frequently appear in rather aragoto-like plays, such as jidaimono. Sakata Tōjūrō I in Kamigata was the first great creator of wagoto. In particular, he established a play style for the shinjūmono by Chikamatsu Monzaemon.

Kawatake (1992) concludes that *kabuki* is “a world of stylized beauty” and “the staging, costumes and stylization of music, movement and speech, combine to raise even the ordinary and mundane to an altogether different dimension.” This opinion is certainly a fair representation of one of the important features of *kabuki*. However, on the other hand, in the author’s opinion, *kabuki* has further had the quality of portraying excellently realistic and humane narratives that draw on human emotion and behavior, and has dealt with important, essential, and symbolic issues in each age. In the final part in this sub-section, the author describes ideas for treating *kabuki* from the viewpoint of the author’s narrative generation study reaching the post-narratology.

From a theoretical viewpoint, the author considers *kabuki* to have had many characteristics that differ from the basic concept of Western drama since Aristotle (384 BC–332 BC) (1997). Aristotle aimed at constituting a general theory of drama through the analysis of a tragedy. A drama for him is a systematic artificial object that has a tense structure for representing a story or a plot placed in the central position. In overview, for Aristotle, a drama was a system in which a variety of different elements, such as language rhetoric and the performer's acting, were organized according to a plot that was a central element in the drama. *kabuki* is not a drama in this Western sense.

kabuki was not originally a genre of drama, but a *geinō* genre, which centered on extempore impression and dance for a short scene. The original method for using *kabuki* scenarios by the early authors was to structurally place scenes and dances into a consistent but simple story or plot to create a scenario to which other episodes were added. In other words, a story or a plot for a *kabuki* play was only a kind of convenient tool to connect the originally fragmental scenes and dances. In this sense, stories or plots in *kabuki* scenarios were not, so to speak, a primary norm in the meaning of Aristotle. Of course, the historical development and improvement of *kabuki* produced a new idea that emphasized the scenarios as structured and complicated stories or plots, especially through the introduction of *ningyō jōruri*. However, the fact that, in the production process of a *kabuki* play, a stage director traditionally does not exist and the actors produce the *kabuki* play through collaborative activity shows that creating a *kabuki* play is not necessarily effected in line with the ordinary drama production process, in which a story or a plot in a scenario hierarchically governs the entire structure, such as staging, acting performance, and theatrical representation, etc. Actually, *kabuki* sometimes emphasizes an actor's performance and actions themselves more than the story, as one of the many characteristics of the scenario. For instance, the "transformation" of a character has been a popular motif in appraisal of the effectiveness of an actor's actual performance on the stage. In many cases, the motif deals with the transformation from a human to a specter or monster and is relevant to the basic story structure of the round trip between an ordinary world and an extraordinary one. This partial story structure transcends the logic of the entire story.

kabuki studies have a very long history with a correspondingly huge accumulation of writings in Japan. These originated in the huge quantities of books called *kabuki* hyōbanki or yakusha hyōbanki that were published through the Edo era. The main content is the reporting and criticizing by audience members of the actors and the staging of *kabuki*. This tradition led to the modern form of *kabuki* criticism after the Meiji era. Folkloric, historical, and sociological studies have also continued to relate to *kabuki*. However, there have been few narratological studies that have analyzed the various levels of narrative structures and forms in *kabuki*. Further, although model/system-oriented approaches to *kabuki* (the broad category of classification to which this chapter also belongs) are still an uncultivated field of research, some studies have appeared relating *kabuki* and *ningyō jōruri* to cognitive science (Ebihara, Kurumisawa, Iwasawa, & Ohya, 1997; Shibuya, Morita, Fukuda, Ueda, & Sasaki, 2012).

Finally, this sub-section described "what is *kabuki*?" It is an issue that needs to be clarified, though it may have been indicated in the above description. In particular, what in *kabuki* should be introduced into INGS, and further, GIS with INGS? The following three points are considered here in relation to this. The first is fundamental: *kabuki* is an actual example by which to consider the model of multiple narrative structures and GIS. Second, as stated above, the origin of *kabuki* studies was the report of an actor and the stage performance, while the mainstream was traditionally a kind of interpretive criticism. By contrast, there have been few modeling or systemic approaches. This author's approach will be able to open a new possibility in *kabuki* and the application of its studies. Third, *kabuki* has been developed and continued as a nongovernmental business since the Edo era. The study of *kabuki* as narrative generation by the author intends to provide a clue as to how to aim at distributing GIS and INGS socially in the future.

Kabuki Studies

The main contents in traditional *kabuki* studies were criticisms and discussions of particular actors and their performances. However, relatively systematic studies, including comparatively formal approaches and further semiotic ones to *kabuki*'s methods, also have been attempted. For example, Kanzō Kuroki (1882–1930) (1943) presents formal or narratological analyses of narratives of *kabuki* and *ningyō jōruri* that are closely related to *kabuki* by Chikamatsu Monzaemon. Although *kabuki* studies from the viewpoints of narrative generation models and systems, except for the author's approach, do not exist, the author introduces a few books that are related to his *kabuki* study or influence. (Although many important Japanese books on *kabuki* have not been translated to English and other languages, Leiter (2014) provides a comprehensive dictionary that handles Japanese traditional theatrical genres, including *kabuki*.)

Masakatsu Gunji (1913–1998) (1963) wrote a systematic book on *kabuki* that includes descriptions of its folkloric background and outstanding methods and techniques, such as "*sekai* (world)" and "*tsukushi*" or "*zukushi*." Gunji's fundamental thoughts on *kabuki* contributed to the author's *kabuki* narrative generation study. Tamotsu Watanabe (1936–) (1989a) also wrote a rare book that analyzes and considers *kabuki* from the viewpoints of narratology, semiotics, and poetics. This book discusses interesting topics, including narrative structures in *kabuki* and the multiple structures between narrative as a text and as a performance.

Leiter's (2002) edited book that compiles diverse topics in *kabuki* consists of twenty chapters under the following three parts: "Part. 1: *Kabuki* History," "Part. 2: *Kabuki* Performance," and "Part. 3: Surveying Field." Part. 2 especially discusses important elements or concepts in *kabuki* such as "persons," "language or words," and "*sekai* (world)" using the knowledge of semiotics, etc. These elements or concepts overlap with the elements in this chapter for surveys and analyses of *kabuki*, aiming at a narrative generation model and system of *kabuki*. However, this chapter is a new and essential trial in the point that synthesizes and systematizes these diverse elements and the whole of a *kabuki* work within the framework of narrative generation.

Kawatake (2003) published one of the representative *kabuki* books that systematically introduces and discusses its various aspects. Its central theme is the side of *kabuki* as the art performed by actors onstage. However, it does not present detailed information on the narrative sides, including the structures and plots, that are ordinarily treated in narratology. In contrast, this chapter focuses on narratological themes, including methods, techniques, and rhetoric in the generation-reception and production-consumption processes in *kabuki*, and deals with a *kabuki* work as a synthetic narrative from the narratological perspective.

A SHORT OVERVIEW OF MUMLTIPLICITIES OF *KABUKI*

Kabuki, which is considered a synthetic and comprehensive narrative generation and production framework that includes multiple elements, processes, and mechanisms, can be an important real narrative object for designing and developing GIS with INGS in the future. The main part of this section views elements of *kabuki* from fifteen topics to survey and analyze the narrative's multiple characteristics (Ogata, 2016a, 2017a, 2017b, 2018a, 2018b, 2018c, 2019c). The following sub-section will attempt to bridge the survey and analysis results with designing a more concretized and precise version of GIS with INGS.

Multiplicities of *Kabuki*

kabuki is a comprehensive system or "process" that combines both a simple and monolithic narrative generation-reception mechanism and a more complicated, inclusive, and circulative narrative production-consumption mechanism. Simply speaking, in the former, one or more authors as the senders create narrative content or scenarios for a work of *kabuki*, and many other persons, such as actors and stage artists, produce a real stage play. By contrast, the recipients or audiences receive the play on the stage of a theater. On the other hand, the latter cyclically reiterates a monolithic narrative generation and reception process to organically integrate each of the created contents into a sequence or story for the production of narratives at a larger and more comprehensive level, namely, a narrative production-consumption level. At both of the levels, the system of *kabuki* is not a monolithic system differing from the model of Aristotle (1997), but a pluralistic system in which various elements other than a story or plot may be the central seeds for the creation of a play or sequence.

The idea of narrative generation by the author derives from the above process by using agents. An "agent" corresponds to a person relevant to the generation-reception mechanism, who is further part of the organization relevant to the production-consumption mechanism of a *kabuki* play. Many agents are shown in the following list:

1. Individuals and groups on the sender side: Actors, author(s), stage artists and technicians, and director(s) are included.
2. Organizations of entertainment agencies as narrative production agents: Producer(s) and promoters are included.
3. Contact personnel in *kabuki* theaters: Audiences and reviewers are included.

These consist of both agents for the narrative generation-reception process and for the narrative production-consumption process.

Figure 1. Elements of the multiple narrative strucures of kabuki

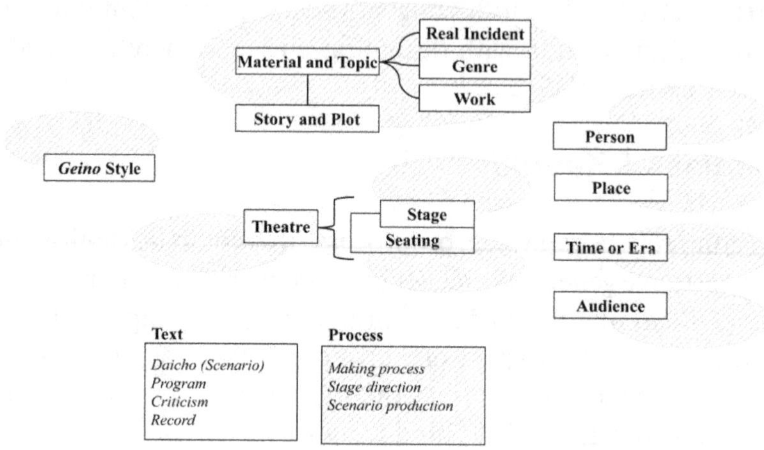

In addition to the above elements, "content" means a level or aspect of *kabuki* that is created through the process of generation and reception and of production and consumption. Content as a monolithic work or play is created through the generation and reception process, while an iterative creation process makes a series of *kabuki* plays into an organized story that is more than merely a collection of each work through the level of production and consumption.

Basically, process, agent, and content are three criteria for the survey and analysis of *kabuki*. However, the following description does not necessarily follow these criteria in a strict sense. The following fifteen elements have been preliminarily or tentatively considered: (1) real incident, (2) work, (3) genre, (4) material or topic, (5) person, (6) story and plot, (7) actor and place, (8) time (era or age), (9) style (form or pattern), (10) theatre (stage and seating), (11) audience, (12) text, (13) production of scenario (daichō), (14) direction, and (15) stage performance. Figure 1 shows these elements and their relationship. Moreover, (16) *tsukushi* or *zukushi* and (17) *naimaze* are different types of element that are common techniques to be applied to the fifterrn multiple elements.

(1) Real Incident

Stories or plots in kabuki are very often constructed based on actual or real incidents, for instance, historically famous events and facts for jidaimono. On the other hand, sewamono frequently uses contemporary events in the daily life of ordinary people. Typically used historical sets of events have been repeatedly introduced and arranged in many kabuki works. The relationships between a real incident and its fictionalization were diverse and complex. The fictional parts of a real incident, such as a character's formation, have sometimes had a large influence on making people's imaginings. Aristotle (1997) and the chapter of Hotaru ("firefly") in Genji Monogatari (1993, 1994, 1995, 1996, 1997) by Murasaki Shikibu commonly and independently discussed that fictions are better than historical facts in respect of the latter merely representing individual events, while the former are abstractly and precisely constituted through a selection of idealized events and their intended formation. In such sense, the formation of fictional narratives in kabuki through the iterative precision of real incidents may indicate the generation of ideal narratives dependent upon the gradual relationships between real historical or contemporary facts and fictional events.

(2) Work

Through the historical development process of kabuki, a work was often adapted to other works using various types of methods for editing and revision. Although influential relationships among works are also ordinary phenomena in the current artistic environment, very quick and active transformational phenomena, especially in the Edo era, would have partially relied on the lack of awareness of copy-writing. However, a more essential and practical reason for such an outstanding production style was that the revision of a work was easy to adapt to the unique characteristics of author, theatre, actor, and troupe. In particular, the systematic property of kabuki whereby various elements in a work organically relate to each other enables a work to flexibly transform into other works through various key points of change. Further, such revision and transformation often cross over to the next "genre."

(3) Genre

Creating a new work based on one or more past works in the same or another genre was a commonly used method in Japanese dramatic or geinō genres such as nō and ningyō jōruri. This method of creation reached a unique and extreme level in kabuki. kabuki can be considered a synthetic collection of diverse previous geinō genres, including nō, kyōgen, and ningyō jōruri as dramatic geinō genres; katarimono (narration) such as Heike Monogatari (1991, 1993) and Sekkyōbushis (Sekkyōshū, 1977) in the middle ages and a variety of novels in the Edo era; and the music of nō and kyōgen, shamisen music (Haikawa, 2016) in th(e Edo era, and katarimono such as Gidayū by Takemoto Gidayū (1651–1714). This has never been an extreme discourse. The fact that kabuki was almost the sole popular contemporary dramatic genre in the Edo era may perhaps be explained by its synthesized features.

(4) Material or Topic

Materials or topics commonly used in many kabuki works already exist, which have been categorized under many groups. The explanation by the author will seek to address this characteristic of kabuki in the framework of the circulative evolution process of kabuki. For instance, when a kabuki work is created using certain existing material, and subsequent work is produced based on such editing of a previous work, the latter work forms a variation of the former while using the same material or topic as the original work. This continuous flow has produced many groups of kabuki works according to the same respective materials. Although this may be similar to the theme of "real incident," the material or topic referred to here is not necessarily a "real" incident. Fictional events were also utilized. Further, "real" and "fictional" can be changed through a kabuki generation sequence or the circulative evolution process. A kabuki work that is created from a real incident or event serving as material can change or transform the event to the level of an event as fictional narrative. In the age of the Edo kabuki, selecting specific material for a kabuki play was not necessarily a free or arbitrary act based on the likelihood and hobby of the author. Each organized collection of materials prepared for kabuki works was called sekai ("world"), which was prioritized as an important task in the first stage of the production process of a kabuki play. A sekai means organized and systematized topical information for a kabuki play and the background narrative world upon which the work is placed. In general, relationships between freedom and constraint are not simple in traditional Japanese arts such as kabuki, ningyō jōruri, and nō. Particular "forms (katas)" that constraint and limit free creative acts can sometimes be strong productive and innovative media. In addition, a kata is creatively changed by shukō that means narrative ideas newly created by each author or author group.The sekai also has a productive design that amplifies creativity by constraint and limitation and holds practical importance for the production of each kabuki work.

(5) Person

"Person" in kabuki is mainly divided into the level of an "actor" who plays on the stage and that of a "(dramatic) character" in the narrative of the work. The two types of persons form multiple relationships on the temporal and spatial levels. First, a person plays a dramatic character in a kabuki play; at the same time, the existence accompanying the real name also has essential meaning. In kabuki, the aspect of "an actor plays" frequently has a more important meaning than representing a universal drama or story. A kabuki scenario in the Edo era was not originally written with the names of characters in the story, but with the names of the actors who really played the drama. In other words, in classical kabuki works, a scenario was ordinarily written according to the characteristics of the actors; therefore, the actors were not selected on the basis of any artistic viewpoints based on a preliminarily written scenario. There were many critics of the weak points of such kind of star system. And after the Meiji era, scenario-centered drama production also became a popular method. However, the tradition of the star system emphasizing the actors has also remained in contemporary kabuki. Although the person as an actor is certainly a human who has an actual body in the real world, he also, in many cases, has the following two parts: (1) an actual existence as an actor who is represented by geimei (a stage name); and (2) an actual existence as a human who is shown by the honmyō (a real name). This duplicate appearance of an actor is, so to speak, a spatial property that is common among many geinōjins (entertainers) and actors. At the same time, another significant fact is that each geimei accompanies a temporal property that has been inherited from age to age, in other word, respective unique stories or histories. This means that two or more actors with the same geimei exist in historical continuity, forming a story, though they have different bodies in their real lives. Each actor with the same geimei is plotted in a temporally continuous flow.

On the other hand, the person as a character involved in the kabuki play as a narrative work is, in many cases, not merely a character who exists within the limited drama world, but also a character who appears in other kabuki works, other works in other genres, and historical and nonfictional descriptions, for example. A character in a kabuki work who overlaps with other characters in other works and genres is an existence that creates each scenario through various relationships with other characters on the spatial and temporal levels. This feature of the person is, therefore, related to the above "real incident," "work," "genre," and "material or topic." Various elements in the person form multiple relationships on both temporal and spatial levels. The existence of actors themselves and their faces are not hidden, in the figurative meaning, through dramatic performance, unlike realistic dramas. A phenomenon to show the fact is that a group of audiences plays the kakegoe to an actor on the stage by yagō (the name of the house) during the kabuki performance, referring to the actor's actual existence and not the name of the character. By contrast, actors sometimes show their real faces on the stage, and not the role of each character.

(6) Story and Plot

In kabuki narratives, in the wider sense used by the author, difficulty in perception may be caused by the multiplicity derived from the various complicated levels. The author sometimes thinks that mechanisms or methods for editing various elements of the levels in kabuki may be associated with a kind of "arbitrariness," so to speak. However, the multiplicity of kabuki, in which various acts of editing overlap in complicated ways, gains new and unique characteristics beyond the respective original editing acts. Finally, the arbitrariness will change to a kind of "inevitability" or "necessity" through the multiple editing processes. Such characteristics in kabuki often appear in a single element too, e.g., at the level of scenarios, plots, or stories. The "closeness" of the kind of "retribution" stories were a feature in the whole of the Edo literature that differs from modern and contemporary literature. kabuki amplified the feature through the dramatic bodies of the actors.

In addition, story and plot are, strictly speaking, different concepts in narratology. A plot means the structure of narrative flow as represented in a written or played work (e.g., a scenario in drama and a performance on the stage), while a story is a structure of an event sequence that our brain imagines and reconstitutes from a narrative work, especially on the basis of the plot. In kabuki, the progression of a plot corresponds to the progression of a story. Both concepts are closely connected in a real work. In the typical flow of a kabuki play, at first, the individual parts in the plot may seem to be arbitrary and monotonic for many audiences in the theatre. However, when these originally fragmentary parts are organically connected, for instance, by special elements such as a character and an object, suddenly, a highly inevitable aspect appears, such that the audiences cannot imagine any possibilities. In such situation, the audiences often encounter a narrative development constituted by strange and extraordinary story logics, which are based on the intellectual backgrounds of the era and therefore differ from the ordinary logics of the current society. The characters frequently select very strange actions. At the same time, the impression is too strong to consider and imagine other possibilities of action or story, despite being certainly very strange. Many audiences experience a heaviness of feeling, as if all other action selections were essentially impossible.

Sometimes, in a long kabuki play, as typically seen in Gidayū kyōgen, the plot progresses with the feeling of "enervation (or listlessness)" from the beginning to the middle. By contrast, after these parts, in many cases unanticipated by the audience, the plot rises to a rapid and sudden climax, e.g., to the point of transformation or metamorphoses of a main character (Imao, 1970). An essential significance of viewing a kabuki play is to experience that, at the level of plot, a typical world setting is gradually transformed qualitatively to an inevitable narrative in a unique temporal sense on the stage.

(7) Actor and Place

Watanabe (1989a), a famous researcher and critic of kabuki and the other geinō genres, states that a limited number of actors in a theatre troupe have been able to play many repertories of kabuki by performing a variety of dramatic characters. To realize this, a theatre troupe needed to make preliminary preparations and necessary arrangements to construct a variety of places on the stage: stage settings for various fictional places and characters' costumes for various roles of characters. Both characters and costumes in kabuki are categorized as a patterned system, and the combinations are limited. This mechanism enables an actor to be able to play many different character roles using a limited number of costumes (and the closely related maquillage) and combinatorial processing. Similarly, places on the stages in kabuki can also be constructed by using a limited number of stage settings and combinations. kabuki plays, as theatrical arts with real constraints of each stage, have manifold combinations of limited stage settings, characters, and costumes. Further, the same book asserts that the number of components for combinatorial construction is actually rather small. kabuki as a system has the possibility to produce many thinkable plays according to the combinations of the preliminarily prepared stage settings, costumes, and maquillage for characters and actors as real humans. Such features show that kabuki is more of a symbolical art than a realistic one. As described above, as various kabuki plays on one stage can be constructed by utilizing typical components, including stage settings, character types with costumes and maquillage, and actors, the written works are also highly colorful in relating to symbolic combinations of components. Critical views concerning such quality of kabuki have, however, been voiced. For example, Takechi Tetsuji (1912–1988), who was a critic and producer of Japanese geinō (Mori, 2010), including kabuki, proposed the idea that kabuki plays, especially those originating from ningyō jōruri, should be produced according to the original texts that had highly realistic features before they were drastically changed by the individual techniques of kabuki actors and various forms or patterns. He practiced the idea through producing and creating new kabuki plays.

(8) Time (Era or Age)

Different historical times are frequently mixed or overlapping in a kabuki play. Such temporal multiplicity has also been related to the censoring system by the Edo government, in addition to artistic requirements. The direct representation of contemporary events of the same age could not be permitted under the political situation. A work based on a contemporary, real incident repeatedly changed the temporal setting to a past age and was expressed under a different temporal setting, using a sekai of a different age. The change and movement of time and era was simultaneously connected to the change of spatial elements, the components for space and place. An interesting phenomenon regarding temporal and spatial change is that it creates an overlapping of spaces and characters from different historical times. For example, various types of costumes and places simultaneously appear in one historical setting. In contrast with the above description, this temporal and spatial multiplicity may not necessarily have been a simple result of the political censoring system. Aesthetic and artistic strategies may have played a hand in these techniques.

(9) Style (Form or Pattern)

"Style" is here used to mean "performing style," in which various styles appeared to be adopted on the actual stages in kabuki plays, such as for dance, music, narration, and action forms or patterns by actors. There were a variety of multiplicities and possibilities dependent upon the constraint of prepared elements and combinations. The music accompanying the actors' actions has included many genres, such as nagauta, kiyomoto, tokiwazu, and ōzatsuma, etc. The narration by a takemoto, corresponding to a tayū in ningyō jōruri, has been transformed for a playing style for human actors instead of one for artificial puppets. While all parts of the scenario of a work are narrated by several tayūs in ningyō jōruri, the scenario of a kabuki work is divided into the parts of katari (narration) by takemoto and serifu (speech) by human actors. The artistic blending of kabuki and ningyō jōruri has given birth to another aesthetic technique called ningyōburi in kabuki, which is a unique performing style in which a human actor imitates or simulates the action of the puppet in ningyō jōruri. This represents one of the interesting and complicated multiple structures in kabuki.

The most significant method to actually control a performing style of an actor is, perhaps, "form (kata)." Each kata is a kind of pattern or norm for actors regarding body actions and the corresponding sets of costumes and maquillage, etc. There are many forms that have been historically inherited by many actors. For an actor, to select a particular kata means determining various elements for constructing the stage of a kabuki play, such as costume, music, narration, and stage setting, as well as their combination. Therefore, a kata is not only an element that constraints an actor, but also an actual framework determining the entire style of a stage or play. The main part of training actors in kabuki is for them to learn and know various katas inherited from its past history. At the same time, the creativity of kabuki has been engendered by maintaining kata or partially or entirely destroying it.

(10) Theatre (Stage and Seating)

According to Kawatake (1989, 2003), traditional Japanese plays of a particular genre incorporate the theater as an essential component. The theater/stage exists as a universal space, and it is not that various genres of plays are performed there, but rather, nō, ningyō jōruri (Bunraku puppet theater), and kabuki are performed in theaters and stages specifically suited to them. The hanamichi, described in the previous sub-section, is a mechanism on the kabuki stage that has multiple usages. Many academic studies have endeavored to determine the origin and meaning of the hanamichi (Suwa, 1991; Kurata, 2013). A narratological feature originating from the hanamichi, or a narratological concept that is symbolic of it, is "fluidity" (Ogata, 2019d) in the relationship of the stage and seating in a theatre. Without an actor entering through the seats directly, by utilizing the hanamichi, kabuki can symbolically represent a situation of multiplicity in which the stage and seats are complicatedly blended in an integral space. Main actors or characters in a kabuki play appear and disappear through the hanamichi. In the seats of kabuki theatre, though these are also related to the next topic of "audience," there are special seats, such as sajiki ("grand tier") and tachimiseki ("standing room"), and various related customs, such as kakegoe from the upper gallery including the tachimiseki

(11) Audience

Kabuki is primarily an object to be viewed in the theatre by the audience. Of course, this does not necessarily mean that only the phenomenon of viewing in the physical space of the theatre can be an absolute experience for the audience. However, it is also certain that kabuki is closely, essentially, and mutually bound with the existence of the theatre. As seen in the examples of the hanamichi and kakegoe, the audience also plays a part in the multiple editing of kabuki. The difference between seeing a kabuki play as a movie and in the theatre is considered to be greater than the difference between seeing a movie on television and at the cinema. The viewpoint in a movie has already been prepared based on the creator's intention and technique. Rather, when an audience views a movie at the cinema, the viewpoint in the work does not change according to the place in which the audience views it. On the other hand, in the case of kabuki, there are various different viewpoints according to the place in the theatre. Unlike the movie film, a kabuki play on the stage does not have any fixed viewpoints which correspond to the camera work in the movie. Therefore, the action of an audience viewing the stage itself plays a role in the viewpoint as one of the narratological techniques. Certainly, the above description may be a common feature for all theatrical dramas, but the stage of kabuki containing the hanamichi, seriagari, and music players, etc., is more complicated than the stage styles of other many genres of drama. This fact brings a variety of influences to the action of viewing by the audience. On another topic, problems for the audience may be considered to be linked with cconomic problems related to money and its association with the hierarchy of the audience seats. However, in contemporary kabuki theaters, for example, ordinary invitees often sit in expensive seats on the first floor, and there are also many regular patrons including critics and researchers sitting in cheap seats on the third floor, so it is not a simple division that reflects the economic circumstances and the social stratum of the audience.

(12) Text Reading

Or receiving a novel or a poem is an experience involving a person reading words that are physically represented on the paper, and the text in such case means the enumeration of the words themselves. By contrast, receiving a drama on the stage is, in the primary sense, to perceive humans' physical actions around the stage settings and decors. Unlike a book, a drama on the stage is not a representation solely based on the enumeration of words, but is also a collection of body actions and sometimes the stage itself accompanying these actions. kabuki has a tradition of essays and criticisms originating from the yakusha hyōbanki in the Edo era (Taguchi, 2003), which was a series of books that commented on the kabuki actors. The style has been persistently dependent upon the fact that an audience really views a kabuki play on the stage in a theatre. This fact indicates that the text in kabuki has been the individual stage performance as a collection of body actions by actors and stage settings. This tradition has not yet been corrupted on the present kabuki scene. For example, the works of critique, which are theoretical or academic texts, by Watanabe (2012a) are based on his viewpoint and experience as an audience member of kabuki. However, as Watanabe also says in his book, it is a fact that kabuki works have another level of text similar to a scenario that differs from the text as a stage performance by actors. Although the rise of scenarios as reading material is a relatively new phenomenon, it is currently possible for people to read and appreciate the scenarios of classical kabuki works in the form of books. In short, the text of kabuki has a multiplicity consisting of the "first (level) text" as a stage performance by actors performing on the stage and the "second (level) text" as the scenario of words written by authors. Further, as an additional feature to the above multiplicity of texts, there have been multiple relationships among different genres, such as the multiplicity and transformational relationships between nō and kabuki, kyōgen and kabuki, and ningyō jōruri and kabuki, etc. In the case that a scenario of ningyō jōruri has been adapted to kabuki according to its narratological or dramatic structure, the resulting kabuki scenario is equivalent to a multiple text that retains elements of the original ningyō jōruri.

(13) Production of Scenario ("Daichō")

The production of actual performance scenarios, which in the book of a scenario is called daichō, was very simple in the initial days of kabuki because the typical works, which had only a single structure containing but a few scenes, were not completed dramas but an assortment of entertainments and dances. However, kabuki works gradually came to have large-scale and complicated structures. The production mechanisms also developed to become more systematic, thanks to excellent authors, especially from ningyō jōruri, such as Chikamatsu Monzaemon and Kino Kaion. As a result, a new type of scenario production style was invented through which several authors created a scenario working collaboratively. According to Imao (2009), a hierarchy existed among a group of authors in a theatre. The top level was occupied by the tatesakusha, who planned the entire structure of a work and wrote the most important parts. This collaborative production style continued until the Meiji era, when the system of tatesakusha ended.

(14) Direction

The property of kabuki plays whereby various different historical ages and places of a narrative are mixed rather than representing a realistic standard influences the production of each actual stage. Various styles of blending or overlapping of costumes, language of speech, music, etc., sometimes exceed ordinary situations. Maquillage or the make-up and costumes of characters are designed according to the form (kata) or kind of traditional norm to be used for each typical character. Such very symbolic and formal features frequently ignore the realistic consistency and naturalness of the entire staging. One of the reasons may certainly be the disordered confusion of forms through historical accumulation. Innovative plans to return realistic styles of play to kabuki have actually been proposed by some actors, such as Ichikawa Danjūrō IX in the Meiji era. However, since over-realistic and symbolic styles form a tense tradition and also a unique beauty involving the audience, kabuki grew into a more heterogeneous and synthesized dramatic genre than one of pure realism. Danjūrō also returned to the traditional style of kabuki (Watanabe, 2018). Another topic that differs from the issue of stage representation in the above sense is that kabuki has not traditionally had a person playing a managerial and integrating role similar to that of a director on the stage. The role of a director of a kabuki play was conducted according to the naturally collaborative work of the actors who really performed the play. This making style is partially relevant to the symbolic features of kabuki, which arose from the characteristic of materials of various levels held by the system being freely combined and edited to complete a work. In addition, a traditional training style of kabuki, in which the actors accumulated the knowledge of diverse playing katas over a long time, was also one factor in enabling collaborative work among actors to direct an actual stage performance.

(15) Stage Performance

Kabuki in the Edo era was a contemporary drama of that age. Therefore, each kabuki play was basically performed in the style of tōshi-jōen, which means that all of the parts are played continuously, in one or two days. However, in truth, the actual form was not necessarily simple as there were complex rules or performing systems. A basic pattern was that two types of newly created works of jidaimono and sewamono were mutually played in the program for a series of kabuki performances. For example, when Tsuruya Nanboku IV's Tōkaidō Yotsuya Kaidan debuted in the year 1825 (Bunsei 8), it was alternately performed along with Kanadehon Chūshingura over two days. By contrast, after the Meiji era, the midori-jōen (a new performing style in which mainly the classical works of Edo kabuki were repeatedly performed) gradually spread and became popular, though many new works (scenarios), including the sub-genre of shin-kabuki ("new kabuki"), were also written by many modern authors. The midori-jōen is a stage performance style consisting of different works through which a kabuki program is constituted. The works include, for instance, the following types of play: sdan (a part cut by the jōshikimaku) and a scene in a dan in long, classical, pure kabuki or Gidayū kyōgen works, a part or a scene in shin-kabuki, a kabuki dance, and a nō and kyōgen-based play, etc. These more or less fragmental works were collected to form a program. The audience who is interested in the entire story in a play contained in a midori-jōen cannot comprehend it adequately only on the basis of the part that they are really viewing on the stage. The perceived performance is in a form in which only a part of the piece of work (e.g., editing one stage or a scene or several stages or scenes) is performed, with several parts coming together to form a full performance. According to Yasutomi (2012), the early midori-jōen was performed by Takeda Shibai (Takeda Troupe) in Osaka in the Edo Period's Hōreki Era (1751–64) and it spread in popularity. In addition, in the Meiji era, the major producer of Ichimuraza theatre, Tamura Nariyoshi (1851–1920) (1922), established the midori-jōen's regular format of "first act: jidaimono, intermission: dance (buyō), second act: sewamono" dependent upon the system of "first act, intermission, second act, and final act." Many audience members try to understand the part that they are viewing through a preliminary and parallel reading of the explanation in which the entire story is described. On the other hand, other audience members may also abandon an understanding of the story to experience an ecstasy of happiness by enveloping themselves in the formal beauty of the stage. Also, certain audience members will repeatedly view the same kabuki play performed by different actors to gradually and naturally come to understand and learn a story. For good or ill, as midori-jōen is at present usually structured as a well-known play, the audience frequently encounters the same plays in their continuing experience of kabuki viewing. However, editing and alteration at the level of a scenario is sometimes undertaken in preparation of its performance as contemporary kabuki. Since it is difficult to directly aim at actually realizing the original scenarios of classical kabuki plays on the stage in contemporary performances, the temporal, spatial, and thematic conditions or constraints of the scenarios are in many cases adapted for current society. Finally, one of the characteristics of kabuki related to the topic of stage performance, as described by Watanabe (2012a) and other critics, is that it has been a kind of star system or actor-centric system with one or more central actors often writing, revising, or editing a scenario. Producing and directing have also in many cases been conducted by the actors.

(16) Tsukushi or Zukushi

Ogata (2016b) considered the problem of "tsukushi" or "zukushi" in the context of kabuki. The term tsukushi or zukushi is not a general term of kabuki and literature. In kabuki vocabulary, there is "kyōgen monomane zukushi," which is a kabuki stage made from realistic or mimetic performances. In Torigoe (2000) and Takei (2002), it was used for the performance style that centers on realistic or mimetic dramatic elements in contrast to the style that centers on dancing elements in the first period of kabuki. Moreover, although it originally may mean the style performed only in nō and kyōgen, it became to be considered as a principle of scenario production and kabuki staging in the yarō kabuki (kabuki by adult men and the direct origin of current kabuki) after the prohibition of wakashu kabuki (kabuki by young men and a form in the first period's kabuki) by the Edo government. kabuki came to have relatively fixed scenarios and rich narratives by introducing scenarios of ningyō jōruri, which has created fixed and great scenarios. Actually, kabuki has continued to have dancing elements throughout its long history. In the current kabuki, one or more dance scenes frequently appear in kabuki works, and on a kabuki stage, dramatic plays that have strong narrativity and dance plays are performed in turns.

(17) Naimaze

Naimaze in kabuki has been known as a narrative creation method or a group of techniques that combines components in existing kabuki works or in the narratives from other genres for making a new work. Various elements such as stories, plots, characters, and places can be used as components in naimaze. Although the narrative and theatrical techniques of naimaze were systematically established by Tsuruya Nanboku IV, one of the greatest kabuki writers, naimaze was organized systematically through meetings and collaborations among various participants of a kabuki production. Ogata (2018d) showed the systematic production process of a kabuki work using naimaze. First, when a new kabuki work is created, a sekai (world) that is a set of materials as the framework of the work is selected or determined. A sekai contains one or more old legends, geinō genres, literary works, narratives, and so on in which dramatic characters and historical people, who are friendly to the period's audiences, appear. In detail, a sekai also includes historical backgrounds, places, names and characteristics of characters, human relationships, and macro patterns of stories. After a sekai is decided, various techniques (shukō) are added to the framework of sekai. The shukō is the originality or creativity that is unique to each kabuki writer. In kabuki, this was a popular characteristic using both narrative frameworks that many audiences commonly held and literary and artistic originality and creativity. In the techniques of shukō, patterned scenes or episodes also exist, such as migawari (substitute or sacrifice), koroshi (killing), and enkiri (dissolution). As well, contemporary, vivid events such as killing, burglary, and shinjū (double suicide), are frequently used in a sekai. Here, naimaze means a special technique or a group of techniques through which several sekais are simultaneously combined each other.

KABUKI AS DOCUMENTS

The author shows related references in various places in this chapter. However, in this section, the author attempts to systematically list more *kabuki*-related references. This attempt is related to the fact that *kabuki* has had many important relationships with documents from its first period. Thus, this part extends the literary survey of *kabuki* studies, especially toward future development in this research. *kabuki* studies have been supported by critical references from the first period during the Edo era, and this was an important and significant characteristic in the pragmatic genre of *kabuki*. In particular, in the Edo era, *kabuki* has held documents called yakusha nendaiki ("chronicle of actors") and shibai nendaiki ("chronicle of plays") that mainly criticize body performances and the patterns or forms (*katas*) of many actors. These documents also covered the descriptions of a variety of areas in *kabuki* from stories to stage mechanisms. Thus, *kabuki* has been the subject of numerous studies and critique. Especially, "*kabuki* critique," beyond studies in the narrow sense, began from the yakusha hyōbanki ("commentary of actors" or "performers' reputation") (a kind of yakusha nendaiki) in the Edo era. Nakano (1985) reports that in the period from Hōreki (1751–1763) to Bunka (1804–1817) in the Edo era, diverse meibutsu hyōbanki ("commnetary of special products") were generated; a pioneering series was yūjo hyōbanki ("commentary of prostitutes or geishas") and yakusha hyōbanki. In Matsuzaki (2000) and Taguchi (2003), yakusha hyōbankis refer to books that criticized the arts and performances of *kabuki* actors and continued from the last half of the seventeenth centuries to the twentieth centuries (1890s in the Meiji era). The main writer was the famous novelist (ukiyo-zōshi writer) Ejima Kiseki (1666–1735). Ukiyo-zōshi was a popular literary genre that portrayed realistic stories of everyday life in the Edo era. From the Meiji era to the present time, new types of *kabuki* critique have been also practiced to report and pass on the artistic performance patterns by actors (Miki, 2004; Shino, 1991; Watanabe, 2004, 2013b). The researches and critique in *kabuki* have formed a unique field and tradition. As these examples show, because *kabuki* required the audience to be practical, both the researches and critique has been closely connected to each other.

Interestingly, the development of scenarios in *kabuki* was delayed compared to the scenarios of ningyō jōuri. Original texts or source texts were also ambiguous in *kabuki* plays. From its beginning, settings and roles in *kabuki* have not been performed based only on the fixed scenarios written. The above was rather an interesting characteristic in *kabuki*. At the same time, *kabuki* has not been established as a *geinō* genre, having a close relationship to "body" instead of language and texts. The tradition of *kabuki* has been that language and texts play a large and important role. For example, there have been critique and reports of performances and forms (*katas*) by actors, of stage setteings and other related elements, and of advertisements. Therefore, collecting, classifying, archiving, and documenting the world of *kabuki* using language is also an important task in *kabuki* studies.

Table 3, dependent on the above consciousness, the author categorizes the *kabuki* references to collect papers and books that the author has refered in the papers so far. (This list is also a small reading guide to the first-class student of *kabuki*.) This table (Table 3) also shows a rough and tentative classification of literatures on *kabuki* for the future in this *kabuki* study. From the broad perspective, documents on *kabuki* include two directions: diachronic references in books, reports, essays, papers, and general writings since the Edo era and synchronic references in the present time. However, this table provides a small list of the paradigmatic references. Before the author presents his idea, here is a general example, a classification of recent *kabuki*-related documents (books) by Hattori, Tomita, and Hirosue (2000): (1) general books, encyclopedias, and dictionaries, (2) history-related books, (3) work-related books, (4) author(writer)-related books, (5) actor-related books, (6) direction-related books, (7) reception-related books, (8) theater-related books, (9) reprinted books, (10) books in Western languages, and (11) movie contents. Tomizawa and Fujita (2012) offers a magazine list, including historical magazines. Additionally, there two classifications based on "method" and "object" and both academic and general documents. Dependent on several document classifications, the author tentatively attempts the following list in Table 1.

KABUKI'S MULTIPLE NARRATIVE STRUCTURES SEEN IN CONCRETE EXAMPLES

Figure 2 tentatively plots the knowledge of *kabuki*. Although this figure has same elements as Figure 1, Figure 2 rather classifys *kabuki*'s multiple elements according to the framework of narrative production of a *kabuki* stage. In particular, a variety of narrative knowledge elements concerns in the generation of a *kabuki* play on the stage, in a theatre that has the seating and the audiences. According to the multiple elements in *kabuki*, shown in Table 1 and Figure 2, this section extends the descriptions by stating concrete examples in each element. In addition, several elements are expanded by supplementary information.

Table 1. Documents related to kabuki

Category	Sub-category	Overview	Reference
Researches	General and Introductory Studies	This category includes documents that contain general and introductory studies on *kabuki* and is classification by method.	Kawatake (1989, 1992, 2003); Keene (1990); Taguchi (2011); Toita (1949)
	Studies Focusing on Actors	This category contains biographical studies of actors and is classification by object.	Baba (1998); Imao (2009); Mori (2010); Nakagawa (2010, 2013a, 2013b); Watanabe (1989b, 1993, 2006); Yamamoto (2013)
	Studies Focusing on Authors	This category contains biographical studies of authors and is classification by object.	Imao (2009); Kitani (2007); Nakamura (2004); Suwa (2005); Watanabe (1997)
	Studies Focusing on Works or Plays	This category collects studies that surveys and analyzes *kabuki* works from various perspectives and is classification by object.	Inumaru (2005); Nakamura (2004); Watanabe, 1986)
	Studies Focusing on Music	This category of studies focused on the music and sound of *kabuki* and is classification by method.	Haikawa (2016)
	Studies Focusing on Theaters and Environments	This is a variety of studies concerning theaters, performances of *kabuki*, and other external or environmental elements and is classification by object.	
	Folkloric and *geinō* studies	This category is related to *kabuki* studies that have strong folkloric characteristics and is classification by method.	
	Historical Studies	This is a methodological classification that focuses on *kabuki*-related histories.	Kurata (2013); Nakamura (2004); Nakano (1985); Takano (2005); Tamura (1922); Watanabe (2009b, 2012b, 2013a); Watanabe & Takaizumi (2012)
	Sociological Studies	This is a research category that surveys and analyzes diverse *kabuki*-related concepts using sociological methodologies and techniques and has a strong methodological consciousness.	Yanai (2011)
	Literary, Artistic, Aesthetic, and Critical Studies	This category contains texts that have a strong methodological consciousness, more than academic researches in the narrow sense, and is classification by method.	Asanuma (2007); Gunji (1963); Hara (2013); Imao (1970, 1979, 2010); Kamimura (1997, 2003); Kawatake (1989); Matsui (2002); Nakagawa (2013a); Suwa (1991); Shino (1991); Watanabe (1965, 1974, 1989a, 2004, 2009a, 2012a, 2013b); Yamamoto (2015)
	Fringe Studies	This category treats other genres relating to *kabuki*, for instance, *nō*, *kyōgen*, *ningyō jōruri*, Shinpa, Shingeki, and others. This is classification by both object and method.	Hashimoto (2012); Kurata (2013); Taguchi (2003)
	Informational and Cognitive Studies	This methodological classification is a research area that surveys and analyzes objects and concepts relating to *kabuki* using methods and techniques of informatics, information technologies, and cognitive science.	

continued on following page

Table 1 Continued

Category	Sub-category	Overview	Reference
	Dictionaries and Encyclopedias	This category includes a wide range of dictionaries and encyclopedias covering broad topics relating to *kabuki*.	Hattori, Tomita, & Hirosue (2000); Kanazawa (2009); Leiter (2014); Nojima (2002); Tomizawa & Fujita (2012)
Dramatic critics	Dramatic Critics (From Early Modern to Modern) <Synchronic>	This category contains dramatic critics in a variety of media since yakusha nendaiki in the Edo era.	Miki (2004)
	Dramatic Critics (Contemporary) <Diachronic>	This refers to contemporary dramatic critics in recent books, magazines, newspapers, and blogs.	
Reports	Reports (a)	In this category, the author refers to pamphlets that we can buy at the reception areas of theaters and shops. A pamphlet explains stories, actors, and other information about a specific *kabuki* stage. These pamphlets are stored in libraries as important documents.	*kabukiza* (2015a, 2015b); Kokuritsu Gekijō (2014) ; Zenshinza (2015)
	Reports (b)	These reports contain other documents such as ukiyoe selections, picture books, and movie collections.	Ōkura & Kamimura (2001)
	Magazines	Since the Meiji era, magazines have been an important medium that reports, explains, and inherits *kabuki*-related information.	
	Actor's Talks	It refers to fixed talks and speeches narrated or written by *kabuki*-related people such as actors.	Bandō (2015); Bandō (1976); Onoe (2014); Onoe (1989); Sugiyama (2004)
Works	Scenarios (Daichō) in *kabuki* and Related Genres	This category also contains selected series of *kabuki* scenarios.	Mayama (1982); Mishima (1956); Yūda (1965)

Real Incident

One of the most famous examples that illustrates a mutual relationship between social fact and fictional arrangement is a work based on the so-called social incident of "the forty-seven samurais of Akō" or "Chūshingura." *Kanadehon Chūshingura* by Namiki Sōsuke, Takeda Izumo II, and Miyoshi Shōraku (dates of birth and death are unknown) was the greatest work of a variety of similar works written by gathering materials regarding this historical event by many authors. Many works of shinjūmono by Chikamatsu Monzaemon, such as Sonezaki Shinjū and Shinjū Ten no Amijima, were also created on the basis of actual stories that men and women really experienced and people had heard of in the same age. Adauchimono, such as works based on The Soga Brother (Soga Jūrō Sukenari (1172–1193) and Soga Gorō Tokimune (1174–1193)), including Kotobuki Soga no Taimen, and works based on a historical event of Araki Mataemon (1599–1638), including *Igagoe Dōchū*

Figure 2. Overview of kabuki production using multiple narrative structures
Source: Ogata (2016a)

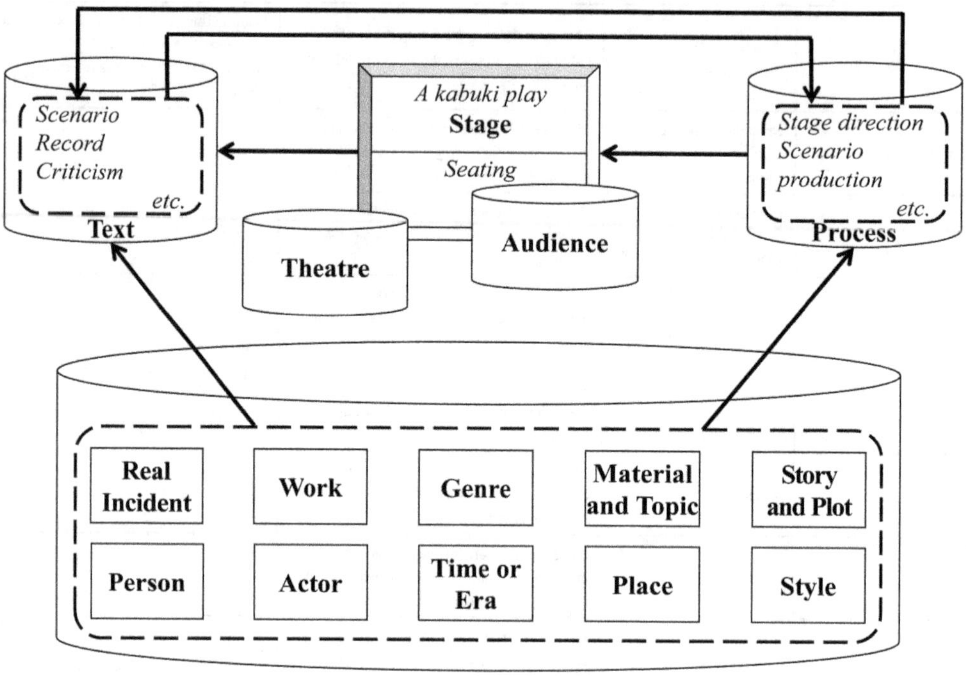

Sugoroku by Chikamatsu Hanji and *Igagoe Norikake Gappa* by Nagawa Kamesuke, were also such examples. These were repeatedly depicted in many works while adding many fictional adaptations. In many cases within these narratives, fictional characters and historical figures are mixed up and difficult to distinguish.

Work

The fictional dramatization of a historical event by Akō samurais can also be considered from the viewpoint of the problem of influence and inheritance from a work to another work. Until *Kanadehon Chūshingura* was completed, a series of works, including Onikage Musashi Abumi40 by Azuma Sanpachi (1701–1730) and Goban Taiheiki41 by Chikamatsu Monzaemon, were continuously being created. Even after *Kanadehon Chūshingura*, many related works, including TōkaidōYotsuya Kaidan and Kamikakete Sango Taisetsu42 by Tsuruya Nanboku VI, were created in the Edo era. Further, after the Meiji era, works such as Genroku Chūshingura43 by Mayama Seika (1878–1948) have been written until now. Together with Genji Monogatari (1993, 1994, 1995, 1996, 1997) by Murasaki Shikibu (the middle of the Heian era) and Heike Monogatari (1991, 1993), these Chūshingura works form a huge group of works of Japanese literary and entertainment history.

Genre

As a supplemental description, "genre" is also related to the next "material or topic." For example, new works were frequently created by going beyond a certain genre of a material. It is a development which transcended the genre of *kabuki*. And even if dealing with the same subject matter, *kabuki* has developed its own structural features. The mixing of various genres in *kabuki* is one of the most distinctive aspects of the genre. However, if only that were the case, *nō* would also be an entertainment genre created by compiling the various elements of prior public entertainment and narratives. However, in the case of *nō*, in its development, production, and transmission process, the works were fixed and classical, new works were only rarely written, and the quality of the performance was keenly oriented toward purity. More than anything, it was incorporated into the system during the Edo era as a form of shogunate ceremony (Watanabe, 2009b, 2009c). On the other hand, *kabuki* is a less pure form of entertainment. It was a genre that evolved and expanded by actively or indiscriminately absorbing various elements such as traditional as well as current entertainment genres, stories, narratives, dances and music. Although the power of classicalization has strengthened since the Meiji era, it is true that many new works, including the group of works collectively referred to as "Shin-*kabuki* (New *kabuki*)" have been written and performed (Nakamura (2004) carried out a detailed description centered on the *kabuki* artists and works since the Meiji Era). In recent years, even though it has been both praised and criticized in equal measure, it is a fact that some attempts at "impure" *kabuki* like "Super *kabuki*" by Ichikawa En'nosuke III (Ichikawa En' ō II now) (1939–), "Cocoon *kabuki*" and "Heisei Shin-*kabuki*" by Nakamura Kanzaburō XVIII (1955–2012) exist and have been commercially established.

More broadly, there are, for example, many films that visualize literary works and other forms of entertainment. Even contemporary novels, especially popular novels, often use past works and their parts of materials. In many cases, transformations between genres in *kabuki* has often been attributed to commercial convenience or the strategy of greedily incorporating popular works from different fields. In that sense, the multiplicity of genres and inter-genre transformations in *kabuki* are not particularly rare. It can be said that it is only an embodied example of intertextuality (Kristeva, 1980; Genette, 1982), which is a universal phenomenon in literature and art in general. However, the characteristic of the inter-genre transformation of *kabuki* is that the form and constituent features of the original genre remain in the *kabuki* that incorporated it, but rather that the composition and form of *kabuki* are clearly influenced and changed. When *ningyō jōruri* was introduced into *kabuki*, the style that the narrator (tayū) told the story and words as an accompaniment to the shamisen performance was also adopted fairly directly. None of those elements, namely *ningyō jōruri*'s elements, were reduced in the style of *kabuki*. Even if they were introduced organically, they were never purified. The inter-genre transformation seen in *kabuki* is not simply the transformation of themes and materials at the level of stories and characters, but rather the introduction of new styles into the resulting *kabuki*. Although it is not the case that such phenomena cannot be seen at all in converting novels to movies or mangas (comics) to movies, it does not appear to be as obvious as in *kabuki*.

As mentioned above, the most typical and large-scale representation of that phenomenon was the introduction of various aspects of *ningyō jōruri* (Bunraku puppet theater). In *ningyō jōruri*, the use of puppets allowed characters to have abilities that went beyond what actual humans could perform, and this aspect was introduced into *kabuki*. This contributed to the massive expansion of the range of human actors' potential for physical activity and for dramatic production. This, however, is largely related to the nature of *kabuki*'s original exhibition. In addition, for *ningyō jōruri*, which had the special characteristic of "narration from writing", the script was completely finished as a narrative work by one or more powerfully artistic author(s). Some distinguished authors included Chikamatsu Monzaemon, Ki-no Kaion, Takeda Izumo I and his son (Takeda Izumo II/Jr.), Namiki Sōsuke (or Namiki Senryū), Miyoshi Shōraku, and Chikamatsu Hanji. From a certain period in the Edo era, *kabuki* began to borrow solid, large, and complicated stories and narratives from the framework of *ningyō jōruri* puppet theater. The form of *kabuki* itself was altered, creating a great genre called Gidayū *kyōgen*. In Gidayū *kyōgen*, there is a phenomenon of duality of text in "a *ningyō jōruri* script and the script of *kabuki* itself based on the *ningyō jōruri* script." In *ningyō jōruri*, the script is told by a narrator (tayū) along with a shamisen accompaniment, while ningyō-tsukais (puppeteers) manipulate the puppets for the performance. On the other hand, in Gidayū *kyōgen*, the narration portion and the speech of the actors are dissected from the *ningyō jōruri* script, and a new dramatic script is created. On the stage, takemotos (narrators) narrate, while a shamisen plays, and actors perform and speak their lines. Of course, other elements (e.g., sound—music and sound effects) that are not found in *ningyō jōruri* puppet theater are also used in *kabuki*. Although human actors can perform more life-like movements when compared to puppets, they sometimes perform body movements called ningyōburi (puppet-motion simulations) that imitate the formal and mechanical movements of the puppets. People who replaced puppets did not simply pursue reality. Scenes when ningyōburi appears are during the climax and important parts of the work, when the tie with realism is severed, and this intensifies the dramatic nature of *kabuki* as a whole.

As examples, as mentioned in the previous part, introducing *ningyō jōruri* into *kabuki*, of which the adaptation called Gidayū *kyōgen* was the lower genre, was the most innovative factor in the evolution of *kabuki*. These are known as maruhonmono, Gidayū *kyōgen*, inponmono, dendenmono, Takemoto-geki, and maruhon *kabuki* (Toita, 1949). In particular, blending with scenarios of *ningyō jōruri* contributed to introducing strong narrativity into the dramas of *kabuki*, which had previously been very arbitrary in their points of narrative, as originally constructed to be centered upon characteristics of dance and as a kind of entertainment. On the other hand, the *kabuki* works based on the form of *nō* are called matsubamemono. As a general tendency, according to the properties of the original genres, many scenarios of Gidayū *kyōgen* have very long and complicated stories compared to matsubamemono works, which are formed by short and clear stories. The big three works that are praised in classic *kabuki*, Sugawara Denju Tenarai Kagami, Yoshitsune Senbon Zakura, and *Kanadehon Chūshingura* are all Gidayū *kyōgen*. As a famous *kabuki* work which incorporates *nō* styles, matsubamemo, Kanjinchō based on the *nō* masterpiece Ataka exists.

Material or Topic

A variety of *sekais* (worlds) were prepared in Edo *kabuki*, and each of them was used to produce a series of works. For instance, adauchimono based on the materials, topics, and themes of adauchi, meaning a kind of revenge story, has a popular *sekai* in classical *kabuki* and *ningyō jōruri*. In addition, the *sekai* of adauchimono hierarchically contains various specific lower-genres according to the characteristics of the corresponding materials, such as Soga-kyōdaimono, Chūshinguramono, and Igagoemono. These form each *sekai*. Different works based on the same *sekai* have been written by different authors. For example, by adopting the same *sekai* of Igagoemono, Chikamatsu Hanji and Nagawa Kamesuke, who were known as genius writers, respectively wrote Igagoe Dōchū Sugoroku44 as a *ningyō jōruri* work and Igagoe Norikake Gappa45 by as a *kabuki* work. These two works are similar in their macro level plots, though there are also many remarkable differences. This example is not in the order of *ningyō jōruri* ® *kabuki*. Between the two genres, this kind of mutual exchange was gradually taking place with the advent of powerful *kabuki kyōgen* artists such as Nagawa Kamesuke. In the first place, there was also a long period of time when Chikamatsu Monzaemon was a *kabuki* writer, and his experience was utilized in later *ningyō jōruri* puppet shows. By contrast, other authors wrote quite different works from the original story for a *sekai*. For instance, Tōkaidō Yotsuya Kaidan by Tsuruya Nanboku VI developed a surprising novel narrative using the *sekai* of Chūshinguramono, whose typical use is seen in *Kanadehon Chūshingura*. Even if we take one author, it is common to make several works based on the same subject matter, and even Chikamatsu Monzaemon wrote many period dramas (jidaimonos) about the revenge of the Soga brothers (Soga Jūrō Sukenari and Soga Gorō Tokimune) when looking at his complete collection. In addition to materials at the level of event, many popular historical characters, such as Minamoto no Yoshitsune (1159–1189), for example Yoshitsune Senbon Zakura, Ichinotani Futaba Gunki, and Kanjinchō, and Akechi Mitsuhide (1528–1582), for example Ehon Taikōki, and Toki ha Ima Kikyō no Hataage,46 also appear as universal materials as part of a *sekai* for producing and imagining the works of *kabuki*.

Person

First, the author describes a supplement. When adding a person, the first is a male actor dressed as a woman. *kabuki* originated with yūjo (prostitute) *kabuki* (Takano, 2005), but later shifted from *wakashu* (adolescent male) *kabuki* to *yarō* (men's) *kabuki* and became a play of only men. The modern style of *kabuki* is based on *yarō kabuki*. *kabuki* is a play in which no real women appear, and males instead take on the role of female characters. Males in female form may be the biggest factor that produces the appeal of the specific feeling of *kabuki*. Therefore, this was the focus of discussion when Watanabe (1965) wrote an essay regarding "actresses in *kabuki*." At the Zenshinza public performance at the National Theater of Japan (Kokuritsu Gekijō) in May of 2015, actresses Ayami Imamura, Kazuko Nishikawa, and Kazuko Tsumakura appeared in Banchō Sarayashiki47 written by Okamoto Kidō (Zenshinza, 2015). It's not the case that originally, there were no women involved in *kabuki* plays at all. Bandō Mitsugorō X (1956–2015) (2015) describes a performer (kyogen mentor called okyōgenshi) who taught and showed plays to women of a Daimyō family during the Edo era. Mitsugorō himself says the first teacher he learned from when he was a child was a woman who inherited the tradition. In terms of multiplicity, in March, 2016, Nakamura Shibajaku VII (1955–) succeeded Nakamura Jakuemon V. In addition to the multiplicity of the actor's name, the multiplicity as a living human being, the continuity from the father Nakamura Jakuemon IV (1920–2012) (see Watanabe (2006) for the story of how war affected Jakuemon IV), by taking over the female form established and attained by Nakamura Jakuemon III (1875–1927) and IV, the male and female multiplicity are also carried over their whole lives.

As for "person," the motif of "metamorphosis" represented by the word of "jitsuha (truly)" (Imao 1970) plays a vital role in *kabuki*'s story. Many characters in *kabuki* are multiplexed through "jitsuha" and create a complex image. In many cases, this "metamorphosis through jitsuha" takes place when a regular townsperson becomes a famous historical figure. For example, in Yoshitsune Senbon Zakura's third dan (act,) "Sushiya (Sushi shop)," Yasuke's character is the disguised Taira no Koremori (1118–1181) in a modest sushi shop in the mountains, but the noble appearance of the scene gives the impression that the whole stage is enveloped in a brilliant glow. *kabuki* has the concepts of sewamono (ordinary story, narrative, or drama) and jidaimono (historical story, narrative, or drama), and although the elements and characters of the sewamono are sometimes mixed in the overall jidaimono, "metamorphosis through jitsuha" often consists of sewamono moving to become jidaimono. These concepts can also appear on a more micro-scale when acting. For instance, actors sometimes employ old-fashioned gestures and utterances, namely like jidaimono, in a sewamono drama. This is also a concept that shows the types of acting, namely, the realism (sewamono) and the style (jidaimono), with the multiple rhythms of putting style in realism and putting realism in style creating the characteristic feel of *kabuki*.

There are various multiple persons and narratives, such as the actor as a character who appears in a drama and narrative; present and near past events and the narratives performed by the actor; the present and near past narrative or story performed by the actor as a real human; and actors and the narratives of the past ages until that of the present actor. The multiplicity of Bandō Tamasaburō V (1950–), who is one of the most excellent on'nagata in contemporary *kabuki*, can be considered partially as follows:

- Bandō Tamasaburō is the stage name of the actor.
- His real name is Shin'ichi Morita.
- The name of Bandō Tamasaburō has a history and accompanying narratives:
    - A female dancer also existed in the Meiji era.
    - Morita Kan'ya XII (1846–1897), who is a great grandfather of Tamasaburō though Tamasaburō was really an adopted son in the Morita family, was one of the greatest theatrical promoters.
- Tamasaburō has various narratives and stories as an actor, creator, and artist:
    - He is a *kabuki* actor.
    - He is a movie director of Gekashitsu (Bandō, 1992) based on a novel, Gekashitsu, by Izumi Kyōka (1873–1939) (1942a), Tenshu Monogatari (Bandō, 1995) based on a scenario, Tenshu Monogatari, by Izumi Kyōka (1942b), etc.
    - He is a creator, director, and actor of new *kabuki* plays: famous attempts are the adaptation of some literary works by Kyōka for *kabuki*.
    - He is a performer of works that transcends genres, such as dancing, classical Chinese operas, and recitations.
    - Recently, he is an educator of young *kabuki* players being performed together in many cases.

Story and Plot

In an example of two-fold and three-fold plot transformation in the episode of Okaru and Kampei in *Kanadehon Chūshingura*, in its own light, it may seem like a successful editing technique, but if we look back at the unexpected accumulation of circumstances from the very beginning, the storyline's editorial complexity does not prevent the audience from getting the feeling that the events are inevitable and existential. In Daimotsunoura in Yoshitsune Senbon Zakura, when a ferryman transforms to Taira no Tomomori (1152–1185), who is really a great head of a distinguished family, Heike, the narrative rapidly develops to a battle with Minamoto no Yoshitsune, who is the head of another distinguished family, Genji. The clear contrast of a peaceful, wagoto- and sewamono-like scene in the first half with that of the furious, dramatic, aragoto- and jidaimono-like scene make a very strong impression on the audiences. For example, Sukeroku (Yukari no Edo Zakura) produces a sharp and surprising impression on the audiences. Hanakawado Sukeroku appears on the stage as a beautiful young man and comical scenes continue. Then, Sukeroku shows his true character as Soga no Gorō, who is the main character of a classical *sekai* (world) in an adauchimono, in the final part. The multiplicity of the levels of story and plot represented by both works simultaneously shows the multiplicity of historical ages or worlds. In addition, Kamikakete Sango Taisetsu by Tsuruya Nanboku IV is also often subject of discussion as an example of multi-layered story development (Inumaru, 2005; Asanuma, 2007). The nature of the compilation is not sufficiently realistic throughout these storylines. At the time of its occurrence, it seems that inevitability and existence were guaranteed to a considerable extent by exterior contextual and circumstantial knowledge, but they remain even today when we lack those contexts. We must consider the meaning of this.

Actor and Place

Date no Jūyaku by Tsuruya Nanboku is the most extreme example in which only one main actor plays ten or more characters' roles by changing costumes and maquillages on a stage that continues to alter to different stage settings. The true title, Haji Momiji Ase no Kaomise,48 indicates not the narrative content but the difficulty for the chief actor. One actor plays the role of ten characters—Nikki Danjō, the spirit of Akamatsu Manyū (Mitsusuke) (1381 or 1373–1441), Ashikaga Yorikane, Hosokawa Katsumoto (1430–1473), Masaoka, Kinugawa Yoemon, Arajishi Otokonosuke, Kasane, Takao, Dote no Dōtetsu, but changing acting patterns helps in this case. For example, the setting of the stage at the Inase River in Kamakura is often used instead of the Sumida River when it comes to place (stage space). (The Inase River, which flows out from Yuigahama Beach in Kamakura, is not really like the Sumida River.) When you continuously watch the *kabuki* stage, you notice that different works sometimes share the same painted backdrops. For example, the showy place of Yoshiwara Nakanochō is used in different works, such as Kagotsurube Sato no Eizame49 by Kawatake Shinshichi III (1842–1901) and the final scene Nakanochō Sayaate no Ba in Tsuruya Nanboku IV's Ukiyozuka Hiyoku no Inazuma.50 Concerning place, for instance, the same stage setting at the entrance space of Yoshiwara Nakanochō, one of the most luxuriant places in Edo, has commonly been used for many different *kabuki* plays, such as Kagotsurube Sato no Eizame and Yoshiwara Suzume.51

Audiences who have seen many *kabuki* plays would encounter a same stage setting for different ones.

Time (Era or Age)

Although the main material of *Kanadehon Chūshingura* was a real incident that occurred in Genroku era (1688–1707) in the Edo era, the temporal positioning of the story was in the Muromachi (or Ashikaga) era (1338–1573). However, the characters' styles and the stage settings in these different times were mixed and overlapping in the play. Although the temporal setting of a story was in the Ashikaga era, for example, the standard for the stage space was clearly the Genroku era. As another example, Sugawara Denju Tenarai Kagami set the story time in the age of Sugawara Michizane (845–903), who was the actual model, in the Heian era. However, there existed various historical contradictions. For instance, while Sugawara Michizane and Fujiwara Shihei (Tokihira) (871–909), who was a villain hero, wore the costumes of the Heian era, Takebe Genzō, who was a teacher of a terakoya (the Edo-style school) as a fictional character, wore the costume of samurai in the Edo era. Various costumes in different ages were mixed in all scenes of this work. This is also related to the two most representative forms of *kabuki* performance: jidaimono and sewamono. In other words, while Michizane's and Shihei's performances have an exaggerated style that is timeless, Genzō and others are based on the realistic form of the same period, and both coexist gracefully on the same stage. Yukio Mishima (1956) took characters derived from *nō*, and placed them into the Japan of the latter twenteeth century within his Five Modern *Nō* Plays. In *kabuki*, people from modern Japan and those from the Muromachi era might coexist within the setting of a house from the Genroku era. Although it is not a *kabuki* play, the novel Shin Karamazov no Kyōdai [New Brothers Karamazov] (Kameyama, 2015a, 2015b) converts all of Dostoevsky's settings and characters into those of modern Japan. In *kabuki*, it is possible for the setting of the stage to remain in the Russia of the nineteeth century, with Russians of the nineteeth century and Japanese of the twenteeth century able to coexist. However, in the present day when the censorship system is not as blatant as in the Edo era, it might be said that the problem is that there is not enough motivation to do that kind of work.

Style (Form or Pattern)

With respect to the third act (dai sandan), Igami no Gonta in Yoshitsune Senbon Zakura, the line dealing with Gonta from Edo is a (Onoe) Kikugorō-gata production and performance style or pattern and the line dealing with a rogue of Yamato is a (Ichikawa) Danzō-gata production and performance style or pattern (Imao, 2010). Regarding the third act (dai sandan), Kumagai Jinya of Ichinotani Futaba Gunki by Namiki Sōsuke and others, the (Nakamura) Shikan-gata style or pattern is handed down from Nakamura Utaemon III (1778–1838), to Nakamura Shikan IV (1831–1899), and the (Ichikawa) Danjurō style started with Ichikawa Danjurō VII and IX and was handed down to Nakamura Kichiemon I (1886–1954). In this way, actors' styles or patterns have been handed down from generation to generation (and in cases where a specific actor belongs to a specific family, these are sometimes called like "ie no gei (house arts)" (Matsui, 2002). There have been different types of forms for Satō Tadanobu (1153–1186) as a fox in Yoshitsune Senbon Zakura. For instance, Ichikawa En'nosuke III created a bold form to represent a fox's actions with magical powers. By contrast, the form of Onoe Kikugorō VI (1885–1949) was more humble and realistic. On the other hand, the audiences get to see ningyōburi especially in Gidayū *kyōgen*. Ningyōburi in which an actor imitates a puppet in *ningyō jōruri* is also one form of performance. There, a complicated multiple-level structure is realized in which a living human performs in a drama that originates from *ningyō jōruri* further imitates a puppet again. The author would like to introduce the more complex examples, in the last half of the scene of Dōmyōji [the Dōmyōji Temple] in Sugawara Denju Tenarai Kagami (Dōmyōji is one of the most famous and excellent *ningyō jōruri* scenario written by Miyoshi Shōraku). In Dōmyōji, a wooden statue appears and walks over to help the main character, Sugawara Michizane, as a "scapegoat." In contemporary *ningyō jōruri*, a puppet operated by three ningyō-tsukai plays a wooden statue that is an acting entity in this scene. In a more complicated *kabuki* version of the play, a real actor plays a wooden statue, like a puppet. This is also a kind of ningyōburi. Kataoka Nizaemon XV (1945–) in contemporary *kabuki* is also known for excellent plays of Dōmyōji in addition to other works. Nizaemon represents multiple bodies with his real body through the circulative transformation of a wooden statue to a puppet, then to a man, and then back to a puppet and a wooden statue again on an impressive stage. Moreover, onstage in December 2018 at *kabuki*za, Bando Tamasaburō performed two different characters, the heroine Akoya and Iwanaga Saemon, in the dan of Akoya in Dan'noura Kabuto Gunki by changing every day. Iwanaga was an enemy of Akoya and treated her cruelly. Further, Iwanaga was played as a puppet like *ningyō jōruri*. In particular, Tamasaburō performed opposing characters, both a human and a puppet.

Theatre (Stage and Seating)

The hanamichi primarily and pragmatically functions as the entrance and exit for the stage and backstage. However, the hanamichi is also a place for dance and narration, especially narrations of reminiscence and self-introduction. An example is the appearance of the five shiranamis (thieves) in fourth act (dai yondan) of Inasegawa Seizoroi no Ba in Kawatake Mokuami's Aoto Zōshi Hana no Nishikie. Similarly, in Sukeroku Yukari no Edo Zakura, the chief character, Sukeroku, performs a very long, stylistic, and impressive dance and shigusa (action) on the hanamichi in the first appearance. Moreover, the hanamichi can also be a passage to strange worlds through which ghosts and monsters appear and disappear. In Meiboku Sendai Hagi by Nagawa Kamesuke, Nikki Danjō, who is a strange entity, namely a big mouse, hatching a bad political plan to destroy his lord's house and usurp it, appears from the place called suppon (snapping turtle) in the hanamichi in a cloud of white smoke. Originally, the hanamichi was a place where an actor would receive flowers from the audience (Suwa, 1991). The current hanamichi still retains the function of the platform on which a character greets the audience as an actor. Although the greeting on the hanamichi is sometimes criticized in the point that the smooth flow of a *kabuki* play is obstructed, for the audience, the friendship with actors (and not characters) is also an attractive element that is part of the pleasure of *kabuki* plays. An example in which the whole of a theatre containing the stage and seats becomes a large and inclusive stage is the scene of Numazu in Igagoro Dōchū Sugoroku. Two unique characters go down to the seats from the stage and walk around them while in conversation with each other. Further, they sometimes talk to the audience. In such situation, the kakegoe from the seats will also blossom. At such times, the fictional space and real space are multiply blended within the same theatre. In accordance with the characteristics of these boundaries, "characters" of *kabuki* are able to flexibly switch back and forth among characters as the people in fictional stories, characters with actual bodies as actors, and real-world characters. Conversely, the multiplicity of *kabuki*'s characters is also supported by these structural features of the theater.

Audience

The author makes a supplementary argument from the point of view of the multiplicity of "viewers."

First of all, we can think of "multiplicity in a single audience member." Just like there is a spatial phase and a temporal phase within actors, these phases exist within individual audience members as well. The spatial phase relates to the seat from which you view the play in a theater. If an audience member watches several performances of the same play from different seats, the viewing experience will take on a layer of multiplicity. In the current *kabukiz*a, although we are not really talking about which is better, the reception from the audience members may obviously vary when we compare the views of audience in the third- and fourth-floor who can only see a part of the hanamichi, thought the whole stage can be seen, with the views of audience near the hanamichi in the first-floor. Of course, each has its own benefits. Further, this seating may also be intertwined with economic problems.

Next, the author compares watching a *kabuki* stage with watching a movie in a "movie" theater. Although it is the same in both cases that your view of the material differs depending on your vantage point, in the case of movie, the scenes where events unfold are cut off in advance by a certain point of view (camera) and this one process is added when compared to play theaters. The difference is that, in *kabuki* theaters, audinences are able to select a viewing perspective. Although the essential element to making a movie may be the presence of the camera's viewpoint, play theaters leave that process to the audience's side rather than to the creater's side like a movie. The stage of the theater, of course, is set up so that the audience members can see the front. In a sense, this can be thought of as corresponding to the point of view in the case of movies. However, at the very least, in current *kabuki* theaters, it is impossible to realize the technique of enlarging images of specific objects frequently appearing on-screen in a movie. For the author who do not simply affirm the viewpoint of the camera in movies, from the placement of relatively large objects on stage in theaters, in other words, from particular audience positions, sweeping scenes taken by video cameras do appear to be more fascinating. The name of Kenji Mizoguchi (1898–1956) may come up, but the author will set aside that argument for another day. When adding the element of time-lapse to elements such as the location and place inside theaters, each audience member has a history of each life and the intersection between the theater and the theater's scene. Within the audience, there are potentially a huge number of stories being developed——even if they are only related to *kabuki*. For audience who has been in theaters that have been performing *kabuki* for some time, the history of that *kabuki* is formed by the accumulation of contact points with the occasional spatial circumstances of what kind of seats they were in when they viewed the plays.

Here, "viewer" does not equal to "audience" or "spectator." If the theater itself is also considered to be part of *kabuki*, you can directly distinguish *kabuki* actors and *kabuki* viewers. The actors also see the audience from the stage. Then, recognizing the current state of the theater, it will be affected to a greater or lesser extent. Such phenomena are talked about in various "geidan" (art discussion) (Kamimura, 1997). There are the geidan of Ichimura Uzaemon XV (1874–1945) introduced by Kamimura (1997) and various geidans by Baba (1998), along with various other geidans by Onoe Shōroku II (1813–1989) (1989) and Bandō Mitsugorō X (2015). There is no direct communication between the actors and the audience members. In the sixth act (rokudanme) known as Numazu of *Igagoe Dōchū Sugoroku* by Chikamatsu Hanji et. al., the characters Jūbē and Heisaku go down from the stage into the audience and a peaceful scene develops in which they talk to the audience. This is likely a technique already incorporated into the production. As the actors' reactions to the audience appear realistic, it's certain that there's still communication happening that's indirect but face-to-face. In a story like a novel, whether you are a reader or a listener who is assumed to be inside the story, there's no face-to-face communication between the narrator or the character and the reader. The relation between them is always delayed. In this way, the viewer's viewpoint in some cases results in a kind of communication problem between the performer (actor) and the viewers (audience). This, of course, does not mean that this kind of communication is encouraged to become more direct. In *kabuki*, which has a mechanism known as the hanamichi, the proximity of an actor to the audience under certain circumstances is definitely one of those characteristics. As can be seen in the above example of Igagoe Dōchū Sugoroku, the stage and the audience are, nevertheless, sharply defined as mutually distinct spaces.

In addition, the audience can be thought of as being both a lot of individual spectators and a single group. The audience as a group acts as if it were a single organism, and kakegoe certainly come from individual audience members. However, looking at the entirety of the theater space, the audience as a group function as a whole. The audience getting noisier bit by bit is known in the industry as jiwa, but this is still one kind of collective phenomenon. Actors do not deal with spectators as individuals in many cases, but rather, they deal with the audience collectively, that is, as a whole. We can only guess, but the actors who see the audience from the stage are probably influenced more or less daily by a subtly different set of reactions. (This guess is partly based on the author's own experience of teaching in classrooms and similar lecture spaces.)

Furthermore, spectators in the audience can be categorized as "amateurs" and "professionals." The latter's extreme form in the Edo era would be an audience member who became a major writer about yakusha hyōbanki at the time, and in the present age it can be a theater critic. Written commentaries and criticisms are delayed reactions instead of real-time reactions from within the theater, but shape communication with the audience as amateur spectators. If an actor comes into contact with it somehow, it will still produce a kind of communication (indirect communication) there. Actors may read the criticism of a playwright and later speak directly to the author about it. On the contrary, various types of critics do not give feedback to the actors in directly after a performance. After all, if you disregard the content, it is the same for amateurs, with amateurs indirectly describing words similar to their impressions, and currently, these can be published on the web, or they can be said directly to the actors after a performance. Therefore, paying attention only to the form of this phenomenon, the direct and indirect expression of critiques does not vary much between so-called amateurs and the professionals, and it can be understood collectively as direct and indirect communication through critiques. However, in spite of this, in *kabuki*, "statements from experts" have continued to be very important. In order for *kabuki* to maintain its vitality, it can be said that a strong tense relationship had to exist between criticism from experts and the actors and writers. Here, although the experts are not subordinate to the creator, having said that, they cannot also just give any random opinions without any restrictions. There exists a theoretical (in a broad sense) type of independence. Traditionally, in the world of *kabuki*, communication has been based on an exceedingly antagonistic relationship between the receiver's side and the sender's side, rather than being based on a relaxed harmony.

For considerding audience, the author describes examples in the previous *kabuki*za, the fourth version of *kabuki*za (the original *kabuki*za theater was opened as a Western-style theater in Meiji 22 (1889), but in Meiji 44 (1911), it was rebuilt into a Japanese-style palatial building. Later, after it was destroyed by a fire and then an earthquake, it was again rebuilt, this time in the style of the Momoyama period, opening again in the Taishō 14 (1925). It was yet again destroyed in the World War II, but rebuilt in the Shōwa 26 (1951).

Finally, in Heisei 22 (2010), this structure was demolished and the theater was rebuilt once more (Abe, 2000), the audience could not see the hanamichi from the back of the second, third, and fourth floors. The audience in these seats guessed and imagined an actor's actions on the hanamichi based on the voice and situational atmosphere. In Sukeroku (Yukari no Edo Zakura), a very long and flowery dance played by Sukeroku on the hanamichi creates a very fast climax. During this time, the audience gazes at invisible things using only their imagination. Also, in several seats at the back side of the first floor by the hanamichi of the previous *kabuki*za, the audience could not see a part of the stage due to a large and red hole. Although the above problems were solved in the currently used new style of *kabuki*za, the large difference in viewing *kabuki* by the location of the seats is still an interesting property of this theatre. On the third or fourth floor, when a person closest to an audience plays the kakegoe, the feeling of excitement and pleasure also increases for the audience. By contrast, when the sound of plastic bags and voices of speech create a terrible atmosphere, the multiplicity of the play on the stage and the strange play by the seats remains in the audiences' mind as a harsh but impressive memory. As a recent memorable example, in March 2015, the author viewed the scene of Dōmyōji in Sugawara Denju Tenarai Kagami with a historical performance by Kataoka Nizaemon XV at the forward side of the second floor in *kabuki*za. The sound of plastic bags by an audience member nearby was overlapping with the narration by Kanshōjo (Nizaemon), who was playing the climax scene on the stage. This was certainly an impressive phenomenon in the traditional theatre, though an unpleasant experience. Including the possibility of such bad experiences, the audience in the actual theatre is an important factor of viewing a *kabuki* play.

Text

Regarding the element of "text," there is a supplement. Watanabe (2004) proposed the concept of "first text" and "second text" in *kabuki*. The first text corresponds to what is called a performance script. The theatrical script is called a play, and as something that should be read like a novel or poem, it is in the form of a book. *kabuki* scripts, even if many have been lost, can also be read in the same way as plays. Of the 68 volumes of Japanese plays, *kabuki* comprises 50 of them (Shun'yodo in Tokyo, pusblished through 1928–1933 compiled by Seitaro Atsumi (1892–1959)), 25 volumes of *kabuki* masterpieces (Tokyo Sōgensha, 1968–1973), 26 volumes of *kabuki* on stage (Hakusuisha in Tokyo, 1985–2008), 45 volumes of aggregated *kabuki* daichōs (*kabuki* Daichō Kenkyūkai, Bensei Shuppan in Tokyo, 1983–2003) are some of the published examples. However, the peculiarity of *kabuki* is that the faithfulness to the play as a script is not necessarily an important criterion, and it is normal for the script itself to continue to change because the "production" is mainly actor-oriented. What remains in *kabuki* is the play itself, which is the "second text," not the script (the "first textA") that it was based on. In *kabuki* the second text has naturally come to be regarded as essential. The kind of improvisation that some avant-garde plays performed experimentally were already quite established in *kabuki* (Tsukamoto & Suzuki, 2010). This has been revised on the basis of modern plays, and furthermore, so-called avant-garde theater has made a conscious return back to its ancestor. However, the point of being supported by "*kata* (form or pattern)" is a special characteristic of improvisation, in a broad sense, in *kabuki*.

The second text that Watanabe (2004) mentions does not simply mean a recording of an actual performance of a certain work. It means a kind of pattern that serves as a norm for actual individual performances. This pattern also includes such things as the stage equipment, props, and actors' costumes. Practically speaking, "*kata* of an actor name" (for instance, (Nakamura) Shikan no *kata* = Shikangata) means a kind of pattern in which the stage setting and the actors' costumes are naturally determined. However, Watanabe more narrowly regards the term "second text" as a record of certain patterns of acting in a performance by an actor = *kata* (called "fū" in Osaka). In classical *kabuki* performances, there is a representative *kata* for each actor, and later actors act in a particular *kata* depending on various circumstances and constraints, even if there realistically exist some ambiguous areas such as differences in degree of dependence on specific *katas*. Therefore, later actors learn the *kata* of the preceding actors in some way, with this being the exemplary method by which the techniques are passed down in *kabuki*. Even if it is not explicitly written down, the second text in this sense can always potentially exist. However, it is sometimes written down. In this sense, as described above, we cannot simplistically call recordings of performances of each individual play "second texts." Even so, now that video technology has developed, it may be possible to link the video recording of each performance to the identification of the second text as an actor's type. The second text is a means to transcend criticism as one's simple impression, to overcome the limits of the one-time nature of plays, and to function as a medium to transmit traditions.

Can a video be considered as a possibility for a second text? A *kabuki* video that is currently on sale or screening at a movie theater, when seen from the perspective of it being a technical record of a particular actor's performance (*kata*), is quite incomplete, as it is obstructed by the contrivance of camera work. Nevertheless, critiques on actors' performances based on video have started to emerge (Yamamoto, 2013, 2015). If it is possible to visualize a chain of actions fully focused on a particular actor, it could lead to a record of that actor's *kata*. It seems that, in the sense of accurately recording an external event at that exact point, an image has an advantage over words. However, even in this case, the weakness of video lies in the viewpoint of the camera work, which is regarded as its own characteristic. Even if, for argument's sake, the camera tracked one individual actor, even if the filming angle varies, it is impossible for the camera to also get a view of the surroundings, and conversely, if the camera is focused on the entire stage, it is difficult to observe the small details within the scene. There may be possibilities for using hypertext-like systems that may compensate for those weaknesses. Also, although language is a seemingly inconvenient medium, because we can record information beyond directly captured visual and auditory information, we should strive to unite language information and visual information.

Now, the second text is a kind of style record, a transmission medium, and a pattern through which that later generations can learn. For the many *kabuki* criticisms and the *kabuki* studies that are existential, the actor's *kata* has come as a major aim in describing the second text through Takeji Miki (1867–1908), (2004), Yōtarō Shino (1913–2009) (1991), Tamotsu Watanabe (2004, 2013b), and other writers. That is not a so-called impressionistic criticism, but rather, it is what you would call a technical critique. If compared to the critique of a novel, it would correspond to the detailed record and analysis of the writing style and structure. In critiques of novels, such things are rarely done, and in many cases it is neglected. Narratology as a technical critique is not popular, and the mainstream, as always, is still only a content-oriented broad ideological criticism or essays that does not go beyond impressionistic criticism. Beginning especially in modern literature, value is placed on the negation and transcendence of previous works, and the imitation and transmission of traditions were at least outwardly despised and dismissed. Or, it was made to be dormant. Naomi Watanabe's (1952–) (2012) attempt is a rare exception. In the case of *kabuki*, on the contrary, value was placed on the follow-ups and imitations of the preceding model. It can also be said that inevitably, technical criticisms and texts had to be developed.

kabuki has multiplicity of several levels of text. Transformation between different genres such as *ningyō jōruri* and *kabuki* mentioned in the multiplicity of "genres" can also be regarded as multiplicity in the text if paying attention to the complex nature of the two. This can be regarded as multiplicity in texts. If this multiplicity is a concept that takes into consideration the manipulation of focus, that is, the pluralism among them, beyond just the existence of various elements in a multilayered manner, the most active of all multiplexed text depends on what the performer or viewer is paying attention to at the time. Conversely, in a *kabuki* work as a whole, out of the multiplicity of texts, the kinds of text brought into the foreground and the kinds of text pushed into the background depend on the circumstances at the time. It is placed in a fluent or fluid situation. However, it remains the case that either of these texts can work in some way, and if we take the discussion from Ogata (2018a), it will play a major role in the development of *kabuki*'s existential character.

Regarding texts, the author would like to add one more thing to the significance of "first text" analysis and research. It is said that a "play's history" does not exist in *kabuki* because the play is often created mainly centered on actors, rather than based on a written script. Especially in early *kabuki*, it is said that the actor himself created a play, rather than a professional writer. In addition, modern-day *kabuki* performances are based on "midori-jōen," which covers only a part of the entire work. Even if an entire "toshi-jōen" play is performed, many things are both added and subtracted from it, and not all of the past scripts left behind are performed as they were originally written. These unique circumstances of *kabuki* as above lead to the fact that analysis and research of entire large plays are not very advanced. However, only a small portion of the vast collection of works has been covered and performed, and if you want to know about the entire *kabuki* play, it is also true that there is no alternative but to look at the script. When we dare to see *kabuki* as a play, in today's world, with regards to the works of the best writers, which are not often performed, besides analyzing and examining at the level of a single play, we have not been given the art of deducing the true value of these works. For example, works that are highly structured and are considered to be of great value, such as Nagawa Kamesuke's *Igagoe Norikake Gappa* may be important objects of narrative analysis.

In addition, the analysis of narratological commonalities and differences by comparison with *Igagoe Dōchū Sugoroku* by Chikamitsu Hanji et. al. by with the same subject matter, with a focus on the group of *kabuki* plays that form an intertextual network as a whole, due to the connection from plays that are performed more often, this also leads to an attempt to analyze works only survive as plays. The narratives of *kabuki* and *ningyō jōruri* are often very complicated and difficult to understand, but although there are many concrete forms of implementation, created by a variety of interesting narrative techniques—routine "kase"s (restrictions) and limitations, a multiplicity of character patterns (the sacred and profane, jidaimono-like and sewamno-like), multiple discourse types and complicated or overlapped worlds (*naimaze*, mixing different narrative elements, often appeared in Tsuruya Nanboku's works), a story's various transformations, etc. They are full of symbolic, arbitrary editing and fluidity. In this sense, they can be thought of as having a more familiar personality than modern or contemporary literature. (However, mentally separating modern and contemporary literature as completely different from Edo literature, including *kabuki*, can be a problem.)

Furthermore, the author thinks the description of the "second text" itself can also be subjected to meta-analysis. The second text itself is a kind of analysis of the first text that accompanies the script. It is a sort of analysis of an analysis, but there is not much narrative analysis of the second text itself. (Research on composition of the *kata* of the plays themselves, and of the historical trends has been conducted by Yanai (2011).) However, this second text has accumulated a lot. In addition to the audience as described above, analyses of the detailed records that actors left behind can also be part of these narrative analyses. In other words, analyzing the relation between the first text and the second text, and how the abstract descriptions of events and the concrete actions and words are linked are an interesting question regarding the concrete ways in which actions and words vary.

As examples, the scenario of a *kabuki* work, e.g., Kagotsurube Sato no Eizame, can be considered the first text(s), though, strictly speaking, differences exist according to the version. By contrast, in the second texts that are really performed, the forms of Sano Jirōzaemon, who is the hero in the work, have been played still recently by various *kabuki* actors, such as Nakamura Kanzaburō XVIII, Nakamura Kichiemon II (1944–), and Onoe Kikugorō VII (1942–), who have given different renditions and impressions to the audiences. A reason why people repeatedly view *kabuki* plays of the same title is related to the level of the second text. For instance, a group of people wants to see or confirm how a current actor plays the second text depending upon the *kata* created by a past actor, which has been historically inherited by his successors. Moreover, people hope to encounter the present scene or place in which a new second text comes into being through the excellent rendition of an actor or actors. Moreover, there are many differences between a scenario of *ningyō jōruri* and the version that is adapted to *kabuki*. All parts of a scenario in the former are written for the narration by narrators (tayūs). In the real play, a tayū precisely narrates all of the descriptions and serifu (speech or words), while its adaptation to the stage of a *kabuki* play requires varied edits and revisions from the *ningyō jōruri* scenario, such as the parts of the narration by a takemoto corresponding to the tayū in *ningyō jōruri*; the parts of body actions by human actors; the parts of speech or serifu by actors; and music in the *kabuki* style. Imao (2010) presents a precise analysis of transformational methods between respective scenarios of *ningyō jōruri* and *kabuki*.

Production or Scenario (Daichō)

Chikamatsu Monzaemon, a great writer who is familiar with *ningyō jōruri* and *kabuki*, in the modern-day critiques and research from the genealogy of Sōkichi Tsuda (1873–1961) (1977–1978) to Donald Keene (1922–2019) (1977), from a realistic point of view, such as realistic and natural outpouring of emotion, it became very highly valued. It is thought that the high evaluation of so-called shinjū (double suicides) in the modern-day is strongly related to this. It's true that Chikamatsu excels at stories about shinjū. Especially, the force of circumstances leading to the shinjū (or its attempt) is unusual. However, the modern-day label of double suicide Chikamatsu does not accurately reflect the actual state of his works. In fact, if you read the 17 volumes of the Chikamatsu Monzaemon Collection (Complete Work of Chikamatsu, Iwanami shoten, 1985–1994), you will find many jidaimono works. And especially, works taking place in the same world such as stories about the Soga brothers, there are many works in which there is a lot of overlap in the theme. There are also abnormal works that have almost nothing to do with reality, such as Shutendōji Makurakotoba (Chikamatsu, 1987). Of course, Chikamatsu's works have been highly regarded since the Edo era. Even from the place where Hozumi Shigeaki, the son of Hozumi Ikan (1692–1769) who was the creator of Naniwa Miyage (Shuzui, Ōkubo, 1959), which contained Chikamatsu's literary theory, used the pen name Chikamatsu Hanji—one could see how that was a name that would be looked up to. However, according to Kurata (2013), the script by Chikamatsu, which had hardly been replayed at the premiere, was re-staged several decades later. Advances in techniques and technologies related to puppet performances greatly contributed to it. This differs from the revivals due to people acknowledging the value of it in modern times. In any case, the high evaluation of Chikamatsu in modern times led to a delay in studying of other great writers and their works.

It seems that such an imbalance is also related to the fact that Chikamatsu has many "independent works." The days when Chikamatsu was creating *ningyō jōruri* and *kabuki* were still before collaboration system had become widespread. In the history of *ningyō jōruri* and *kabuki*, the collaborative system ("gassakusei") was new. Rather, it was this era of collaboration that is linked to a kind of systematization in production that brought about the golden age in the history of *ningyō jōruri* and *kabuki* plays. The collaborative system is a powerful technique and method for the edited production of a scenario. The standard of literary evaluations today is based on the originality and novelty of individual authors (also related to copyright), and rather than *kabuki* and *ningyō jōruri* puppet shows, which were created by teams instead of by individuals, we can presume that Chikamatsu was a more convenient subject to be evaluated in modern times.

The above-mentioned emphasis on Chikamatsu also brings about clear negative effects that lead to the stagnation of research about plays and authors such as Namiki Sōsuke and Chikamatsu Hanji, who are very frequently featured in performances of works these days. It is necessary to correct these defects in research and criticism of the past and investigate, in particular, the production system known as the collaborative system from a proactive point of view.

As examples, in a *ningyō jōruri* work, Sugawara Denju Tenarai Kagami, a common theme through the famous three dans (parts) (Dōmyōji, Igami no Gonta, and Terakoya) was "separation (betsuri or wakare)." According to Watanabe (2009a), another common theme, "scapegoat (*migawari*)," was also added to the first theme. Each dan was written by three different great authors, respectively: Miyoshi Shōraku, Namiki Sōsuke, and Takeda Izumo I. These works are frequently performed even in *ningyō jōruri* and *kabuki* and all of these three were great writers. Although Takeda Izumo I's theory of determining the author used to be the most influential, it seems that in recent years, there has been a strong rise in popularity of Namiki Sōsuke's theory (Uchiyama, 2009). Kawai (2000) also takes this position and concludes that the main author of the famous Terakoya in Sugawara Denju Tenarai Kagami is also Namiki Sōsuke. Although there has been some discussion over who the "tatesakusha (main writer)" was in case of this miracle work, the difference among the tateakusha and other level of authors was usually clear. For each author, there was a history of rising from the lower to the higher level of the hierarchy and finally to reaching the utmost rank of the tatesakusha, at least in the happy cases. It is a famous case that Tsuruya Nanboku IV, who was known as one of the greatest and most famous *kabuki* authors, rose to become a tatesakusha in his fifties, but continued to be active as the first author in the *kabuki* world for over 25 years until his death (Suwa, 2005).

Direction

kabuki's direction or production is also deeply involved with multiplicity and pluralism. A multiplicity of direction is pursued because the "standard as a work" in *kabuki* is pluralistic. There is not a single standard for creating, performing, and directing works. Instead, these standards can be said to be fluid. Aristotle (1997), who represents the central idea of the narrative, in Poetics, based on the tragedy Oedipus Rex (Sophocles, 1972), the elements of the play were classified hierarchically, and the "structural" importance was ranked. This goal-oriented idea tries to interpret and organize plays with a focus on a "plot" and "story." In this sense, it is the theory that centrally placed the narrative in its narrow meaning (the narrative from the author does not mean necessarily this narrative in the narrow sense, but is the narrative in a broader sense.) The narrative of a play, as described above, is not only a work (second text) performed on the stage. It has scripts as its foundation (first text) and is also the subject of reading as a genre of literature, similarly to novels and poems. It was this level that Aristotle considered the most important. However, despite the centralized and hierarchical ordering process, as Aristotle himself showed, the play has several other key components, which are eventually integrated into the work itself as a stage performance. Positioned as one of the least important in Aristotle's ranking, appearances such as costumes are not necessary in order for a play to be put on. However, in the *kabuki* narratives, they are extremely essential elements and should not necessarily be subordinated under the story. In many cases, characters and stories or plots are integrated with costumes, and it is difficult to separate them. While Aristotle's (1997) systemic thinking is not entirely contrary to the real nature of *kabuki*, narrative in *kabuki* is a multi-structured, multi-dimensional system that does not fall into a single, hierarchical order of importance. A *kabuki* work may be said to be a momentary fixation in a fluid situation.

In Aristotle's way of thinking, characters that are essential but embedded in a narrative (in a narrow sense) are, in the end, elements that are embedded in the developmental logic of the story, and even if they stand out in particular, they stand out only in the story, and it is not permissible to transcend the logic of their development using out-of-narrative logic. However, in *kabuki*, characters frequently go beyond the developmental logic of the story itself. As described above, not only does the existence of "actor" itself stand in parallel with the characters in an antagonistic existence, but it is a situation that occurs often. For instance, it is normal for a character to cross the boundaries of a story when acting or dancing. Below are some examples:

- Nakamura Tokizō III (1895–1959) was gleefully performing the ningyōburi (*ningyō jōruri* puppet movements) during the sad scenes of stories (Watanabe, 1989a, p. 314). It is a common phenomenon in *kabuki* for gestures of a *kata* in a story stand out beyond the logical development of the story.
- Side stories and side characters are sometimes transformed into main stories and characters, mainly by changing clothes and makeup. For example, Nakamura Nakazō I was only a supporting actor in *Kanadehon Chūshingura*, and was killed the second he appeared on stage. Then, he changed the image or role of a villain Ono Sadakurō in this drama who only speaks one word, and constructed the style that is most popularly passed down these days. Although Ono Sadakurō in the story was only a hackneyed villain, he became an important, and ikina (stylish) character in the story played by star actors.
- In the midst of a play's story, a character can instantly transform into an actor. For example, the final part of a play (such as the scene of a duel between the main character and the villain) may be suddenly brought to an end. Entire performances forced to end with a phrase such as "Honjitsu ha koregiri (That's all for today)" are quite common.
- A very short scene in the story is positioned as a very important scene by using costumes and gestures. In the case of Kagotsurube Sato no Eizame, the instant gestures and expressions of the heroine in gorgeous clothes determine all of the tragedies that follow. This scene itself only shows the oiran-dōchū (procession of courtesans).
- It is important to show scenes and people as stationary pictures, and these pictures stand independent from the temporal flow of the story. We sincerely hope that for that brief period of time before the end of the story when the curtain closes, we can behold the beauty of a single painting that could never appear in this world. (One complaint about *kabuki* videos is that they do not allow people to fully enjoy the beauty of this picture.)

The narratives of *kabuki* are neither a tight integration in Aristotle's sense nor a multimodal fluctuation as a basis for existence, both from the structure and from the performance. Editing and producing *kabuki* narratives have a superficial affinity with the character of arbitrariness. In *kabuki*, there is a sense that there is something at least as important as the rigor and integration of the plot and the story, such as antagonism and rivalries. Classic *kabuki* scripts are originally written for specific actors. Therefore, in the play scripts, the serifu (words or lines) of a character was described under the name of an actor rather than the name of the character. The actor did not act according to the story, but the story was made for a specific actor. Although this trend has changed somewhat since the Meiji Era, this methodology, which has been cultivated for many years, has not completely been overtaken by the Aristotelian play structure. Even today, the *kabuki* narratives exist as a highly impure mixture. Liquidity and arbitrary editing seem to exist well, being underpinned by multiplicity and pluralism.

As mentioned with regard to the topic of time (era or age), for Sugawara Denju Tenarai Kagami, the historical setting is the Heian era. The hero, Kan Shōjō, modeling Sugawara Michizane, and the anti-hero, Fujiwara Shihei, modeling Fujiwara Tokihira, appear on the stage wearing the costumes of the Heian era, and one of the major characters, Takebe Genzō, appears in the garb of a samurai in the Edo era, while the main setting is a terakoya (school) in the mountains of Kyoto in the Edo style. Further, the makeup of the face of Fujiwara Shihei does not represent a human but undoubtedly that of a monster. The characteristic of having no director has also been inherited by the modern form of *kabuki*. When Yukio Mishima produced the last *kabuki* play, Chinsetsu Yumihari Zuki52 (Mishima, 2002) in Kokuritsu Gekijō (the National Theatre of Japan) in 1969, in his final days, he came to despair of the selfishness of the actors who ignored direction and wanted to make the stage and play by their own self-direction (Mishima, 2004). Therefore, although he intended to rewrite *ningyō jōruri* with a strict script, its completion was not fulfilled by the time of the author's death. Only the first volume was recorded in his complete collection (Mishima, 2002). According to the explantion by Tanaka (2002), in the "yukahon" script performed during the Kokuritsu Gekijō (National Theater of Japan) in November, 1971, the second and third volumes written by Shōichi Yamada (1925–) were also recorded together with this first volume.

Furthermore, although the following is related to various elements including person and story, we can consider it as a theme of *kabuki* direction (moreover, it is associated with a new topic, the problem of recognitions of various types of persons related to *kabuki*). Oikawa and Ogata (2012) and Ogata (2018a) take Yukio Mishima's (1963) novel *Gogo no Eikō* to suggest the idea of narrative as a model for characters' mutual perception. Mishima was a writer that was substantially influenced by *kabuki* throughout his life, and he created and staged several pseudoclassical kabiki works, such as Iwashiuru Koi no Hikiami53 and Chinsetsu Yumiharizuki (See Kitani (2007) and Nakamura (2004, pp. 355-362) for more details.) His novels are also frequently theatrical. The composition of *Gogo no Eiko* itself is extremely theatrical and even *kabuki*-esque. In particular, the movement of the children and (the fallen) hero, Ryuji, to the final scene very much gives the impression of imitating *kabuki* 's *michiyuki*. The narrative development can also be understood as based on discrepancies between the multiple characters' mutual perceptions. The aiso-zukashi (spiteful words) in Kagotsurube Sato no Eizame by Kawatake Shinshich III also follow this pattern. Aiso-zukashi or *enkiri* (severing connection) is a pattern for story development in Japanese *ningyō jōruri* and *kabuki* and usually refers to a woman intentionally acting cold against her actual feelings in order to distance herself from a man for some reason. It is characterized by the contradiction between the woman's internal emotional information and her external actions. The man does not notice her feelings and simply falls in the trap of her external actions, so that his feelings become directly connected with his external actions. In this way, aiso-zukashi is established through the division between the characters' internal and external as well as their interpersonal relations. This is a technique in *kabuki* narratives that constitutes a driving force for story development. The mutual perceptional inconsistencies seen here happens between the characters on stage as well as between the characters and the spectators in the guest seating. The misunderstanding between the two characters on stage develops the drama, while the spectators watch the drama from another viewpoint knowing that it is a misunderstanding. However, as with novels and movies, it is also possible to act it out so that the spectators will not be able to recognize the misunderstanding by only giving them access to the internal perspective of a certain character. For example, in Ichinotani Futaba Gunki, the device of Kumagai Naozane sacrificing his son Kojirō to save Taira no Atsumori is for some time not recognized by either the other characters or the audience. Alternatively, in plays, all the audience sees is basically just the external actions of the characters, so that even when everything is expressed on stage, the recognition of any misunderstandings can only ever be conjecture on the part of the spectators' perception. From this, it is possible think of the narrative form of *kabuki* as based on an external omniscient perspective, but all that the narrator provides is a scene and some circumstances, while the spectators' actions of watching and interpreting have a major and substantial influence on the narrative itself. We should rather think of stage art like *kabuki* as a manifestation of plural recognition that involves also the receiver and is not necessarily objective. An interesting task for the future is trying to take theatrical narratives like The Sailor Who Fell from Grace with the Sea and change them into a theatrical structure for interpretation and analysis).

Stage Performance

For example, the composition of the *kabuki*za performance in September 2015 (*kabuki*za, 2015b) is—(1) the first scene, the ba (scene) of Ukamuse no Ba and Shin-kiyomizu no Ba in Futatsu Chōchō Kuruwa Nikki54 by Namiki Sōsuke and others, (2) Momijigari, and (3) Hadekurabe Ise Monogatari55 by Nagawa Kamesuke, in the afternoon, and then, from Meiboku Sendai Hagi, Jomaku (the first act) Kamakura Hanamizubashi no Ba, Nimakume (the second act) Ashikaga-ke Takenoma no Ba, Mimakume Daiichiba (the first scene of the third act) Ashikaga-ke Okuden no Ba, Mimakume Dainiba (the second scene of the third act) Ashikaga-ke Yukashita no Ba, Yomakume (the fourth act) Monchūjo Taiketsu no Ba, and Ōzume (the final act) Monchūjo Ninjō no Ba, in the evening. (According to Matsui (2000), in the story of turmoil within the Date clan of Sendai, *kabuki* is famous for Meiboku Sendai Hagi by Nagawa Kamesuke and for Datekurabe Okuni *kabuki*56 by Sakurada Jisuke I and Kasanui Sensuke (1772–1800), but using these as a base a *ningyō jōruri* dramatizastion of Meiboku Sendai Hagi by Matsu Kanshi I (?–1798), et. al. was created. Dependent on these preceding works, Meiboku Sendai Hagi as a synthetic work has been played from the middle of the Meiji era until today.) The afternoon section is a midori-jōen, and is composed of (1) Gidayū *kyōgen*, (2) a dance performance, and (3) pure *kabuki* (but the play takes a Gidayū *kyōgen* form). This is a rather special combination, as it is called the "Shūzansai (Shūzan Festival)" to commemorate Nakamura Kichiemon I. Meanwhile, the evening section is an entire play (tōshi-jōen). In addition, (1) of the afternoon section was edited and revised ("hotei") by Toyoshige Imai (1964–), a *kabuki* playwright enrolled in Shōchiku, and which is not exactly the same as the original *ningyō jōruri* script. In addition, the hotei of original works of *kabuki* and *ningyō jōruri* is also conducted on at the National Theater of Japan (in a department designated as the "Bungeika (literature department or creative section)." This approaches to a "production based on a certain work or production edited by a work itself", that is, an "adaptation." An example is seen in Chikamatsu Monzaemon (Chikamatsu Zenshū [The Complete Works of Chikamatsu, 1985–1994, Tokyo: Iwanami Shoten), in the form of a part of a scenario sometimes being used for part of another work with another title. There are various examples of such type of adaptation or editing in Chikamatsu. It corresponds to an editing style of *kabuki* according to which there is gradual development through incremental changes among works. There is an interesting example concerning the midori-jōen, which is partially the practice in which a book was edited according to the form of midori-jōen (Yūda, 1965). Similar to real stage performances by a midori-jōen, this book collected only the parts of some long scenarios. Reading this book is therefore similar to the experience of viewing a *kabuki* program on stage.

Tsukushi or *Zukushi*

Tsukushi, or *zukushi*, is used as a method for collectively and concisely representing objects in a category in *kabuki* and other literatures. For example, there are comical "*sakana* (fish) *zukushi*" appearing in *Yoshitsune Senbon Zakura*, adapted in a *kabuki* work from a *ningyō jōruri* work, and a short story, *Hashi-zukushi* by Mishima (1969), including Miyoshibashi, Tsukijibashi, Irifunebashi, Akatsukibashi, Sakaibashi, and Bizenbashi). In many current television dramas, various places appear (a kind of *basho-* (place) *zukushi*). Rabelais's (1991a, 1991b, 1991c, 1991d, 1991e) novels contain various parts showing strong *tsukushi*. Therefore, *tsukushi* may be a universal phenomenon in literatures and narratives in the world. In Japanese traditional literatures and narratives especially, we frequently see many scenes showing a character's positional movement in few pages. For instance, in *Woguri* (2017), a famous work in a sekkyōbushi that was one of the original genres of *ningyō jōruri*, the hero, Oguri, was transformed to the figure of an ugly *gakiami* and went to Yunomine in Kumano through the country of Sagami, the rivers Sakawa, Odawara, Hakone, the foot of Mount Fuji, and the country of Mikawa. Oguri narrates the long journey. Additionally, a novel from the Edo era, *Tōkaidōchū Hizakurige* [Travel Cronicles of Yaji and Kita] (1995) by Jippensha Ikku (1765–1831) inserts relatively long stories into positional movement and variable towns. A positional movement typically appears in the *michiyuki* in *kabuki*. Each place of a *michiyuki* is frequently where one or more actors are dancing on the stage. A *kakiwari* (stage setting) behind one or more dancing actors is gradually or suddenly changed to another *kakiwari* to show positional change or movement. Yukio Mishima (1963), who was influenced by *kabuki* and Japanese classical literatures, shortly described in the final chapter of the novel *Gogo no Eikō*, a *michiyuki* scene where boys move after killing a hero, Ryūji, from Yamate in Yokohama to a hill in Sugita where they can see the sea.

Thus, a typical type or pattern of *tsukushi* is related to places. For example, the first scene of Sonezaki Shinjū by Chikamatsu Monzaemon is a scene of *kan'non-meguri* (*kan'non* tour; *kan'non* is a kind of god in Buddhism) by the heroine, Ohatsu. Ohatsu walks from the Daiyū temple, the first *fudasho* (each temple or shrine on the route of a pilgrim), to the Goryō shurine, the thirty-third *fudasho*. In *Michiyuki Tabiji no Yomeiri* [*Michiyuki* Her Wedding's Travel] in *Kanadehon Chūshingura*, a mother and her daughter, Tonase and Konami, who were the wife and daughter of Kakogawa Honzō, walked through Yui and Kanbara to Yamashina in Kyoto to attend Konami's wedding to Ōishi Chikara (1688–1703) before the predestined raid of the Kira mansion by Chikara, who was a member of the raid. Additionally, *Yoshinoyama Michiyuki no Ba* [Scene of *Michiyuki* at River Yoshino] (in *ningyō jōruri*, *Michiyuki Hatsune no Tabi* [*Michiyuki* the First Sound's Travel]) in Yoshitsune Senbon Zakura drew a small movement at a place in Mount Yoshino. Although, in *kabuki*, the scene is placed in the middle of spring during cherry blossom season, the season in ningyō jōruuri is the first spring with no cherry blossoms. In a *ningyō jōruri* work, *Igagoe Dōchū Sugoroku* by Chikamatsu Hanji, the narrative places gradually move according to the narrative progression from Kamakura to Numazu, Fujikawa, Okazaki, and Fushimi in Kyoto. These places are post towns in Tōkaidō. However, in a previous *kabuki* work, *Igagoe Norikake Gappa* by Nagawa Kamesuke, positional movement is not clear.

Naimaze

In an example by Tsuruya Nanboku IV, *Sumidagawa Hana no Goshozome* (*On'na Seigen*)57 was the *naimaze* of the following three *sekais* (worlds): Sumidagawa *sekai*, Seigen Sakurahime [*Sakurahime Azume Bunsho*, 58etc.] *sekai*, and Kagamiyama *sekai*.

As one type of narratology that can be evaluated is the *sekai* of *kabuki* as a playwriting method. In *kabuki*, the *sekai* is a form of a narrative or a story that is used in the production of *kabuki*. It is a comprehensive and practical concept that includes the main components of a story—the characters (name, character, relationship), historical background and events, story development patterns, and scenes. *kabuki* productions are the presence of this *sekai* that enables production from something rather than from nothing. *kabuki* has always been produced as a commercial entertainment or play from the beginning to the present, so the speed and efficiency of production and performance are emphasized, and this has produced a systematic method called the collaborative production and *katas*. The *sekai* has adopted a lot of Japanese history and traditions that were familiar to the people of the Edo era, but the *sekai* does not necessarily understand the meaning of these original books. Many *sekais* have been performed repeatedly in medieval forms such as performing arts and *ningyō jōruri* puppet dramas. In that sense, the *sekai* was boiled down together with the times, and it also changed and transformed with the times. Also, it was impossible to exclude the existence of people who shared a certain *sekai* as recognition, knowledge, and culture. For example, in the *sekai* of *Kanadehon Chūshingura*, the *sekai* of Taiheiki (1960a, 1960b, 1961) was used to avoid censorship by the Edo Shogunate at that time. Many *sekais* exist, including "Mikurai Arasoi" (Ariwara no Narihira (825–880), Ariwara no Yukihira (818–893)), "Zen (previous) Taiheiki" (Minamoto no Yorimitsu (1159–1189) and Shiten'nō (four heros)), "Ōshūzeme" (Minamoto no Yoshiie (Hachiman Tarō) (1039–1106), Abe no Sadatō (1019–1062)), "Hachinoki" (the Kamakura era), "Higashiyama" (Fuwa Banzaemon, Nagoya Sanzaburō (1572 or 1576–1603)), "Shusse Yakko" (Toyotomi Hideyoshi (1537–1598)).

Also, some examples of vengeance that relate to the *sekai*: Oiesodomono (household troubles), Kaidanmono (ghost stories), Chūshinguramono, Igauchimono (Iga revenge), Sugawara Michizane mono, Minamoto no Yoshitsune mono, Sogamono (the Soga brother), Akechi Mitsuhide mono, and such other works. The *kabuki* playwriting method (or the *ningyō jōruri* playwriting method) adds "*shukō*" (each author's unique technqiues) to this *sekai*, showing the individuality of each author. These *shukōs* seem to mean the whole of rhetoric at a smaller level than the *sekai*. In addition, the domain of rhetorical flourish encompasses rhetoric in the narrative, rhetoric in the narrative discourse, and rhetoric in the presentation of the narrative (language, image, and music).

As a *sekai*-related document, *Sekai* Kōmoku (1916) is known. This was written in the Edo era and actually *kabuki*-related people, especially the authors, used when creating *kabuki* works. In this sense, it was a pragmatic book. However, general people could not read for a long while. *Sekai* Kōmoku are divided into the following four parts:

1. The part of *kabuki* jidai-*kyōgen sekai*: This part means the world of historical plays. 58 *sekais* are contained in this category. "Yamato Takeru no Mikoto," "Heike Monogatari," and "Kōga Saburō" are examples.
2. The part of katakiuchi or tekiuchi in oie-*kyōgen sekai*: Oie-*kyōgen* means family feud tales and katakiuchi or tekiuchi means the revenge for the honor. 7 *sekais* are contained in this category. There are "Gishiden" and "Hinin Tekiuchi."
3. The part of *kabuki* sewa-*kyōgen sekai*: This means the world of contemporary and realistic plays. 74 *sekais* are contained in this category. The part includes the sekais such as "Ichidai-otoko", "Shinodazuma," "Dōjōji (temple)," and "Shakushi Meimoku."
4. The part of jingi (gods): Jingi means hevenly gods and gods on the ground. 3 *sekais* are contained in this category. "Awashima" and "Takasago" are examples.

Each *sekai* consists of the following four items:

1. The name of the *sekai*
2. Yakumei: This official title is a list of the main characters that appear in the *sekai*.
3. Hikigaki: The reference is a list of old documents in which the *sekai* appears.
4. Gidayū jōruri: Gidayū Jōruri refers to the name of the work of the ningyō jōruuri in which the *sekai* was used.

For example, the *sekai* of "Oguri" consists of the three elements:

1. Yakumei: contains 14 characters, including Oguri Hangan and Terute Hime.
2. Hikigaki: 2 references are shown, Oguri Jikki and Shinpen Kamakurashi.
3. Gidayū jōruri: 5 Gidayū jōruris are shown, including Oguri Hangan and Onikage Musashiabumi.

Furthermore, according to Sakusha Shikihō Kezairoku (1972) written by Nyūgatei Ganyū or Namiki Shōza II (?–1809), which forms one type of a *kabuki* playwriting method, according to the schema of "Sakusha Kokoroe no Koto, Goka Juyō no Den (Author's Knowledge, The Method of Five Flowers and Ten Leaves)," focusing on "*sekai* (world)," "*shukō* (technique)," and "shikumi or shigumi (structure)," Nyūgatei Ganyū placed the five types of methods— "keiyō," "yama," "yusuri," "ōkuruwa," and "kama ireru" upon the *sekai* (these show the five-stage composition of a play):

1. Keiyō: This is made to be the same as the description, which means the description of a thing is shown in the opening act.
2. Yama: This marks the "climax" of a play.
3. Yusuri: This is the part that shows trembling, with the meaning of "action."
4. Ōkuruwa: This is the place of great development.
5. Kama ireru: This means the harvest, that is, the conclusion.

Moreover, under *shukō* and shigumi, the "jo (opening)," the "ha (middle)," and the "kyū (emergency) are structured and below these, there is the "ki (cause)," the "shō (second line)," the "ten (turnaround)," and the "gō (outcome)."

TOWARD INTRODUCING *KABUKI* KNOWLEDGE INTO GIS WITH INGS: FROM ANALYSIS TO GENERATION

Based on the above framework or components, the author surveyed and analyzed several *kabuki* elements, including "person," "story," "*tsukushi* or *zukushi*," and "*naimaze*.". Although person and story were included in the fifteen elements, *tsukushi* and *naimaze* were not included. They belong to another group in the author's framework of *kabuki*'s narrative generation study. The latter elements are understood as a group of universal narrative techniques that can be applied to various elements. In particular, the fifteen *kabuki* elements themselves are not *kabuki*'s concrete narrative techniques, but the elements for constructing a *kabuki* phenomenon. They can contain various narrative techniques employed in *kabuki*. However, *tsukushi* and *naimaze* are respectively a narrative technique or a set of narrative techniques that can be used in various elements of *kabuki*. For example, the narrative technique or techniques of *naimaze* can be used in person (character), story, and place. Although the analyzed narrative techniques for person, story, *tsukushi*, and *naimaze* are not organically incorporated into INGS and GIS or GIS with INGS, they will be positioned as important knowledge components of the unified system. Therefore, the following section shows the macro level framework of GIS with INGS.

Figure 3. Expanded narrative generation by INGS and GIS

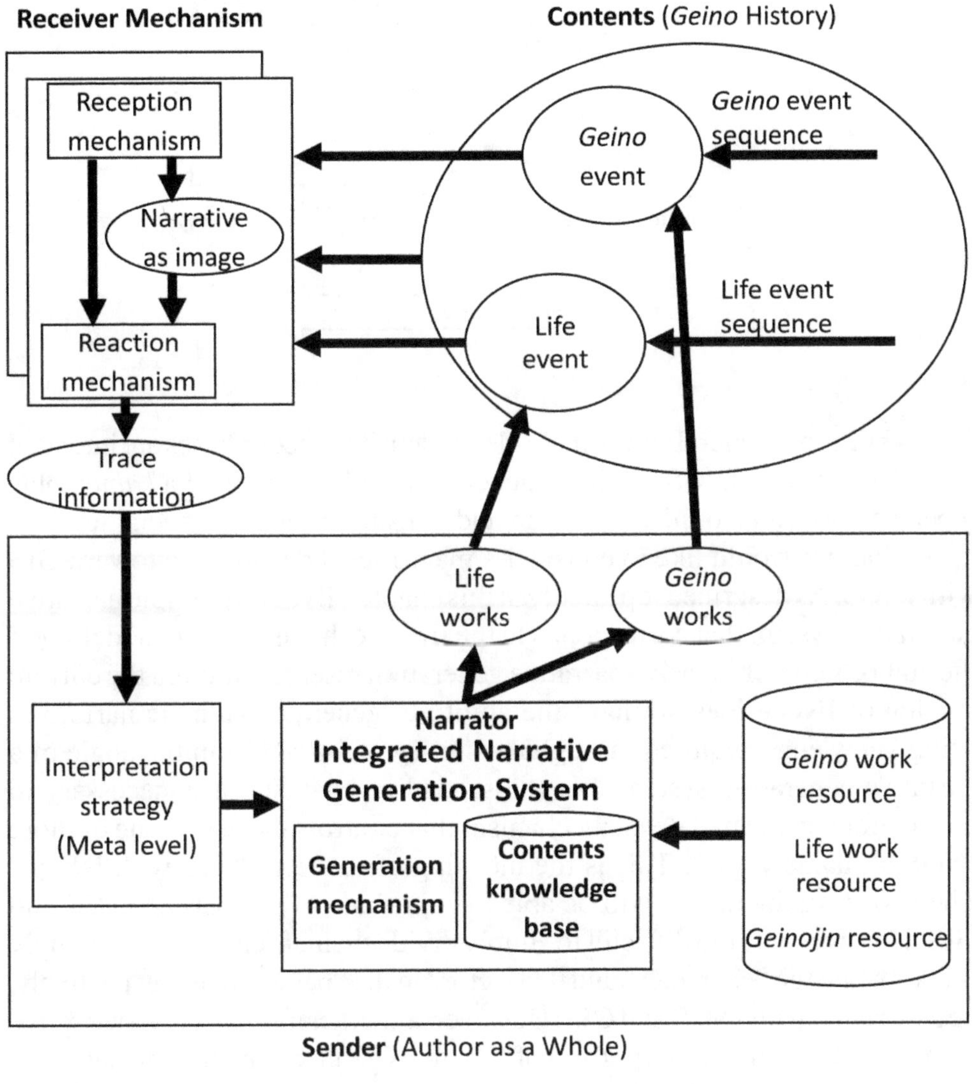

A Macro Framework of GIS With INGS

A *kabuki* play and sequence are produced through a comprehensive narrative generation and production mechanism, the GIS with INGS in the author's model, using various types of *kabuki* knowledge, as stated in the previous section. These knowledge elements correspond to each knowledge base, storing historical, cultural, and other information regarding *kabuki*. The *kabuki*-related domain specific knowledge bases are aimed to be included in

Figure 4. Multiplicity of persons and narratives

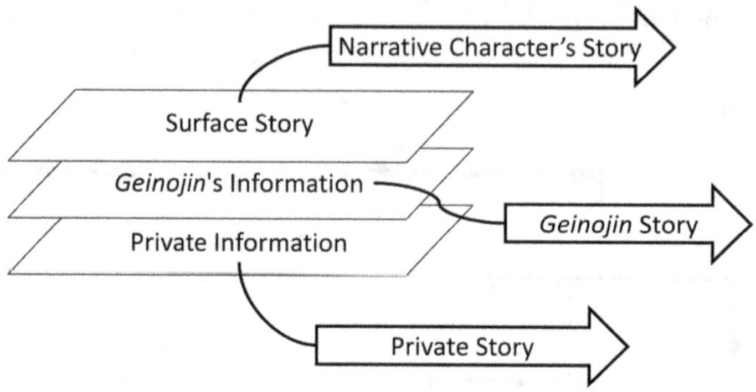

GIS, while more general and domain-dependent knowledge bases are intended to be developed in INGS. A single generation and reception of a *kabuki* play repeatedly extends to the production and consumption at the social level.

The author would like to consider a macro-level design plan toward GIS with INGS. As described repeatedly in this chapter, INGS is a simple narrative generation system that corresponds to the first mechanism in the modeling of the author's comprehensive narrative generation research, though the current version of INGS does not have the ability to generate synthetic narratives using knowledge acquired through the survey of *kabuki* from the angle of a narrative generation system. For the second mechanism, it is necessary to concentrate on the metalevel mechanism that controls the use of one or more INGS(s), namely, GIS. This is the meaning of the term "GIS with INGS." The above survey results will be able to contribute to any part of INGS and GIS. However, it is too difficult to simply divide them, for instance, by simply thinking that the "character" and the "actor" in the "person" can, respectively, be associated with INGS and GIS. However, a character is partially a narrative element to be stored into a portion of GIS in its connection with the narrative world (*sekai*). Regarding the framework of the relation between INGS and GIS, as shown in Figure 3, the INGS plays narrative generation as a simple subject. By contrast, the GIS produces sequential narratives on a broad level that includes each narrative work and narratives of geinōjins (entertainment performers) as ordinary persons, actors, and characters.

In addition, INGS is not necessarily a narrative generation model at the individual level. For example, a movie production requires many participants such as a producer, scenario writer, director, actors, and composers. As INGS is a model of the entire movie narrative production process, it is related to the narrative production and consumption process beyond a simple narrative generation and reception process. In the model of GIS with INGS, the role of GIS is not to directly produce a specific narrative work. Although GIS is also related to a narrative production process in the point that, for instance, it provides various narrative resources, including actors, GIS focuses on the aspect of the relationships among a produced narrative work and other works. In other words, an essential role is to generate a story at a different level through the narrative works produced by INGS. This role is related to the concept of circulative narrative control mentioned in Ogata (2019d). In this circulative narrative generation process, each narrative work corresponds to an event in each narrative work produced by INGS. This means that the author states GIS is a macro-level narrative production and consumption mechanism. Further, regarding the aspect of consumption, the narrative reception in INGS is mainly intended to connect to the relatively internal function of a narrative work itself. In contrast, the consumption of GIS implicates a more socially opened narrative reception through a business content service system, in many cases.

Two Multiple Elements and Two Technical Concepts: Person, Story, *Naimaze*, and *Tsukushi*

Based on the previous studies (Ogata, 2016a, 2018a, 2018c, 2019c), this section summarizes and supplements two types of multiple elements, person and story, and two types of technical concepts, *naimaze* and *tsukushi* or *zukushi*. These studies showed the multiple structures of *kabuki*'s person or people (Figure 4). In *kabuki*, there are various kinds of "persons," such as "actors (or performers)," "(dramatic) characters," and "real persons." These temporally and spatially form and possess multiple existences and relationships.

First, an actor performs a character on a *kabuki* stage. The actor him or herself has an important and essential value; in *kabuki*, "an actor plays" sometimes has a meaning that is more important than in universal dramas that focus on characters. Originally, through the Edo era, a *kabuki* scenario (daichō) was not written using the characters' names but using the actors' names. In other words, many *kabuki* scenarios were written to emphasize and strengthen the personalities and characteristics of actors or performers. Next, although the actor is a human who has a real body, a real person in *kabuki* is further divided into two parts: the real actor who is represented by his "stage name," and the real human who is symbolized by his "real name."

These correspond to the spatial forms of a person. Another unique point is the fact that a stage name, in many cases, has a temporal tradition and a history or story that is inherited over generations. This means that actors that retain stage names through a variety of different performances exist at each point in continuity with other "real" persons. An actor is located in a temporal and historical continuity. Thus, the existences of various persons and stories can be multiplied in various ways. Finally, a "person" as a character is not limited to a single *kabuki* play. A certain character in a work, for instance, can occasionally appear in other *kabuki* plays, other works in other genres, or non-fictional histories. A character is, thus, an existence that represents a particular scenario in relation to the other appearances in the temporal and spatial situations of *kabuki*. A character overlaps with his or her appearances in other *kabuki* plays and other genres.

Moreover, the author describes interesting elements regarding person or people of *kabuki*, such as "on'nagata" and "transformation (henshin) of characters," in previous papers.

Although Ogata (2018a) described characteristics of *kabuki*'s story in detail, in this section, the author simply considers the story of *kabuki* in relation to the above person or people. In a narrative generation process of *kabuki*, when a set of persons is chosen or generated by the narrative generation mechanism, one of the necessary tasks is to design, define, and form their aspects. In particular, each person as an actor (geinōjin) is used from the geinōjin resource in GIS. Although this person's level as an actor (geinōjin) is closely related to the level of private individuals, they are two overlapping but different narrative roles in GIS. Further, each person as a character is selected from the resource for narrative characters and other knowledge.

Moreover, the formation of these types of persons is associated with various kinds of stories. This shows that stories in a narrative work are not limited to one story; it may be a collection of many sub-stories. Furthermore, a narrative work has several stories corresponding to the above three types of persons. Namely, although persons as dramatic characters have respective different stories, the stories have strong relationships with the story of the narrative work itself. As many narratives are developed based on the sequence of an actor's actions or several actors' actions, the story of a narrative work itself and the actor's or actors' story or stories overlap at the level of the action sequence. Next, persons as actors or geinōjins also have respective different stories from the story of a narrative work itself and the actor-level story. Focusing on person and story in the multiple elements in *kabuki*, we know that a narrative work is constructed as the collection of many types of stories that may be associated with different types of people.

Furthermore, if the above stories are internal stories for the related persons and narrative work, external stories also exist that have indirect relationships with the persons and narrative work. For example, the story related to an actor is typically the actor's life history, or their life course that corresponds to the internal story, in this case. By contrast, the external stories in this case include the stories of other past actors who share their name. For example, although the great contemporary *kabuki* actor Bandō Tamasaburō V was the son-in-law of Morita Kan'ya XIV (1907–1975), Kan'ya was originally Bandō Tamasaburō IV and the son of Bandō Tamasaburō III (1883–1905). Further, Bandō Tamasaburō III's real name was Morita Miki; she was a female *kabuki* actress and was the fifth child of Morita Kan'ya XII, who established Shintomiza near Tsukiji and was an excellent *kabuki* producer of the Meiji era. Bandō Tamasaburō III performed as a dancer at the American World Exposition at Meiji 37 (1904) and died in New York during her training period. Thus, the story of the *kabuki* actor Bandō Tamasaburō V is connected to other Tamasaburōs' stories and, further Morita Kan'yas' stories. Moreover, a person as a private individual is also connected to external stories. For instance, the real name of Bandō Tamasaburō V is Morita Shin'ichi, a name that does not come from the world of *kabuki*. Morita Shin'ichi has unique personal stories that do not overlap with the *kabuki* world.

A very important perspective regarding internal and external stories in the above sense appears in relation with the person as a character. The internal story that is directly related to the person as a character is, of course, the story of the narrative work in which the character appears. As stated above, this level of the story overlaps with the story as a scenario of the narrative work itself. However, in *kabuki*, the concept of story and the concept of narrative are clearly different in the point that many common stories are repeatedly used in various different narrative works; the internal story in a narrative work is naturally associated and connected with the external stories. As many *kabuki* characters, which are closely related to stories or are a part of stories, are also common existences in character resources or knowledge bases, characters themselves are also divided into internal characters and external characters.

Furthermore, the problem of story related to the person as a character also has a close relationship with the concept of *naimaze* introduced above. Although, as shown in the previous section, *naimaze* was a unique concept in *kabuki* that has been explored and practiced since the Edo era, Ogata (2018a, 2019c) considered the possibilities of *naimaze* from a more general viewpoint. In particular, the author considers possible "objects" of *naimaze* in the following aspects. The elements of the multiple narrative structures of *kabuki* that can be considered the objects of *naimaze* in the ordinary sense are "real incident," "work," "material and topic," "story and plot," "person," "time or era," and "place." Furthermore, "text," "genre," "actor," and "style" may also become the objects. An important point here is that, as the story is one of the most synthetic element that contains other various multiple elements, the *naimaze* of stories involves other elements.

Ogata (2018a) analyzed an example of the *naimaze* of stories and related other elements based on research by Inumaru (2005) that demonstrated the multiplicity of *kabuki*'s stories. This book by Inumaru treats a synthetic *kabuki* work at Kawarasakiza (in Edo) in Tenpō 11 (1840) called Kibakazari Chūshingura.59 As a sequence of plays in Kibakazari Chūshingura, Sayogoromo Oshidori no Tsurugiba60 by Namiki Sōsuke, this makes up the first part, and Tōkaidō Yotsuya Kaidan and Kamikakete Sango Taisetsu by Tsuruya Nanboku IV follow. Ogata showed that this synthetic *kabuki* work was structured as the multiplicity, namely *naimaze*, of the following diverse elements comprising a story:

1. Real Event: Taiheiki (1960a, 1960b, 1961) has been treated as a historical book, and the events of Akō have been recognized as historical fact.
2. Work: The literary sequence from one work to another work and the continuation are seen.
3. Genre: There are variations on genres, such as Taiheiki as a historical narrative, Sayogoromo Oshidori no Tsurugiba and *Kanadehon Chūshingura* as *ningyō jōruri* plays, and Godairiki Koi no Fūjime by Namiki Gohei I, Tōkaidō Yotsuya Kaidan, and Kamikakete Sango Taisetsu, originally written as *kabuki* plays.
4. Material or Topic: For example, Chūshinguramono served as setting material, and the magic word Godairiki served as a motif used in several works.
5. Person: In particular, both Kō no Moronao (Kira Kōzukenosuke Yoshihisa (1641–1703) and En'ya Hangan (Asano Takuminokami Naganori (1667–1701) are regarded as actual historical persons. At the same time, in *Kanadehon Chūshingura*, Kira Kōzukenosuke and Asano Takuminokami are largely different from their actual historical selves. In addition, this *kabuki* play was, in the Kawarasakiza in Edo, performed by Ichikawa Ebizō V (previously Ichikawa Danjūrō VII) (1791–1859),

Onoe Eizaburō III (later Onoe Kikugorō IV) (1808–1860), Ichikawa Danzaburō V (?–1854), and other actors.

6. Time (Era): As an ordinary style of *kabuki*, the Muromachi era (past) and Edo era (now) were created by mixing the sequence of the plays.
7. Theater (Stage and Seats): Kawarasakiza in Edo, which was a theater that had an extraordinary and imperial meaning, was used.
8. Production, Direction, and Stage Play of Scenarios (Daichōs): Larger and comprehensive plays for actual stages were fluidly produced to contain several plays or parts.

Moreover, we can consider the concept of *tsukushi* or *zukushi* in the associations with the person and story. In previous papers (Ogata, 2016b, 2019c), the author mentioned the following as a basic standpoint on *tsukushi*. First, *tsukushi* is not necessarily a group of *kabuki*-oriented techniques and can been seen broadly in Japanese classical literary and narrative works. However, the author introduces them into the methods of multiplicity. For example, the *tsukushi* of places, which is a typical type of *tsukushi*, is connected to historical narrative memories and geographical information that has no direct relationships with a narrative. *Tsukushi* for various objects overlaps with the outside of the narrative through the introduction of encyclopedic information. As mentioned above, the ordinary multiple narrative structures in *kabuki* correspond to the methods that perform the vertical editing in a narrative. By contrast, *tsukushi* techniques are conducted on the literal editing in a narrative through which similar objects are listed, collected, and condensed in a narrative scene. Through both vertical and literary editing, more macro and large-scale multiplicity is realized in a narrative.

Compared to places and objects, considering persons and stories as the objects of *tsukushi* is not easy. However, thinking generally and theoretically, physical objects, characters, places, times, abstract objects, and all noun concepts can be the objects of *tsukushi*. As the examples of the *tsukushi* of persons, the author would like to focus on the phenomenon of the multiplicity of a character. For example, although Kyōkanoko Musume Dōjoji by Fujimoto Tobun (the middle of the Edo era) and Kineya Yasaburō I (?–1758) and Yugiri [Kuruwa Bunshō]61 were originally *kabuki* works played by a heroine, various transformational versions, such as Ninin Musume Dōjōji (Kyōkanoko Musume Dōjōji by two heroines), Gonin Musume Dōjōji (Kyōkanoko Musume Dōjōji by five heroines), and Ninin Yūgiri (Yūgiri by two heroines), appeared in which several heroines perform the same role in essence. Further, although the main character of Kirare Yosa by Segawa

Figure 5. Property information acquisition algorithm
Source: Ogata (2019c)

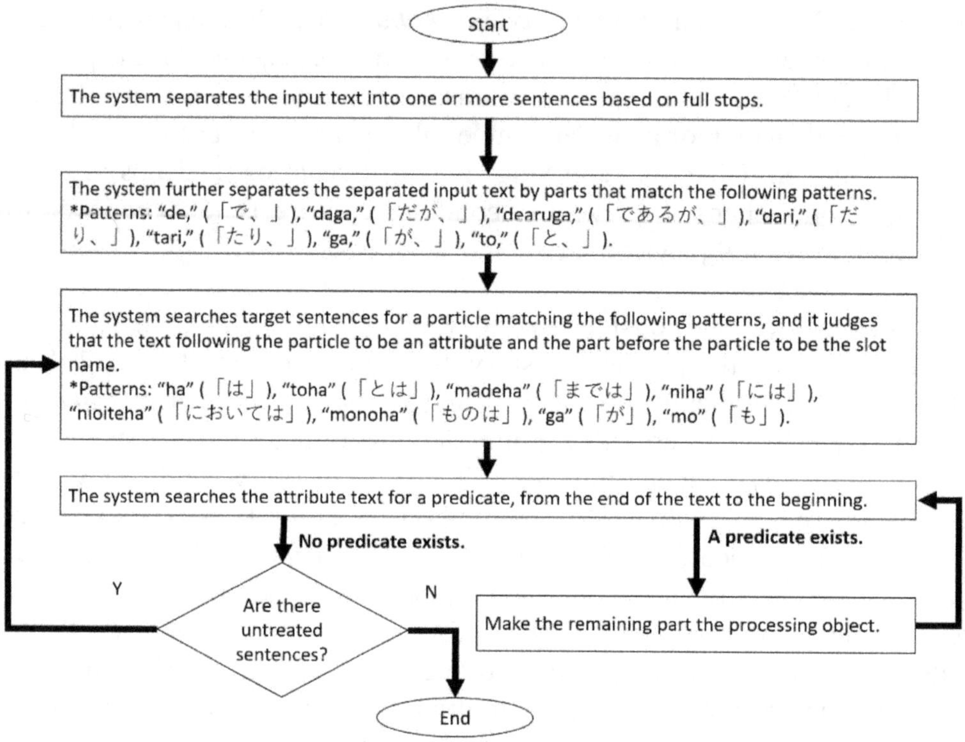

Jokō III is a man, the main character of Kirare Otomi [Musumegonomi Ukina no Yokogushi],62 by Kawatake Mokuami which is based on the same plot line with Kirare Yosa and is a kind of parody of Kirare Yosa, is a woman. In these works, the main character is the same. However, a kind of *tsukushi* effect occurs in the mind of audiences who viewed both *kabuki* works. As another example, Sugawara Denju Tenarai Kagami is a famous *kabuki* play in which respectively contrasting, unique, and attractive main characters play the central roles in the drama. It can be also interpreted as *tsukushi* of persons in a broad sense.

Next, another way to approach the problem of *tsukushi* of stories is to concretely consider how *tsukushi* is realized in a story. The story's *tsukushi* refers to how various types of elements or objects are used in a story. The direct objects of *tsukushi* typically include persons, places, and objects. For instance, in the positioning of *tsukushi* in a narrative, an object can be inserted into a narrative. The types are divided into the collective insertion into a micro

Figure 6. The description form of knowledge base
Source: Ogata (2019c)

```
(person
 (geino
  (g-base <data> <data> ...)
  (evaluation <data> <data> ...)
  (g-event <data> <data> ...)
  (work <data> <data> ...)
  (prize <data> <data> ...))
 (life
  (l-base <data> <data> ...)
  (l-event <data> <data> ...)
(family <data> <data> ...)))
```

and narrow range and are distributed in a broad range. The former is a case in which the *tsukushi* of objects is collectively inserted into a narrative place, and the latter is a case in which the objects cover a relatively broad range or appear throughout the narrative. This is part of the method of *tsukushi*. By contrast, there is a different method through which the entire structure of a story is constructed based on the principle of *tsukushi*. In such instances, basic story lines are structured by *tsukushi*. For example, the story lines of *Igagoe Norikake Gappa* and *Igagoe Dōchū Sugoroku* are developed through the *tsukushi* of places or towns along Tōkaidō, which was a main road in the Pacific side in the Edo era. An interesting feature of this narrative was its introduction of Tōkaidō and the towns by the *tsukushi* of places throughout the narrative.

Figure 7. Input and output examples of property acquisition mechanism
Source: Ogata (2019c)

Input	十一代目 市川 海老蔵(いちかわ えびぞう、1977年12月6日 -)は、日本の歌舞伎役者、俳優。歌舞伎名跡「市川海老蔵」の当代。屋号は成田屋。定紋は三升、替紋は杏葉牡丹。 ...<omitted>... 古典の大役に挑み、初役を多くつとめ、高い評価を得ている。「海老さま」と人気のあった十一代目市川團十郎に重ね合わせるファンもいる。現代の歌舞伎を担う若手スターの一人。また、NHK大河ドラマ『武蔵 MUSASHI』で主役をつとめたほか、現代劇にも挑戦している。 ...<omitted>...
Output	((("は" "歌舞伎役者、俳優。" (")は" "十一代目市川海老蔵(じゅういちだいめいちかわえびぞう、1977年12月6日 -")) (nil "nil")) (("は" "当代。" ("」の" "歌舞伎名跡「市川海老蔵")) (nil "nil")) (("は" "杏葉牡丹。" ("は" "三升、替紋" ("は" "成田屋。定紋" ("は" "屋号")))) (nil "nil")) ...<omitted>... ((("は" "。" ("を多くつとめ、高い評価を得ている" "初役" ("に挑み" "大役" ("の" "古典")))) (nil "nil")) (("は" "。" ("もいる" "ファン" ("に重ね合わす" "十一代目市川團十郎" ("のあっ" "人気" ("」と" "「海老さま"))))) (nil "nil")) (("は" "若手スターの一人。" ("を担う" "歌舞伎" ("の" "現代"))) (nil "nil")) (("は" "。" ("にも挑戦している" "ほか、現代劇" ("をつとめ" "主役" ("』で" "また、NHK大河ドラマ『武蔵 MUSASHI")))) (nil "nil")))) ...<omitted>...

Figure 8. Ichikawa Ebizō XI's property information acquired from Japanese Wikipedia
Source: Ogata (2018a)

Geino information

●Basic information

Juichidaime (the eleventh) Ichikawa Ebizo (born June 12 1977) is a kabuki actor and actor in other media. His kabuki name is "Ichikawa Ebizo" as the head of the Ichikawa clan. His acting house name is Naritaya. His family crest is Mimasu and his alternate personal crest is Gyoyobotan. He is affectionately known as "Ebi sama and "Ebi chan."

●Critical reception

He has taken on the challenge of the major roles in the kabuki classics and also played many roles for the first time, and his performances have been highly praised.

There are many fans of Ebi sama" who are also fans of the popular Juichidaime Ichikawa Danjuro. He is one of the young stars of contemporary kabuki theater. He played the leading role in Musashi, NHK's Taiga (long-running) drama, and also taken on the challenge of performing in contemporary dramas.

Preparatory System Development Based on the Analysis of *Kabuki*

This section discusses the preparatory system development using techniques acquired by previous *kabuki* analysis by the author. An objective is to consider concretely and in detail narrative functions, mechanisms, and techniques regarding *kabuki* to introduce into GIS and INGS in the future. Therefore, the proposed systems are relatively simple, although they are based on a consistent method: "generation based on editing." In particular, the systems use actual *kabuki* texts and edit them using narrative techniques acquired through the author's *kabuki* analysis. As Ogata (2018a, 2019c) showed the basic method, this section revises and expands the description based on the summarization.

First, the author explains the technique for acquiring property information of *kabuki* through existing texts. Although Ono and Ogata (2017) presented a property acquisition system from the existing texts using geinōjin information of *kabuki* actors that were analyzed and structuralized based on the above method for the useful form in computer systems, this method can be applied to other types of information, such as *kabuki* works. However, this first explanation is based on *kabuki* actors' texts. Here, the properties refer to a collection of the descriptions of characteristics of each *kabuki* actor. One of the examples can be acquired by formally describing a Wikipedia item in the following algorithm. Each unit of structured and formalized properties can be placed on a part in GIS. Figure 5 is the detailed algorithm of the property acquisition system. Further, the description form of the geinōjin information of *kabuki* actors is shown in Figure 6.

As an example, Figure 7 shows a part of the result that the algorithm transformed from the Japanese Wikipedia of Ichikawa Ebizō XI.

A SEQUENTIAL GENERATION EXAMPLE

In this section, the author shows a narrative generation process based on the connection of several generations using the techniques of person, story, *naimaze*, and *tsukushi*.

Analysis and Generation of Person's Information

The author considers how acquired, described, and structuralized property information of *kabuki* actors are connected to the GIS and INGS. Figure 3 included the conceptual figure of the GIS itself. The geinōjin information and life information of *kabuki* actors are stored in the *geinō* work resource (or *geinō* resource) and life work resource (or life resource). At the same time, Figure 3 indicated that the GIS with INGS, generates and produces from a narrative to the sequence of narratives and receives and consumes them. Moreover, this figure also showed the circulation of these processes.

Figure 9. The structure in the geinō information of Ichikawa Ebizō XI
Source: Morita and Ogata (2018)

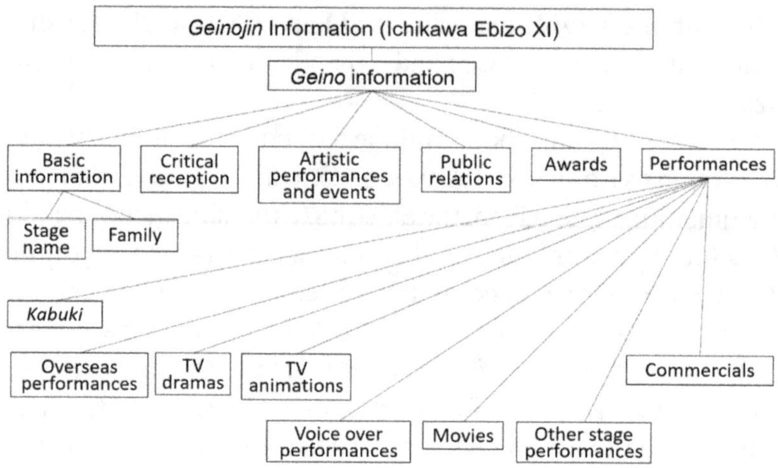

Figure 10. The structure in the life information of Ichikawa Ebizō XI
Source: Morita and Ogata (2018)

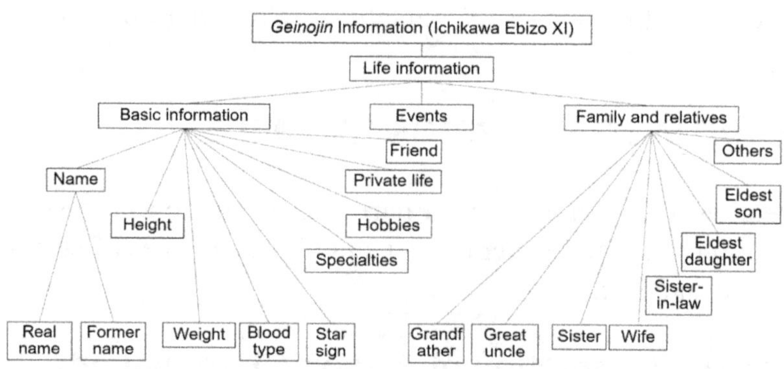

In addition, in the above system, the author basically treated geinōjin information and private information in person information. However, personal information as characters also exists. Especially in the Edo era, historical characters and fictional characters whom many people knew as general knowledge frequently and repetitively appeared in many works. They created another level of narrative, in particular, narratives of specific characters. *kabuki* has also had specific characters who frequently and repetitively appear

Figure 11. The structure in the geinōjin information of kabuki actors
Source: Morita and Ogata (2018)

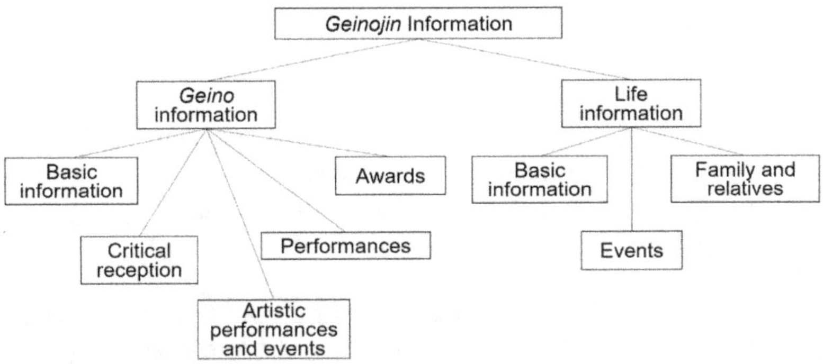

in different works. This kind of character information will be stored in a part of *geinō* resource in the GIS in the future.

In this section, for developing a research regarding the multiplicity in *kabuki*'s people, the author divided *kabuki* actors' text descriptions in Japanese Wikipedia into *geinō* resources and life resources and determined respective hierarchical structural forms. The author formalized the text data into property information that can be utilized in the author's narrative generation system, namely the GIS and INGS.

As more a detailed example, the author shows Ichikawa Ebizō XI's property information acquired from Japanese Wikipedia based on Ogata (2018a). Figure 8 shows the first part of the data and APPENDIX 2 (1) shows the complete version. Through the following process, this information will be transformed to other forms using computer programs.

Next, the author presents a simple method for treating the element of "persons" in *kabuki* by computer program. This program is designed for use in the INGS and GIS. First, the author determines the knowledge description structures of *kabuki* actors (geinōjins) through the survey and analysis of several actors. As an example, the author focuses on Ichikawa Ebizō XI and analyzes the hierarchical structure of his Japanese Wikipedia and the items are re-categorized. In particular, these items are divided into two types of information, "*geinō* information" and "life information." Further, this hierarchical structure has subcategories under the *geinō* information and life information. In the *geinō* information, the author means the information including *geinō* activities themselves and the relating information. Life information contains the private lives and families of actors. For the categorization, for example, information regarding performances and family can be directly categorized in the *geinō*

Figure 12. Property data of Ichikawa Ebizō XI

```
(setq *Ebizou*
  '(geinojinInfo
    (geino
     (g-base
      ((("は" "歌舞伎役者、俳優。" (")は" "十一代目市川海老蔵(じゅういちだいめいちかわえびぞう、1977年12月6日-")) (nil "nil"))
      (("は" "当代。" ("」の" "歌舞伎名跡「市川海老蔵")) (nil "nil"))
      (("は" "杏葉牡丹。" ("は" "三升、替紋" ("は" "成田屋。定紋" ("は" "屋号")))) (nil "nil"))
      (("は" "。" ("」がある" "海老さま」、「海老ちゃん" ("に" "愛称"))) (nil "nil"))
      (("は" "た。" ("がよくあっ" "こと" ("で揉め" "時には演出のこと" ("に上がっ" "舞台" ("などで一緒に" "健在であった当時、海外公演" ("が" "父・團十郎")))))) (nil "nil"))
      (("は" "。" ("はよくなかったという" "ときの家庭内の空気" ("もあっ" "とき" ("の意見を聞き入れて許してくる" "海老蔵" ("が" "團十郎"))))) (nil "nil"))
      (("は" "。" ("を述べている" "團十郎に常々感謝の言葉" ("にずっと優しく見守ってくれてい" "だが、反発していたとき"))) (nil "nil"))
      (("は" "た。" ("」とコメントしてい" "ん" ("は同情でき" "今回" ("がなってい" "んだから、自己管理" ("を空け" "穴" ("はさすがに激怒し" "時" ("を降板し" "歌舞伎公演" ("をし" "父・團十郎も、2007年に海老蔵が風呂場でケガ" ("を見せてい" "際に寛大な態度" ("(非嫡出子)が発覚し" "婚外子")))))))))) (nil "nil"))
      (("は" "のではないか。" ("の)けいこに入れ" "京都南座での顔見世興行" ("にも再度激怒し「(予定されてい" "際" ("を受け" "被害" ("に発生し" "2010年"))))) (nil "nil"))
      (("は" "た。" ("を突き放してい" "海老蔵" ("。憤りを感じている" "わざ" ("がなせる" "大変失礼な行動で、当人の自覚のなさ" ("では、出演すべきではないと思っている」と前置きし" "責任の上" ("を作ったと言う" "事件の原因" ("に巻き込まれたとは言え" "(事件"))))))) (nil "nil"))
      (("は" "。" ("に耐え切れず反発を繰り返していたという" "家柄・伝統の重責" ("から続く" "幼少時" ("は" "海老蔵" ("の" "頃" ("の" "10代")))))) (nil "nil"))
      (("は" "。" ("に感動したからと語っている" "祖父・十一代目市川團十郎の芸の美しさと勇姿" ("をつとめる" "弁慶" ("で見" "のは、偶然フィルム" ("を立ち直らせ" "そんな彼"))))) (nil "nil"))
      (("は" "。" ("も、もがき続けているという" "現在" ("を追うがゆえに苦しみ" "理想" ("に掲げている" "目標" ("を" "祖父"))))) (nil "nil"))
      (("は" "。" ("に感じるという" "今はプライベートの方が苦痛" ("になっ" "よう" ("に打ち込む" "ことが苦痛であったが、歌舞伎に真剣" ("に上がる" "舞台" ("は" "以前")))))) (nil "nil"))
      (("は" "。" ("よ」と返され、安心したと述べている" "ところ、「あら、それは当たり前" ("に相談し" "五代目坂東玉三郎" ("かと悩ん" "異様なこと" ("が" "これ"))))) (nil "nil"))
```

Figure 13. Generated sentences of Ichikawa Ebizō XI

十一代目市川海老蔵（じゅういちだいめいちかわえびぞう、1977年12月6日-）は歌舞伎役者、俳優。
歌舞伎名跡「市川海老蔵」の当代。
屋号は成田屋。定紋三升、替紋杏葉牡丹。
愛称に海老さま」、「海老ちゃん。
父・團十郎が健在であった当時、海外公演舞台時には演出のことことた。
團十郎が海老蔵ときときの家庭内の空気。
だが、反発していたときにずっと優しく見守ってくれてい團十郎に常々感謝の言葉。
婚外子（非嫡出子）が発覚し際に寛大な態度父・團十郎も、2007年に海老蔵が風呂場でケガ歌舞伎公演時穴んだから、自己管理今回んた。
2010年に発生し被害際京都南座での顔見世興行のではないか。
（事件に巻き込まれたとは言え事件の原因責任の上大変失礼な行動で、当人の自覚のなさわざ海老蔵た。
10代の頃海老蔵幼少時家柄・伝統の重責。
そんな彼を立ち直らせのは、偶然フィルム弁慶祖父・十一代目市川團十郎の芸の美しさと勇姿。
祖父を目標理想現在。
以前は舞台ことが苦痛であったが、歌舞伎に真剣よう今はプライベートの方が苦痛。
これが異様なこと五代目坂東玉三郎ところ、「あら、それは当たり前。
また、坂東玉三郎とことこと。
本格的な筋力トレーニングやヨガ雑誌『ターザン』などの表紙こと。
成田屋のお家芸「歌舞伎十八番うち、今日演目の復活上演を積極的。
古典の大役初役。
「海老さま」と人気十一代目市川團十郎ファン。
現代の歌舞伎若手スターの一人。
また、NHK大河ドラマ『武蔵MUSASHI』で主役ほか、現代劇。
1983年5月-歌舞伎座『源氏物語』の春宮お目見得。
1985年5月-歌舞伎座『外郎売』の坊七代目市川新之助襲名。
2000年歌舞伎座『源氏物語』の光君会場は連日満員、チケット人気五代目尾上菊之助、二代目尾上辰之助とともに「平成の三之助新たな歌舞伎ブーム。
2003年2月-日置明子との間女児、当時1歳ことが判明。子ことを会見で説明。
2004年4月4日-成田山新勝寺で襲名披露、奉告の練り。京成本線京成上野駅-京成成田駅間列車「海老蔵号京成AE100形電車を使用）。
2004年5月歌舞伎座『助六由縁江戸桜』の花川戸助六、『暫鎌倉権五郎ほか十一代目市川海老蔵襲名。
2004年6月30日-成田山大阪別院明王院で京阪本線天満橋駅-香里園駅間列車「海老蔵号京阪8000系電車使用）。

Figure 14. Edited description of Ichikawa Ebizō XI (structural lisp desctiption)

```
(setq *Ebizou*
 '(geinojinInfo
  (geino
   (g-base
    ((("は" "歌舞伎役者、俳優。" (")は" "十一代目市川海老蔵(じゅういちだいめいちかわえびぞう、1977年12月6日-")) (nil "nil"))
     (("は" "当代。" ("」の" "歌舞伎名跡「市川海老蔵")) (nil "nil"))
     (("は" "杏葉牡丹。" ("は" "三升、替紋" ("は" "成田屋。定紋" ("は" "屋号")))) (nil "nil"))))
   (work
    ((("は" "・歌舞伎") (nil "nil"))
     (("は" "弁慶、富樫左衛門" (", 勧進帳の" "精" ("の" "獅子" ("に" "小姓弥生後" ("の" "花川戸助六, 鏡獅子" ("の" "光君, 助六由縁江戸桜" ("の" "鳴神上人, 源氏物語" ("の" "片岡直次郎, 鳴神" (")の" "侍" ("と" "天衣紛上野初花「雪暮夜入谷畦道」(三千歳")))))))))) (nil "nil"))))
   (g-event
    ((("は" "お目見得。" ("で" "春宮" ("』の" "1983年5月-歌舞伎座『源氏物語"))) (nil "nil"))
     (("は" "襲名。" ("を" "七代目市川新之助" ("で" "坊" ("』の" "1985年5月-歌舞伎座『外郎売")))) (nil "nil"))
     (("は" "。" ("を起こす" "新たな歌舞伎ブーム" ("」と呼ばれ" "五代目尾上菊之助、二代目尾上辰之助とともに「平成の三之助" ("を得る" "人気" ("は発売初日に完売する" "会場は連日満員、チケット" ("をつとめ" "光君" ("』の" "2000年歌舞伎座『源氏物語"))))))) (nil "nil"))
     (("は" "ことを会見で説明。" ("は認知して養育費を支払っている" "ことが判明。子" (")がいる" "女児、当時1歳" ("に" "間" ("との" "2003年2月-日置明子"))))) (nil "nil"))
     (("は" "京成AE100形電車を使用)。" ("」に乗車し" "列車「海老蔵号" ("で運行され" "練り。京成本線京成上野駅-京成成田駅間" ("で襲名披露、奉告の" "2004年4月4日-成田山新勝寺")))) (nil "nil"))
     (("は" "襲名。" ("を" "十一代目市川海老蔵" ("で" "鎌倉権五郎ほか" ("』の" "花川戸助六、『暫" ("』の" "2004年5月歌舞伎座『助六由縁江戸桜"))))) (nil "nil"))
     (("は" "使用)。" ("を" "京阪8000系電車" ("」が" "列車「海老蔵号" ("で" "京阪本線天満橋駅-香里園駅間" ("で" "2004年6月30日-成田山大阪別院明王院"))))) (nil "nil"))
     (("は" "。" ("を呼ぶ" "現地でも大きな話題" ("』を持っている」となどと報じ" "一族の出身」「26歳のスター」「若く美男で特に『必殺の視線" ("をとどろかせてき" "17世紀末以来の日本の伝統的演劇の中で名" ("を使っ" "ル・モンド』紙が一面全面" ("も観劇し" "トップ" ("の" "市川新之助改メ十一代目市川海老蔵襲名披露フランス公演」。同国政財界" ("で" "シャイヨ宮劇場" ("の" "2004年10月-フランス・パリ"))))))))) (nil "nil"))
     (("は" "クイズ$ミリオネア』(フジテレビ)に出演。" ("』の宣伝も兼ね" "海" ("の" "2006年9月7日-映画『出口"))) (nil "nil"))
     (("は" "。" ("を驚かせる" "観客" ("で行っ" "中央通路" ("に見立て" "六方を花道" ("の飛び" "引っ込み" ("の絶賛を浴びる" "パリ" ("」の舞いを披露し" "紅葉狩" ("で" "弁慶" ("とともに実現。『勧進帳』の" "父" ("を" "歌舞伎公演" ("の" "初" ("で" "2007年3月-パリ・オペラ座")))))))))))) (nil "nil"))))
)))
```

Figure 15. Edited description of Ichikawa Ebizō XI (sentence representation)

十一代目市川海老蔵(じゅういちだいめいちかわえびぞう、1977年12月6日-)は歌舞伎役者、俳優。

歌舞伎名跡「市川海老蔵」の当代。

屋号は成田屋。定紋三升、替紋杏葉牡丹。

愛称に海老さま」、「海老ちゃん。

・歌舞伎は

天衣紛上野初花「雪暮夜入谷畦道」(三千歳と侍片岡直次郎, 鳴神鳴神上人, 源氏物語光君, 助六由縁江戸桜花川戸助六, 鏡獅子小姓弥生後獅子精弁慶、富樫左衛門

1983年5月-歌舞伎座『源氏物語』の春宮お目見得。

1985年5月-歌舞伎座『外郎売』の坊七代目市川新之助襲名。

2000年歌舞伎座『源氏物語』の光君会場は連日満員、チケット人気五代目尾上菊之助、二代目尾上辰之助とともに「平成の三之助新たな歌舞伎ブーム。

2003年2月-日置明子との間女児、当時1歳ことが判明。子ことを会見で説明。

2004年4月4日-成田山新勝寺で襲名披露、奉告の練り。京成本線京成上野駅-京成成田駅間列車「海老蔵号京成AE100形電車を使用)。

2004年5月歌舞伎座『助六由縁江戸桜』の花川戸助六、『暫鎌倉権五郎ほか十一代目市川海老蔵襲名。

2004年6月30日-成田山大阪別院明王院で京阪本線天満橋駅-香里園駅間列車「海老蔵号京阪8000系電車使用)。

2004年10月-フランス・パリのシャイヨ宮劇場市川新之助改メ十一代目市川海老蔵襲名披露フランス公演」。同国政財界トップル・モンド』紙が一面全面17世紀末以来の日本の伝統的演劇の中で名一族の出身」「26歳のスター」「若く美男で特に『必殺の視線現地でも大きな話題。

2006年9月7日-映画『出口の海クイズ$ミリオネア』(フジテレビ)に出演。

2007年3月-パリ・オペラ座で初歌舞伎公演父弁慶紅葉狩パリ引っ込み六方を花道中央通路観客。

information. On the other hand, for elements that overlap in *geinō* information and life information, new groups or categories are created. Figure 9 and 10 respectively show the structures of the *geinō* information and life information of Ichikawa Ebizō XI.

In these syntheses, the common or general structural form of geinōjin information is defined (Figure. 11).

Additionally, as full-scale examples, for Wikipedia documents of Ichikawa Ebizō XI, Figure 12 shows conceptual structures using the program explained in the above section and Figure 13 is natural lanauuge transformed by a developed sentence generation program. Figure 12 shows the partial data and APPENDIX 2 (2) corresponds to the complete version. APPENDIX 2 (3) generats sentences of Ichikawa Ebizō XI based on APPENDIX 2 (2).

Table 2. Synopses, property data, and generated sentences [Sukeroku, Narukami, Kanjinchō]

	Synopsis
Sukeroku	花川戸助六という侠客に姿をやつした曾我五郎は、源氏の宝刀「友切丸」を探すため吉原に通っている。様々な男が集まる吉原で、遊客にわざと喧嘩を吹っ掛けて刀を抜かせようというのである。そこに助六を情夫にしている花魁の揚巻と、揚巻に言い寄る髭の意休が登場。意休が友切丸を持っていると勘づいた助六は刀を抜かせようとするが、なかなかうまくいかない。そこへ白酒売に身をやつした兄の曾我十郎がやってきて弟に意見するが、助六の真意を知った十郎は自らも喧嘩を売る稽古を始める。やがて揚巻が一人の侍を伴って再登場。助六はその侍に喧嘩を売ろうとするが、驚いたことにその侍は、兄弟を心配してやってきた母の満江であった。満江は助六に破れやすい紙子の衣を着せて、激しい喧嘩を戒めると十郎とともに帰ってゆく。舞台には再び意休が登場。意休は実は助六が曾我五郎と見抜いており、友切丸を抜いて源氏を裏切ることをそそのかす。助六はもちろん応じず、意休（実は平家の残党・伊賀平内左衛門）を斬り、友切丸を取り返して吉原を抜け出す。
	[Soga no Gorō, who appears as a guest named Hanakawado Sukeroku, goes to Yoshiwara (an entertainment district in Edo) to search for Genji's treasured sword "Tomokirimaru." In Yoshiwara, which is a place where various men gather, he intentionally tries to start fights with men so that they will draw their swords, revealing if they have the sword he is searching for. Agemaki is a courtesan who is frequented by Sukeroku, and the magnificently bearded Ikyū also appears with Agemaki. Sukeroku thinks that Ikyū might have Tomokirimaru and tries to get him to draw his sword, but he is unable to. Sukeroku's older brother, Soga no Jūrō, than appears as a seller of white sake, and on hearing the opinion of his younger brother, he himself starts trying to pick fights. Soon Agemaki returns to the stage accompanied by a samurai. Sukeroku tries to pick a fight with the samurai, but surprisingly that samurai is his mother, Mankō, in disguise, who has come because she is worried about the two brothers. Mankō dresses Sukeroku in an easily breakable paper kimono and warns him against a fierce fight, and they return to Jūrō. Ikyū once again appears on the stage. Ikyū sees that Sukeroku is actually Soga no Gorō, draws Tomokirimaru and tries to persuade him to betray Genji. Sukeroku of course does not agree to this, cuts-down Ikyū, retrieves Tomokirimaru, and leaves Yoshiwara.]
Narukami	世継ぎのない天皇からの依頼をうけて、鳴神上人は寺院建立を約束に皇子誕生の願をかけ、見事これを成就させる。しかし当の天皇が寺院建立の約束を反故にしたため、怒った上人は呪術を用いて、雨を降らす竜神を滝壷に封印してしまう。それからというもの雨の降らぬ日が続き、やがて国中が旱魃に襲われ、民百姓は困りはててしまった。そこで朝廷では女色をもって上人の呪術を破ろうと、内裏一の美女・雲の絶間姫を上人の許に送り込む。姫の色仕掛けにはさすがの上人も抗しきれず、思わずその身体に触れたが最後、とうとう戒律を犯し、さらには酒に酔いつぶれて眠ってしまう。その隙を見計って姫が滝壷に張ってある注連縄を切ると封印が解け、竜神がそこから飛び出すと一天にわかにかき曇ってやがて豪雨となり、姫はその場を逃げ去る。雨の音に飛び起きた上人はやっと騙されたことに気づき烈火のごとく怒り、髪は逆立ち着ている物は炎となって姫を逃さじと、その後を追いかける。
	[The rain has stopped falling and the whole country is suffering from a drought. This is because Narukami Shōnin (a priest) holds a grudge against the Emperor for not keeping his promise, and has caged the dragon god of rain behind a waterfall. A beautiful young woman appears, heading for Shōnin's hermit cave in the northern mountain. She says that she wants to wash the clothes of her dead husband in which her feelings for him linger, but because there is no water, she has come to the waterfall to find the water to wash the clothes. To say farewell to her husband, the woman crosses the river toward Shōnin, and lifts up the hem of her skirt to reveal her white legs, and without thinking, Shōnin bends forward toward her. To the suspicious Narukami Shōnin, the woman turns and says that she wants to become a nun and his disciple. Shōnin sends out his two disciples, Hakuunbō (the white cloud bonze) and Kōkuunbō (the black cloud bonze) to bring her nun garments, and then the woman suddenly becomes sick. The surprised Shōnin attends to her, and while doing so he touches her breast. He has never touched a woman before and he is gradually unable to control his desire due to the softness of her skin. Seizing her chance, the woman brings out to drinking cups and gets Shōnin drunk, and in his drunkenness he reveals that he keeps the dragon god imprisoned behind a sacred rope across the waterfall. As Shōnin sleeps drunkenly, the woman creeps away and cuts the rope and releases the dragon god, and immediately it begins to rain heavily. The woman was in fact Kumonotaemahime (Princess), the most beautiful woman at the court, sent by the Emperor to break the power of Narukami. When he revives and becomes aware of what happened, his hair stands on end and he becomes enraged, he turns into flames, he fights and knocks down the disciples who attempt to stop him, and he chases after Kumonotaemahime.]

continued on following page

Table 2. Continued

	Synopsis
Kanjinchō	源頼朝の怒りを買った源義経一行が、北陸を通って奥州へ逃げる際の加賀国の、安宅の関での物語。義経一行は武蔵坊弁慶を先頭に山伏の姿で通り抜けようとする。辿り着いた関で、弁慶は焼失した東大寺再建のための勧進を行っていると言う。しかし、関守の富樫左衛門の元には既に義経一行が山伏姿であるという情報が届いており、山伏は通行罷りならぬと厳命する。これに憤慨した弁慶は仲間と富樫調伏の呪文を唱え、疑いを晴らそうとする。感心した富樫は先の弁慶の言葉を思い出し、勧進帳を読んでみるよう命じる。弁慶はたまたま持っていた巻物を勧進帳であるかのように装い、朗々と読み上げる。なおも疑う富樫は山伏の心得や秘密の呪文について問い質すが、弁慶は淀みなく答える。富樫は通行を許すが、部下の一人が強力に疑いをかけた。弁慶は主君の義経を金剛杖で叩き、その疑いを晴らす。危機を脱出した義経は弁慶の機転を褒めるが、弁慶はいかに主君の命を助けるためとは言え無礼を働いたことを涙ながらに詫びる。それに対して義経は優しく弁慶の手を取り、共に平家を追った戦の物語に思いを馳せる。そこへ富樫が現れ、先の非礼を詫びて酒を勧める。それに応じて、弁慶は酒を飲み、舞を披露する。舞いながら義経らを逃がした弁慶は、笈を背負って富樫に目礼。主君の後を急ぎ追いかける。
	[The story of Kaga no kuni at Ataka no Seki of Kaga when he got the anger of Minamoto no Yoritomo and fled through Hokuriku to Ōshū. Yoshitsune and others will go through Yamabushi with Musashibō Benkei at the head. Having arrived at Ataka no Seki, it is said that Benkei has been promoting to rebuild the destroyed Tōdaiji Temple. However, the information that Yoshitsune and his group are in the form of a mountain is already delivered to Togashi left guard gate of Ataka no Seki ward, and Yamabushi is strict as not to pass. Benkei, who is envyed to this, casts a Togashi Fushin spell with his friends and tries to clear up his doubts. Impressive Togashi recalls the previous Benkei words and commands him to read the Kanjinchō. Benkei dressed up as if it were a scroll book and happened to read scrolls. Togashi still doubts questions about Yamabushi's mastery and secret spells, but Benkei answers with enthusiasm. Togashi allowed the passage, but one of his subordinates strongly questioned. Benkei knocks on the Lord's Yoshitsune with a gong stick and clears the suspicion. Yoshitsune who escaped from the crisis gives up the opportunity of Benkei, but Benkei apologizes with tears that he has worked with rudeness, though to save the life of the Lord. On the other hand, Yoshitsune gently takes Benkei's hand and reminds me of the story of the war that followed Heike. Togashi appears there and apologizes for the rudeness and encourages you to drink. In response, Benkei drinks and shows dance. Benkei, who missed Yoshitsune while dancing, carried a bag on her back and gave thanks to Togashi. Benkei hurries after the lord.]
Property data	
Sukeroku	((“” “”) ((“” “”) (“と” “花川戸助六”)) ((“は” “曽我五郎” (“をやつし” “いう侠客に姿”)) (“探すため吉原通っている” “宝刀「友切丸」” (“の” “、源氏”))) ((“” “”) (“で、” “吉原” (“が集まる” “様々な男”))) ((“” “”) (“にわざと” “遊客”)) ((“” “”) (“を抜かせようと” “喧嘩を吹っ掛けて刀”)) ((“” “”) (“のである” “いう”))) ((“” “”) ((“” “”) (“と” “花魁の揚巻” (“にしている” “情夫” (“を” “助六” (“に” “そこ”))))) ((“” “”) (nil “髭の意休が登場” (“に言い寄る” “、揚巻”))) ((“” “”) (“を持っていると” “友切丸” (“が” “意休”))) ((“は” “勘づいた助六”) (“を抜かせようと” “刀”)) ((“” “”) (“、” “するが”)) ((“” “”) (nil “なかなかうまくいかない”)) ((“” “”) (“するが、” “弟に意見” (“がやってき” “兄の曽我十郎” (“をやつし” “身” (“に” “白酒売” (“へ” “そこ”)))))) ((“は” “十郎” (“を知っ” “真意” (“の” “助六”))) (“始める” “稽古” (“も喧嘩売る” “自ら”)))) ((“” “”) ((“” “”) (nil “て再登場” (“を伴っ” “侍” (“の” “一人” (“が” “やがて揚巻”))))) ((“は” “助六”) (“を売ろうと” “喧嘩” (“に” “その侍”))) ((“” “”) (“、” “するが”)) ((“” “”) (“こと” “驚いた”)) ((“は” “にその侍”) (“であった” “母の満江” (“を心配してやってき” “、兄弟”))) ((“は” “満江”) (“を戒めると” “て、激しい喧嘩” (“を着せ” “紙子の衣” (“に破れ” “助六”)))) ((“” “”) (“と” “十郎”)) ((“” “”) (nil “と”)) ((“” “”) (nil “もに帰ってゆく”)) ((“には” “舞台”) (nil “登場” (“が” “再び意休”))) ((“は” “意休”) (“と” “曽我五郎” (“が” “実は助六”))) ((“” “”) (“を裏切ること” “源氏” (“を抜い” “見抜いており、友切丸”))) ((“” “”) (nil “をそそのかす”)) ((“は” “助六”) (“を抜け出す” “吉原” (“を取り返し” “友切丸” (“）を斬り” “もちろん応じず、意休（実は平家の残党· 伊賀平内左衛門”)))))
Narukami	((“は” “鳴神上人” (“からの依頼をうけ” “天皇” (“の” “世継ぎ”))) ((“は” “鳴神上人” (“からの依頼をうけ” “天皇” (“の” “世継ぎ”))) (“させる” “願かけ見事これ成就” (“の” “皇子誕生” (“に” “寺院建立約束”)))) ((“は” “上人” (“、怒っ” “ため” (“にし” “反故” (“の約束を” “寺院建立” (“が” “しかし当の天皇”))))) (“てしまう” “竜神滝壷封印” (“降らす” “雨” (“用い” “呪術”)))) ((“” “”) (nil “それからと”)) ((“は” “民百姓” (“に襲われ” “国中が旱魃” (“が続き” “日” (“の降ら” “いうもの雨”)))) (nil “困りはててしまった”))) ((“” “”) ((“” “”) (“を破ろうと” “上人の呪術” (“をもっ” “女色” (“では” “そこで朝廷”)))) ((“” “”) (“に送り込む” “許” (“の” “上人” (“を” “絶間姫” (“の” “美女· 雲” (“の” “、内裏一”)))))) ((“には” “色仕掛け” (“の” “姫”)) (“、と” “最後” (“に触れ” “身体” (“も抗しきれ” “さすが上人”)))) ((“” “”) (nil “うと”)) ((“は” “、さらに” (“を犯し” “う戒律”)) (“に酔いつぶれて眠ってしまう” “酒”)) ((“” “”) (“を切ると” “注連縄” (“に張ってある” “姫が滝壷” (“を見計っ” “その隙”)))) ((“” “”) (“がそこから飛び出すと” “封印が解け、竜神”)) ((“” “”) (“と” “豪雨” (“にかき曇っ” “一天にわか”))) ((“は” “なり、姫”) (“を逃げ去る” “その場”)) ((“は” “上人” (“に飛び起き” “音” (“の” “雨”))) (nil “やっと”)) ((“” “”) (“こと” “騙された”)) ((“” “”) (“のごと” “に気づき烈火”)) ((“は” “く怒り、髪”) (“と” “物炎” (“着ている” “逆立ち”))) ((“” “”) (“を逃さじと” “なって姫”)) ((“” “”) (“、その後を追いかける” “”)))

Table 2. Continued

	Synopsis
Kanjinchō	((“は” “nil”) ((“は” “nil”) (“が、” “源義経一行” (“を買っ” “怒り” (“の” “源頼朝”)))) ((“は” “nil”) (nil “際の加賀国の、安宅の関での物語” (“へ逃げる” “奥州” (“を通っ” “北陸”))))) ((“は” “義経一行”) ((“は” “義経一行”) (“で通り抜けようと” “姿” (“の” “山伏” (“に” “先頭” (“を” “武蔵坊弁慶”))))) ((“は” “nil”) (nil “する”)) ((“は” “nil”) (“で、” “辿り着いた関”)) ((“は” “弁慶”) (“を行っていると” “焼失した東大寺再建のための勧進”)) ((“は” “nil”) (nil “言う”)) ((“には” “元” (“の” “富樫左衛門” (“の” “しかし、関守”))) (“であると” “山伏姿” (“が” “既に義経一行”))) ((“は” “山伏” (“が届いており” “いう情報”)) (“罷りならぬと” “通行”)) ((“は” “nil”) (“する” “厳命”)) ((“は” “弁慶” (“に憤慨し” “これ”)) (“と” “仲間”)) ((“は” “nil”) (“を晴らそうと” “疑い” (“を唱え” “呪文” (“の” “富樫調伏”)))) ((“は” “nil”) (nil “する”))) ((“は” “感心した富樫”) ((“は” “感心した富樫”) (“を読んでみるよう命じる” “、勧進帳” (“を思い出し” “言葉” (“の” “弁慶” (“の” “先”))))) ((“は” “弁慶”) (“と” “朗々” (“に装い” “たまたま持っていた巻物を勧進帳であるかのよう”))) ((“は” “nil”) (nil “読み上げる”)) ((“は” “なおも疑う富樫”) (“について問い質すが、” “呪文” (“の” “秘密” (“や” “心得” (“の” “山伏”))))) ((“は” “弁慶”) (“なく答える” “淀み”))) ((“は” “富樫”) ((“は” “富樫”) (“を許すが、” “通行”)) ((“は” “nil”) (“をかけた” “疑い” (“に” “強力” (“が” “一人” (“の” “部下”))))) ((“は” “弁慶”) (“を晴らす” “疑い” (“で叩き” “金剛杖” (“を” “義経” (“の” “主君”)))))) ((“は” “義経” (“を脱出し” “危機”)) ((“は” “義経” (“を脱出し” “危機”)) (“褒めるが、” “機転” (“の” “弁慶”))) ((“は” “弁慶”) (“を助けるためと” “命” (“の” “いかに主君”))) ((“は” “”) (“を働いたこと” “言え無礼”)) ((“は” “nil”) (“ながらに詫びる” “を涙”)) ((“は” “義経” (“に対して” “それ”)) (“を馳せる” “優しく弁慶の手を取り、共に平家を追った戦の物語に思い”)) ((“は” “nil”) (“を勧める” “酒” (“を詫び” “先の非礼” (“が現れ” “富樫” (“へ” “そこ”))))) ((“は” “弁慶” (“に応じ” “それ”)) (“する” “舞を披露” (“を飲み” “酒”))) ((“は” “弁慶” (“を逃がし” “舞いながら義経ら”)) (nil “富樫に目礼” (“背負っ” “、笈”))) ((“は” “nil”) (“の後を急ぎ追いかける” “主君”)))
Generated sentences	
Sukeroku	花川戸助六という侠客に姿をやつし曾我五郎は、源氏の宝刀「友切丸」探すため吉原通っている様々な男が集まる吉原で、遊客にわざと喧嘩を吹っ掛けて刀を抜かせようというのである 情夫にしている花魁の揚巻と、揚巻に言い寄る髭の意休が登場。 意休が友切丸を持っていると勘づいた助六は刀を抜かせようとするが、なかなかうまくいかない 兄の曾我十郎がやってき弟に意見するが、真意を知っ十郎は自らも喧嘩売る稽古始める 侍を伴って再登場。助六はその侍に喧嘩を売ろうとするが、驚いたことにその侍は、兄弟を心配してやってき母の満江であった 満江は紙子の衣を着せて、激しい喧嘩を戒めると十郎とともに帰ってゆく 舞台には再び意休が登場。 意休は実は助六が曾我五郎と見抜いており、友切丸を抜い源氏を裏切ることをそそのかす 助六は友切丸を取り返し吉原を抜け出す
Narukami	天皇からの依頼をうけ鳴神上人は皇子誕生の願かけ見事これ成就させるため、怒った上人は雨降らす竜神滝壷封印てしまう それからと国中が旱魃に襲われ民百姓は困りはててしまった 女色をもっ上人の呪術を破ろうと上人の許に送り込む 姫の色仕掛けには身体に触れ最後、とうとう戒律を犯し、さらには酒に酔いつぶれて眠ってしまう 姫が滝壷に張ってある注連縄を切ると封印が解け、竜神がそこから飛び出すと一天にわかにかき曇っ豪雨となり、姫はその場を逃げ去る 音に飛び起き上人はやっと騙されたことに気づき烈火のごとく怒り、髪は逆立ち着ている物炎となって姫を逃さじと、その後を追いかける
Kanjinchō	怒りを買っ源義経一行が、奥州へ逃げる際の加賀国の、安宅の関での物語。 義経一行は山伏の姿で通り抜けようとする。 辿り着いた関で、弁慶は焼失した東大寺再建のための勧進を行っているとは言う。富樫左衛門の元には既に義経一行が山伏姿であるという情報が届いており山伏は通行罷りならぬと厳命する これに憤慨し弁慶は仲間と呪文を唱え疑いを晴らそうとする。 感心した富樫は言葉を思い出し、勧進帳を読んでみるよう命じる弁慶はたまたま持っていた巻物を勧進帳であるかのように装い朗々と読み上げる。なおも疑う富樫は秘密の呪文について問い質すが、弁慶は淀みなく答える 富樫は通行を許すが、強力に疑いをかけた 弁慶は金剛杖で叩き疑いを晴らす。 危機を脱出し義経は弁慶の機転褒めるが、弁慶はいかに主君の命を助けるためとは言え無礼を働いたことを涙ながらに詫びる。 それに対して義経は優しく弁慶の手を取り、共に平家を追った戦の物語に思いを馳せる。 先の非礼を詫び酒を勧める。それに応じ弁慶は酒を飲み舞を披露する。舞いながら義経らを逃がし弁慶は、笈背負っ富樫に目礼。主君の後を急ぎ追いかける

Table 3. Explanations, property data, and generated sentences [Sukeroku, Narukami, Kanjinchō]

	Explanation
Sukeroku	『助六』は、歌舞伎の演目の一つの通称。本外題は主役の助六を務める役者によって変わる。江戸の古典歌舞伎を代表する演目のひとつ。「粋」を具現化した洗練された江戸文化の極致として後々まで日本文化に決定的な影響を与えた。歌舞伎宗家市川團十郎家のお家芸である歌舞伎十八番の一つで、その中でも特に上演回数が多く、また上演すれば必ず大入りになるという人気演目である。『助六』は歌舞伎の形式上「曾我もの」の演目。そのため侠客の助六が「実ハ曾我五郎」で、白酒売りは「実ハ五郎の兄曾我十郎」という設定である。助六のモデルではないかと考えられている人物は三人いる。江戸浅草の米問屋あるいは魚問屋の大店に大捌助六あるいは戸澤助六という若旦那がいたという説、京・大坂でその男気をもって名を馳せた助六という侠客だとする説、そして江戸・蔵前の札差で、粋で気前のいい文化人として知られた大口屋暁雨だとする説である。このうち、史家の多くは第一の助六を否定する。その理由の一つが「助六」という名。これは上方でならありそうな名だが、江戸の「粋」の感覚からはどうにも野暮な名なのだという。
	[Sukeroku is one of the kabuki's popular names. This extra topic changes with the actors who play the protagonist. It is one of the works representing Edo classical kabuki. As the ultimate in refined Edo culture that embodies "iki (sui)," it has had a decisive influence on Japanese culture until later. It is one of the Kabuki Jūhachiban, one of the family arts of the kabuki's main family, Ichikawa family, with a particularly high number of performances among them, and it is a popular program that will always be a large show if it is performed. Suketoku is a form of Sogamono in the form of kabuki. For this reason, the customer's supporting party is "Jitsu ha (his true name is) Soga no Gorō", and the white liquor seller is set as "Jitsu ha (his true name is) Gorō's big brother Soga no Jūrō." There are three people who are believed to be a supporting model. One is a master of Edo Asakusa's rice wholesaler or fish wholesaler's large store. There was a young husband named Ōsabaki Sukeroku or Tozawa Sukeroku. The second is the theory that it is a customer named Sukeroku who has a distinguished name with the masculinity in Kyoto and Osaka. The third is Ōguchiya Gyōu. In Edo, it is the theory that it is who is known as a smart and generous cultural person by the tag difference of the storehouse. Among these, many historians deny the first aid. One of the reasons is the name "Suketoku". This is a name that seems likely to be above, but it is said that it is a wild name from Edo's sense of "iki".]
Narukami	『鳴神』とは、歌舞伎十八番のひとつ。この『鳴神』は貞享元年（1684年）の正月、初代市川團十郎が三升屋兵庫の名で台本を書き江戸中村座の『門松四天王』（かどまつしてんのう）において上演したものが濫觴であるが、現行で上演されているものは、寛保2年（1742年）に大坂で上演された『雷神不動北山桜』がもとになっている。『雷神不動北山桜』は、その三幕目が『毛抜』、四幕目がこの『鳴神』(正式には、雷神不動北山桜北山岩屋の場)、そして五幕目大切が『不動』となっていて、今日では『毛抜』と『鳴神』は独立した芝居として上演されることが多い。そしてそのいずれもが七代目市川團十郎によって歌舞伎十八番に撰ばれたが、その後『鳴神』は嘉永4年（1851年）に八代目團十郎が演じて以降、九代目團十郎は自分の柄にあわないとして演じなかったので上演が絶えていた。その後、明治43年 (1910年) に二代目市川左團次が岡鬼太郎と提携し、演出を改めて上演に漕ぎ着けた。なお二代目左團次はこの前年、前述の『毛抜』も復活上演しており、現在ではともに歌舞伎の人気演目のひとつになっている。
	[Narukami is one of the Kabuki Jūhachiban. Narukami was staged in the first 1684, when Ichikawa Danjūrō I wrote a script in the name of Mimasuya Hyōgo and wrote in the name Kadomatsu Shiten'nō for Edo Nakamuraza. What is currently being staged is based on the Narukami Fudō Kitayama Zakura which was staged at Osaka in Kanpō 2 (1742). The third act of the Narukami Fudō Kitayama Zakura is Kenuki, the fourth act is this Narukami (officially, the place of the Narukami Fudō Kitayama Zakura, Kitayama Iwaya no Ba), and the fifth act is Fudō. Nowadays, Kenuki and Narukami are often staged as independent plays. And all of them were enshrined by the seventh Ichikawa Danjūrō in Kabuki Jūhachiban but after that Narukami was played by Kayō 4 (1851) and after the eighth Danjūrō played, the nineth Danjūrō did not played by his own. In particular, the presentation was over because he did not play as he did not match the handle. After that, the second Ichikawa Sadanji left a tie-up with Oka Onitarō in 1910, and started directing for a while again. The second Sadanji has a revival stage in the previous year, Kenuki, mentioned above, and Narukami are now popular kabuki performances together.]
Kanjinchō	原形は初代市川團十郎が元禄15年（1702年）2月初演の『星合十二段』に取り入れたのが最初。これを五代目市川海老蔵（七代目團十郎）が能の様式を取り入れ、天保11年（1840年）3月に江戸の河原崎座で初演された。配役は海老蔵の弁慶、八代目市川團十郎の義経、二代目市川九蔵（後の六代目團蔵）の富樫左衛門。作詞は三代目並木五瓶、作曲は四代目杵屋六三郎だった。ただ、この上演では富樫の番卒が時代物の軍兵姿であり、弁慶の水衣も棒縞であるなど（公開されている錦絵「源義経と武蔵坊弁慶と富樫左衛門」を参照）、今日の舞台とはいろいろな違いがあった。現在見られる型は明治時代、九代目市川團十郎により完成されたもので、これ以降、番卒は狂言風の衣装に、弁慶の水衣も黒地に金糸の梵字をあしらったものへと改められた。かつては市川宗家（團十郎家）のお家芸として、他家の役者が勝手に上演できなかった。しかし、九代目團十郎の没後、宗家に有力な後継者がいなかった事情もあって、他の役者が弁慶を演じるようになっていく。以後『勧進帳』は、弁慶・義経・富樫の三役を歴代の看板役者が生涯に一度は演じるという、歌舞伎の代表作の一つとなった。特に昭和初期の七代目松本幸四郎の弁慶・六代目尾上菊五郎の義経・十五代目市村羽左衛門の富樫は近来の白眉とされ、1943年（昭和18年）歌舞伎座で上演された舞台は映画にも記録された。

continued on following page

Table 3. Continued

	Explanation
	[It is the first time that the original form was adopted by Ichikawa Danjurō in the Hoshiawase Jūnidan, which premiered in February of Genroku 15 (1702). The fifth Ichikawa Ebizō (seventh Danjūrō) adopted the style of nō, and premiered in March in Tenpō 11 (1840) at Kawarasakiza in Edo. Casts are Benkei in Ebizō, Yoshitsune in the eighth Danjūrō, Togashi Saemon in the second Ichikawa Kuzō (later the sixth Ichikawa Danzō). The lyrics were the third Namiki Gohei, and the composition was the fourth Kineya Rokusaburō. However, in this performance, Togashi's warrior is a soldier's figure of the period, and Benkei's Mizugoromo (water cloth) is a bar stripe (see the published nishikie, Minamoto no Yoshitsune and Musashibō Benkei and Togashi Saemon). There were many differences from the stage. The type currently seen was completed by the ninth Ichikawa Danjurō during the Meiji era, and since that time, the guard has changed into a kyōgen style costume, and the Benkei's aquatic coat to a black ground with a gold thread knurled. In the past, the actors of other family members could not perform as their own art of Ichikawa Main Family (Danjūrō's family). However, after the death of the ninth Danjūrō, other actors began to play Benkei, partly because the Main Family had no leading successor. Since then, Kanjinchō has become one of kabuki's masterpieces, with successive signboard actors playing the three roles of Benkei, Yoshitsune, and Togashi once in a lifetime. Especially in the early Shōwa period, seventh Matsumoto Kōshirō's Benkei, sixth Onoe Kikugorō's Yoshitsune, fifteenth Ichimura Uzaemon's Togashi is regarded as a recent excerent stage, and it was staged in Kabukiza in 1943 (Shōwa 18) was also recorded in the movie.]
Property data	
Sukeroku	(("は" "『助六』") (("は" "『助六』") (nil "通称" ("の" "一つ" ("の" "演目" ("の" "、歌舞伎"))))) (("は" "本外題") ("によって変わる" "役者" ("を務める" "助六" ("の" "主役"))))) (("は" "nil") (("は" "nil") (nil "演目のひと" ("を代表する" "古典歌舞伎" ("の" "江戸")))) (("は" "nil") (nil "つ")) (("は" "nil") ("と" "「粋」を具現化した洗練された江戸文化の極致")) (("は" "nil") ("を与えた" "して後々まで日本文化に決定的な影響")) (("は" "nil") ("で、" "一つ" ("の" "お家芸である歌舞伎十八番" ("の" "歌舞伎宗家市川團十郎家")))) (("は" "nil") ("になると" "大入り" ("でも特に上演回数が多く、また上演すれ" "その中"))) (("は" "nil") ("である" "いう人気演目"))) (("は" "『助六』") (("は" "『助六』") (nil "演目" ("もの」の" "形式上「曾我" ("の" "歌舞伎")))) (("は" "nil") ("で、" "八曾我五郎」" ("が" "助六" ("の" "そのため侠客")))) (("は" "白酒売り") ("と" "兄曾我十郎」" ("の" "「実八五郎"))) (("は" "nil") ("である" "いう設定"))) (("は" "nil") (("は" "nil") ("ではないかと" "モデル" ("の" "助六"))) (("は" "考えられている人物") ("いる" "三人")) (("は" "nil") ("と" "捌助六あるいは戸澤助六" ("に" "店" ("の" "米問屋あるいは魚問屋" ("の" "江戸浅草"))))) (("は" "nil") ("がいたと" "いう若旦那")) (("は" "nil") ("と" "助六" ("を馳せ" "名" ("をもっ" "いう説、京・大坂でその男気")))) (("は" "nil") ("だと" "いう侠客")) (("は" "nil") ("で、" "札差" ("の" "する説、そして江戸・蔵前"))) (("は" "nil") ("と" "文化人" ("の" "気前" ("で" "粋")))) (("は" "nil") ("だと" "して知られた大口屋暁雨")) (("は" "nil") ("である" "する説"))) (("は" "多く" ("の" "このうち、史家")) (("は" "多く" ("の" "このうち、史家")) ("する" "否定" ("を" "第一助六"))) (("は" "nil") ("と" "助六」" ("が" "一つ" ("の" "その理由")))) (("は" "nil") (nil "いう名")) (("は" "これ") ("だが、" "そうな名" ("でならあり" "上方"))) (("は" "nil") ("なのだと" "野暮な名" ("からは" "感覚" ("」の" "粋" ("の" "江戸"))))) (("は" "nil") (nil "いう"))) (("は" "nil") (("は" "nil") ("と" "『鳴神』")) (("は" "") (nil "ひと" ("の" "、歌舞伎十八番"))) (("は" "nil") (nil "つ")))
Narukami	(("は" "この『鳴神』") (("は" "この『鳴神』") ("であるが、" "ものが濫觴" ("してんのう）において上演し" "江戸中村座の門松四天王』（かどまつ" ("を書き" "台本" ("で" "名" ("の" "三升屋兵庫" ("が" "正月、初代市川團十郎" ("）の" "貞享元年（1684年")))))))) (("は" "もの" ("で上演されている" "現行")) (nil "もと" ("』が" "坂た『雷神不動北山桜" ("）に" "、寛保2年（1742年")))) (("は" "nil") (nil "になっている"))) (("は" "『雷神不動北山桜』") (("は" "『雷神不動北山桜』") ("が不動と" "場)、そして五幕目大切" ("の" "雷神不動北山桜北山岩屋" ("には" "鳴神(正式" ("が" "毛抜』、四幕目" ("が" "、その三幕目")))))) (("は" "nil") ("と" "なっていて、今日では『毛抜』")) (("は" "『鳴神』") ("と" "独立した芝居")) (("は" "nil") ("されること" "して上演")) (("は" "nil") (nil "が多い")) (("は" "nil") ("に撰ばれたが、" "歌舞伎十八番" ("によって" "七代目市川團十郎" ("もが" "そしてそのいずれ")))) (("は" "その後『鳴神』") ("にあわないと" "以降、九代目團十郎自分の柄" ("が演じ" "八代目團十郎" ("）に" "嘉永4年（1851年")))) (("は" "nil") ("が絶えていた" "して演じなかったので上演")) (("は" "nil") ("と" "岡鬼太郎" ("が" "二代目市川左團次" ("に" "その後、明治43年(1910年)")))) (("は" "nil") ("に漕ぎ着けた" "提携し、演出を改めて上演"))) (("は" "なお二代目左團次") (("は" "なお二代目左團次") ("しており、現在ではと" "復活上演" ("』も" "毛抜" ("、前述の" "この前年")))) (("は" "nil") (nil "ひと" ("の" "人気演目" ("の" "もに歌舞伎")))) (("は" "nil") (nil "つになっている")))

continued on following page

Table 3. Continued

	Explanation
Kanjinchō	((“は” “nil”) ((“は” “nil”) (“が、” “源義経一行” (“を買っ” “怒り” (“の” “源頼朝”)))) ((“は” “nil”) (nil “際の加賀国の、安宅の関での物語” (“へ逃げる” “奥州” (“を通っ” “北陸”))))) ((“は” “義経一行”) ((“は” “義経一行”) (“で通り抜けようと” “姿” (“の” “山伏” (“に” “先頭” (“を” “武蔵坊弁慶”))))) ((“は” “nil”) (nil “する”)) ((“は” “nil”) (“で、” “辿り着いた関”)) ((“は” “弁慶”) (“を行っていると” “焼失した東大寺再建のための勧進”)) ((“は” “nil”) (nil “言う”)) ((“には” “元” (“の” “富樫左衛門” (“の” “しかし、関守”))) (“であると” “山伏姿” (“が” “既に義経一行”))) ((“は” “山伏” (“が届いており” “いう情報”)) (“罷りならぬと” “通行”)) ((“は” “nil”) (“する” “厳命”)) ((“は” “弁慶” (“に憤慨し” “これ”)) (“と” “仲間”)) ((“は” “nil”) (“を晴らそうと” “疑い” (“を唱え” “呪文” (“の” “富樫調伏”)))) ((“は” “nil”) (nil “する”))) ((“は” “感心した富樫”) ((“は” “感心した富樫”) (“を読んでみるよう命じる” “、勧進帳” (“を思い出し” “言葉” (“の” “弁慶” (“の” “先”))))) ((“は” “弁慶”) (“と” “朗々” (“に装い” “たまたま持っていた巻物を勧進帳であるかのよう”))) ((“は” “nil”) (nil “読み上げる”)) ((“は” “なおも疑う富樫”) (“について問い質すが、” “呪文” (“の” “秘密” (“や” “心得” (“の” “山伏”))))) ((“は” “弁慶”) (“なく答える” “淀み”))) ((“は” “富樫”) ((“は” “富樫”) (“を許すが、” “通行”)) ((“は” “nil”) (“をかけた” “疑い” (“に” “強力” (“が” “一人” (“の” “部下”))))) ((“は” “弁慶”) (“を晴らす” “疑い” (“で叩き” “金剛杖” (“を” “義経” (“の” “主君”)))))) ((“は” “義経” (“を脱出し” “危機”)) ((“は” “義経” (“を脱出し” “危機”)) (“褒めるが、” “機転” (“の” “弁慶”))) ((“は” “弁慶”) (“を助けるためと” “命” (“の” “いかに主君”))) ((“は” “”) (“を働いたこと” “言え無礼”)) ((“は” “nil”) (“ながらに詫びる” “を涙”)) ((“は” “義経” (“に対して” “それ”)) (“を馳せる” “優しく弁慶の手を取り、共に平家を追った戦の物語に思い”)) ((“は” “nil”) (“を勧める” “酒” (“を詫び” “先の非礼” (“が現れ” “富樫” (“へ” “そこ”))))) ((“は” “弁慶” (“に応じ” “それ”)) (“する” “舞を披露” (“を飲み” “酒”))) ((“は” “弁慶” (“を逃がし” “舞いながら義経ら”)) (nil “富樫に目礼” (“背負っ” “、笈”))) ((“は” “nil”) (“の後を急ぎ追いかける” “主君”)))
Generated sentences	
Sukeroku	『助六』は一つの通称。 本外題は助六を務める役者によって変わる古典歌舞伎を代表する演目のひと。 「粋」を具現化した洗練された江戸文化の極致として後々まで日本文化に決定的な影響を与えたお家芸である歌舞伎十八番の一つで、その中でも特に上演回数が多く、また上演すれ大入りになるという人気演目である。 『助六』は形式上「曾我もの」の演目。 助六が八曾我五郎」で、白酒売りは「実八五郎の兄曾我十郎」という設定である。 助六のモデルではないかと考えられている人物は三人いる 店に捌助六あるいは戸澤助六という若旦那がいたと名を馳せ助六という侠客だとする説、そして江戸・蔵前の札差で、は気前の文化人として知られた大口屋暁雨だとする説である。 このうち、史家の多くは第一助六を否定する。 一つが助六」という名これは上方でならありそうな名だが、感覚からは野暮な名なのだという。
Narukami	『鳴神』とは、歌舞伎十八番のひと。 この『鳴神』は江戸中村座の門松四天王』(かどまつしてんのう)において上演しものが濫觴であるが、現行で上演されているものは『雷神不動北山桜』がもとになっている。 『雷神不動北山桜』は雷神不動北山桜北山岩屋の場)、そして五幕目大切が不動となっていて、今日では『毛抜』と『鳴神』は独立した芝居として上演されることが多い。 七代目市川團十郎によって歌舞伎十八番に撰ばれたが、その後『鳴神』は八代目團十郎が演じ以降、九代目團十郎自分の柄にあわないとして演じなかったので上演が絶えていた。 二代目市川左團次が岡鬼太郎と提携し、演出を改めて上演に漕ぎ着けた。 なお二代目左團次は毛抜』も復活上演しており、現在では人気演目のひとつになっている。
Kanjinchō	原形は星合十二段』に取り入れのが最初。様式を取り入れ天保11年3月に江戸の河原崎座で初演された配役は義経、二代目市川九蔵の富樫左衛門。作詞は三代目並木五瓶、作曲四代目杵屋六三郎だった 時代物の軍兵姿であり、弁慶の水衣も棒縞であるなど、今日舞台とはいろいろな違いがあった 現在見られる型は明治時代、九代目市川團十郎により完成されたもので、これ以降、番卒は金糸の梵字をあしらったものへと改められた。 かつては市川宗家のお家芸として、他家の役者が勝手に上演できなかった有力な後継者がい事情もあって他役者弁慶を演じるようなっていく 以後『勧進帳』は生涯に一度演じるという、歌舞伎の代表作の一つとはなった。 義経・十五代目市村羽左衛門の富樫は近来白眉とされ、1943年歌舞伎座で上演され舞台は映画にも記録。

Figure 16. Basic construction of the advertising stories of kabuki

Episodes Related to the Stage of Ebizō XI		
Sukeroku's Explanation	*Narukami's* Explanation	*Kanjinchō's* Explanation
Original Story of *Sukeroku*	Originanl Story of *Narukami*	Originanl Story of *Kanjinchō*
Mixed Story of *Sukeroku, Narukami and Kanjinchō* Using *Naimaze* and *Tsukushi*		

Furthremore, the author can transform the original structure of the above Ebizō's information according to the purposes of use. In this senction, the author wants to use for the advertising stories regarding Ebizō and his stage performance in the context of GIS. Figure 14 and 15 respectively the structural and sentence representations that were newly edited and restructured from the above data.

Selecting and Editing Stories

Ogata (2018a, 2019c) selected two stories on *kabuki* works, Sukeroku and Narukami, that Ichikawa Ebizō XI frequently played. In this section, Kanjinchō is added as another example. The three *kabuki* works have relatively simple and clear scenarios. Table 2 contains these scenarios' synopses (from Japanese Wikipedia), their property data, and generated sentences. For translation from Japanese to English in Tables 2 and 3 and Figures 17 and 18, the author used Google Translate and added the necessary revisions, such as to names and objects.

(((“” “”) ((“” “”) (“を破ろうと” “上人の呪術” (“をもっ” “女色” (“では” “そこで朝廷”)))) ((“” “”) (“に送り込む” “許” (“の” “上人” (“を” “絶間姫” (“の” “美女・ 雲” (“の” “、内裏一”)))))) ((“には” “色仕掛け” (“の” “姫”)) (“、と” “最後” (“に触れ” “身体” (“も抗しきれ” “さすが上人”)))) ((“” “”) (nil “うと”)) ((“は” “、さらに” (“を犯し” “う戒律”)) (“に酔いつぶれて眠ってしまう” “酒”)) ((“” “”) (“を切ると” “注連縄” (“に張ってある” “姫が滝壷” (“を見計っ” “その隙”)))) ((“” “”) (“がそこから飛び出すと” “封印が解け、竜神”)) ((“” “”) (“と” “豪雨” (“にかき

Figure 17. An example of synthesized Sukeroku, Narukami, and Kanjinchō by naimaze techniques

Structural Lisp Desctiption

```
(s継起
(event 依頼する (agent 天皇) (counter-agent 鳴神))
(event 約束する (agent 天皇))
(event 祈祷する (agent 鳴神))
(event 生まれる (agent 王子))
(event 破る (agent 天皇))
(event 怒る (agent 鳴神))
(event 封印する (agent 鳴神) (counter-agent 竜神))
(event 乾く (agent 国))
(event 困る (agent 百姓))
(event 探す (agent 助六) (object 友切丸))
(event 通う (agent 助六) (to 吉原))
(event 喧嘩する (agent 助六) (counter-agent 遊客))
(event 現れる (agent 揚巻))
(event 現れる (agent 意休))
(event 持つ (agent 意休) (object 友切丸))
(event 知る (agent 助六))
(event 帰る (agent 揚巻))
(event 帰る (agent 意休))
(event 現れる (agent 十郎))
(event 練習する (agent 十郎) (object 喧嘩))
(event 現れる (agent 揚巻))
(event 現れる (agent 満江))
(event 喧嘩する (agent 助六) (counter-agent 満江))
(event 与える (agent 満江) (object 紙子の衣) (to 助六))
(event 怒る(agent 満江))
(event 帰る (agent 満江))
(event 帰る (agent 十郎))
(event 現れる (agent 意休))
(event 見抜く (agent 意休) (object 助六の正体))
(event そそのかす (agent 意休) (object 裏切り)
(couner-agent 助六))
(event 奪う (agent 助六) (object 友切丸))
(event 切る (agent 助六) (counter-agent 意休))
(event 去る (agent 助六) (from 吉原))
(event 怒る (agent 頼朝))
(event 逃げる (agent 助六) (to 奥州))
(event 到着する (agent 助六) (to 安宅の関))
(event 知る (agent 富樫))
<to right>
```

```
<from left>
(event 禁止する (agent 富樫) (object 通行))
(event 唱える (agent 弁慶) (object 呪文))
(event (agent 富樫) (object (event 読む (agent 弁慶)
(object 勧進帳))))
(event 読む (agent 弁慶) (object 巻物))
(event 聞く (agent 富樫) (object 心得))
(event 答える (agent 弁慶))
(event 許す (agent 富樫) (object 通行))
(event 疑う (agent 部下) (counter-agent 助六))
(event 叩く (agent 弁慶) (counter-agent 助六)
(instrument 金剛杖))
(event 許す (agent 富樫) (counter-agent 助六))
(event ほめる (agent 助六) (object 機転))
(event 詫びる (agent 弁慶))
(event 取る (agent 助六) (object 手))
(event 現れる (agent 富樫))
(event 進める (agent 富樫) (object 酒))
(event 飲む (agent 弁慶))
(event 披露する (agent 弁慶) (object 舞))
(event 背負う (agent 弁慶) (object 笈))
(event 目礼する (agent 弁慶) (counter-agent 富樫))
(event 追う (agent 弁慶) (conter-agent 助六))
(event 送る (agent 天皇) (counter-agent 揚巻))
(event たぶらかす (agent 揚巻) (counter-agent 鳴神))
(event 破る (agent 鳴神))
(event 酔う (agent 鳴神))
(event 眠る (agent 鳴神))
(event 切る (agent 揚巻) (object しめ縄))
(event 解放する (agent 揚巻) (counter-agent 竜神))
(event 飛ぶ (agent 竜神))
(event 振る (agent 豪雨))
(event 逃げる (agent 揚巻))
(event 起きる (agent 鳴神))
(event 怒る (agent 鳴神))
(event 追う (agent 鳴神) (counter-agent 揚巻))
)
```

Generated Sentences

天皇が鳴神へ依頼する。天皇が約束する。鳴神が祈祷する。皇子が生まれる。天皇が約束を破る。鳴神が怒る。鳴神が龍神を封印する。国が乾く。百姓が困る。助六が友切丸を探す。助六が吉原に通う。助六が遊客と喧嘩する。揚巻が現れる。意休が現れる。意休が友切丸を持つ。助六は知る。揚巻が帰る。意休が帰る。十郎が現れる。十郎が喧嘩を練習する。揚巻が現れる。満江が現れる。助六が満江と喧嘩する。満江が紙子の衣を助六に与える。満江が怒る。満江が帰る。十郎が帰る。意休が現れる。意休が助六の正体を見抜く。意休が裏切りを助六にそそのかす。助六が友切丸を奪う。助六が意休を切る。助六が吉原から去る。頼朝が怒る。助六が奥州へ逃げる。助六が安宅の関に辿り着く。富樫は正体を知る。富樫は通行を禁止する。弁慶は呪文を唱える。富樫は勧進帳を読むと命じる。弁慶は巻物を読み上げる。富樫は心得を聞く。弁慶は答える。富樫は通行を許す。部下が助六を疑う。弁慶は助六を金剛杖で叩く。富樫は助六を許す。助六は機転を褒める。弁慶は詫びる。助六は手を取る。富樫が現れる。富樫は酒を勧める。弁慶は酒を飲む。弁慶は舞を披露する。弁慶は笈を背負う。弁慶は富樫に目礼する。弁慶は助六を追う。天皇が揚巻を送る。揚巻が鳴神をたぶらかす。鳴神が戒律を破る。鳴神が酒に酔う。鳴神が眠る。揚巻がしめ縄を切る。揚巻が龍神を解放する。龍神が飛び出す。豪雨が降る。揚巻が逃げる。鳴神が起きる。鳴神が怒る。鳴神が揚巻を追う。[The emperor makes a request to Narukami. The Emperor promises. Narukami prays. A prince is born. The emperor breaks his promise. Narukami gets angry. Narukami seals Ryūjin. The country dries up. Farmers suffer from lack of food. Sukeroku looks for *Tomokirimaru*. Sukeroku goes to Yoshiwara. Sukeroku hesitates with the guests. Agemaki appears. Ikyū appears. Ikyū has *Tomokirimaru*. Sukeroku knows. Agemaki comes back. Ikyū goes back. Jūrō appears. Jūrō practices acupuncture. Agemaki appears. Mankō appears. Sukeroku reminisces with Mankō. Mankō gives a paper clothe to Sukeroku. Mankō is angry. Mankō is back. Jūrō comes home. Ikyū comes to mind. Ikyū finds out the identity of Sukeroku. Ikyū is an endorsement of the betrayal. Sukeroku robs *Tomokirimaru*. Sukeroku breaks his mind. Sukeroku leaves from Yoshiwara. Yorritomo gets angry. Sukeroku escapes to Ōshū.

Figure 18. Sukeroku, Narukami, and Kanjinchō synopsis using tsukushi technique

東京駅で助六が奥州へ逃げる。東京駅で助六が鰺の押寿しを食べる。盛岡駅で助六が安宅の関に辿り着く。青山駅で富樫は正体を知る。厨川駅で富樫は通行を禁止する。巣子駅で弁慶は呪文を唱える。滝沢駅で富樫は勧進帳を読むと命じる。渋民駅で弁慶は巻物を読み上げる。好摩駅で富樫は心得を聞く。岩手川口駅で弁慶は答える。いわて沼宮内駅で富樫は呪文を尋ねる。御堂駅で弁慶は答える。奥中山高原駅で富樫は通行を許す。小繋駅で部下が助六を疑う。小鳥谷駅で弁慶は助六を金剛杖で叩く。一戸駅で富樫は助六を許す。富樫は一戸駅でロースカツ弁当を見つける。二戸駅で助六は機転を褒める。斗米駅で弁慶は詫びる。金田一温泉駅で助六は手を取る。目時駅で富樫が現れる。三戸駅で富樫は酒を勧める。諏訪ノ平駅で弁慶は酒を飲む。諏訪ノ平駅富樫が南部どり釜飯を出す。剣吉駅で弁慶は舞を披露する。苫米地駅で弁慶は笈を背負う。北高岩駅で弁慶は富樫に目礼する。八戸駅で弁慶は助六を追う。東京駅で天皇が揚巻を送る。神田駅で揚巻が鳴神をたぶらかす。秋葉原駅で鳴神が戒律を破る。御徒町駅で鳴神が酒に酔う。御徒町駅で鳴神がシウマイ弁当を食べる。上野駅で鳴神が眠る。鶯谷駅で揚巻がしめ縄を切る。日暮里駅で揚巻が龍神を解放する。西日暮里駅で龍神が飛び出す。田端駅で豪雨が降る。駒込駅で揚巻が逃げる。巣鴨駅で鳴神が起きる。大塚駅で鳴神が怒る。池袋駅で鳴神が揚巻を追う。
[Sukeroku escapes to Ōshū at Tokyo station. At the Tokyo station, Sukeroku eats *Aji no Oshizushi Bentō*. At Morioka station, Sukeroku arrives at Ataka no Seki. At Aoyama station, Togashi knows his true identity. Togashi prohibits the passage at Kuriyagawa station. Benkei casts a spell at Sugo station. At Takizawa station, Togashi orders to read the *kanjinchō* book. Benkei reads the scroll at Shibutami station. Togashi listens to the information at Kōma station. Benkei answers at Iwate-Kawaguchi station. Togashi asks for a spell at Iwate-Numakunai station. Benkei answers at Midō station. Togashi allows traffic at Okunakayama-kōgen station. The subordinates doubt Sukeroku at Kotsunagi station. At Kozuya station, Benkei taps Sukeroku with a gong stick. At Ichinohe station, Togashi grants support. Togashi eats *Rosukatsu Bentō* at Ichinohe station. At Ninohe station, Sukeroku gives up on the move. Benkei apologizes at Tome station. Sukeroku takes a hand at Kindaichi-onsen station. Togashi appears at Medoki station. Togashi is recommended to drink *sake* at San'nohe station. Benkei drinks *sake* at Suwanodaira station. At Suwanodaira station, Togashi shows *Nanbudori Kamameshi*. Benkei shows dance at Ken'yoshi station. Benkei bears a bag at Tomabechi station. Benkei at Kitatakaiwa Station pays tribute to Togashi. At Hachinohe station, Benkei follows Sukeroku. The emperor sends Agemaki at Tokyo station. At Kanda station, a pilgrimage sets off the narcissus. Narukami breaks the discipline at Akihabara station. Narukami gets drunk at Okachimachi station. Narukami eats *Shiumai Bentō* at Okachimachi station. Narukami sleeps at Ueno station. At Uguisudani station, Agemaki is cut off. Agemaki releases Ryūjin at Nippori station. Ryūjin jumps out at Nishi-nippori station. It is heavy rain at Tabata station. Agemaki escapes at Komagome station. Narukami comes up at Sugamo station. Narukami gets angry at Ōtsuka station. Narukami follows Agemaki at Ikebukuro station.]

曇っ" "一天にわか"))) (("は" "なり、姫") ("を逃げ去る" "その場")) ((" は" "上人" ("に飛び起き" "音" ("の" "雨"))) (nil "やっと")) (("" "") ("こと" "騙された")) (("" "") ("のごと" "に気づき烈火")) (("は" "く怒り、髪") ("と" "物炎" ("着ている" "逆立ち"))) (("" "") ("を逃さじと" "なって姫")) (("" "") ("、その後を追いかける" "")))

(("は" "nil") (("は" "nil") ("が、" "源義経一行" ("を買っ" "怒り" (" の" "源頼朝")))) (("は" "nil") (nil "際の加賀国の、安宅の関での物語" ("へ逃げる" "奥州" ("を通っ" "北陸")))))

(("は" "義経一行") (("は" "義経一行") ("で通り抜けようと" "姿" (" の" "山伏" ("に" "先頭" ("を" "武蔵坊弁慶"))))) (("は" "nil") (nil "する")) (("は" "nil") ("で、" "辿り着いた関")) (("は" "弁慶") ("を行っていると" "焼失した東大寺再建のための勧進")) (("は" "nil") (nil "言う")) ((" には" "元" ("の" "富樫左衛門" ("の" "しかし、関守"))) ("であると" "

Figure 19. An example of advertising story in GIS as a synthesis

十一代目市川海老蔵（じゅういちだいめいちかわえびぞう、1977年12月6日-）は歌舞伎役者、俳優。

歌舞伎名跡「市川海老蔵」の当代。

屋号は成田屋。定紋三升、替紋杏葉牡丹。

愛称に海老さま」、「海老ちゃん。

・歌舞伎は

天衣紛上野初花「雪暮夜入谷畦道」（三千歳と侍片岡直次郎，鳴神鳴神上人，源氏物語光君，助六由縁江戸桜花川戸助六，鏡獅子小姓弥生後獅子精弁慶、富樫左衛門

1983年5月-歌舞伎座『源氏物語』の春宮お目見得。

1985年5月-歌舞伎座『外郎売』の坊七代目市川新之助襲名。

2000年歌舞伎座『源氏物語』の光君会場は連日満員、チケット人気五代目尾上菊之助、二代目尾上辰之助とともに「平成の三之助新たな歌舞伎ブーム。

2003年2月-日置明子との間女児、当時1歳ことが判明。子ことを会見で説明。

2004年4月4日-成田山新勝寺で襲名披露、奉告の練り。京成本線京成上野駅-京成成田駅間列車「海老蔵号京成AE100形電車を使用）。

2004年5月歌舞伎座『助六由縁江戸桜』の花川戸助六、『暫鎌倉権五郎ほか十一代目市川海老蔵襲名。

2004年6月30日-成田山大阪別院明王院で京阪本線天満橋駅-香里園駅間列車「海老蔵号京阪8000系電車使用）。

2004年10月-フランス・パリのシャイヨ宮劇場市川新之助改メ十一代目市川海老蔵襲名披露フランス公演」。同国政財界トップル・モンド』紙が一面全面17世紀末以来の日本の伝統的演劇の中で名一族の出身」「26歳のスター」「若く美男で特に『必殺の視線現地でも大きな話題。

2006年9月7日-映画『出口の海クイズ$ミリオネア』（フジテレビ）に出演。

2007年3月-パリ・オペラ座で初歌舞伎公演父弁慶紅葉狩パリ引っ込み六方を花道中央通路観客。

<table>
<tr>
<td>『助六』は一つの通称。

本外題は助六を務める役者によって変わる古典歌舞伎を代表する演目のひと。

「粋」を具現化した洗練された江戸文化の極致として後々まで日本文化に決定的な影響を与えたお家芸である歌舞伎十八番の一つで，その中でも特に上演回数が多く、また上演すれ大入りになるという人気演目である。

『助六』は形式上「曾我もの」の演目。

助六がハ曾我五郎」で、白酒売りは「実ハ五郎の兄曾我十郎」という設定である。

助六のモデルではないかと考えられている人物は三人いる

店に捌助六あるいは戸澤助六という若旦那がいたと名を馳せ助六という侠客だとする説、そして江戸・蔵前の札差で、は気前の文化人として知られた大口屋暁雨だとする説である。

このうち、史家の多くは第一助六を否定する。

一つが助六」という名これは上方でならありそうな名だが、感覚からは野暮な名なのだという。</td>
<td>『鳴神』とは、歌舞伎十八番のひと。

この『鳴神』は江戸中村座の門松四天王』（かどまつしてんのう）において上演しものが濫觴であるが、現行で上演されているものは『雷神不動北山桜』がもとになっている。

『雷神不動北山桜』は雷神不動北山桜北山岩屋の場)、そして五幕目大切が不動となっていて、今日では『毛抜』と『鳴神』は独立した芝居として上演されることが多い。

七代目市川團十郎によって歌舞伎十八番に撰ばれたが、その後『鳴神』は八代目團十郎が演じ以降、九代目團十郎自分の柄にあわないとして演じなかったので上演が絶えていた。

二代目市川左團次が岡鬼太郎と提携し、演出を改めて上演に漕ぎ着けた。

なお二代目左團次は毛抜』も復活上演しており、現在では人気演目のひとつになっている。</td>
<td>原形は星合十二段』に取り入れのが最初。様式を取り入れ天保11年3月に江戸の河原崎座で初演された配役は義経、二代目市川九蔵の富樫左衛門。作詞は三代目並木五瓶、作曲四代目杵屋六三郎だった

時代物の軍兵姿であり、弁慶の水衣も棒縞であるなど、今日舞台とはいろいろな違いがあった

現在見られる型は明治時代、九代目市川團十郎により完成されたもので、これ以降、番卒は金糸の梵字をあしらったものへと改められた。

かつては市川宗家のお家芸として、他家の役者が勝手に上演できなかった有力な後継者がい事情もあって他役者弁慶を演じるようなっていく

以後『勧進帳』は生涯に一度演じるという、歌舞伎の代表作の一つとはなった。

義経・十五代目市村羽左衛門の富樫は近来白眉とされ、1943年歌舞伎座で上演され舞台は映画にも記録。</td>
</tr>
<tr>
<td>助六が友切丸を探す。助六が吉原に通う。助六が遊客と喧嘩する。揚巻が現れる。意休が現れる。意休が友切丸を持つ。助六は知る。揚巻が帰る。意休が帰る。十郎が現れる。十郎が喧嘩を練習する。揚巻が現れる。満江が現れる。助六が満江と喧嘩する。満江が紙子の衣を助六に与える。満江が怒る。満江が帰る。十郎が帰る。意休が現れる。意休が助六の正体を見抜く。意休が裏切りを助六にそそのかす。助六が友切丸を奪う。助六が意休を切る。助六が吉原から去る。</td>
<td>天皇が鳴神上人へ依頼する。天皇が約束する。鳴神上人が祈祷する。皇子が生まれる。天皇が約束を破る。鳴神が怒る。鳴神が龍神を封印する。国が乾く。百姓が困る。天皇が姫を送る。姫が鳴神をたぶらかす。鳴神が戒律を破る。鳴神が酒に酔う。鳴神が眠る。姫がしめ縄を切る。姫が龍神を解放する。龍神が飛び出す。豪雨が降る。姫が逃げる。鳴神が起きる。鳴神が怒る。鳴神が姫を追う。</td>
<td>源頼朝が怒る。義経が奥州へ逃げる。義経が安宅の関に到着する。富樫は正体を知る。富樫は通行を禁止する。弁慶は呪文を唱える。富樫は勧進帳を読むと命じる。弁慶は巻物を読む。富樫は心得を聞く。弁慶は答える。富樫は通行を許す。部下が義経を疑う。弁慶は義経を金剛杖で叩く。富樫は義経を許す。義経は機転を褒める。弁慶は詫びる。義経は手を取る。富樫が現れる。富樫は酒を勧める。弁慶は酒を飲む。弁慶は舞を披露する。弁慶は笈を背負う。弁慶は富樫に目礼する。弁慶は義経を追う。</td>
</tr>
</table>

東京駅で助六が奥州へ逃げる。東京駅で助六が鯵の押寿しを食べる。盛岡駅で助六が安宅の関に辿り着く。青山駅で富樫は正体を知る。厨川駅で富樫は通行を禁止する。巣子駅で弁慶は呪文を唱える。滝沢駅で富樫は勧進帳を読むと命じる。渋民駅で弁慶は巻物を読み上げる。好摩駅で富樫は心得を聞く。岩手川口駅で弁慶は答える。いわて沼宮内駅で富樫は呪文を尋ねる。御堂駅で弁慶は答える。奥中山高原駅で富樫は通行を許す。小繋駅で部下が助六を疑う。小鳥谷駅で弁慶は助六を金剛杖で叩く。一戸駅で富樫は助六を許す。富樫は一戸駅でロースカツ弁当を見つける。二戸駅で助六は機転を褒める。斗米駅で弁慶は詫びる。金田一温泉駅で助六は手を取る。目時駅で富樫が現れる。三戸駅で富樫は酒を勧める。諏訪ノ平駅で弁慶は酒を飲む。諏訪ノ平駅富樫が南部どり釜飯を出す。剣吉

山伏姿" ("が" "既に義経一行"))) (("は" "山伏" ("が届いており" "いう情報")) ("罷りならぬと" "通行")) (("は" "nil") ("する" "厳命")) (("は" "弁慶" ("に憤慨し" "これ")) ("と" "仲間")) (("は" "nil") ("を晴らそうと" "疑い" ("を唱え" "呪文" ("の" "富樫調伏")))) (("は" "nil") (nil "する")))

(("は" "感心した富樫") (("は" "感心した富樫") ("を読んでみるよう命じる" "、勧進帳" ("を思い出し" "言葉" ("の" "弁慶" ("の" "先"))))) (("は" "弁慶") ("と" "朗々" ("に装い" "たまたま持っていた巻物を勧進帳であるかのよう"))) (("は" "nil") (nil "読み上げる")) (("は" "なおも疑う富樫") ("について問い質すが、" "呪文" ("の" "秘密" ("や" "心得" ("の" "山伏"))))) (("は" "弁慶") ("なく答える" "淀み")))

(("は" "富樫") (("は" "富樫") ("を許すが、" "通行")) (("は" "nil") ("をかけた" "疑い" ("に" "強力" ("が" "一人" ("の" "部下"))))) (("は" "弁慶") ("を晴らす" "疑い" ("で叩き" "金剛杖" ("を" "義経" ("の" "主君"))))))

(("は" "義経" ("を脱出し" "危機")) (("は" "義経" ("を脱出し" "危機")) ("褒めるが、" "機転" ("の" "弁慶"))) (("は" "弁慶") ("を助けるためと" "命" ("の" "いかに主君"))) (("は" "") ("を働いたこと" "言え無礼")) (("は" "nil") ("ながらに詫びる" "を涙")) (("は" "義経" ("に対して" "それ")) ("を馳せる" "優しく弁慶の手を取り、共に平家を追った戦の物語に思い")) (("は" "nil") ("を勧める" "酒" ("を詫び" "先の非礼" ("が現れ" "富樫" ("へ" "そこ"))))) (("は" "弁慶" ("に応じ" "それ")) ("する" "舞を披露" ("を飲み" "酒"))) (("は" "弁慶" ("を逃がし" "舞いながら義経ら")) (nil "富樫に目礼" ("背負っ" "、笈"))) (("は" "nil") ("の後を急ぎ追いかける" "主君")))

花川戸助六という侠客に姿をやつし曾我五郎は、源氏の宝刀「友切丸」探すため吉原通っている様々な男が集まる吉原で、遊客にわざと喧嘩を吹っ掛けて刀を抜かせようというのである

情夫にしている花魁の揚巻と、揚巻に言い寄る髭の意休が登場。

意休が友切丸を持っていると勘づいた助六は刀を抜かせようとするが、なかなかうまくいかない

兄の曾我十郎がやってき弟に意見するが、真意を知っ十郎は自らも喧嘩売る稽古始める

侍を伴って再登場。助六はその侍に喧嘩を売ろうとするが、驚いたことにその侍は、兄弟を心配してやってき母の満江であった

満江は紙子の衣を着せて、激しい喧嘩を戒めると十郎とともに帰ってゆく

舞台には再び意休が登場。

意休は実は助六が曾我五郎と見抜いており、友切丸を抜い源氏を裏切ることをそそのかす

助六は友切丸を取り返し吉原を抜け出す

天皇からの依頼をうけ鳴神上人は皇子誕生の願かけ見事これ成就させせるため、怒った上人は雨降らす竜神滝壷封印てしまう

それからと国中が旱魃に襲われ民百姓は困りはててしまった

女色をもっ上人の呪術を破ろうと上人の許に送り込む

姫の色仕掛けには身体に触れ最後、とうとう戒律を犯し、さらには酒に酔いつぶれて眠ってしまう

姫が滝壷に張ってある注連縄を切ると封印が解け、竜神がそこから飛び出すと一天にわかにかき曇っ豪雨となり、姫はその場を逃げ去る

音に飛び起き上人はやっと騙されたことに気づき烈火のごとく怒り、髪は逆立ち着ている物炎となって姫を逃さじと、その後を追いかける

怒りを買っ源義経一行が、奥州へ逃げる際の加賀国の、安宅の関での物語。

義経一行は山伏の姿で通り抜けようとする。

辿り着いた関で、弁慶は焼失した東大寺再建のための勧進を行っているとは言う。富樫左衛門の元には既に義経一行が山伏姿であるという情報が届いており山伏は通行罷りならぬと厳命する

これに憤慨し弁慶は仲間と呪文を唱え疑いを晴らそうとする。

感心した富樫は言葉を思い出し、勧進帳を読んでみるよう命じる弁慶はたまたま持っていた巻物を勧進帳であるかのように装い朗々と読み上げる。なおも疑う富樫は秘密の呪文について問い質すが、弁慶は淀みなく答える

富樫は通行を許すが、強力に疑いをかけた

弁慶は金剛杖で叩き疑いを晴らす。

危機を脱出し義経は弁慶の機転褒めるが、弁慶はいかに主君の命を助けるためとは言え無礼を働いたことを涙ながらに詫びる。

それに対して義経は優しく弁慶の手を取り、共に平家を追った戦の物語に思いを馳せる。

先の非礼を詫び酒を勧める。それに応じ弁慶は酒を飲み舞を披露する。舞いながら義経らを逃がし弁慶は、笈背負っ富樫に目礼。主君の後を急ぎ追いかける

Kanjinchō was originally based on Ataka of nō (Muromachi era). In Ataka, Minamoto no Yoshitsune and his vassals, on their way to Ōshū (Tōhoku region), are stopped at the Border of Ataka by Togashi Saemon, the border guard barrier keeper. Musashibō Benkei (?–1186), a vassal of Yoshitsune, reads a false kanjinchō (subscription list) and escapes from there. Although the framework of this story originally relied on previous narratives, such as Azumakagami (2007–2015), Genpei Seisuiki (1988–1991), and Gikeiki

(2000), the kanjinchō reading episode was considered the author's idea. In other *kabuki* works, Sakurada Jisuke I, a famous *kabuki* writer in the Edo era, wrote Gohiiki Kanjinchō63, a kind of parody of Ataka. In the world of movies, Akira Kurosawa directed Tora no O wo Fumu Otokotachi (Kurosawa, & Ito, 1952), a musical based on Kanjinchō in *kabuki*.

In this example, as a narrative work in the context of GIS, the author aimed to make a kind of advertising story, including explanations of Ichikawa Ebizō XI and the three *kabuki* works, Sukeroku, Narukami, and Kanjinchō. Further, the synopsis of a scenario applied *naimaze* and *tsukushi* techniques. The presentation on the stage by Ebizō of newly constructed *kabuki* work was planned in the virtual world in GIS. Table 3 gives the explanations of Sukeroku, Narukami, and Kanjinchō, property data, and generated sentences. Figure 16 preliminarily shows the basic construction of the advertising stories.

((“は” “この『鳴神』”) ((“は” “この『鳴神』”) (“であるが、” “ものが濫觴” (“してんのう)において上演し” “江戸中村座の門松四天王』(かどまつ” (“を書き” “台本” (“で” “名” (“の” “三升屋兵庫” (“が” “正月、初代市川團十郎” (“)の” “貞享元年(1684年”)))))))) ((“は” “もの” (“で上演されている” “現行”)) (nil “もと” (“』が” “坂た『雷神不動北山桜” (“)に” “、寛保2年(1742年”)))) ((“は” “nil”) (nil “になっている”)))

((“は” “『雷神不動北山桜』”) ((“は” “『雷神不動北山桜』”) (“が不動と” “場)、そして五幕目大切” (“の” “雷神不動北山桜北山岩屋” (“には” “鳴神(正式” (“が” “毛抜』、四幕目” (“が” “、その三幕目”)))))) ((“は” “nil”) (“と” “なっていて、今日では『毛抜』”)) ((“は” “『鳴神』”) (“と” “独立した芝居”)) ((“は” “nil”) (“されること” “して上演”)) ((“は” “nil”) (nil “が多い”)) ((“は” “nil”) (“に撰ばれたが、” “歌舞伎十八番” (“によって” “七代目市川團十郎” (“もが” “そしてそのいずれ”)))) ((“は” “その後『鳴神』”) (“にあわないと” “以降、九代目團十郎自分の柄” (“が演じ” “八代目團十郎” (“)に” “嘉永4年(1851年”)))) ((“は” “nil”) (“が絶えていた” “して演じなかったので上演”)) ((“は” “nil”) (“と” “岡鬼太郎” (“が” “二代目市川左團次” (“に” “その後、明治43年(1910年)”)))) ((“は” “nil”) (“に漕ぎ着けた” “提携し、演出を改めて上演”)))

((“は” “なお二代目左團次”) ((“は” “なお二代目左團次”) (“しており、現在ではと” “復活上演” (“』も” “毛抜” (“、前述の” “この前年”))))

((“は” “nil”) (nil “ひと” (“の” “人気演目” (“の” “もに歌舞伎”)))) ((“は” “nil”) (nil “つになっている”)))

((“は” “nil”) ((“は” “nil”) (“が、” “源義経一行” (“を買っ” “怒り” (“の” “源頼朝”)))) ((“は” “nil”) (nil “際の加賀国の、安宅の関での物語” (“へ逃げる” “奥州” (“を通っ” “北陸”)))))

((“は” “義経一行”) ((“は” “義経一行”) (“で通り抜けようと” “姿” (“の” “山伏” (“に” “先頭” (“を” “武蔵坊弁慶”))))) ((“は” “nil”) (nil “する”)) ((“は” “nil”) (“で、” “辿り着いた関”)) ((“は” “弁慶”) (“を行っていると” “焼失した東大寺再建のための勧進”)) ((“は” “nil”) (nil “言う”)) ((“には” “元” (“の” “富樫左衛門” (“の” “しかし、関守”))) (“であると” “山伏姿” (“が” “既に義経一行”))) ((“は” “山伏” (“が届いており” “いう情報”)) (“罷りならぬと” “通行”)) ((“は” “nil”) (“する” “厳命”)) ((“は” “弁慶” (“に憤慨し” “これ”)) (“と” “仲間”)) ((“は” “nil”) (“を晴らそうと” “疑い” (“を唱え” “呪文” (“の” “富樫調伏”)))) ((“は” “nil”) (nil “する”)))

((“は” “感心した富樫”) ((“は” “感心した富樫”) (“を読んでみるよう命じる” “、勧進帳” (“を思い出し” “言葉” (“の” “弁慶” (“の” “先”))))) ((“は” “弁慶”) (“と” “朗々” (“に装い” “たまたま持っていた巻物を勧進帳であるかのよう”))) ((“は” “nil”) (nil “読み上げる”)) ((“は” “なおも疑う富樫”) (“について問い質すが、” “呪文” (“の” “秘密” (“や” “心得” (“の” “山伏”))))) ((“は” “弁慶”) (“なく答える” “淀み”)))

((“は” “富樫”) ((“は” “富樫”) (“を許すが、” “通行”)) ((“は” “nil”) (“をかけた” “疑い” (“に” “強力” (“が” “一人” (“の” “部下”))))) ((“は” “弁慶”) (“を晴らす” “疑い” (“で叩き” “金剛杖” (“を” “義経” (“の” “主君”))))))

((“は” “義経” (“を脱出し” “危機”)) ((“は” “義経” (“を脱出し” “危機”)) (“褒めるが、” “機転” (“の” “弁慶”))) ((“は” “弁慶”) (“を助けるためと” “命” (“の” “いかに主君”))) ((“は” “”) (“を働いたこと” “言え無礼”)) ((“は” “nil”) (“ながらに詫びる” “を涙”)) ((“は” “義経” (“に対して” “それ”)) (“を馳せる” “優しく弁慶の手を取り、共に平家を追った戦の物語に思い”)) ((“は” “nil”) (“を勧める” “酒” (“を詫び” “先の非礼” (“が現れ” “富樫” (“へ” “そこ”))))) ((“は” “弁慶” (“に応じ” “それ”)) (“する” “舞を披露” (“を飲み” “酒”))) ((“は” “弁慶” (“を逃がし” “舞いながら義経ら”)) (nil “富樫に目礼” (“背負っ” “、笈”))) ((“は” “nil”) (“の後を急ぎ追いかける” “主君”)))

『助六』は一つの通称。

本外題は助六を務める役者によって変わる古典歌舞伎を代表する演目のひと。

「粋」を具現化した洗練された江戸文化の極致として後々まで日本文化に決定的な影響を与えたお家芸である歌舞伎十八番の一つで、

その中でも特に上演回数が多く、また上演すれ大入りになるという人気演目である。

『助六』は形式上「曾我もの」の演目。

助六が八曾我五郎」で、白酒売りは「実ハ五郎の兄曾我十郎」という設定である。

助六のモデルではないかと考えられている人物は三人いる

店に捌助六あるいは戸澤助六という若旦那がいたと名を馳せ助六という侠客だとする説、そして江戸・蔵前の札差で、は気前の文化人として知られた大口屋暁雨だとする説である。

このうち、史家の多くは第一助六を否定する。

一つが助六」という名これは上方でならありそうな名だが、感覚からは野暮な名なのだという。

『鳴神』とは、歌舞伎十八番のひと。

この『鳴神』は江戸中村座の門松四天王』(かどまつしてんのう)において上演しものが濫觴であるが、現行で上演されているものは『雷神不動北山桜』がもとになっている。

『雷神不動北山桜』は雷神不動北山桜北山岩屋の場)、そして五幕目大切が不動となっていて、今日では『毛抜』と『鳴神』は独立した芝居として上演されることが多い。

七代目市川團十郎によって歌舞伎十八番に撰ばれたが、その後『鳴神』は八代目團十郎が演じ以降、九代目團十郎自分の柄にあわないとして演じなかったので上演が絶えていた。

二代目市川左團次が岡鬼太郎と提携し、演出を改めて上演に漕ぎ着けた。

なお二代目左團次は毛抜』も復活上演しており、現在では人気演目のひとつになっている。

原形は星合十二段』に取り入れのが最初。様式を取り入れ天保11年3月に江戸の河原崎座で初演された配役は義経、二代目市川九蔵の富樫左衛門。作詞は三代目並木五瓶、作曲四代目杵屋六三郎だった

時代物の軍兵姿であり、弁慶の水衣も棒縞であるなど、今日舞台とはいろいろな違いがあった

現在見られる型は明治時代、九代目市川團十郎により完成されたもので、これ以降、番卒は金糸の梵字をあしらったものへと改められた。

かつては市川宗家のお家芸として、他家の役者が勝手に上演できなかった有力な後継者がい事情もあって他役者弁慶を演じるようなっていく

以後『勧進帳』は生涯に一度演じるという、歌舞伎の代表作の一つとはなった。

義経· 十五代目市村羽左衛門の富樫は近来白眉とされ、1943年歌舞伎座で上演され舞台は映画にも記録。

Using *Naimaze*

The dramaturgy of *kabuki* is constructed based on *kata* and *shukō*. "*Kata*" refers to the typical patterns of various elements in *kabuki*, including characters, places or scenes, and stories or plots. The tradition of *kabuki* has stored diverse *kata*. It is also related to the concept of "*sekai* (world)" in *kabuki*. The author provided explanations of *sekai* in a post-narratology chapter (Ogata, 2019b) and an INGS chapter (Ogata, 2019a) in a previous work (Ogata & Akimoto, 2019). In the history of *kabuki*, the systematic description (as a kind of database or knowledge base) of *sekai* is given in the document *Sekai* Kōmoku (1916) in the Edo era. *Sekai* Kōmoku provides actual works in many narrative genres, including *nō*, *ningyō jōruri*, and novels, famous characters, and typical stories or plots. In the production of a *kabuki*'s work, one more *sekai* would be selected for deciding the basic *kata* to produce a work efficiently. However, *kabuki* writers did not simply apply the *kata* to the work. The second concept, *shukō*, is equal to special and individual narrative techniques for transforming the *kata* to make the work more interesting. *Naimaze*, which fuses different stories (as *kata*) to make a new story, is also a type of *shukō*. Applying a *naimaze* method or technique to a story or plot is the most general usage. *Naimaze* is also related to the element of a "character" as a "person." In particular, although it is possible for *naimaze* to mix "stories" or "plots" of two or more narratives, the use of the *naimaze* of "stories" or "plots" is related to the *naimaze* of "characters" of different works. Similarly, introducing "characters" from different works into a new work enables the associative relation of the new work to past works. In addition, Ogata (2019c) presented an approach for designing the narrative techniques of *naimaze* for INGS (and GIS) by applying the method for combining "moves," as explained in the narratological study of Vladimir Propp (1895–1970), Morphology of the Folktale (Propp, 1968). However, the following example is simpler. Figure 17 synthesizes the synopses of Sukeroku, Narukami, and Kanjinchō into a new scenario using *naimaze* techniques. The author used the conceptual structures of Sukeroku, Narukami, and Kanjinchō. This form of a conceptual structure is the story representation form that is used in INGS explained in detail in Chapter 1. Therefore, the narrative data described here can be utilized in the synthesis with INGS. In particular, first, the author transformed the sentences in each story shown in Table 4 into a collection of simple sentence using a semi-automatic script generation program developed by the authors (Arai, Ono, & Ogata, 2016), and, second, each transformed sentence is transformed by hand using an input format. Using these input data, in Figure 17, stories of Sukeroku and Kanjinchō are inserted into the story of Narukami as the base. Sukeroku and Kanjinchō are connected with Narukami in the points in which large special changes appear. These connecting stories cut the natural flow of Narukami's story. It is a type of story-cutting rhetoric that has been studied by Ogata and Kanai (2010a, 2010b). Sukeroku, the hero in Sukeroku, reaches new places after the events of "Narukami is angry" or "Yoritomo is angry." Kumonotaemahime, the heroine in Narukami, is changed to Agemaki, who is the heroine in Sukeroku, based on the common property of being female.

Using *Tsukushi*

Tsukushi is comprehended from the perspective of techniques through which objects respectively relate, meaning that objects are inserted or sprinkled into a narrative work or a part. Methods of *tsukushi* may correspond to lateral techniques for expanding objects. In the basic methods of *tsukushi*, methods for connecting objects to one another are included and principally divided into methods by semantic and non-semantic relations. The former, for instance, includes the geographical relationships of places and the conceptual relationships of object. The latter contains the relationships by dajare (puns) and linguistic sound. Many spatial objects can be inserted into a narrative using geographical movement and temporal change. The typical structure is *michiyuki*.

Here, the synthesized story of Sukeroku, Narukami, and Kanjinchō is extended by the use of *tsukushi* techniques, as shown in Figure 18. Using the story in Figure 17 generated based on the *naimaze* technique in *kabuki*, characters (Sukeroku and Agemaki in Sukeroku, Togashi and Benkei in Kanjinchō, and Narukami and Ryūjin in Narukami) in this figure play a type of *michiyuki* (as a dynamic pattern of *tsukushi*). In particular, the characters move to Ōshū (Tōhōku region) from Tokyo Station via the Tōhōku Shinkansen and travel to local areas via the Iwate Ginga Tetsudō (Railway). Afterward, they return to Tokyo Station and get on the Yamate line in Tokyo. The characters eat ekiben in several stations, according to the ekiben data in Wikipedia.

A Final Example and the Positioning in GIS With INGS

According to the story construction in Figure 18, the final story structures were generated, including the generated explanation sentences of Ichikawa Ebizō XI, generated explanation of the three *kabuki* works (i.e., Sukeroku, Narukami, and Kanjinchō), the works' generated synopses, and synthesized and expanded *kabuki* synopses applying *naimaze* and *tsukushi* techniques. A generated example is shown in Figure 19.

FUTURE RESEARCH DIRECTIONS

The main short-term plans concerning *kabuki*'s survey and analysis are presented.

1. The modeling and design of GIS with INGS based on the introduction of elements and knowledge of *kabuki*: The first objective, followed by a preliminary attempt, is to design a more detailed GIS with the INGS model, extended by a survey of *kabuki*. This also draws on the next theme, namely, deepening the survey and analysis of *kabuki* itself. Regarding this direction, the author conducted a further literature survey on *kabuki* to describe in detail *kabuki* as Documents in BACKGROUND. Furthermore, the author newly surveyed and analyzed "person," "story," "*tsukushi*" and "*naimaze*" in *kabuki* in detail in a new section, TOWARD INTRODUCING *KABUKI* KNOWLEDGE INTO GIS WITH INGS: FROM ANALYSIS TO GENERATION. In the future, to produce deeper and broader surveys and analyses of these elements, the author would like to extend the study's range to other elements and components in *kabuki*.
2. Deepening the survey and analysis of *kabuki* itself as an information system: This *kabuki* analysis, along with the following system design and implementation, forms a mutual process in that the *kabuki* analysis is progressed toward the systematization of GIS with INGS or directed by it, and the system design and implementation require a more detailed, precise, and systematic *kabuki* analysis. Dependent on the analysis, this chapter showed experimental programs using person, story, *naimaze*, and *tsukushi*. In the future, the author will conduct a design and experiment

for incorporating the proposed program on persons and other elements into the INGS and GIS.

3. An experimental implementation and the verification of GIS with INGS: Although this topic accompanies the expansion and precision of INGS that is the chief part of GIS, it is possible to use the current, tentative version in the GIS experiment system because it can act as an entire system. One of the future issues is to develop a meta-level mechanism to control the mutual relationship between the two mechanisms to make a sequential story of narrative works from the narrative content. In addition, multiple narrative structures can be seen in each constituent element in the narrative generation-reception and production-consumption mechanisms. INGS will be able to tackle this issue by enhancing knowledge bases for elements including persons and other features.
4. The applicable system design and implementation, including social distribution experiments: A current ambitious goal of GIS with INGS is to perform experiments at the social level, such as the internet-based distribution of generated narratives or of the generation mechanism itself. To carry out such socially open experiments, application systems of GIS with INGS need to be developed by limiting and constraining direct objectives. Further, as stated in detail in Ogata (2019d) and Chapter 4 in this book, a narratological concept in this respect represents the concept of fluidity and fixation in the expanded literary theory (Ogata, 2014). For the social distribution of GIS with INGS, the fluid aspect of narrative generation is connected to the generation and distribution on the online system, and the fixed aspect is associated with publishing the processes and results of the narrative generation in the form of a book. Moreover, by social distribution, the author also means a new type of business model for social applications using the web for the narrative generation system itself, or for a content system including narrative generation systems. The author anticipates directions for future research on the social distribution of the GIS with INGS through a greater understanding of *kabuki* and its multiple narrative structures model. Although this part has also left the stage of planning or idea formation, the author has an idea that models the GIS with INGS as a kind of a theatrical troupe or *geinō* production (entertainment agency).

CONCLUSION

This chapter was a synthesis of several papers written on *kabuki* addressing the narrative generation of *kabuki* or *kabuki* as narrative generation. This chapter first provided a focused description of *kabuki*, which has been presented in the author's previous papers, and presented the synthetic characteristics of *kabuki* as narrative generation in BACKGROUND.

Next, in the section *KABUKI*'S MULTIPLE NARRATIVE STRUCTURES SEEN IN CONCRETE EXAMPLES, the author describes the fifteen main narrative elements or techniques and several cross-sectional elements or techniques beyond conceptual description. In this chapter, the author extended respective explanations using concrete examples because *kabuki* is a pragmatic art that has been developed by many actors through its long history. However, as the author feels that the systematic consideration has several weak points, for example, the description of musical elements is weak, considering *kabuki*'s elements more comprehensively and deeply is required in the future. Next, from the framework of the multiple narrative structures, INGS described in Chapter 1 was the mechanism that performs the narrative generation-reception at the simple substance or personal level. In the integrated approach of narrative generation or post-narratology, a fundamental concept of the modeling is "multiplicity," or multiple narrative structures. The author aimed to explore *kabuki* narratives from the viewpoint of the multiple structures of the narrative generation-reception process and narrative production-consumption process. In contrast to INGS, GIS mentioned in this chapter corresponds with the narrative production-consumption process. The main role of GIS is not to generate a single narrative work, but to execute narrative generation as the outer system for making sequences of narrative generation using a simple substance. In this chapter, through two types of systems, including INGS and GIS, the author presented an entire image and design of narrative generation mechanisms in *kabuki*.

A *kabuki* play is a collection of narratives built around original scenarios, authentic and related histories, and actors' private scenes. It was intended to be connected with plans for a future narrative generation system comprising several other narrative generation systems. In TOWARD INTRODUCING *KABUKI* KNOWLEDGE INTO GIS WITH INGS: FROM ANALYSIS TO GENERATION, the author showed the design and implementation of experimental programs to concretize *kabuki*'s narrative generation. In particular, using program modules of "person," "story," "*naimaze*," and "*tsukushi* or zukshi," this chapter presented the simulation of the process that gradually generates more complex narratives according to the mutual relationships of the above modules. The presented simple programs are experiments for concretely considering the introduction methods to INGS and GIS. For instance, the techniques of *naimaze* and *tsukushi* will be used in the narrative content knowledge bases, especially the story content knowledge base, in INGS. The *naimaze* technique can be defined as a type of intertextual narrative generation method, and the technique of *tsukushi* is also a narrative method of repetition and change. At the same time, more detailed descriptions of *naimaze* and *tsukushi* are suitable to be stored in more specific narrative resources in GIS. Similarly, concerning the elements of person and story, although more general knowledge should be added to the various content knowledge bases, more specific and concrete knowledge should be stored in various resources. From the above viewpoint, this chapter focused on the elements of "person," "story," "*tsukushi*," and "*naimaze*," to present ideas of concrete design and implementation of *kabuki*-related experimental narrative generation mechanisms.

ACKNOWLEDGMENT

This chapter's research was supported by JSPS KAKENHI Grant Number 18K18509, Kayamori Foundation of Information Science Advancement Research Grant, The Telecommunication Advancement Foundation Research Grant, and AOI TYO Holdings Inc.

REFERENCES

Abe, Y. (2000). kabukiza. In Y. Hattori, T. Tomita, & T. Hirosue (Eds.), Shinpan kabuki jiten [New edition kabuki dictionry] (pp. 124–125). Tokyo, Japan: Heibonsha.

Arai, T., Ono, J., & Ogata, T. (2016). Semi-automatic generation of events sequence knowledge for narrative generation: The use in an integrated narrative generation system. In Proceedings of the 30th Annual Conference of the Japanese Society for Artificial Intelligence (3P1-1in2). Tokyo, Japan: The Japanese Society for Artificial Intelligence.

Aristotle. (1997). Poetics (M. Heath, Trans.). London, UK: Penguin Classics. (Original work published c. 335 BC)

Asanuma, K. (2007). Monogatari toha nanika—Tsuruya Nanboku to Fujisawa Shūhei no shudai ni yoru capricchio [What is narratives? Capricchio by the themes of Tsuruya Nanboku and Fujisawa Shūhei]. Tokyo, Japan: Suiseisha.

Azumakagami. (2007–2015). In F. Gomi & K. Hongō (Eds.), Gendaigo-yaku Azumakagami, 1–16 [Modern Japanese translation Azumakagami, Vols. 1-16]. Tokyo, Japan: Yoshikawa Kōbunkan. (Original work publishued the Kamakura era)

Baba, J. (1998). Hito to geidan—Sakigaketa haiyū tachi [Artists and talks: Pioneering actors]. Tokyo, Japan: Engeki Shuppansha.

Bandō, M. (2015). kabuki no tanoshimi [kabuki's pleasures]. Tokyo, Japan: Iwanami Shoten.

Bandō, T. (1976). Mayonaka no note [A notebook in the midnight]. Tokyo, Japan: Sanrio.

Bandō, T. (Director) (1992). Gekashitsu [Operating room] (Motion picture). Tokyo, Japan: Shōchiku.

Bandō, T. (Director) (1995). Tenshu monogatari [Castle tower story] (Motion picture). Tokyo, Japan: Shōchiku.

Chikamatsu, M. (1987). Shutendōji makura-kotoba. In Chikamatsu zenshū, 6 [Complete collection of Chikamatsu Monzaemon, 6] (pp. 1–91). Tokyo, Japan: Iwanami Shoten. (Original work played 1710)

Ebihara, K., Kurumisawa, J., Iwasawa, S., & Ohya, J. (1997). Virtual KABUKI system. In IEICE Technical Report (Image Engineering), 96(507) (pp. 61–67). Tokyo, Japan: The Institute of Electoronics, Information and Communication Engineers.

Genette, G. (1982). Palimpsestes: La littérature au second degré. Paris: Seuil.

Genji Monogatari. (1993). In S. Yanai, S. Murofushi, Y. Ōasa, H. Suzuki, S. Fujii, & Y. Imanishi (Eds.), Shin nihon koten bungaku taikei, 19 [New Japanese classic literature collection, 19]. Tokyo, Japan: Iwanami Shoten. (Original work published early 11th century, the Heian era)

Genji Monogatari. (1994). In S. Yanai, S. Murofushi, Y. Ōasa, H. Suzuki, S. Fujii, & Y. Imanishi (Eds.), Shin nihon koten bungaku taikei, 20 [New Japanese classic literature collection, Vol. 20]. Tokyo, Japan: Iwanami Shoten. (Original work published early 11th century, the Heian era)

Genji Monogatari. (1995). In S. Yanai, S. Murofushi, Y. Ōasa, H. Suzuki, S. Fujii, & Y. Imanishi (Eds.), Shin nihon koten bungaku taikei, 21 [New Japanese classic literature collection, Vol. 21]. Tokyo, Japan: Iwanami Shoten. (Original work published early 11th century, the Heian era)

Genji Monogatari. (1996). In S. Yanai, S. Murofushi, Y. Ōasa, H. Suzuki, S. Fujii, & Y. Imanishi (Eds.), Shin nihon koten bungaku taikei, 22 [New Japanese classic literature collection, Vol. 22]. Tokyo, Japan: Iwanami Shoten. (Original work published early 11th century, the Heian era)

Genji Monogatari. (1997). In S. Yanai, S. Murofushi, Y. Ōasa, H. Suzuki, S. Fujii, & Y. Imanishi (Eds.), Shin nihon koten bungaku taikei, 23 [New Japanese classic literature collection, Vol. 23]. Tokyo, Japan: Iwanami Shoten. (Original work published early 11th century, the Heian era)

Genpei Seisuiki. (1988–1991). In Mizuhara, H. (Ed.), Shintei Genpei seisuiki, 1–6 [New authentic edition Geipei seisuiki, Vols. 1–6]. Tokyo, Japan: Shin Jinbutsu Ōraisha. (Original work published the Kamakura era)

Gikeiki. (2000). In M. Kajiwara (Ed., & Trans.), Shinpen nihon koten bungaku zenshū, 62 [New edition Japanese classic literature collection, Vol. 62]. Tokyo, Japan: Shōgakukan. (Original work publishued the Kamakura era or the Muromachi era)

Gunji, M. (1963). kabuki no bigaku [Aesthetics of kabuki]. Tokyo, Japan: Engeki Shuppansha.

Haikawa, M. (2016). *kabuki* no ongaku, oto [Music, sound of *kabuki*]. Tokyo, Japan: Ongaku no Tomosha.

Hara, M. (2013). Chikamatsu jōruri no sakugekihō [The dramaturgy of Chikamatsu's jōruris]. Tokyo, Japan: Yagi Shoten.

Hashimoto, O. (2012). Jōruri wo yomō [Let's read jōruris]. Tokyo, Japan: Shinchōsha.

Hattori, Y., Tomita, T., & Hirosue, T. (Eds.). (2000). Shinpan kabuki jiten [New edition kabuki dictionry]. Tokyo, Japan: Heibonsha.

Heike Monogatari. (1991). In M. Kajiwara & H. Yamashita (Eds.), Shin nihon koten bungaku taikei, 44 [New Japanese classic literature collection, Vol. 44]. Tokyo, Japan: Iwanami Shoten. (Original work published the Kamakura era)

Heike Monogatari. (1993). In M. Kajiwara & H. Yamashita (Eds.), Shin nihon koten bungaku taikei, 45 [New Japanese classic literature collection, Vol. 45]. Tokyo, Japan: Iwanami Shoten. (Original work published the Kamakura era)

Imao, T. (1970). Henshin no shisō [The theory of metamorphosis]. Tokyo, Japan: Hosei Daigaku Shuppankyoku.

Imao, T. (1979). kabuki wo miru hito no tame ni [For people who view kabuki]. Tokyo, Japan: Tamagawa Daigaku Shuppankyoku.

Imao, T. (2009). Kawatake Mokuami—Moto no mokuami to naran [Kawatake Mokuami]. Kyoto, Japan: Minerva Shobō.

Imao, T. (2010). kabuki—<Tsūsetsu> no kenshō [kabuki: The verification of ‘common views]. Tokyo, Japan: Hosei Daigaku Shuppankyoku.

Inumaru, O. (2005). Tempō jūichinen no chūshingura—Tsuruya Nanboku Kamikakete Sango Taisetsu wo yomu [Chūshingura at Tempō 11: Reading Tsuruya Nanboku’s Kamikakete Sango Taisetsu]. Tokyo, Japan: Yūzankaku.

Izumi, K. (1942a). Gekashitsu [Operating room]. In Izumi Kyōka zenshū, 2 [Complete collection of Izumi Kyōka, Vol. 2] (pp. 11–28). Tokyo, Japan: Iwanami Shoten. (Original work published 1888)

Izumi, K. (1942b). Tenshu monogatari [Castle tower story]. In Izumi Kyōka zenshū, 2 [Complete collection of Izumi Kyōka, Vol. 2] (pp. 449–499). Tokyo, Japan: Iwanami Shoten. (Original work published 1917)

*kabuki*za. (2015a). Shōchiku sōgyō 120 shūnen 3 gatsu ō-*kabuki* [Shōchiku 120 year in business march great *kabuki*]. Tokyo, Japan: *kabuki*za.

*kabuki*za. (2015b). Shōchiku sōgyō 120 shūnen Shūzansai 9 gatsu ō*kabuki* [Shōchiku 120 year in business Shūzan festival september great *kabuki*]. Tokyo, Japan: *kabuki*za.

Kameyama, I. (2015a). Shin Karamazov no kyōdai, 1 [The new brothers Karamazov, Vol. 1]. Tokyo, Japan: Kawade Shobō Shinsha.

Kameyama, I. (2015b). Shin Karamazov no kyōdai, 2 [The new brothers Karamazov, Vol. 2]. Tokyo, Japan: Kawade Shobō Shinsha.

Kamimura, I. (1997). kabuki no jōkei [Scenes of kabuki]. Tokyo, Japan: Engeki Shuppansha.

Kamimura, I. (2003). Jidai no naka no kabuki—Kindai kabuki hihyōkaron [kabuki in ages: Modern kabuki critics]. Tokyo, Japan: Keiōgijuku Daigaku Shuppankai.

Kanazawa, Y. (2009). kabuki meisaku jiten (shinsōban) [kabuki masterpieces dictionary]. Tokyo, Japan: Seiabō. (Original work published 1959)

Kawai, M. (2000). Gassaku jōruri ni okeru tate-sakusha no suitei [Study on a leader author in joint jōruri works]. In Kinsei bungaku no kōryū—Engeki to shōsetsu [Interactions of early modern literature: Dramas and novels] (pp. 47–65). Tokyo, Japan: Seibundō Shuppan.

Kawatake, T. (1989). kabuki-bi ron [On the beauty of kabuki]. Tokyo, Japan: Tokyo Daigaku Shuppankai.

Kawatake, T. (1992). kabuki: Its beauty and tradition (C. Holmes, Trans.). Tokyo, Japan: Japan Arts Council.

Kawatake, T. (2003). kabuki: Baroque fusion of the arts. Tokyo, Japan: The International House of Japan.

Keene, D. (1977). Nihon bungakushi kinsei hen, 2 [The history of Japanese literature the part of early modern period, Vol. 2] (T. Tokuoka, Trans.). Tokyo, Japan: Chūōkōronsha.

Keene, D. (1990). Nō and bunraku: Two forms of Japanese theatre. Columbia University Press.

Kitani, M. (2007). Mishima Yukio to kabuki [Mishima Yukio and kabuki]. Tokyo, Japan: Kanrin Shobō.

Kokuritsu Gekijō. (2014). Dai 292 kai Heisei 26 nen 12 gatsu kabuki kōen [The 292th Heisei 26 december kabuki stage]. Tokyo, Japan: Kokuritsu Gekijō (Japan National Theatre).

Kristeva, J. (1980). Desire in language: A semiotic approach to literature and art. Columbia University Press.

Kurata, Y. (2013). Bunraku no rekishi [A history of bunraku]. Tokyo, Japan: Iwanami Shoten.

Kuroki, K. (1943). Jōruri-shi [The history of jōruri]. Tokyo, Japan: Seijisha.

Kurosawa, A. (Director), & Ito, M. (Producer). (1952). Tora no o wo fumu otokotachi [The men who tread on the tiger's tail] (Motion picture). Japan: Tōhō.

Leiter, S. L. (2014). Historical dictionary of Japanese traditional theatre, 2n'd Edition. Lanham, MD: Rowman & Littlefield.

Leiter, S. L. (Ed.). (2002). A kabuki reader: History and performance. London, UK: Routledge.

Matsui, S. (2000). Meoboku sendai hagi. In Y. Hattori, T. Tomita, & T. Hirosue (Eds.), Shinpan kabuki jiten [New edition kabuki dictionry]. (pp. 395–396). Tokyo, Japan: Heibonsha.

Matsui, S. (2002). kabuki—Ie no gei [kabuki: Families' arts]. Tokyo, Japan: Engeki Shuppansha.

Matsuzaki (2000). Yakusha hyōbanki [Performers' reputation]. In Y. Hattori, T. Tomita, & T. Hirosue (Eds.), Shinpan *kabuki* jiten [New edition *kabuki* dictionry] (pp. 404–405). Tokyo, Japan: Heibonsha.

Mayama, S. (1982). Genroku chūshingura, 1, 2. Tokyo, Japan: Iwanami Shoten. (Original work published 1934–1941)

Miki, T. (2004). Kangeki gūhyō [Critique of kabuki performances]. Tokyo, Japan: Iwanami Shoten.

Mishima, Y. (1956). Kindai nōgaku shū. Tokyo, Japan: Shinchōsha.

Mishima, Y. (1963). Gogo no eikō. Tokyo, Japan: Kōdansha.

Mishima, Y. (1969). Hashi-zukushi. [The seven bridges]. In S. Ōoka & Y. Mishima (Eds.), Ōoka Shōhei, Mishima Yukio shū (pp. 360–368). Tokyo, Japan: Chikuma Shobō.

Mishima, Y. (2002). Chinsetsu yumiharizsuki. In Ketteiban Mishima Yukio zenshū, 25 [Complete works of Mishima Yukio (definitive edition), Vol. 25] (pp. 99–160). Tokyo, Japan: Shinchōsha. (Original work published 1969)

Mishima, Y. (2004). Bunraku chinsetsu yumihari zuki. In Ketteiban Mishima Yukio zenshū, 40 [Complete works of Mishima Yukio (definitive edition), Vol. 40] (pp. 161–177). Tokyo, Japan: Shinchōsha. (Original work published 1970)

Mori, A. (2010). Takechi Tetsuji to iu geijutsu—Amarini contemporary na [The art of Takechi Tetsuji: Very contemporary]. Tokyo, Japan: Shin'yōsha.

Morita, M., & Ogata, T. (2018). Character information acquisition of *kabuki* actors for the geinō information system. In Proceedings of the 55th Special Interest Group on Language Sense Processing Engineering (pp. 1–5). Tokyo, Japan: The Japanese Society for Artificial Intelligence.

Nakagawa, Y. (2010). Bandō Tamasaburō—kabukiza tateoyama he no michi [Bandō Tamasaburō: His road to a leading female role actor in kabukiza]. Tokyo, Japan: Gentōsha.

Nakagawa, Y. (2013a). Jūhachidaime Nakamura Kanzaburō—Zen kiseki [Nakamura Kanzaburō XVIII: All tracks]. Tokyo, Japan: Asahi Shimbun Shuppan.

Nakagawa, Y. (2013b). kabuki—Ie to chi to gei [kabuki: Families, blood, and arts]. Tokyo, Japan: Kōdansha.

Nakamura, T. (2004). kabuki no kindai—Sakka to sakuhin [The modern age of kabuki: Authors and works]. Tokyo, Japan: Iwanami Shoten.

Nakano, M. (1985). Edo meibutsu hyōbanki an'nai [A guide to written commentaries on notable objects and events in the Edo era]. Tokyo, Japan: Iwanami Shoten.

Nojima, J. (Ed.). (2002). Shintei zōho kabuki jinmei jiten [New and supplement edition kabuki biographical dictionary]. Tokyo, Japan: Nichigai Associates.

Ogata, T., & Akimoto, T. (Eds.). (2019). Post-narratology through computational and cognitive approaches. Hershey, PA: IGI Global. doi:10.4018/978-1-5225-7979-3 doi:10.4018/978-1-5225-7979-3

Ogata, T., & Kanai, A. (2010a). Story to shōhin = brand—Seisei ni muketa kōkoku no bunseki [Stories and products = brands: An analysis of advertisements toward narrative generation]. In T. Ogata & A. Kanai (Eds.), Monogatariron no jōhōgaku josetsu—Monogatari seisei no shisō to gijutsu wo megutte [An introduction to informatics of narratology: Toward the thoughts and technologies of narrative generation] (pp. 42–115). Tokyo, Japan: Gakubunsha.

Ogata, T., & Kanai, A. (2010b). Monogatari seisei to shite no kōkōku seisei no tame ni—Danpenteki kōsatsu [For advertisement generation as narrative generation: A fragmental consideration]. In T. Ogata & A. Kanai (Eds.), Monogatariron no jōhōgaku josetsu—Monogatari seisei no shisō to gijutsu wo megutte [An introduction to informatics of narratology: Toward the thoughts and technologies of narrative generation] (pp. 384–447). Tokyo, Japan: Gakubunsha.

Ogata, T. (2014). Expanded literary theory for automatic narrative generation. In Proceedings of Joint Conference of the 7th International Conference on Soft Computing and Intelligent Systems and the 15th International Symposium on Advanced Intelligent Systems (pp. 1558–1563). Tokyo, Japan: Japan Society of Fuzzy Theory and Intelligent Informatics.

Ogata, T. (2016a). kabuki as multiple narrative structures. In T. Ogata & T. Akimoto (Eds.), Computational and cognitive approaches to narratology (pp. 400–431). Hershey, PA: IGI Global. doi:10.4018/978-1-5225-0432-0.ch016 doi:10.4018/978-1-5225-0432-0.ch016

Ogata, T. (2016b). Narrative generation of "zukushi"—Towards its systematization through surveying *kabuki* and developing a proper noun conceptual dictionary in INGS. In Proceedings of the 30th Annual Conference of the Japanese Society of Artificial Intelligence (1K5-OS-06b-2). Tokyo, Japan: The Japanese Society for Artificial Intelligence.

Ogata, T. (2017a). Multiple narrative structures of *kabuki* and the geinō information system. In Proceedings of the 31st Annual Meeting of the Association for Natural Language Processing (1D2-OS-29a-4). Tokyo, Japan: The Japanese Society for Artificial Intelligence.

Ogata, T. (2017b). Analyzing multiple narrative structures of kabuki based on the frameworks of narrative generation systems. Journal of Robotics, Networking and Artificial Life, 4(3), 221–226. doi:10.2991/jrnal.2017.4.3.10 doi:10.2991/jrnal.2017.4.3.10

Ogata, T. (2018a). An integrated approach to narrative generation: From Mishima and kabuki to narrative generation systems. In T. Ogata & S. Asakawa (Eds.), Content generation through narrative communication and simulation (pp. 49–147). Hershey, PA: IGI Global. doi:10.4018/978-1-5225-4775-4.ch002 doi:10.4018/978-1-5225-4775-4.ch002

Ogata, T. (2018b). kabuki ni mukete (1)—Shiisei to henshūsei no monogatari kara tajūsei to jitsuzon no monogatari he [Toward kabuki (1): From narratives of arbitrariness and editing to narratives of multiplicity and real existence]. In T. Ogata, Y. Kawamura, & A. Kanai (Eds.), Jōhō monogatariron—Jinkōchinō, ninchi, shakai katei to monogatari seisei [Informational narratology: Artificial intelligence/cognition/social process and narrative generation] (pp. 187–208). Tokyo, Japan: Hakutō Shobō.

Ogata, T. (2018c). kabuki ni mukete (2)—Tajū monogatari kōzō no shosō [Toward kabuki (2): Aspects of multiple narrative structures]. In T. Ogata, Y. Kawamura, & A. Kanai (Eds.), Jōhō monogatariron—Jinkōchinō, ninchi, shakai katei to monogatari seisei [Informational narratology: Artificial intelligence/cognition/social process and narrative generation] (pp. 209–244). Tokyo, Japan: Hakutō Shobō.

Ogata, T. (2018d). A method of naimaze of narratives based on *kabuki* analyses and Propp's move techniques for an automated narrative generation system. In Proceedings of the 2018 International Conference on Artificial Life and Robotics (pp. 668–674). Ōita, Japan: International Society of Artificial Life and Robotics. doi:10.5954/ICAROB.2018.OS17-310.5954/ICAROB.2018.OS17-3

Ogata, T. (2019a). A computational, cognitive, and narratological approach to narrative generation. In T. Ogata & T. Akimoto (Eds.), Post-narratology through computational and cognitive approaches (pp. 1–84). Hershey, PA: IGI Global. doi:10.4018/978-1-5225-7979-3.ch001 doi:10.4018/978-1-5225-7979-3.ch001

Ogata, T. (2019b). Toward a post-narratology or the narratology of narrative generation. In T. Ogata & T. Akimoto (Eds.), Post-narratology through computational and cognitive approaches (pp. 85–142). Hershey, PA: IGI Global. doi:10.4018/978-1-5225-7979-3.ch002 doi:10.4018/978-1-5225-7979-3.ch002

Ogata, T. (2019c). kabuki as multiple narrative structures and narrative generation. In T. Ogata & T. Akimoto (Eds.), Post-narratology through computational and cognitive approaches (pp. 192–275). Hershey, PA: IGI Global. doi:10.4018/978-1-5225-7979-3.ch005 doi:10.4018/978-1-5225-7979-3.ch005

Ogata, T. (2019d). Theoretical or philosophical considerations for an integrated narrative generation approach. In T. Ogata (Ed.), Toward an integrated approach to narrative genertion: Emerging research and opportunities (pp. 315–403). Hershey, PA: IGI Global.

Ōkura, S., & Kamimura, I. (2001). kabuki—kabuki today: The art and tradition [Photographs]. Tokyo, Japan: Kōdansha International.

Ono, J., & Ogata, T. (2017). Attribute information acquisition in a conceptual dictionary by using Wikipedia and the use in an automatic narrative generation game. In Proceedings of the 31st Annual Conference of the Japanese Society for Artificial Intelligence (1D3-OS-29b-2in2). Tokyo, Japan: The Japanese Society for Artificial Intelligence.

Onoe, B. (2014). On'nagata no koto [On on'nagata]. Tokyo, Japan: Chūōkōronsha. (Original work published 1944)

Onoe, S. (1989). Shōroku geiwa [Shōroku's artistic talk]. Tokyo, Japan: Kōdansha.

Propp, V. Y. (1968). Morphology of the folktale (L. Scott, Trans.). Austin, TX: University of Texas Press. (Original work published 1928)

Rabelais, F. (1991a). Gargantua monogatari [La vie très horrifique du grand Gargantua, père de Pantagruel] (Watanabe, K., Trans.). Tokyo, Japan: Iwanami Shoten. (Original work published 1534 or 1535)

Rabelais, F. (1991b). Pantagruel monogatari, 1 [Horribles et épouvantables faits et prouesses du très renommé Pantagruel] (Watanabe, K., Trans.). Tokyo, Japan: Iwanami Shoten. (Original work published c. 1532)

Rabelais, F. (1991c). Pantagruel monogatari, 2 [Le tiers livre des faicts et dicts héroïques du noble Pantagruel] (K. Watanabe, Trans.). Tokyo, Japan: Iwanami Shoten. (Original work published 1546)

Rabelais, F. (1991d). Pantagruel monogatari, 3 [Le quart livre des faicts et dicts héroïques du bon Pantagruel] (K. Watanabe, Trans.). Tokyo, Japan: Iwanami Shoten. (Original work published 1552)

Rabelais, F. (1991e). Pantagruel monogatari, 4 [Le cinquième et dernier livre des faicts et dicts héroïques du noble Pantagruel] (K. Watanabe, Trans.). Tokyo, Japan: Iwanami Shoten. (Original work published 1564)

Sakusha Shikihō Kezairoku [Authors' Ceremonial Methods Kezairoku]. (1972). In M. Gunji (Ed.), Kinsei geidōron (Nihon shisō taikei, 61) [Theories of early modern times' arts (Japanese classic phirosophical thoughts collection, Vol. 61)] (pp. 493–532). Tokyo, Japan: Iwanami Shoten. (Original work published 1801)

Sekai Kōmoku. (1916). In Sekai kōmoku, shibai nenjū gyōji, gekikai chinwa (Chinsho kankōkai sōsyo, 9) [Compendium of narrative worlds, annual events related to kabuki plays, rare tales in theatrical circles (Rare books publication society series, Vol. 9)] (pp. 6–46). Tokyo, Japan: Chinsyo kankōkai. (Written in the late Edo period (before 1791))

Sekkyōshū. (1977). In Y. Muroki (Ed.), Shinchō nihon koten shūsei [Shinchō Japanese classic collection]. Tokyo, Japan: Shinchōsha (Sekyōbushis were narrated from the end of the medieval period to the early modern period in Japan)

Shibuya, Y., Morita, Y., Fukuda, H., Ueda, K., & Sasaki, M. (2012). Asynchronous relation between body action and breathing in bunraku: Uniqueness of manner of breathing in Japanese traditional performing arts. Cognitive Studies, 19(3), 337–364.

Shino, Y. (1991). kabuki—Kata no denshō [kabuki: Transmission of forms]. Tokyo, Japan: Engeki Shuppansha.

Shuzui, K., & Ōkubo, T. (1959). Chikamatsu jōruri shū (Nihon koten bungaku taikei, 50) [Chikamatsu jōruri collection (Japanese classic literature collection, Vol. 50)]. Tokyo, Japan: Iwanami Shoten. (Original works published from the late 17th century to the early 18th century in the Edo era)

Sophocles. (1972). Oedipus the king (A. Burgess, Trans.). Minneapolis, MN: University of Minnesota Press. (Original work first performed around 429 BC)

Sugiyama, S. (2004). Jōruri shirōto kōshaku, 1, 2 [Joruri amateur's lectures, Vol. 1, 2]. Tokyo, Japan: Iwanami Shoten. (Original work published 1926)

Suwa, H. (1991). kabuki no hōhō [Methods of kabuki]. Tokyo, Japan: Bensei Shuppan.

Suwa, H. (2005). Tsuruya Nanboku—Kokkei wo konomite, hito wo warawasu koto wo gō to su [Tsuruya Nanboku]. Tokyo, Japan: Minerva Shobō.

Taguchi, A. (2003). kabuki to ningyō jōururi [kabuki and ningyō jōruri]. Tokyo, Japan: Yoshikawa Kōbunkan.

Taguchi, A. (2011). kabuki kara Edo wo yominaosu [Rereading the Edo from kabuki]. Tokyo, Japan: Yoshikawa Kōbunkan.

Taiheiki. (1960a). In Nihon koten bungaku taikei, 34 [Japanese classic literature collection, Vol. 34]. Tokyo, Japan: Iwanami Shoten. (Original work published the Muromachi era)

Taiheiki. (1960b). In Nihon koten bungaku taikei, 35 [Japanese classic literature collection, Vol. 35]. Tokyo, Japan: Iwanami Shoten. (Original work published the Muromachi era)

Taiheiki. (1961). In Nihon koten bungaku taikei, 36 [Japanese classic literature collection, Vol. 36]. Tokyo, Japan: Iwanami Shoten. (Original work published the Muromachi era)

Takano, T. (2005). Yūjo kabuki [Prostitute's kabuki]. Tokyo, Japan: Kawade Shobō Shinsha.

Takei, W. (2002). Kyōgen zukushi [Kyōgen devotion]. In Y. Tomizawa & H. Fujita (Eds.), Saishin kabuki daijiten [The newest kabuki dictionary] (p. 200). Tokyo, Japan: Kashiwa Shobō.

Tamura, N. (1922). Zokuzoku kabuki nendaiki, Ichikan [A sequel to a kabuki chronicle, Volume I]. Tokyo, Japan: Ichimuraza.

Tanaka, M. (2002). Kaidai [Bibliography]. In Ketteiban Mishima Yukio zenshū, 25 [Complete works of Mishima Yukio (definitive edition), Vol. 25] (pp. 825–847). Tokyo, Japan: Shinchōsha.

Toita, Y. (1949). Maruhon kabuki. Tokyo, Japan: Kōdansha.

Tōkaidōchū Hizakurige. (1995). In Y. Nakamura (Ed. & Trans.), Shinpen nihon koten bungaku zenshū, 81 [New edition Japanese classic literature collection, Vol. 81]. Tokyo, Japan: Shōgakukan. (Original work publishued 1802–1814)

Tomizawa, Y., & Fujita, H. (Eds.). (2012). Saishin kabuki daijiten [The newest kabuki dictionary]. Tokyo, Japan: Kashiwa Shobō.

Torigoe, B. (2000). Monomane kyōgen zukushi [Miming kyōgen devotion]. In Y. Hattori, T. Tomita, & T. Hirosue (Eds.), Shinpan kabuki jiten [New edition kabuki dictionry] (p. 399). Tokyo, Japan: Heibonsha.

Tsuda, S. (1977–78). Bungaku ni arawaretaru waga kokumin shisō no ken'kyu, 1–8 [Study on Japanese thoughts appearing in literature, Vols. 1–8]. Tokyo, Japan: Iwanami Shoten. (Original work published 1917–1921)

Tsukamoto, M., & Suzuki, M. (2010). Zen'ei toha nanika? kōei toha nanika?—Bungakushi no kyokō to kindaisei no jikan [What is avant-garde? what is rear guard?: The fiction of literary histories and the time of modanity]. Tokyo, Japan: Heibonsha.

Uchiyama, M. (2009). Enkyō san'nen kara Kan'en san'nen no Namiki Sōsuke—Jōruri gassakusha mondai [Namiki Sōsuke from Enkyō 3 to Kan'en 3: Jōruri joint work problem]. In Namiki Sōsuke ten—Jōruri no ōgon jidai [Namiki Sōsuke exhibition: The golden age of jōruri] (pp. 67–70). Waseda Daigaku Tsubouchi Hakushi Kinen Engeki Hakubutsukan (Waseda University Dr. Tsubouchi Memorial Theatre Musium).

Wanatabe, T. (1986). Musume dōjōji. Tokyo, Japan: Shinshindō Shuppan.

Watanabe, N. (2012). Nihon shōsetsu gijutsushi [A history of technologies of Japanese novels]. Tokyo, Japan: Shinchōsha.

Watanabe, T., & Takaizumi, A. (2012). Shōwa engeki daizenshū [Shōwa theatre large collection]. Tokyo, Japan: Heibonsha.

Watanabe, T. (1965). kabuki ni joyū wo [Actresses for kabuki]. Tokyo, Japan: Maki Shoten.

Watanabe, T. (1974). On'nagata no unmei [The destiny of an on'nagata]. Tokyo, Japan: Kinokuniya Shoten.

Watanabe, T. (1989a). kabuki—Kajō naru kigō no mori [kabuki: The forest of excessive signs]. Tokyo, Japan: Shin'yōsha.

Watanabe, T. (1989b). Nakamura Kanzaburō. Tokyo, Japan: Kōdansha.

Watanabe, T. (1993). Nizaemon no fūkaku [Nizaemon's dignity]. Tokyo, Japan: Kawade Shobō Shinsha.

Watanabe, T. (1997). Mokuami no Meiji Ishin [The Meiji restoration of Mokuami]. Tokyo, Japan: Shinchōsha.

Watanabe, T. (2004). kabuki—Kata no miryoku [kabuki: Charm of forms]. Tokyo, Japan: Kadokawa Shoten.

Watanabe, T. (2006). Meioyama Jakuemon [A great on'nagata Jakuemon]. Tokyo, Japan: Shinchōsha.

Watanabe, T. (2009a). kabuki no mikata [kabuki's viewpoints]. Tokyo, Japan: Kadokawa Gakugei Shuppan.

Watanabe, T. (2009b). Edo engekishi, 1 [The history of theatre in the Edo era, Vol. 1]. Tokyo, Japan: Kōdansha.

Watanabe, T. (2009c). Edo engekishi, 2 [The history of theatre in the Edo era, Vol. 2]. Tokyo, Japan: Kōdansha.

Watanabe, T. (2012a). Watashi no kabuki henreki—Aru gekihyō-ka no kokuhaku [My kabuki wandering: Confession of a drama critic]. Tokyo, Japan: Engeki Shuppansha.

Watanabe, T. (2012b). Meiji engekishi [The history of theatre in the Meiji era]. Tokyo, Japan: Kōdansha.

Watanabe, T. (2013a). Chūshingura—Mouhitotsu no rekishi kankaku [Chūshingura: Another historical sense]. Tokyo, Japan: Kōdansha.

Watanabe, T. (2013b). kabuki—Kata no shinzui [kabuki: The essence of kata]. Tokyo, Japan: Kadokawa Gakugei Shuppan.

Watanabe, T. (2018). Kudaime Danjūrō [Danjūrō IX]. Tokyo, Japan: Engeki Shuppansha.

Woguri. (2017). In Y. Muroki (Ed.), Shinchō nihon koten shūsei Sekkyōshū [Shinchō Japanese classic literature collection Sekkyōshū] (pp. 209–298). Tokyo, Japan: Shinchōsha. (Original work narrated the end of the medieval period)

Yamamoto, K. (2013). Jūhachidaime Nakamura Kanzaburō no gei—Apollon to Dionysus [The arts of Nakamura Kanzaburō XVIII: Apollon and Dionysus]. Tokyo, Japan: Alpha-beta-sha.

Yamamoto, K. (2015). On'nagata no bigaku—Taoyameburi no senryaku [The aesthetics of a onnagata: The strategy of taoyameburi]. Tokyo, Japan: Alpha-beta-sha.

Yanai, K. (2011). Meiji no kabuki to shuppan media [Meiji's kabuki and the media for publication]. Tokyo, Japan: Pelican-sha.

Yasutomi, J. (2012). Midori. In Y. Tomizawa & H. Fujita (Eds.), Saishin kabuki daijiten [The newest kabuki dictionary] (pp. 72–73). Tokyo, Japan: Kashiwa Shobō.

Yūda, Y. (Ed.). (1965). Bunraku jōruri shū [Collection of bunraku and jōruri works]. Tokyo, Japan: Iwanami Shoten.

Zenshinza. (2015). Zenshinza go gatsu Kokuritsu Gekijō kōen Banchō sarayashiki, Ninjōbanashi Bunshichi mottoi [Zenshinza may National Theatre performace Banchō sarayashiki and Ninjōbanashi Bunshichi mottoi]. Tokyo, Japan: Zenshinza Eihō Sendenbu.

KEY TERMS AND DEFINITIONS

***Geinō* Information System (GIS):** This means a system model of *geinō* production and consumption. GIS has been considered a framework in which various levels of narrative generation processes are driven by writers, receivers, characters, and actors. Furthermore, GIS is a model for circulative narrative generation that repeats a narrative generation process. In contrast, in the author's narrative generation study, the "Integrated Narrative Generation System: INGS" corresponds to a single-level narrative generation system.

Integrated Narrative Generation System (INGS): INGS is a synthetic narrative generation system architecture integrating the previous studies by the author. From the broadest perspective, INGS is divided into two types of part, knowledge and procedure. INGS aims at a kind of narrative synthesizer that integrates a variety of narrative techniques, methods, rhetoric, and knowledge into an organic and dynamic generation framework.

Multiple Narrative Structures Model: This is a conceptual model by which a narrative text is multiply constituted, and the narrative generation process multiply executed. This is not a special narrative phenomenon, but the multiplicity is positioned as an essentially important concept for considering and designing narrative generation mechanisms in the author's narrative generation study.

"*Naimaze*" in *Kabuki*: *Naimaze* is a narrative creation method or a group of techniques that combines components in existing *kabuki* works or in the narratives from other genres for creating a new work. Components that can be used for *naimaze* include stories, plots, characters, and places. It is a representative narrative technique for the multiple narrative structures in *kabuki*.

"Person" in *Kabuki*: This is divided into the level of an "actor" who performs onstage and that of a "(dramatic) character" in the narrative of the work. Moreover, the existence accompanying the real name also has essential meaning. The multiplicity of "person" in *kabuki* shows an example of the multiple narrative structures. On the *kabuki* stage, these three levels or sides simultaneously multiply exist through a person's body.

"Story" in *Kabuki*: This is related to the multiple narrative structures model in the point that a story played onstage frequently overlaps with one or more other stories in many cases. The principle of overlapping stories adds depth and thickness to narratives of *kabuki*.

"*Tsukushi*" or "*Zukushi*" in *Kabuki*: This is a rhetorical method for collectively and concisely representing objects in a category. Although *tsukushi* or *zukushi* is not a technique limited to *kabuki*, diverse *tsukushi*s are used in *kabuki*, such as "yama (mountains)-*zukushi*," "*sakana* (fishes)-*zukushi*," and "kane (bells)-*zukushi*" to ensure effectiveness and create strong and deep impressions on the audiences related to narrative multiplicity.

APPENDIX 1

Table 4. kabuki works list

	English Title	First Performance		Notes
		Year	Theatre (City)	
1	The Treasury of Loyal Retainers	1748	Takemotoza (Osaka)	Originally, *ningyō jōruri*
2	Sugawara's Secrets of Calligraphy	1746	Takamotoza (Osaka)	Originally, *ningyō jōruri*
3	Yoshitsune and the Thousand Cherry Trees	1747	Takamotoza (Osaka)	Originally, *ningyō jōruri*
4	The Ghost Stories at Yotsuya on the Tōkaidō	1825	Nakamuraza (Edo)	
5	The Scandalous Love of Osome and Hisamatsu	1813	Moritaza (Edo)	Formal name: Osome Hisamatsu Ukina no Yomiuri
6	Scarface Yosa	1853	Nakamuraza (Edo)	Formal name: Yo ha Nasake Ukina no Yokogushi
7	Legend of the Righteous Man of Sakura	1851	Nakamuraza (Edo)	Formal name: Higashiyama Sakura Sōchi
8	The Three Kichisas and the New Year's First Visit to the Pleasure Quarters	1860	Ichimuraza (Edo)	
9	The Glorious Picture Book of Aoto's Exploits	1862	Ichimuraza (Edo)	
10	Sukeroku: Flower of Edo	1811	Ichimuraza (Edo)	Formal name: Sukeroku Yukari no Edo Zakura
11	The Soga Confrontation	1676	Nakamuraza (Edo)	Formal name: Kotobuki Soga no Taimen [The Felicitous Soga Encounter]
12	Benten the Thief			In Aoto Zōshi Hana no Nishikie [The Glorious Picture Book of Aoto's Exploits]
13	Mount Imo and Mount Se: An Exemplary Tale of Womanly Virtue	1771	Takemotoza (Osaka)	Originally, *ningyō jōruri*
14	Chronicle of the Battle of Ichinotani	1751	Toyotakeza (Osaka)	Originally, *ningyō jōruri*
15	Kumagai's Battle Camp			In Ichinotani Futaba Gunki [Chronicle of the Battle of Ichinotani]
16	Japan's 24 Paragons of Filial Piety	1766	Takemotoza (Osaka)	Originally, *ningyō jōruri*
17	The Picture Book of the Taikō	1799	Toyotakeza (Osaka)	Originally, *ningyō jōruri*
18	The Precious Incense and Autumn Flowers of Sendai	1777	Naka no shibai (Osaka)	
19	Mirror Mountain	1782	Gekiza (Edo)	
Formal name: Kagamiyama Kokyō no Nishikie [Mirror Mountain: A Women's Treasury of Loyalty]	Originally, *ningyō jōruri*	20	The Love Suicides at Sonezaki	1703

continued on following page

Table 4. Continued

	English Title	First Performance		Notes
		Year	Theatre (City)	
Takemotoza (Osaka)	Originally, *ningyō jōruri*	21	The Love Suicides at Ten no Amijima	1720
Takemotoza (Osaka)	Originally, *ningyō jōruri*	22	Five Great Powers That Secure Love	1794
Nishi no shibai (Kyoto)		23	The Balladeer's New Tale	1780
Takemotoza (Osaka)	Originally, *ningyō jōruri*	24	Five Male Bandits	
	Formal name: Aoto Zōshi Hana no Nishikie [The Glorious Picture Book of Aoto's Exploits]	25	The Maiden at the Dōjōji Temple	1753
Nakamuraza (Edo)		26	Viewing the Autumn Foliage	1887
Shintomiza (Tokyo)	Originally, *nō* (by Kanze Kojirō Nobumitsu (1435-1516))	27	The Subscription List	1840
Kawarasakiza (Edo)	Originally, *nō* (Ataka)	28	Earth Spider	1881
Shintomiza (Tokyo)	Originally, *nō*	29	Benkei aboard Ship	1885
Shintomiza (Tokyo)	Originally, *nō*	30	A Substitute for Meditation	1910
Ichimuraza (Tokyo)	Originally, *kyōgen* (Hanago)	31	Tied to a Pole	1916
Ichimuraza (Tokyo)	Originally, *kyōgen*	32	The Barrier Gate	1784
Kiriza (Edo)	Formal name: Tsumoru Koi Yuki no Sekinoto [Love and Deep Snow at the Mountain Barrier]	33	Echigo Lion	1811
Nakamuraza (Edo)		34	The Ferry	1843
Ichimuraza (Edo).		35	The Three-Shrine Festival	1832
Nakamuraza (Edo)	Formal name: Yayoi no Hana Asakusa Matsuri	36	Water-Seller	1813
Moritaza (Edo)		37	Just a Minute!	1697
Nakamuraza (Edo)		38	The Whisker Tweezers	
	In Narukami Fudō Kitayama Zakura [Saint Narukami, the God Fudō, and Mount Kita's Cherry Blossoms]	39	The Thunder God	1742
Sadoshima Chōgorōza (Osaka)		Formal name: Narukami Fudō Kitayama Zakura [Saint Narukami, the God Fudō, and Mount Kita's Cherry Blossoms]	(Narukami [The Thunder God] is also a scene of Narukami Fudō Kitayama Zakura.)	40

continued on following page

Table 4. Continued

	English Title	First Performance		Notes
		Year	Theatre (City)	
Onikage Musashi Stirrup	1710	Shinozukaza (Osaka)		41
Chronicle of the Great Peace on a Go Board	1710	Takemotoza (Osaka)	Originally, *ningyō jōruri*	42
Absolutely Sango is Important				43
The Treasury of Loyal Retainers: Genroku Version	1934	*kabukiza* (Tokyo)		44
Through Iga Pass with the Tōkaidō	1783	Takamotoza (Osaka)	Originally, *ningyō jōruri*	45
Through Iga Pass with Riding Kappa	1776	Arashiza (Osaka)		46
Now is the Time to Raise the Bellflower Flag	1808	Ichimuraza (Edo)		47
The Mansion of Plates	1916	Hongōza (Tokyo)		48
Ashamed to Show a Sweaty Face as Red as the Autumn Leaves	1815	Kawarasakiza (Edo)		49
The Sword Kagotsurube and Sobering Up the Brothel District	1888	Chitoseza (Tokyo)		50
The Floating World Sword Hilt and the Conjoined Thunderbolts	1825	Nakamuraza (Edo)		51
The Sparrows of Yoshiwara	1768	Ichimuraza (Edo)		52
Anothr Version of a Brave Archer's Tale	1969	Kokuritsu Gekijō (Tokyo)		53

continued on following page

Table 4. Continued

	English Title	First Performance		Notes
		Year	Theatre (City)	
The Sardine Seller Draws the Net of Love	1954	*kabukiza* (Tokyo)		54
Chōgorō and Chōkichi: A Diary of Two Butterflies in the Pleasure Quaters	1749	Takemotoza (Osaka)	Originally, *ningyō jōruri*	55
A Colorful Rivalry: Tales of Ise	1775	Naka no Shibai (Osaka)		56
The Date Rivalry and Okuni *kabuki*	1778	Nakamuraza (Edo)		57
Sumida River Flower Palace Dyeing	1814	Ichimuraza (Edo)		58
The Scarlet Princess of Edo	1817	Kawarasakiza (Edo)		59
The Treasury of Loyal Retainers: Kibakazari Version	1840	Kawarasakiza (Edo)		60
Small Night Clothes Mardarin Ducks' Swords' Edges	1739	Toyotakeza (Osaka)	Originally, *ningyō jōruri*	61
Love Letter from the Licensed Quarter	1808	Nakamuraza (Edo)		62
Scarface Otomi	1864	Moritaza (Edo)	Formal name: Musumegonomi Ukina no Yokosushi	63

Everyone's Favorite Kanjinchō
1773
Nakamuraza (Edo)

APPENDIX 2

Ichikawa Ebizō XI's Property Information Acquired From Japanese Wikipedia

(The author indicates the parts used in figures in this chapter based on Source: Ogata, 2018a)

Geino information
Basic information

Jūichidaime (the eleventh) Ichikawa Ebizō (born June 12 1977) is a kabuki actor and actor in other media. His kabuki name is "Ichikawa Ebizō" as the head of the Ichikawa clan. His acting house name is Naritaya. His family crest is Mimasu and his alternate personal crest is Gyoyobotan. He is affectionately known as "Ebi sama and "Ebi chan."

Critical reception

He has taken on the challenge of the major roles in the kabuki classics and also played many roles for the first time, and his performances have been highly praised.

There are many fans of Ebi sama" who are also fans of the popular Jūichidaime Ichikawa Danjūrō. He is one of the young stars of contemporary kabuki theater. He played the leading role in Musashi, NHK's Taiga (long-running) drama, and also taken on the challenge of performing in contemporary dramas.

At the time his father, Danjūrō was alive, they frequently came into conflict when they performed together on stage, such as during overseas performances. There were times when Danjūrō listed to and accepted Ebizō's opinion, but the atmosphere within their home at that time was not good. However, Ebizō has frequently expressed his gratitude to Danjūrō, who always watched kindly over him even when they were in conflict.

While his father Danjūrō had shown a tolerant attitude when it was discovered that Ebizō had a child born outside of marriage (an illegitimate child), he was furious with Ebizō in 2007 when his son was injured in a bathroom and his kabuki performances has to be cancelled, commenting that "He has made this absence in his public performances, so he has no self-control. I cannot forgive him this time." He was furious again when he was injured during a violent incident that occurred in 2010, and he commented that "Will he be able to join (the performance with star cast scheduled to take place at the Kyoto Minami-za Theater)? (While Ebizō was caught up in the incident, it was said that it occurred because of his behavior) As it was his responsibility, I think he should not perform," and he harshly criticized Ebizō, saying "His actions were extremely rude and he must take a strong look at himself. I feel angry."

When Ebizō was in his teens, he repeatedly rebelled against the rigorous training that he had undergone continually since early childhood and also the heavy responsibility of his family name and tradition. It is said that what made him reform his attitude was when he was greatly moved by the beauty and the courage of a performance by his grandfather, Jūichidaime Ichikawa Danjūrō, playing the role of Benkei in Kanjinchō, which he saw by chance on film. With his grandfather's performance as his goal, he has suffered in his pursuit of this ideal and even today continues to struggle to achieve it.

In the past he found it difficult to appear on stage, but today, when he is devoting himself diligently to kabuki, he feels it is more difficult to be a private person. He was worrying about the strangeness of this and discussed it with Bandō Tamaburō V, who told Ebizō that "It is natural you feel that way," after which he became more relaxed about it. He also said that he was very happy that Bandō Tamasaburō was able to empathize with him about his concerns.

He engages in full-scale strength training and yoga and has appeared on the front of magazines, such as Tarzan.

The Naritaya acting house of artists are actively reviving and performing those works in the kabuki Jūhachiban (the repertoire of 18 major kabuki plays) that today are not frequently performed.

Artistic performances and events

December 6, 1977 – Born in Tokyo. The eldest son of his father Jūnidaimei (the twelfth) Ichikawa Danjūrō and his mother Kimiko (maiden name, Shoji). His grandfather was "Ebi sama," the Jūichidaime Ichikawa Danjūrō.

May 1983 – His gave his first public performance as Togu in The Tale of Genji kabuki play

May 1985 - He assumed the name of shichidaime (the seventh) Ichikawa Shinnosuke for the performance of Kikanbo in the Uiro Uri kabuki play

March 1996 – After attending the kindergarten, primary, secondary, and high school departments of Aoyama Gakuin, he transferred to Horikoshi High School

March 1997 – He graduated from Horikoshi High School

2000 – He played the role of Hikaru Kimi in The Tale of Genji kabuki play, and the venue was full on each night and the tickets were so popular as to be sold out on the first day they went on sales. Together with Godaime (the fifth) Onoe Kikunosuke and Nidaime (the second) Onoe

Tatsunosuke, he became of the "Heisei no san Suke" (the three "Sukes" of the Heisei period) and helped create a new boom in kabuki.

February 2003 – It was discovered that he had an illegitimate child with Akiko Hioki (a girl, 1 year old at that time)."At a press conference, he explained that he recognized the child was his and was paying child support.

April 4 2004 – He gave the performance for the succession of his stage name at the Naritasan Shinshoji Temple and the parade for the report to the deity was held. A train named "Ebizō" began operations between the Keisei Main Line Keisei Ueno Station and Keisei Narita Station (a Keisei AE 100-type train was used).

May 2004 – He assumed the name Jūichidaime Ichikawa Ebizō on performing various kabuki roles including Hanakawado Sukeroku in Sukeroku Yukari no Edozakura and Kamakuranogongorou in Shibaraku.

June 30, 2004 – he gave the performance for the assumption of the name and the notification ceremony was held at the Myoo-in Temple of Naritasan Osaka Betsui. The train "Ebizō" began operations between the Keihan Main Line Tanmabashi Station and Kariyosen Station (a Keihan 8000 series train was used).

October 2004 – At the Palais de Chaillot Theater in Paris France, he gave a public performance of the performance he gave when his name was changed from Ichikawa Shinnosuke to Jūichidaime Ichikawa Ebizō. The performance was attended by leading members of the political and business worlds, and the newspaper Le Monde gave an entire page for its coverage of the performance, reporting that "he is a member of family that has been distinguishing itself in traditional Japanese theater since the end of the 17th century," that "he is a star at 26 years old," and that he is "a young and handsome man who in particular possesses 'the gaze of death'"(the nirami glare used in kabuki), and his performance attracted a lot of attention locally.

September 7 2006 – While promoting the movie Deguchi no nai Umi (The Sea with No Exit), he appeared in the Quiz $ Millionaire (Fuji Television).

March 2007 – He performed with this father for the first ever kabuki performance at the Theatre National de l'Opera de Paris. He performed the Momijigari dance of Benkei in Kanjinchō, and earned the praise of Paris. He surprised the audience by jumping into the central aisle to perform the Tobiroppo (the flying exit) along the hanamichi (the flower road) passage from the stage.

July 13 2007 – He fell over in the bathroom of a dressing room and suffered a serious injury, cutting the back of his right foot on broken door glass, and he had to withdraw from the July Great kabuki at Osaka Shochiku-za Theater.

2009 – October 14 2007, a documentary on the life of Ebizō was broadcast "Born into fame; the story of the fate and struggles of Ichikawa Ebizō" (Fuji Television-affiliated) and it received the Gold Award in the Arts Section of the New York Festival.

May 23, 2009 – He appeared in BS's serialized drama MR.BRAIN. It was the first time he had appeared in a commercial serialized drama and it was his first appearance in a contemporary drama.

Early in the morning of November 25, 2010, he injured his head and face in an assault incident in a restaurant in Nishi-Azabu (the Jūichidaime Ichikawa Ebizō assault incident). After returning home, his wife, Mao, reported the incident to the emergency services and to the police, and he was rushed to Toranomon Hospital. It seems that after eating and drinking with his friends, who were kabuki actors, a group of a few people joined with Ebizō, and when they want home, he was assaulted by a stranger whose group he had joined, and the Metropolitan Tokyo Meguro Police Station started an investigation into the incident in which he was injured. However, there was criticism of Ebizō when it was discovered that he went out drinking on the night of November 24, even though he had earlier cancelled a press conference, citing ill health, for the Early Spring Flower kabuki (to be performed in January 2011 at Le Theater Ginza, Tokyo) that was scheduled to be held in the afternoon of November 25. On the evening of November 25, after the Naritaya acting house hurriedly convened a meeting of the supporters' association, his father, Danjūrō, held a press conference and criticized him, saying "I feel angry at his lack of self-awareness." Due to his injuries, he had to withdraw from the festive kaomise performance of the Tozaigodo kabuki, which was scheduled to be held from November 30 at the Minzami-za Theater, Kyoto. Ebizō himself gave a press conference at a hotel in Tokyo on December 7, and apologized that "I think my daily behavior invited this incident." In addition, the Shochiku Theater Company President, who attended the press conference, announced that Ebizō would not be appearing in any productions for an indefinite period. Alongside this punishment for an indefinite period, the Early Spring Flower kabuki was cancelled, and he withdrew from the February Great kabuki to be performed at the Misono-za Theater, Nagoya (in February 2011).

July 2, 2011 – His punishment was ended and he returned to the stage for the July Great kabuki at Shinbashi Enbujō. He released a comment that "the warm cheers and applause from the audience inspired me. I am grateful for being able to return to the stage and every day, I will take on the challenge of performing with all of my heart and soul."

New Year 2013 – On the first day of the New Year Asakusa kabuki, in which he was performing for the first time in 14 years, prior to this performance, he held a mirror opening together with his co-starts in front of the Asakusa Public Hall, Tokyo.

January 30, 2014 – He announced on his blog that he plans to create his own official website.

July 29, 2015 – He was appointed as the "tour guide" tourism ambassador of Narita City, Chiba Prefecture.

November 2015– The first performance of his eldest son Kangen in Edo no Hana Narita no Omokage in the festive kaomise Performance of the Great kabuki in the 11th Generation Ichikawa Danjūrō Fifteenth Festival (kabuki theater).

He has served as the ambassadorial 'tour guide' of Narita City since 2015.

Public relations

He started a blog on April 9, 2013, and he updates it very frequently, sometimes more than 10 times a day. He received the Ameba MVB award.

In 2009, His father, Ichikawa Danjūrō, was appointed public relations ambassador to France.

New Year 2013. It was announced that if Ichikawa Ebizō glared at you in the New Year Asakusa kabuki, you would enjoy good health.

Awards

1995 Best newcomer at the Mayama Seika Awards for his role of Kiyoshi Hamamoto in Kanayakome1995 Shochiku Chairman's Award for his performance in Kagamijishi

1997 Shochiku Chairman's Award for his performance in Yoshiwarasuzume

2000 Yomiuri Theatrical Grand Awards Haruko Sugimura Award for Benkei in Kanjinchō and for Zusho Nosuke in Tenshu Monogatari

2001 Best newcomer prize in the Matsuo Prize of Entertainment. Best newcomer award in the Minister of Education Award for Fine Arts

2007 Bestowed the title of Chevalier of the L'Ordre des Arts et des letter by France

2007 Special award in the Best Dressed Awards

2014 Best actor award in the 37th Japan Academy Awards

2014 Yahoo! Search Grand Prize 2014 in the actor category

Performances

・ kabuki

Naojiro Kataoka in Kumo ni Magou Ueno no Hatsuhana Yukinoyubeiiriyanoazemich (3,000 years old and Naozamurai),Narukami Shonin in Narukami,Hikaru Kimi in The Tale of Genji,Hanakawado Sukeroku in Sukeroku Yukari no Edozakura,the spirit of a lion after Kosho Yayoi in Kagamijishi,Benkei in Kanjinchō, Saemon Togashi

・ Overseas performances

kabuki performance in Paris of the performance he gave on the assumption of his name of Jūichidaime Ichikawa Ebizō (the National Palais de Chaillot Theater, Paris) (October 2004),kabuki performances in a tour of Europe (at the Saddlers Wells Theater, London, and the Stadsschouwburg Theater, Amsterdam) (May and June 2006),the Theatre National de l'Opera de Paris Shochiku Great kabuki (the Theatre National de l'Opera de Paris, Palais Garnier) (March 2007),Shochiku Great kabuki Monaco performance (the Opéra de Monte-Carlo, Salle Garnier) (September 2009),Shochiku Great kabuki London performance (the Saddlers Wells Theater, London) (June 2010),Shochiku Great kabuki Rome performance Shochiku Grand kabuki featuring Ebizō Ichikawa XI (Teatro-Argentina, Rome) (June 2010),Ichikawa Ebizō Singapore performance "Invitation to the Classics (Izana)" JAPAN THEATER "Produced and headlined by EBIZŌ ICHIKAWA XI (Marina Bay Sands MasterCard Theater, Singapore) (November 2014),EBIZŌ ICHIKAWA XI's JAPAN THEATER 2015(Marina Bay Sands Grand Theater, Singapore) (October 2015), Ichikawa Ebizō GRAND JAPAN THEATER Fujairah performance (Fujairah Miramar, the Al-Acka Beach Resort) (February 2016), Ichikawa Ebizō GRAND JAPAN THEATER New York performance (Carnegie Hall, New York) (March 2016)

・ TV dramas

Taiga drama (NHK)

Hanna no Ran (1994) – role of Ashikaga Yoshimasa (in his youth),MUSASHI (2003) – lead role of Miyamoto Musashi,Onna Joshu Naotora (2017) – the role of Oda Nobunaga,12-jikan Cho-wide Drama (TV Tokyo),Toyotomi Hideyoshi Unifier of a Country! (1995) – role of Kipposhi,Honoo no Bugyou Ookaechizennokami (1997) – role of Oka Ichijūrō, The 8th Era Novel Grand

Prix Special "Nakazo Kyouran" (Asahi Broadcasting) – role of Nakamura Nakazo

MR.BRAIN episodes 1, 7, and 8 (2009, TBS) – role of Kohei Takei

Drama Special on the 100 year anniversary of the birth of Matsumoto Seichou, "Kiri no Hata" (2010, Nippon Television) – lead role of Kinya Otsuka

Friday Prestige "Hasegawa Machiko Monogatari – the day that Sazae san was born -" (2013, Fuji Television) –role of the 9th Ichikawa Ebizō

Saturday Drama "Yowakutemo Katemasu: Aoshi Sensei to Heppoko Kōkō Kyūji no Yabō" (2014, Nippon Television) – role of Kentaro Yachida

Ishikawa Gouemon (2016, TV Tokyo) – lead role of Ishikawa Gouemon

- TV animation

Detective Conan Conan and the Ebizō kabuki Jūhachiban Mystery (January 9 to 16, 2016 Yomiuri Television) – role of Ichikawa Ebizō (himself)

- Voice over performances

All is Lost – the Last Letter (2014) – Role of 'our man" (Robert Redford)

- Movies

The Sea with No Exit (2006) – lead role of Koji Namiki,Harakiri: Death of a Samurai (2011) – lead role of Hanshirô Tsugumo, Ask This of Rikyu (2013) – the lead role of Sen no Rikyu, Over Your Dead Body (2014) – lead role of Kousuke Hasegawa, Blade of the Immortal (2017) – role of Eiku Shizuma

- Other stage performances

Lyle (1990) – role of Josh, Stand by Me (1991) – lead role of Chris,Tenshu Monogatari (1999) – role of Zusho Nosuke (Bandō Tamasaburō V, joint performance with Rie Miyazawa),Kaijin Besso – role of young nobleman (joint performance with Bandō Tamasaburō),On the Wings of the Wind (2002) – role of Suzuki Inohira,Miyamoto Musashi (2003) – the lead role of Miyamoto Musashi, Nobunaga (2006) - the lead role of Oda Nobunaga, Tenshu Monogatari (2006) – the role of Zusho Nosuke (joint performance with Bandō Tama saburō, Nidaime (the second) Junen Ichikawa,Kaijin Besso (2006) – role of young nobleman (joint performance with Bandō

Tamasaburō),Dracula (2007)- the lead role of Ray (joint performance with Rie Miyazawa))

・ Commercials

Itoen O–Io-Cha (2000 -),Yamaki (2000 - 2010),Meiji Seika Chocolate Cool (1995),Mitsui Life Insurance (1998),NEC Express Workstation (1998),Mizuno Superstar (2003),TSUTAYA DISCAS (August 2009 -) moderator role in a TV shopping program, Iida Group Holdings (2014 -),Lion Luxury Care (2015),JTEKT (2016)

Life information

Basic information

His real name is Takatoshi (Tomisei) Horikoshi. His former name is Takatoshi Horikoshi (written using different kanji characters).

His wife was TV newscaster and celebrity Mao Kobayashi (died in June 2017). Height, 176cm. Weight, 80kg. Blood type, AB. Star sign, Sagittarius.

Hobbies and likes

He plays basketball and was a member of the basketball club when a student. He enjoys watching sumo. His is a big fan of the anime movies of Studio Ghibli. When he appeared in the program Oshareism, he was presented with a figure from Ponyo that was not available for sale and the pen that Miyazaki Hayao had used to draw this anime, which moved him to tears of joy. He also talked with Mr. Miyazaki, and he learned that Mr. Miyazaki also practices a health technique using a brush on the body that he himself practices, and he was moved by this point in common between them. He had previously lived an unhealthy life, eating sushi up to 50 times a month, smoking 4 packets of cigarettes a day, and drinking until he was drunk. But he had a change of heart and gave up cigarettes in 2004 when he assumed the name of Ebizō, saying that "When I perform kabuki, I think I must be aware of giving praise to God." Therefore, from the idea of "purifying his mind and body," from around November 2008 he gave up his irregular eating habits, and also gave up alcohol, meat, and fish. But he began to again drink alcohol and eat meat and fish after his marriage to Ma

Articles

National Institute of Informatics' Collection of Articles, National Institute of Informatics

He was a childhood friend and classmate of Godaime (the fifth) Onoe Kikunosuke. When he was an elementary school student, he was invited to his friend's residential villa, and he was scolded by Junko Fujimoto, the mother of Kikunosuke, for being mischievous. This was his first experience of being scolded by someone outside of his family.

His second cousins (male) include Dainime (the second) Onoe Shōroku, Shichidaime (the seventh) Ichikawa Somegorō, and his second cousins (female) are Matsumo Tokio (the sister of Shichidaime Somegorō) and Matsu Takako (the same). He is close friends with Hideaki Ito and they call each other "Hideaki" and "Takatoshi" (his real name is Takatoshi Horikoshi).

Events

November 19, 2009 – It was reported in Nikkan Sports that he was dating the TV newscaster Mao Kobayashi. On the same day, it was announced by fax that they had been dating on the premise of marriage toward them becoming engaged. After the New Year, on January 29, 2010, they held a press conference in which they officially announced their engagement. They submitted the marriage notification on March of the same year, and the wedding ceremony and reception took place on July 29.

November 25, 2010 Early in the morning he injured his head and face in an assault incident in a restaurant in Nishi-Azabu (the Jūichidaime Ichikawa Ebizo assault incident). After returning home, his wife, Mao, reported the incident to the emergency services and to the police, and he was rushed to Toranomon Hospital. It seems that after eating and drinking with his friends, who were kabuki actors, a group of a few people joined with Ebizo, and when they want home, he was assaulted by a stranger whose group he had joined, and the Metropolitan Tokyo Meguro Police Station started an investigation into the incident in which he was injured. However, there was criticism of Ebizo when it was discovered that he went out drinking on the night of November 24, even though he had earlier cancelled a press conference, citing ill health, for the Early Spring Flower kabuki (to be performed in January 2011 at Le Theater Ginza, Tokyo) that was scheduled to be held in the afternoon of November 25. On the evening of November 25, after the Naritaya acting house hurriedly convened a meeting of the supporters' association, his father, Danjūrō, held a press conference and criticized him, saying "I feel angry at his lack of self-awareness." Due to his injuries, he had to withdraw from the festive kaomise performance of the Tozaigodo kabuki, which was scheduled to be held from November 30 at the Minzami-za Theater, Kyoto. Ebizo himself gave a press conference at a hotel within Tokyo on December 7, and apologized that "I think my daily behavior invited this incident." In addition, the Shochiku Theater Company President, who attended the press conference, announced that Ebizo would not be appearing in any productions for an indefinite period. Alongside this punishment for an indefinite period, the Early Spring Flower kabuki was cancelled, and he withdrew from the February Great kabuki to be performed at the Misono-za Theater, Nagoya (in February 2011).

July 25, 2011, Reika, his eldest daughter and his first child with his wife Mao, is born.
February 3, His father Danjūrō XII, dies. He inherits from his father the title of the head of the Ichikawa clan.
March 22, 2013, Kangen, his eldest son and second child with his wife, Mao is born
February 4, 2015, He announces on his blog that he has changed his real name from Takatoshi Horikoshi to Takatoshi Horikoshi (Tomisei) (written with different kanji characters).
June 9, 2016, It is announced that his wife Mao is suffering from breast cancer.
June 22, 2017 His wife, Mao, dies. Her death is announced in a special press conference the following day.

Family and relatives

Great grandfather – Shichidaime Matsumoto Koshiro,Grandfather – Jūichidaime Ichikawa Danjūrō, great uncle – Hachidaime (the eighth) Matsumoto Koshiro,great uncle - Nidaime (the second) Onoe Shōroku,father – Junidaime Ichikawa Danjūrō, aunt – Nidaime (the second) Ichikawa Kōbai,sister – Sandaimei (the third) Ichikawa Botan,wife - Mao Kobayashi, sister-in-law - Maya Kobayashi

January 29, 2010, He holds a press conference to announce his engagement to freelance TV announcer Mao Kobayashi. After that, they submitted the wedding notification on March 3 and held the wedding ceremony.
July 25, 2011 His eldest and first child, Reika, is born
March 22, 2013, His second child and eldest son, Kangen, is born
June 9 2016, It is announced at a press conference that Mao is suffering from breast cancer.
June 23, 2017, On his own blog, he posts that this is "the day I have cried the most in my life," implying Mao's death. On the same day, after a stage performance at 2.30pm, he holds a press conference and reports that Mao died at home on the evening of the previous day, June 22. On the day of Mao's death, Ebizo was rehearsing on stage, and he returned home suddenly upon receiving a LINE message on his mobile phone informing him that Mao's condition had deteriorated. Together with his

sister-in-law Maya Kobayashi, they spent the final moments together with Mao. Mao's final words were "I love you."

Property Data of Ichikawa Ebizō XI

(The following description was artially deleted and edited based on Source: Ogata, 2018a)

(setq *Ebizou*

'(geinojinInfo

(geino

(g-base

((("は" "歌舞伎役者、俳優。" (")は" "十一代目市川海老蔵(じゅういちだいめいちかわえびぞう、1977年12月6日-")) (nil "nil"))

(("は" "当代。" ("」の" "歌舞伎名跡「市川海老蔵")) (nil "nil"))

(("は" "杏葉牡丹。" ("は" "三升、替紋" ("は" "成田屋。定紋" ("は" "屋号")))) (nil "nil"))

(("は" "。" ("」がある" "海老さま」、「海老ちゃん" ("に" "愛称"))) (nil "nil"))

(("は" "た。" ("がよくあっ" "こと" ("で揉め" "時には演出のこと" ("に上がっ" "舞台" ("などで一緒に" "健在であった当時、海外公演" ("が" "父・團十郎")))))) (nil "nil"))

(("は" "。" ("はよくなかったという" "ときの家庭内の空気" ("もあっ" "とき" ("の意見を聞き入れて許してくる" "海老蔵" ("が" "團十郎"))))) (nil "nil"))

(("は" "。" ("を述べている" "團十郎に常々感謝の言葉" ("にずっと優しく見守ってくれてい" "だが、反発していたとき"))) (nil "nil"))

(("は" "た。" ("」とコメントしてい" "ん" ("は同情でき" "今回" ("がなってい" "んだから、自己管理" ("を空け" "穴" ("はさすがに激怒し" "時" ("を降板し" "歌舞伎公演" ("をし" "父・團十郎も、2007年に海老蔵が風呂場でケガ" ("を見せてい" "際に寛大な態度" ("(非嫡出子)が発覚し" "婚外子")))))))))) (nil "nil"))

(("は" "のではないか。" ("の)けいこに入れ" "京都南座での顔見世興行" ("にも再度激怒し「(予定されてい" "際" ("を受け" "被害" ("に発生し" "2010年"))))) (nil "nil"))

(("は" "た。" ("を突き放してい" "海老蔵" ("。憤りを感じている" "わざ" ("がなせる" "大変失礼な行動で、当人の自覚のなさ" ("では、出演すべきではないと思っている」と前置きし" "責任の上" ("を作ったと言う" "事件の原因" ("に巻き込まれたとは言え" "(事件"))))))) (nil "nil"))

(("は" "。" ("に耐え切れず反発を繰り返していたという" "家柄・伝統の重責" ("から続く" "幼少時" ("は" "海老蔵" ("の" "頃" ("の" "10代")))))) (nil "nil"))

(("は" "。" ("に感動したからと語っている" "祖父・十一代目市川團十郎の芸の美しさと勇姿" ("をつとめる" "弁慶" ("で見" "のは、偶然フィルム" ("を立ち直らせ" "そんな彼"))))) (nil "nil"))

(("は" "。" ("も、もがき続けているという" "現在" ("を追うがゆえに苦しみ" "理想" ("に掲げている" "目標" ("を" "祖父"))))) (nil "nil"))

(("は" "。" ("に感じるという" "今はプライベートの方が苦痛" ("になっ" "よう" ("に打ち込む" "ことが苦痛であったが、歌舞伎に真剣" ("に上がる" "舞台" ("は" "以前")))))) (nil "nil"))

(("は" "。" ("よ」と返され、安心したと述べている" "ところ、「あら、それは当たり前" ("に相談し" "五代目坂東玉三郎" ("かと悩ん" "異様なこと" ("が" "これ"))))) (nil "nil"))

(("は" "。" ("を嬉しく思ったとも語っている" "こと" ("を共感でき" "こと" ("と" "また、坂東玉三郎")))) (nil "nil"))

(("は" "。" ("もある" "こと" ("を飾っ" "雑誌『ターザン』などの表紙" ("を行っており" "ヨガ" ("や" "本格的な筋力トレーニング"))))) (nil "nil"))

(("は" "。" ("に行っている" "演目の復活上演を積極的" ("ではあまり上演されなくなっ" "うち、今日" ("」の" "お家芸「歌舞伎十八番" ("の" "成田屋"))))) (nil "nil"))))

(evaluation

((("は" "。" ("を多くつとめ、高い評価を得ている" "初役" ("に挑み" "大役" ("の" "古典")))) (nil "nil"))

(("は" "。" ("もいる" "ファン" ("に重ね合わす" "十一代目市川團十郎" ("のあっ" "人気" ("」と" "「海老さま"))))) (nil "nil"))

(("は" "若手スターの一人。" ("を担う" "歌舞伎" ("の" "現代"))) (nil "nil"))

(("は" "。" ("にも挑戦している" "ほか、現代劇" ("をつとめ" "主役" ("』で" "また、NHK大河ドラマ『武蔵MUSASHI")))) (nil "nil"))))

(g-event

((("は" "お目見得。" ("で" "春宮" ("』の" "1983年5月-歌舞伎座『源氏物語"))) (nil "nil"))

(("は" "襲名。" ("を" "七代目市川新之助" ("で" "坊" ("』の" "1985年5月-歌舞伎座『外郎売")))) (nil "nil"))

(("は" "。" ("を起こす" "新たな歌舞伎ブーム" ("」と呼ばれ" "五代目尾上菊之助、二代目尾上辰之助とともに「平成の三之助" ("を得る" "人気" ("は発売初日に完売する" "会場は連日満員、チケット" ("をつとめ" "光君" ("』の" "2000年歌舞伎座『源氏物語"))))))) (nil "nil"))

(("は" "ことを会見で説明。" ("は認知して養育費を支払っている" "ことが判明。子" (")がいる" "女児、当時1歳" ("に" "間" ("との" "2003年2月-日置明子"))))) (nil "nil"))

(("は" "京成AE100形電車を使用)。" ("」に乗車し" "列車「海老蔵号" ("で運行され" "練り。京成本線京成上野駅-京成成田駅間" ("で襲名披露、奉告の" "2004年4月4日-成田山新勝寺")))) (nil "nil"))

(("は" "襲名。" ("を" "十一代目市川海老蔵" ("で" "鎌倉権五郎ほか" ("』の" "花川戸助六、『暫" ("』の" "2004年5月歌舞伎座『助六由縁江戸桜"))))) (nil "nil"))

(("は" "使用)。" ("を" "京阪8000系電車" ("」が" "列車「海老蔵号" ("で" "京阪本線天満橋駅-香里園駅間" ("で" "2004年6月30日-成田山大阪別院明王院"))))) (nil "nil"))

(("は" "。" ("を呼ぶ" "現地でも大きな話題" ("』を持っている」となどと報じ" "一族の出身」「26歳のスター」「若く美男で特に『必殺の視線" ("をとどろかせてき" "17世紀末以来の日本の伝統的演劇の中で名" ("を使っ" "ル・モンド』紙が一面全面" ("も観劇し" "トップ" ("の" "市川新之助改メ十一代目市川海老蔵襲名披露フランス公演」。同国政財界" ("で" "シャイヨ宮劇場" ("の" "2004年10月-フランス・パリ"))))))))) (nil "nil"))

(("は" "クイズ$ミリオネア』(フジテレビ)に出演。" ("』の宣伝も兼ね" "海" ("の" "2006年9月7日-映画『出口"))) (nil "nil"))

(("は" "。" ("を驚かせる" "観客" ("で行っ" "中央通路" ("に見立て" "六方を花道" ("の飛び" "引っ込み" ("の絶賛を浴びる" "パリ" ("」の舞いを披露し" "紅葉狩" ("で" "弁慶" ("とともに実現。『勧進帳』の" "父" ("を" "歌舞伎公演" ("の" "初" ("で" "2007年3月-パリ・オペラ座")))))))))))) (nil "nil"))

(("は" "大阪松竹座「七月大歌舞伎」を降板。" ("を負い" "けが" ("を切る" "戸のガラスで右足の裏" ("で転倒し、割れ" "風呂場" ("の" "2007年7月13日-楽屋"))))) (nil "nil"))

(("は" "こと…?市川海老蔵・宿命と苦悩の物語?』(フジテレビ系)が、ニューヨークフェスティバルのアート部門で金賞を受賞。" ("に生まれる" "ドキュメンタリー番組『名門" ("に密着し" "海老蔵" ("に放送され" "2009年-2007年10月14日")))) (nil "nil"))

(("は" "。" ("となる" "現代劇出演" ("の" "初" ("の" "民放" ("の" "初" ("』に" "MR.BRAIN" ("の" "2009年5月23日-TBS"))))))) (nil "nil"))

(("は" "た。" (")を降板となっ" "名古屋・御園座「二月大歌舞伎」(2011年2月公演" ("」は中止となり" "初春花形歌舞伎" ("に伴い" "期限謹慎" ("を発表し" "こと" ("で自粛する" "松竹の社長が、海老蔵の歌舞伎出演を無期限" ("だと思います」と謝罪した。また、会見に同席し"

"こと"("が招い""自分のおごり"("を行い""海老蔵本人が12月7日、都内のホテルで記者会見"("」を降板し""ため、11月30日から出演予定だった京都・南座「吉例顔見世興行東西合同大歌舞伎"("を感じる」と謝罪し""本人の自覚のなさに怒り"("を開き""ことが発覚。11月25日夜、成田屋後援会主催の「後援会の集い」開催後に急遽、父・團十郎が記者会見" ("から飲みに出てい" "11月24日夜" ("にキャンセルしたにもかかわら""初春花形歌舞伎」(2011年1月公演、東京・ルテアトル銀座)の記者会見を体調不良を理由"("に予定されてい""警視庁目黒署は傷害事件として捜査を開始。しかし、海老蔵は11月25日午後" ("になり殴られたとみられ""数人のグループとは別の数人のグループとトラブル"("に帰っ""歌舞伎俳優の仲間らと飲食後、数人のグループと海老蔵だけが合流、その後、先"("される""虎の門病院に緊急搬送" ("に通報し" "頭部、顔面を負傷(11代目市川海老蔵暴行事件)。帰宅後に妻の麻央が救急と警察"("の飲食店で暴行を受け""2010年11月25日早朝、西麻布"))))))))))))))))))))) (nil "nil"))

(("は""た。"("で挑んでまいります」とのコメントを発表し""日々の舞台に全身全霊" ("に感謝し" "こと" ("に立たせていただける""思い。こうして舞台" ("の引き締まる" "身" ("から温かい声援を頂戴し" "新橋演舞場「七月大歌舞伎」で舞台復帰。「お客様" ("、謹慎を解かれ" "2011年7月2日"))))))) (nil "nil"))

(("は" "た。" ("を行っ" "東京・浅草公会堂前で共演者らと鏡開き" ("を迎え""新春浅草歌舞伎」の初日"("に出演する""2013年正月、14年ぶり")))) (nil "nil"))

(("は" "発表。" ("で" "ブログ" ("を作成予定と" "公式ホームページ" ("の" "2014年1月30日、自身")))) (nil "nil"))

(("は""就任。"("の観光大使「御案内人」に""2015年7月29日、千葉県成田市")) (nil "nil"))

(("は" "初お目見え。" ("が" "長男・觀玄" ("』で" "江戸花成田面影" (")の" "2015年11月、「吉例顔見世大歌舞伎11世市川団十郎五十年祭」(歌舞伎座")))) (nil "nil"))

(("は" "。" ("御案内人を務めている" "成田市" ("より" "2015年"))) (nil "nil"))))

(work

((("は" "・歌舞伎") (nil "nil"))

(("は" "弁慶、富樫左衛門" (",勧進帳の" "精" ("の" "獅子" ("に" "小姓弥生後" ("の" "花川戸助六,鏡獅子" ("の" "光君,助六由縁江戸桜" ("の" "鳴神上人,源氏物語" ("の" "片岡直次郎,鳴神" (")の" "侍" ("と" "天衣紛上野初花「雪暮夜入谷畦道」(三千歳")))))))))) (nil "nil"))

(("は" "・海外公演") (nil "nil"))

((“は”“誘い(いざない)」JAPANTHEATER”ProducedandheadlinedbyEBIZŌICHIKAWAXI(シンガポール・マリーナベイ・サンズマスターカードシアター)(2014年11月),EBIZŌICHIKAWAXI’sJAPANTHEATER2015(シンガポール・マリーナベイ・サンズグランドシアター)(2015年10月),市川海老蔵　GRANDJAPANTHEATER　フジャイラ公演(フジャイラ・ミラマール・アルアッカビーチリゾート)(2016年2月),市川海老蔵　GRANDJAPANTHEATER　ニューヨーク公演(ニューヨーク・カーネギー・ホール)(2016年3月)”(“への”“劇場)(2004年10月),訪欧歌舞伎公演(ロンドン・サドラーズ・ウェルズ劇場、アムステルダム・スタッツスハウブルフ劇場)(2006年5月、6月),パリ　オペラ座　松竹大歌舞伎(パリ・オペラ座　ガルニエ宮)(2007年3月),松竹大歌舞伎　モナコ公演(モナコ・オペラ座　サル・ガルニエ)(2009年9月),松竹大歌舞伎　ロンドン公演(ロンドン・サドラーズ・ウェルズ劇場)(2010年6月),松竹大歌舞伎　ローマ公演　ShochikuGrandkabukifeaturingEbizōIchikawa?(ローマ・アルジェンティーナ劇場)(2010年6月),市川海老蔵シンガポール公演「古典”(“ヨー”“十一代目市川海老蔵襲名披露　歌舞伎パリ公演(パリ・国立シャイ”))) (nil “nil”))

((“は” “・テレビドラマ”) (nil “nil”))

((“は” “大河ドラマ(NHK)”) (nil “nil”))

((“は”“炎の奉行大岡越前守(1997年)-大岡市十郎役,第8回時代小説大賞スペシャル「仲蔵狂乱」(朝日放送)-中村仲蔵役” (“を獲る” “乱(1994年)-足利義政(青年時代)役,武蔵MUSASHI(2003年)-主演・宮本武蔵役,おんな城主直虎(2017年)-織田信長役,12時間超ワイドドラマ(テレビ東京),豊臣秀吉天下” (“の” “花”))) (nil “nil”))

((“は” “MR.BRAIN 第1話・第7話・第8話(2009年、TBS) - 武井公平 役”) (nil “nil”))

((“は” “旗」(2010年、日本テレビ)-主演・大塚欽也役” (“の” “生誕100年記念松本清張ドラマスペシャル「霧”)) (nil “nil”))

((“は” “日?」(2013年、フジテレビ)-9代目市川海老蔵役” (“が生まれ” “金曜プレステージ「長谷川町子物語?サザエさん”)) (nil “nil”))

((“は” “へっぽこ高校球児の野望?」(2014年、日本テレビ)-谷内田健太郎役” (“「弱くても勝て” “土曜ドラマ”)) (nil “nil”))

((“は” “石川五右衛門(2016年、テレビ東京) - 主演・石川五右衛門役”) (nil “nil”))

((“は” “・テレビアニメ”) (nil “nil”))

((“は” “名探偵コナン コナンと海老蔵 歌舞伎十八番ミステリー(2016年1月9日・16日、読売テレビ) - 市川海老蔵(本人) 役”) (nil“nil”))

((“は” “・吹き替え”) (nil “nil”))

(("は" "男役(ロバート・ レッドフォード)" (")-我らの" "手紙?(2014年" ("の" "オール・ イズ・ ロスト?最後"))) (nil "nil"))

(("は" "・ 映画") (nil "nil"))

(("は""2013年)-主演・ 千利休役,喰女-クイメ-(2014年)-主演・ 長谷川浩介役,無限の住人(2017年)-閑馬永空役" ("にたずねよ" "海(2006年)-主演・ 並木浩二役,一命(2011年)-主演・ 津雲半四郎役,利休" ("の" "出口"))) (nil "nil"))

(("は" "舞台" ("の" "・ その他")) (nil "nil"))

(("は" "共演)" ("と" "ドラクル(2007年)-主演・ レイ役(宮沢りえ" ("と" "海神別荘(2006年)-公子役(坂東玉三郎" ("と" "助役(坂東玉三郎、二代目市川春猿" (")-図書之""2002年)-鈴木猪平役,宮本武蔵(2003年)-主演・ 宮本武蔵役,信長(2006年)-主演・ 織田信長役,天守物語(2006年" ("の" "疾風" ("と" "海神別荘-公子役(坂東玉三郎" ("と" "助役(五代目坂東玉三郎、宮沢りえ" (")-図書之" "ライル(1990年)-ジョッシュ役,スタンド・ バイ・ ミー(1991年)-主演・ クリス役,天守物語(1999年"))))))))) (nil "nil"))

(("は" "・ CM") (nil "nil"))

(("は" "お茶」(2000年-),ヤマキ(2000年-2010年),明治製菓「ショコラクール」(1995年),三井生命(1998年),NEC「Expressワークステーション」(1998年),ミズノ「スーパースター」(2003年),TSUTAYADISCAS(2009年8月-)テレビショッピング番組の司会者役,飯田グループホールディングス(2014年-),ライオン「贅沢Care」(2015年),ジェイテクト(2016年)" ("「お?い" "伊藤園")) (nil "nil"))))

(prize

((("は" "眞山青果賞新人賞" ("で" "濱本清" ("』の" "1995年『仮名屋小梅"))) (nil "nil"))

(("は" "松竹会長賞" ("』で" "1995年『鏡獅子")) (nil "nil"))

(("は" "松竹会長賞" ("』で" "1997年『吉原雀")) (nil "nil"))

(("は" "読売演劇大賞杉村春子賞" ("で" "助" ("之" "図書" ("』の" "弁慶、『天守物語" ("『勧進帳』の" "2000年"))))) (nil "nil"))

(("は" "2001年 松尾芸能賞新人賞、芸術選奨文部科学大臣新人賞受賞") (nil "nil"))

(("は" "シュヴァリエ(Chevalier)受章" (")の" "2007年フランス芸術文化勲章(L'OrdredesArtsetdeslettre")) (nil "nil"))

(("は" "2007年 ベストドレッサー賞特別賞") (nil "nil"))

(("は" "受賞" ("を" "2014年第37回日本アカデミー賞優秀主演男優賞")) (nil "nil"))

(("は" "2014年 Yahoo!検索大賞 2014・ 俳優部門") (nil "nil")))))

(life

(l-base
((("は" ")。" ("こしたかとし" "堀越寶世(宝世)(ほり" ("は" "本名")))
(nil "nil"))
(("は" ")。" ("こしたかとし" "堀越孝俊(ほり" ("は" "旧名"))) (nil "nil"))
(("は" "小林麻央(2017年6月死別)。" ("の" "タレント" ("で" "キャスター" ("は" "妻")))) (nil "nil"))
(("は" "身長176cm、体重80kg。") (nil "nil"))
(("は" "AB型。" ("は" "血液型")) (nil "nil"))
(("は" "射手座。" ("は" "星座")) (nil "nil"))
(("は" "相撲鑑賞。" ("は" "バスケットボール、趣味" ("は" "特技")))
(nil "nil"))
(("は" "た。" ("に所属してい" "バスケットボール部" ("は" "学生時代"))) (nil "nil"))
(("は" "ファン。" ("の" "アニメ作品" ("の" "スタジオジブリ"))) (nil "nil"))
(("は" "た。" ("をプレゼントされ、涙ぐんで喜んでい" "鉛筆" ("に使用し" "際" ("を制作する" "際に、『崖の上のポニョ』の非売品の人形と宮崎駿が同作" ("」に出演し" "「おしゃれイズム"))))) (nil "nil"))
(("は" "。" ("に感動したという" "こと" ("がわかり、共通点があっ" "こと" ("も行っている" "健康法を宮崎" ("をなでる" "タワシで体" ("が実践している" "際に自身" ("もあり" "こと" ("と対談し" "宮崎本人"))))))))
(nil "nil"))
(("は" "。" ("は再開している" "麻央との結婚後に酒、肉、魚" ("からやめてい" "思いから不規則な食生活や酒、肉、魚も2008年11月頃" ("を清くし" "心と体" ("も意識しなくちゃいけ" "こと" ("に捧げる" "とき、神様" ("をやる" "歌舞伎" ("にやめ" "心境の変化でタバコは2004年の海老蔵襲名時" ("な生活をしてい" "健康" ("に飲む" "よう" ("をあびる" "酒" ("吸っ" "タバコを1日4箱分" ("食べに行っ" "月50回" ("を" "寿司" ("は" "以前"))))))))))))))) (nil "nil"))
(("は" "同級生。" ("で" "幼馴染" ("とは" "五代目尾上菊之助"))) (nil "nil"))
(("は" "。" ("に叱られた初めての経験だったという" "これが家族以外の人" ("に叱られ" "菊之助の母である富司純子" ("にいたずらをしようとし" "とき" ("に招かれ" "別荘" ("の" "家" ("の" "頃、彼" ("の" "小学生")))))))) (nil "nil"))
(("は" "。" (")がいる" "姉)· 松たか子(同妹" ("の" "松本紀保(七代目染五郎" ("には" "従姉妹" ("が" "四代目尾上松緑、七代目市川染五郎" ("に" "再従兄弟")))))) (nil "nil"))

((“は” “仲。” (”)」と呼び合う” “堀越孝俊” (“の” “英明」、「孝俊(本名” (“を” “お互い” (“で” “親友” (“とは” “伊藤英明”)))))) (nil “nil”))))

(i-event

(((“は” “父・十二代目市川團十郎と母・希実子(旧姓・庄司)の長男。祖父は「海老さま」こと十一代目市川團十郎。” (“に生まれる” “1977年12月6日-東京都”)) (nil “nil”))

((“は” “高等部から堀越高等学校に転校。” (“で過ごし” “青山学院” (“まで” “1996年3月-幼稚園・初等部・中等部・高等部”))) (nil “nil”))

((“は” “卒業。” (“を” “1997年3月-堀越高等学校”)) (nil “nil”))

((“は” “た。” (“を執り行っ” “7月29日に結婚式並びに披露宴” (“を提出し” “同年3月3日婚姻届” (“を執り行っ” “2010年1月29日に正式に婚約発表の記者会見” (“が明け” “ことをファックスで発表。年” (“に交際し婚約する” “同日中に連名で、結婚を前提” (“』で報じられる” “日刊スポーツ” (“との交際が” “小林麻央” (“の” “2009年11月19日-フリーキャスター”))))))))) (nil “nil”))

((“は” “た。” (”)を降板となっ” “名古屋・御園座「二月大歌舞伎」(2011年2月公演” (“」は中止となり” “初春花形歌舞伎” (“に伴い” “期限謹慎” (“を発表し” “こと” (“で自粛する” “松竹の社長が、海老蔵の歌舞伎出演を無期限” (“だと思います」と謝罪した。また、会見に同席し” “こと” (“が招い” “自分のおごり” (“を行い” “海老蔵本人が12月7日、都内のホテルで記者会見” (“」を降板し” “ため、11月30日から出演予定だった京都・南座「吉例顔見世興行東西合同大歌舞伎” (“を感じる」と謝罪し” “本人の自覚のなさに怒り” (“を開き” “ことが発覚。11月25日夜、成田屋後援会主催の「後援会の集い」開催後に急遽、父・團十郎が記者会見” (“から飲みに出てい” “11月24日夜” (“にキャンセルしたにもかかわら” “初春花形歌舞伎」(2011年1月公演、東京・ルテアトル銀座)の記者会見を体調不良を理由” (“に予定されてい” “警視庁目黒署は傷害事件として捜査を開始。しかし、海老蔵は11月25日午後” (“になり殴られたとみられ” “数人のグループとは別の数人のグループとトラブル” (“に帰っ” “歌舞伎俳優の仲間らと飲食後、数人のグループと海老蔵だけが合流、その後、先” (“される” “虎の門病院に緊急搬送” (“に通報し” “頭部、顔面を負傷(11代目市川海老蔵暴行事件)。帰宅後に妻の麻央が救急と警察” (“の飲食店で暴行を受け” “2010年11月25日早朝、西麻布”)))))))))))))))))))))) (nil “nil”))

((“は” “長女・麗禾(れいか)が誕生。” (“となる” “1子” (“との” “2011年7月25日、妻・麻央”))) (nil “nil”))

((“は” “。” (“を引き継ぐ” “家督” (“の” “市川宗家当主” (“が” “2013年2月3日、父・12世團十郎”)))) (nil “nil”))

((“は”“長男・ 勸玄(かんげん)が誕生。”(“となる”“2子”(“との”“2013年3月22日、妻・ 麻央”))) (nil “nil”))

((“は”“ことを、ブログで発表。”(“)と改名し”“堀越寶世(宝世”(“から”“堀越孝俊”(“を”“2015年2月4日、本名”)))) (nil “nil”))

((“は”“た。”(“を公表し”“こと”(“を患っている”“乳がん”(“が”“2016年6月9日、妻・ 麻央”)))) (nil “nil”))

((“は”“た。”(“を公表し”“死”(“の”“妻”(“で”“緊急会見”(“の”“翌日”(“が”“2017年6月22日、妻・ 麻央”))))))) (nil “nil”))))

(family

(((“は”“たん,妻-小林麻央,義姉-小林麻耶”(“ぼ”“曾祖父-七代目松本幸四郎,祖父-十一代目市川團十郎,大叔父-八代目松本幸四郎,大叔父-二代目尾上松緑,父-十二代目市川團十郎,叔母-二代目市川紅梅,妹-三代目市川”)) (nil “nil”))

((“は”“た。”(“に婚姻届を提出し、結婚披露宴を行っ”“その後3月3日”(“と婚約発表会見を行っ”“小林麻央”(“の”“2010年1月29日、フリーアナウンサー”)))) (nil “nil”))

((“は”“長女・ 麗禾が誕生。”(“となる”“2011年7月25日、第一子”)) (nil “nil”))

((“は”“長男・ 勸玄が誕生。”(“となる”“2013年3月22日、第二子”)) (nil “nil”))

((“は”“た。”(“にし”“ことを記者会見にて明らか”(“を患っている”“乳がん”(“が”“2016年6月9日、麻央”)))) (nil “nil”))

((“は”“」だった。”(“は「愛してる”“麻央の最期の言葉”(“を看取っ”“義姉の小林麻耶らとともに最期”(“から急遽帰宅し”“こと”(“があっ”“LINEのメッセージ”(“を知らせる”“麻央の死去当日、海老蔵は舞台の稽古中で、自身の携帯電話に麻央の危篤”(“を報告し”“こと”(“で死去し”“麻央が前日の6月22日夜に自宅”(“を行い”“同日、舞台終演後の14時30分に緊急記者会見”(“を公表し”“文面”(“をほのめかす”“日です」と麻央の死”(“泣い”“一番”(“で”“人生”(“で”“ブログ”(“の”“2017年6月23日、自身”)))))))))))))))) (nil “nil”)))))))

3. Generated Sentences of Ichikawa Ebizō XI Based on APPENDIX 2 (2)

十一代目市川海老蔵(じゅういちだいめいちかわえびぞう、1977年12月6日-)は歌舞伎役者、俳優。

歌舞伎名跡「市川海老蔵」の当代。

屋号は成田屋。定紋三升、替紋杏葉牡丹。

愛称に海老さま」、「海老ちゃん。

父· 團十郎が健在であった当時、海外公演舞台時には演出のことことた。

團十郎が海老蔵とききときの家庭内の空気。

だが、反発していたときにずっと優しく見守ってくれてい團十郎に常々感謝の言葉。

婚外子(非嫡出子)が発覚し際に寛大な態度父· 團十郎も、2007年に海老蔵が風呂場でケガ歌舞伎公演時穴んだから、自己管理今回んた。

2010年に発生し被害際京都南座での顔見世興行のではないか。

(事件に巻き込まれたとは言え事件の原因責任の上大変失礼な行動で、当人の自覚のなさわざ海老蔵た。

10代の頃海老蔵幼少時家柄· 伝統の重責。

そんな彼を立ち直らせのは、偶然フィルム弁慶祖父· 十一代目市川團十郎の芸の美しさと勇姿。

祖父を目標理想現在。

以前は舞台ことが苦痛であったが、歌舞伎に真剣よう今はプライベートの方が苦痛。

これが異様なこと五代目坂東玉三郎ところ、「あら、それは当たり前。

また、坂東玉三郎とこととこと。

本格的な筋力トレーニングやヨガ雑誌『ターザン』などの表紙こと。

成田屋のお家芸「歌舞伎十八番うち、今日演目の復活上演を積極的。

古典の大役初役。

「海老さま」と人気十一代目市川團十郎ファン。

現代の歌舞伎若手スターの一人。

また、NHK大河ドラマ『武蔵MUSASHI』で主役ほか、現代劇。

1983年5月-歌舞伎座『源氏物語』の春宮お目見得。

1985年5月-歌舞伎座『外郎売』の坊七代目市川新之助襲名。

2000年歌舞伎座『源氏物語』の光君会場は連日満員、チケット人気五代目尾上菊之助、二代目尾上辰之助とともに「平成の三之助新たな歌舞伎ブーム。

2003年2月-日置明子との間女児、当時1歳ことが判明。子ことを会見で説明。

2004年4月4日-成田山新勝寺で襲名披露、奉告の練り。京成本線京成上野駅-京成成田駅間列車「海老蔵号京成AE100形電車を使用)。

2004年5月歌舞伎座『助六由縁江戸桜』の花川戸助六、『暫鎌倉権五郎ほか十一代目市川海老蔵襲名。

2004年6月30日-成田山大阪別院明王院で京阪本線天満橋駅-香里園駅間列車「海老蔵号京阪8000系電車使用)。

2004年10月-フランス・パリのシャイヨ宮劇場市川新之助改メ十一代目市川海老蔵襲名披露フランス公演」。同国政財界トップル・モンド』紙が一面全面17世紀末以来の日本の伝統的演劇の中で名一族の出身」「26歳のスター」「若く美男で特に『必殺の視線現地でも大きな話題。

2006年9月7日-映画『出口の海クイズ$ミリオネア』(フジテレビ)に出演。

2007年3月-パリ・オペラ座で初歌舞伎公演父弁慶紅葉狩パリ引っ込み六方を花道中央通路観客。

2007年7月13日-楽屋の風呂場戸のガラスで右足の裏けが大阪松竹座「七月大歌舞伎」を降板。

2009年-2007年10月14日に放送され海老蔵ドキュメンタリー番組『名門こと...?市川海老蔵・宿命と苦悩の物語?』(フジテレビ系)が、ニューヨークフェスティバルのアート部門で金賞を受賞。

2009年5月23日-TBSのMR.BRAIN初民放初現代劇出演。

2010年11月25日早朝、西麻布の飲食店で暴行を受け頭部、顔面を負傷(11代目市川海老蔵暴行事件)。帰宅後に妻の麻央が救急と警察虎の門病院に緊急搬送歌舞伎俳優の仲間らと飲食後、数人のグループと海老蔵だけが合流、その後、先数人のグループとは別の数人のグループとトラブル警視庁目黒署は傷害事件として捜査を開始。しかし、海老蔵は11月25日午後初春花形歌舞伎」(2011年1月公演、東京・ルテアトル銀座)の記者会見を体調不良を理由11月24日夜ことが発覚。11月25日夜、成田屋後援会主催の「後援会の集い」開催後に急遽、父・團十郎が記者会見本人の自覚のなさに怒りため、11月30日から出演予定だった京都・南座「吉例顔見世興行東西合同大歌舞伎海老蔵本人が12月7日、都内のホテルで記者会見自分のおごりこと松竹の社長が、海老蔵の歌舞伎出演を無期限こと期限謹慎初春花形歌舞伎名古屋・御園座「二月大歌舞伎」(2011年2月公演た。

2011年7月2日、謹慎を解かれ新橋演舞場「七月大歌舞伎」で舞台復帰。「お客様身思い。こうして舞台こと日々の舞台に全身全霊た。

2013年正月、14年ぶりに出演する新春浅草歌舞伎」の初日東京・浅草公会堂前で共演者らと鏡開きた。

2014年1月30日、自身の公式ホームページブログ発表。

2015年7月29日、千葉県成田市の観光大使「御案内人」に就任。

2015年11月、「吉例顔見世大歌舞伎11世市川団十郎五十年祭」(歌舞伎座)の江戸花成田面影長男・觀玄初お目見え。

2015年より成田市。

・歌舞伎は

天衣紛上野初花「雪暮夜入谷畦道」(三千歳と侍片岡直次郎,鳴神鳴神上人,源氏物語光君,助六由縁江戸桜花川戸助六,鏡獅子小姓弥生後獅子精弁慶、富樫左衛門

・海外公演は

十一代目市川海老蔵襲名披露　歌舞伎パリ公演(パリ・国立シャイヨー劇場)(2004年10月),訪欧歌舞伎公演(ロンドン・サドラーズ・ウェルズ劇場、アムステルダム・スタッツスハウブルフ劇場)(2006年5月、6月),パリ　オペラ座　松竹大歌舞伎(パリ・オペラ座　ガルニエ宮)(2007年3月),松竹大歌舞伎　モナコ公演(モナコ・オペラ座　サル・ガルニエ)(2009年9月),松竹大歌舞伎　ロンドン公演(ロンドン・サドラーズ・ウェルズ劇場)(2010年6月),松竹大歌舞伎　ローマ公演　ShochikuGrandkabukifeaturingEbizoIchikawa?(ローマ・アルジェンティーナ劇場)(2010年6月),市川海老蔵シンガポール公演「古典誘い(いざない)」JAPANTHEATERProducedandheadlinedbyEBIZŌICHIKAWAXI(シンガポール・マリーナベイ・サンズマスターカードシアター)(2014年11月),EBIZŌICHIKAWAXI'sJAPANTHEATER2015(シンガポール・マリーナベイ・サンズグランドシアター)(2015年10月),市川海老蔵　GRANDJAPANTHEATER　フジャイラ公演(フジャイラ・ミラマール・アルアッカビーチリゾート)(2016年2月),市川海老蔵　GRANDJAPANTHEATER　ニューヨーク公演(ニューヨーク・カーネギー・ホール)(2016年3月)

・テレビドラマは

大河ドラマ(NHK)は

花の乱(1994年)-足利義政(青年時代)役,武蔵MUSASHI(2003年)-主演・宮本武蔵役,おんな城主直虎(2017年)-織田信長役,12時間超ワイドドラマ(テレビ東京),豊臣秀吉天下炎の奉行大岡越前守(1997年)-大岡市十郎役,第8回時代小説大賞スペシャル「仲蔵狂乱」(朝日放送)-中村仲蔵役

MR.BRAIN 第1話・第7話・第8話(2009年、TBS) - 武井公平 役は

生誕100年記念松本清張ドラマスペシャル「霧の旗」(2010年、日本テレビ)-主演・大塚欽也役

金曜プレステージ「長谷川町子物語?サザエさんが生まれ日?」(2013年、フジテレビ)-9代目市川海老蔵役

土曜ドラマ「弱くても勝てへっぽこ高校球児の野望?」(2014年、日本テレビ)-谷内田健太郎役

石川五右衛門(2016年、テレビ東京) - 主演・石川五右衛門 役は

・テレビアニメは

名探偵コナン コナンと海老蔵 歌舞伎十八番ミステリー(2016年1月9日・16日、読売テレビ) - 市川海老蔵(本人) 役は

・吹き替えは

オール・イズ・ロスト?最後の手紙?(2014年男役(ロバート・レッドフォード)

・映画は

出口の海(2006年)-主演・並木浩二役,一命(2011年)-主演・津雲半四郎役,利休2013年)-主演・千利休役,喰女-クイメ-(2014年)-主演・長谷川浩介役,無限の住人(2017年)-閑馬永空役

・その他の舞台

ライル(1990年)-ジョッシュ役,スタンド・バイ・ミー(1991年)-主演・クリス役,天守物語(1999年)-図書之助役(五代目坂東玉三郎、宮沢りえ海神別荘-公子役(坂東玉三郎疾風2002年)-鈴木猪平役,宮本武蔵(2003年)-主演・宮本武蔵役,信長(2006年)-主演・織田信長役,天守物語(2006年助役(坂東玉三郎、二代目市川春猿海神別荘(2006年)-公子役(坂東玉三郎ドラクル(2007年)-主演・レイ役(宮沢りえ共演)

・CMは

伊藤園「お?いお茶」(2000年-),ヤマキ(2000年-2010年),明治製菓「ショコラクール」(1995年),三井生命(1998年),NEC「Express ワークステーション」(1998年),ミズノ「スーパースター」(2003年),TSUTAYADISCAS(2009年8月-)テレビショッピング番組の司会者役,飯田グループホールディングス(2014年-),ライオン「贅沢Care」(2015年),ジェイテクト(2016年)

1995年『仮名屋小梅』の濱本清眞山青果賞新人賞

1995年『鏡獅子』で松竹会長賞

1997年『吉原雀』で松竹会長賞

2000年『勧進帳』の弁慶、『天守物語図書助読売演劇大賞杉村春子賞

2001年 松尾芸能賞新人賞、芸術選奨文部科学大臣新人賞受賞は

2007年フランス芸術文化勲章(L'OrdredesArtsetdeslettre)のシュヴァリエ(Chevalier)受章

2007年 ベストドレッサー賞特別賞は

2014年第37回日本アカデミー賞優秀主演男優賞を受賞

2014年 Yahoo!検索大賞 2014・俳優部門は

本名は堀越寶世(宝世)(ほり)。

旧名は堀越孝俊(ほり)。

妻はキャスタータレント小林麻央(2017年6月死別)。

身長176cm、体重80kg。は

血液型はAB型。

星座は射手座。

特技はバスケットボール、趣味相撲鑑賞。

学生時代はバスケットボール部た。

スタジオジブリのアニメ作品ファン。

「おしゃれイズム」に出演し際に、『崖の上のポニョ』の非売品の人形と宮崎駿が同作際鉛筆た。

宮崎本人と対談しこと際に自身タワシで体健康法を宮崎とことこと。

以前は寿司月50回タバコを1日4箱分酒よう健康心境の変化でタバコは2004年の海老蔵襲名時歌舞伎とき、神様こと心と体思いから不規則な食生活や酒、肉、魚も2008年11月頃麻央との結婚後に酒、肉、魚。

五代目尾上菊之助とは幼馴染同級生。

小学生の頃、彼家別荘とき菊之助の母である富司純子これが家族以外の人。

再従兄弟に四代目尾上松緑、七代目市川染五郎従姉妹松本紀保(七代目染五郎姉)· 松たか子(同妹。

伊藤英明とは親友お互い英明」、「孝俊(本名堀越孝俊仲。

1977年12月6日-東京都に生まれる父· 十二代目市川團十郎と母· 希実子(旧姓· 庄司)の長男。祖父は「海老さま」こと十一代目市川團十郎。

1996年3月-幼稚園· 初等部· 中等部· 高等部まで青山学院高等部から堀越高等学校に転校。

1997年3月-堀越高等学校を卒業。

2009年11月19日-フリーキャスターの小林麻央日刊スポーツ同日中に連名で、結婚を前提ことをファックスで発表。年2010年1月29日に正式に婚約発表の記者会見同年3月3日婚姻届7月29日に結婚式並びに披露宴た。

2010年11月25日早朝、西麻布の飲食店で暴行を受け頭部、顔面を負傷(11代目市川海老蔵暴行事件)。帰宅後に妻の麻央が救急と警察虎の門病院に緊急搬送歌舞伎俳優の仲間らと飲食後、数人のグループと海老蔵だけが合流、その後、先数人のグループとは別の数人のグループとトラブル警視庁目黒署は傷害事件として捜査を開始。しかし、海老蔵は11月25日午後初春花形歌舞伎」(2011年1月公演、東京· ルテアトル銀座)の記者会見を体調不良を理由11月24日夜ことが発覚。11月25日夜、成田屋後援会主催の「後援会の集い」開催後に急遽、父· 團十郎が記者会見本人の自覚のなさに怒りため、11月30日から出演予定だった京都· 南座「吉例顔見世興行東西合同大歌舞伎海老蔵本人が12月7日、都内のホテルで記者会見自分のおごりこと松竹の社長が、海老蔵の歌舞伎出演を無期限こと期限謹慎初春花形歌舞伎名古屋・御園座「二月大歌舞伎」(2011年2月公演た。

2011年7月25日、妻・麻央との1子長女・麗禾(れいか)が誕生。

2013年2月3日、父・12世團十郎が市川宗家当主家督。

2013年3月22日、妻・麻央との2子長男・勸玄(かんげん)が誕生。

2015年2月4日、本名を堀越孝俊堀越寶世(宝世ことを、ブログで発表。

2016年6月9日、妻・麻央が乳がんことた。

2017年6月22日、妻・麻央が翌日緊急会見妻死た。

曾祖父-七代目松本幸四郎,祖父-十一代目市川團十郎,大叔父-八代目松本幸四郎,大叔父-二代目尾上松緑,父-十二代目市川團十郎,叔母-二代目市川紅梅,妹-三代目市川ぼたん,妻-小林麻央,義姉-小林麻耶

2010年1月29日、フリーアナウンサーの小林麻央その後3月3日た。

2011年7月25日、第一子となる長女・麗禾が誕生。

2013年3月22日、第二子となる長男・勸玄が誕生。

2016年6月9日、麻央が乳がんことを記者会見にて明らかた。

2017年6月23日、自身のブログ人生一番日です」と麻央の死文面同日、舞台終演後の14時30分に緊急記者会見麻央が前日の6月22日夜に自宅こと麻央の死去当日、海老蔵は舞台の稽古中で、自身の携帯電話に麻央の危篤LINEのメッセージこと義姉の小林麻耶らとともに最期麻央の最期の言葉」だった。

Chapter 3
An Analysis of Advertisements for Narrative Generation

ABSTRACT

This chapter focuses on an advertisement analysis from the author's previous narrative analyses. The advertisement analysis is based on the idea of "advertisement as a narrative." In particular, this chapter first introduces the previous fundamental attempts for a study of advertisement narrative generation systems. This was planned as an application of a narrative generation study and has continued in advertisement analyses and in the design and implementation of experimental systems from two viewpoints: the macro framework and the methods introducing brands into advertisements. In the next, central part of this chapter, the author describes in detail the analysis of television commercial strategies based on three regular or ordinary strategy types and nine irregular or extraordinary strategy types. This study has been applied to the development of advertisement narrative generation techniques in the subsequent research. The last part of this chapter mentions the systematization's topic.

INTRODUCTION

The objective of this chapter is to analyze how a brand, i.e., a product or a service, is introduced into the events in story-like television commercial works using real commercials. An event in a story-like commercial means an event that is developed through a character's actions. Each brand corresponds to a

DOI: 10.4018/978-1-5225-9943-2.ch003

symbol showing a product or a service and focusing on the semantic aspect of the product or service. An event is simply a set piece including action(s) and the necessary related elements, and it is different from a shot, which is a basic unit in moving films. Events are frequently constructed using several shots. However, shots can also contain several events. The direct objective of this analysis is to invent the algorithms and develop the knowledge systems that introduce a brand into an event in the advertising narrative generation system. These systems are used in the mechanism that constructs the advertising flow.

Although this study can be applied directly to the systems that support advertising planning and production, such analysis may also be able to be applied to other forms of advertising generation in the future. Advertising narratives seen in television commercials have mythic or folkloric characteristics beyond realistic narratives. A mixture of rhetoric regarding the relationships between an action and an object, and the rhetoric of metamorphoses of objects and characters, are also interesting for the mechanism of general narrative generation. Moreover, in the study of a story generation system based on the method of Propp's narrative morphology, Ogata (2007) presented the concept of "narrative generation by destruction and reconstruction." In this concept, ideas destroy and reconstruct existing narrative texts or narratology and literary theories to create new possibilities not seen in existing texts. This idea is related to the concept of intertextuality in the sense of the usage of existing texts (Nakashima & Ogata, 2006; Ogata & Nakashima, 2006). This advertising narrative generation is also an example of the studies regarding the destruction and reconstruction in the point that aims to develop the combinatorial generation possibilities beyond the simulative generation possibilities based on existing texts, though it uses existing advertising texts. In addition, a similar plan is seen in the idea of "liquidation and crystallization" presented by Akaishi (2006) and Hori (2007) in the context of creative aid in AI. The philosophical concept of "fluidity and fixation" by the author (Ogata, 2019e) is a broader concept that includes the above concepts.

The chapter's configuration is as follows. First, **BACKGROUND** introduces many of the previous narrative analysis research by the author and describes the concept of "advertisement as narrative" that regards all advertisement works as contemporary, mythic, or folkloric narratives. This idea extends the range of advertising approaches of the author to the directions to broader general narratives.

Next, in **APPROACHES TO ADVERTISEMENT GENERATION AS NARRATIVE GENERATION** from two viewpoints of the macro framework and the brand introduction methods into advertising narratives, the author introduces a series of studies of advertising narrative generation in which advertising narrative analysis, system design, and implementation were continued in the positioning of an application of narrative generation.

RHETORICAL STRATEGIES FOR ACTIONS AND OBJECTS: A DETAILED ANALYSIS OF TELEVISION ADVERTISEMENT VIDEOS AND THE EXTRACTION AND CLASSIFICATION OF STORY-TYPE BRAND INTRODUCTION STRATEGIES is the central portion of this chapter and describes the developed advertising narrative analyses of the above methods in detail. The author classified the advertising rhetoric or rhetorical strategies in the following twelve types:

- **R1: Rhetorical strategy of manufacturing process**
- **R2: Rhetorical strategy of purchasing process**
- **R3: Rhetorical strategy of core usage method**
- **R4: Rhetorical strategy of deviated use or defamiliarization use**
- **R5: Rhetorical strategy of no usage**
- **R6: Rhetorical strategy of a character as a narrator**
- **R7: Rhetorical strategy of defamiliarization of a character's state**
- **R8: Rhetorical strategy of a character's defamiliarization of the product usage action**
- **R9: Rhetorical strategy of defamiliarization of a product's state**
- **R10: Rhetorical strategy of defamiliarization of the state of an object related to the product**
- **R11: Rhetorical strategy of the product as a subject**
- **R12: Rhetorical strategy of the defamiliarization of a background**

Moreover, the author systematically shows knowledge bases acquired through the advertising narrative analysis, including characters, places, and objects.

The last part in this chapter, **FROM ANALYSIS TO GENERATION**, deals with the systematization's topic.

BACKGROUND

The author introduces previous narrative analyses for narrative generation and the idea that addresses advertisement as a kind of narrative.

Previous Narrative Analyses

Ogata (2018a) introduced narrative analyses for narrative generation by the author from the viewpoints of "genre," "medium," "object," "goal," "result," "relationships with an Integrated Narrative Generation System (INGS)," and "relationships with literary theories." In this section, the author presents only "genre," "medium," "object," "goal," and "result." "Genre" means each narrative group in which an object to be analyzed is contained and "medium" is each representation tool through which an analysis object is represented and socially distributed. Next, "object" shows an actual work or a collection and, by "goal" and "result," the author means the object, intention, or aim for an analysis and the acquired results. Furthermore, the term "relationships with INGS" corresponds to the future possibilities of use in INGS and may include a simple explanation for a previously introduced part. The term "relationships with literary theories" mentions the relations of this issue to literary theories and narratology. It is also related to the "expanded literary theory" by Ogata (2019e).

In addition, the author's narrative analyses used narratological research regarding story, including Vladimir Propp (1895–1970) (1968) theory and various narratological approaches regarding narrative discourse, including Gerald Genette (1930–2018) (1972) theory. Although narrative analyses and narrative theories indicate that narratology and literary theories are related, Table 1 only presents the narrative analyses more than the above types of narrative analyses using theories and methods from narratology and literary theories.

Advertisment as a Narrative

The narrative genre system in Ogata (2019d) treated advertisement as a genre in narrative. From the perspective of the narrative generation study performed by the author, for example, the genres of novel, movie, advertisement, *kabuki*, and social event are equally included as narrative genres and they all can have a common structure and content although their social functions, roles, and representation forms are different. More concretely, as shown by Ogata, Watanabe, Hori, and Ohsuga (1995a), the scenarios of television advertisement works (films or videos) have similar structures as other narratives such as stories. Moreover, the image representation of television advertisement works has the same features as other narrative arts such as movies. Therefore,

Table 1. Narrative analyses by the author

Genre	Medium	Object	Purpose	Result
Advertisement	***Television***	About 100 Television commercial films that were broadcasted in Japan in the first half of the 1990s.	After verifying that stories in television commercials can be analyzed using the similar methods with folk tales and novels, this study applies the analysis results into an advertising scenario generation system using a developing narrative generation system.	The author achieved this goal, as a person (the author) described, using language, the event sequences in television advertising films and represented them using the tree structure form for a story used in a narrative generation system (Ogata & Watanabe, 1994, 1995; Ogata, Watanabe, Hori, & Ohsuga, 1993, 1995a, 1995b; Ogata & Kanai, 2010a, 2010b).
		More than 1,000 television commercials that were broadcasted in Japan from the beginning of the 2000s to about 2010.	This study defines the objective of a television commercial as the proposition and appeal of a product (brand) and analyzes the forms of television commercials as "rhetoric or rhetorical techniques for introducing a product" to develop an experimental system including the above analysis.	This study analyzed the relationships between a product that forms a theme in a television commercial and the event development, the appearance and usage, amongst others, by describing the event development. In particular, this study showed that various types of defamiliarization rhetoric to a product are used as main techniques to extract and classify such techniques. This classification was enabled by what was defamiliarized in an event that represents a product and the usage form (Abe, Ogata, & Hanada, 2008;Zhang, Ono, & Ogata, 2011). The authors developed an experimental system for performing the product defamiliarization using this result (Ono, Zhang, & Ogata, 2012; Zhang, Ono, & Ogata, 2011, 2012).
		More than 1,000 TV commercials in Japan.	The objectives of this study are divided into the following two parts: the first is to extract detailed CM parameters according to about 20 categories, including narrative techniques, world models, tension, and tone for the audiences' reception and the creators' production, based on the precise and comprehensive analysis of actual commercials to understand the characteristics of contemporary TV commercials. The second goal is to design and implement an automated commercial scenario generation system based on the first analysis. Further, these objectives are organized under the direction through which we present and create new commercial works in the context of the marketing and advertisement industry.	The study is currently in progress (Ito, Sasaki, & Ogata, 2018; Ono, Ogata, & Ito, 2019; Ono, Sasaki, Ito, & Ogata, 2018; Ono, Sasaki, & Ogata, 2018, 2019a, 2019b).

continued on following page

Table 1. Continued

Genre	Medium	Object	Purpose	Result
Short Stories and Novels	***Book (paper)***	*Maiko Satsujin Jiken* by Misa Yamamura (1934-1996) (1988)	This study analyzes the methods of mutual transformation between a story and a plot, which is the construction of a work to be represented, using a short detective story by Misa Yamamura, who is a writer of popular fiction in Japan, to introduce the results into the narrative generation systems by the author. A characteristic of detective stories is that the temporal order in a story is variously transformed in the plot. In this story, a preceding murder in the temporal order appears in the last part of the plot, and other plot-making techniques such as omission and imagination, are diversely utilized.	The author wrote a summary of this fiction as a constructive event sequence, namely a plot, and transformed its description into a temporal event sequence of the story. Additionally, the author extracted methods that enable the temporal order transformation between the story and plot, including methods based on characters' viewpoints to define and utilize them as the techniques for the part of story-plot transformation in a narrative generation system that the author was developing (Ogata, Hori, & Ohsuga, 1994, 1995, 1996b).
	Book (paper)	*Mienai Fukushū* by Tōru Miyoshi (1931–) (1993)	This study analyzes (describes) that different stories are produced based on several characters' viewpoints, or more correctly speaking, different stories are generated from a story, from the perspective of a hypertext-like method. This forms a part of considering the narrative discourse mechanism in a narrative generation system by the author.	This analysis described each story by focusing on all characters' viewpoints or perspectives in a detective story, and constructed and visualized by hyper-text structures respectively in independent events and parts where several characters' routes are combined to consider the structure of a story (Ishii & Ogata, 1998).
	Book (paper)	*Gogo no Eikō* by Yukio Mishima (1925–1975) (1963).	Under the hypothesis that characters of a fiction are constructed through the recognitions of both the characters themselves and the narrator, this study analyzes quantitatively and qualitatively the main characters and the narrator from the perspective of their recognitions about themselves and other characters. This analysis was performed for linking to a developmental consideration for modeling each character defined in narrative generation systems.	Throughout the entire story, the authors temporarily listed information related to the characters, including himself or herself, for each character and the narrator. Through this information, a narrative world (regarding the characters) was emerged based on the construction by the characters including the narrator. The criteria of the world construction in this novel involve the viewpoints (perspectives or focalization) through which a character sees or perceives events and, furthermore, how a character recognizes events and characters including the character's self. At last, the dramatic and dynamic characteristics in a novel are generated from especially the mutual recognitions among the characters, such as matching and discrepancy. This study bridged the above results to a developmental idea of the narrative generation systems (Ogata, 2018a).

continued on following page

Table 1. Continued

Genre	Medium	Object	Purpose	Result
	Book (paper)	Several short stories and novels by Ryūnosuke Akutagawa (1892–1927), Kenzaburō Ōe (1935–), Yukio Mishima, and so on.	This study analyzes the Japanese writing method, particularly character notation, in especially novels to utilize these results in a part of the character notation mechanism and the language representation dictionaries in INGS. This is intended, for example, to develop a generation mechanism for making the notation that imitates the visual and visible impressions the character representation of a work by a writer.	In this study, mainly *kanji, hiragana, katakana*, and *Romaji* were used as the main character types for narrative language representation to experiment the mechanisms for controlling the representation according to the evaluation criteria including readability or comprehensiveness and difficulty to read, and narrative analyses and the simulation experiments to incorporate a part of the results into INGS for the usage (Kamada & Ogata, 2013a, 2013b, 2013c).
Movie	*Film (image)*	*Battleship Potemkin* by Sergei Eisenstein (1898–1948) (1925), *The Color of Pomegranates* by Sergei Parajanov (1924–1990) (1968), and several photographs that were taken by the authors (Kanai, Ogata, & Shinohara, 2002, 2003)	A point of a film study with Kanai (Kanai & Niwa, 2008; Kanai, 2013) is the problem of non-story's characteristics in films. Through several cooperative research projects, the author also analyzed, for image narratives, rhetoric of works in which temporal progression is complicated and rhetoric for intervening intentionally a story's sequence. This study was also conducted as a part of these studies originally to clearly analyze the characteristics of story recognition by the receivers and introduce the acquired knowledge into narrative generation systems.	For the author's systematic narrative generation study, the introduction of this analysis field is a future issue. In particular, the author is considering the rhetorical knowledge acquired from the narrative analyses as a group of narrative discourse techniques in the framework of the Integrated Narrative Generation System (INGS).
	Film (image)	*Tokyo Monogatari* by Ozu Yaujirō (1903–1963) (1953)	Although a direct goal was to simulate the camerawork of *Tokyo Monogatari* by Ozu (1953) using an imaging tool, TVML (NHK Science & Technology Research Laboratories, 2017), the long term goal was to make a first step for developing the techniques of viewpoint or camerawork for visualizing the image in a movie generation mechanism for narrative generation.	This study divided all parts of this work into the units of shots to be analyzed and described the camerawork of all shots in the relationships with the narrative content described in the film. As a result, the authors showed the basic patterns regarding the techniques of making a motion picture of a character and two characters, and introducing external scenes. Moreover, the authors showed that the violation and deviation, such as a moving shot, from a typical pattern in Ozu were performed in a kind of a climax scene and the point of transition in the story. Thus, this analysis showed the movement form based on the repetition of basic patterns as the majority and very few deviations. Continuously, the authors have tried to simulate this result using the TVML tool and partially developed an image representation mechanism that automatically simulate or imitate the basic pattern of this film in accordance with the arrangement and the context of the prepared image materials. Other remaining issues include comparing this analysis result of *Tokyo Monogatari* to other movies by Ozu, by generalizing the method of the shot analysis (Ogata, Tachibana, & Tomite, 2009; Tachibana & Ogata, 2009a, 2009b).

continued on following page

Table 1. Continued

Genre	Medium	Object	Purpose	Result
Manga	***Book (paper)***	*Maison Ikkoku* by Rumiko Takahashi (1957–) (2003-2006)	Originally, the authors have studied *manga* (comic) in the context in order to develop a "hyper-comic" that represents a *manga* work using the hyper-comic structure (Endo & Ogata, 2003a, 2004a; Mori & Ogata, 2005; Ogata & Endo, 2004; Ogata & Mori, 2005). Based on various interesting knowledge acquired in this process, the authors have continued a series of studies regarding the analyses of *manga* works. Concurrently, these studies have mixed analysis and development.	The authors called one of the most interesting and exciting methods, for the author's narrative discourse techniques, that is original and unique in *manga* "multilayered narrative discourse." This is a technique through which different narrative discourses are simultaneously represented in a *koma* (frame) in *manga*. For instance, for an event in a *manga* work, various multiple representation techniques are applied including the following: simultaneous representation from different viewpoints in a *koma* (the narrative discourse of focalization), inserting both the past and future of an event into a *koma* (the narrative discourse of temporal order transformation), and arranging both an external action or state and an internal or mental state (Endo & Ogata, 2003b, 2004b).
Kabuki	***Stage***	Many *kabuki* works and documents including academic books, essays, and criticisms.	This study has been continuously addressed. The author's previous paper on *kabuki* (Ogata, 2016, 2018a, 2018b, 2018c, 2019a) undertook a comprehensive survey and analysis of the art form, aiming to explore a narrative generation-reception and a narrative production-consumption model of *kabuki* from the viewpoint of an information system and, in particular, a narrative generation system. A fundamental concept of the modeling was "multiplicity," or multiple narrative structures, and the author associated this model with the concept of the "*Geinō* Information System (GIS)" representing a system model in which multiple narrative generations and production mechanisms or processes are included.	The author has been implementing the first stage of the *kabuki* study from the viewpoint of narrative generation (Ogata, 2016, 2018a, 2018b, 2018c, 2019a). The main results were the representation of the framework of *kabuki*'s multiple elements and the partial precision and extension. The multiple elements include the following themes: (1) real incident, (2) work, (3) genre, (4) material or topic, (5) person, (6) story and plot, (7) actor and place, (8) time (era or age), (9) style (form or pattern), (10) theatre (stage and seating), (11) audience, (12) text, (13) production of scenario (*daichō*), (14) direction, and (15) stage performance. Moreover, (16) *tsukushi* or *zukushi* and (17) *naimaze* are different elements that are common techniques to be applied to the fifteen multiple elements. The newest detailed information is in Chapter 2 in this book. The author shows the results of concrete analyses of *kabuki*'s elements, including "person," "story," "*tsukushi*," and "*naimaze*." A present topic is to introduce the knowledge and techniques of the above elements into GIS with INGS.
Fokltale	***Book (paper)***	Motif collection of folktales	The objective of this study is to implement patterns from folktales or motifs acquired and systematized by the folktale study by Keigo Seki (1899-199019) (1957, 1966, 1977, 1978a, 1978b, 1978c, 1978d, 1978e, 1979a, 1979b, 1979c, 1979d, 1980) and Seki, Nomura, and Ōshima (1979, 1980) as a computer system to incorporate the system into INGS, further GIS, and use it as a type of story content knowledge base in INGS and the work resource in GIS.	This study is currently in progress. Ono, Ogata, & Ito (2019) present the formalization by Common Lisp of folktale motifs and show the direction to utilize the described system in INGS.

movie directors sometimes produce advertisement works using images. In the aforementioned interpretations, advertisement is a narrative genre. The author has continued a series of studies on advertising narrative generation systems as derivations or applications of narrative generation systems studies and developments.

However, the advertising narrative generation, in this case, does not imply studying the system that synthesizes advertising sentences according to the mere combination of surface works. Moreover, although catch-copy is an element of advertisement, advertisement generation systems are not merely systems that make catch-copies by the combination of words by the author. Furthermore, the author's advertising idea is different from that of the system that inserts prepared image and music components into preliminarily patterned scenarios (for instance, Kawamura (2019)).

Narrative generation systems and advertisement generation systems for the author represent the entire system that concretizes a generated narrative or advertising deep structure into the surface structure, i.e. representation structure. Previous narratology and literary theories utilize many of the methods and theories regarding the components of deep structures, including story, plot, event development pattern, internal structure of each event, the ontology for represented narrative content, conceptual dictionaries, character, object, place or space, and time and seasons. However, of course, "narratology and literary theories can be used directly" or there are not narrative theories, and methods for narrative and advertisement generation readily viable based on direct use. A reason for this is that previous narratology and literary theories have focused on the interpretation and reading of literary and narrative works.

By contrast, the author has presented an expanded literary theory or post-narratology to reconsider old narratology and old literary theories from the perspective of "the generation of narratives and advertisements." The use of a "constructive method" facilitates the performance of concrete studies that involve important questions that include the following: which narratology and literary theories can be used? How can the theories and methods in narratology and literary theories be used? Which necessary theories and methods are newly added? What precisions and revisions are required for new approaches?

However, one of the unique features is that generally, advertising narratives are very short. The brevity is extreme and they are shorter than many short movies. Considering this fact, the author has focused on the motifs of folktales as the knowledge for defining story structures and plot structures for comparatively short narratives. As indicated in the background section in

Ogata (2019d), each folktale motif is a short or summarized description of the sequence of main actions by characters and corresponds to a type of summary. For example, in a folktale motif, a character moved from an ordinary world to another strange world and is subjected to a variety of experiences in this world, and finally returns to the original world. We can interpret advertising works and create new advertising works by directly using this motif. Conversely, we can also change the narrative flow of this motif. For instance, in a new narrative pattern, a character in a strange world comes to an ordinary world, disturbs the world, and finally returns to the original world.

In addition, the author does not necessarily state that advertisement narratives are the same as folktales. However, folktales are narratives that represent the emotions and desires of people in an age. Based on this meaning, we can state that advertisements are current folktales. Of course, current advertisements have several differences in form and content compared to past folktales. However, advertisements are similar to past folktales in that they both present short and simple stories, the narratives appeal to people's emotions, etc.

From an academic perspective, folktales refer to oral narratives that appeared before the emergence of artistic and entertaining narratives. As such, advertisements are not folktales. However, as previously mentioned, folktales implicitly or explicitly express people's feelings, emotions, and desires in naïve forms. The narrative genre of folktale is a kind of synthetic and collective narrative genre. Many artistic and entertaining narrative works have various folktale-like parts in the form of stories, narrative developments, character construction, etc. Advertisements are not an exception. Rather, in contemporary society, universal narrative structures and content are starting to appear in advertisements. This viewpoint, namely that advertisements are truly universal narratives in contemporary society, is a basic idea that supports the author's narrative generation study. In this sense, advisements are folktales in the present. Additionally, as explained in the background in Ogata (2019d), in Japanese literary history, folktales were utilized to activate stereotypical old narrative styles in literature because of the Heian era (794–1185 or 1192) to produce fresh and animated narratives, including *Otogi Zōshi* (1958) and *Sekkyōbushi* (Sekkyōshū, 1977) from the Muromachi era to the later ages.

However, in general, contemporary advertisements are intentionally and consciously produced at much expense. Contrastingly, folktales as an oral tradition were naturally produced by people and have been inherited. For this reason, the author hesitates to make a simple comparison between folktales and contemporary advertisements (however, national or other institutional

Figure 1. A generated example of an advertisement story

```
($対照 ($命令-遵守[$order-obey]
        ($命令[$oder] ($関係-目的[$relation-purpose]
                (命令する[order]
                ((AGT TH%異人[foreigner]) (TO AGT&HARADA-TOMOKO) (OBJ OBJ%OCHAZUKE
                (TIM TIM&YONAKA1) (LOC OBJ&HARADA-KE)))
                (与える[give]
                ((AGT AGT&HARADA-TOMOKO) (TO TH%異人[foreigner]) (OBJ OBJ%OCHAZUKE
(TIM NIL)
                (LOC NIL)))))
        ($遵守[$obey] (与える[give] ((AGT AGT&HARADA-TOMOKO) (TO TH%異人[foreigner]) (C
OBJ%OCHAZUKE)
                        (TIM TIM&YONAKA1) (LOC LOC&HARADA-KE))))
        ($結果[$result] ($継起[$sequence]
                (料理する[cook]
                ((AGT TH%異人[foreigner]) (OBJ OBJ%OCHAZUKE) (TIM TIM&YONAKA1)
                (LOC OBJ&HARADA-KE)))
                (食べる[eat]
                ((AGT TH%異人[foreignser]) (OBJ OBJ%OCHAZUKE) (TIM TIM&YONAKA1)
                (LOC OBJ&HARADA-KE))))
            (感謝する[appreciate]
            ((AGT TH%異人[foreigner]) (OBJ AGT&HARADA-TOMOKO) (TIM TIM&YONAKA1)
            (LOC OBJ&HARADA-KE)))
            (与える[give]
            ((AGT TH%異人[foreigner]) (TO AGT&HARADA-TOMOKO) (OBJ OBJ%飲み物[drink]) (TIM
TIM&YONAKA1)
            (LOC OBJ&HARADA-KE)))))
 ($禁止-違反[$interdiction-violation]
  ($禁止[[$interdiction] ($関係-目的[$relation-purpose]
            (禁止する[prohibit]
            ((AGT TH%異人[foreigner]) (TO AGT&MORITA-KENSUKE) (TIM #:G1263)
            (LOC #:G1264)))
            (食べる[eat]
            ((AGT AGT&MORITA-KENSUKE) (OBJ OBJ%OCHAZUKE) (LOC #:G1264)))))
  ($違反[$violation]] (食べる[eat] ((AGT AGT&MORITA-KENSUKE) (OBJ OBJ%OCHAZUKE) (TIM
#:G1263)
                (LOC #:G1264)))))
 ($結果[$result] ($継起[$sequence]
            (生える[grow]
            ((AGT AGT&MORITA-KENSUKE) (OBJ OBJ%OCHAZUKE) (TIM #:G1263)
            (LOC #:G1264)))
            (変わる[transform]
            ((AGT AGT&MORITA-KENSUKE) (OBJ OBJ%動物[animal]) (TIM #:G1263)
            (LOC #:G1264)))
```

systems must have participated to collecting folkloric narrative or oral traditional literature).

From the perspective of multiple narrative structures, the narratives of contemporary advertisements can be interpreted based on the multiplicity and complexity of the sender's level. However, the narrative multiplicity and complexity is not necessarily the advancement and complication of narratives. Furthermore, this kind of intentional and conscious production frequently conspires with the unconscious level of collective illusion (Yoshimoto, 1968). However, this does not mean that there is a crucial difference between advertisements and folktales.

APPROACHES TO ADVERTISEMENT GENERATION AS NARRATIVE GENERATION

According to the above background idea, the author has continued a series of studies on advertisement generation as narrative generation. Although this chapter shows the analysis of advertisement brand-introduction strategies as one of the concrete approaches after the next section, this specific section presents several topics as the premise.

The Basic Framework of an Advertisement Scenario Generation System and its Development

The advertisement narrative generation study began by presenting a system that generates advertisement scenarios as an application of narrative generation systems (Ogata & Watanabe, 1994,1995; Ogata, Watanabe, Hori, & Ohsuga, 1993, 1995a). A structural analysis of television commercial films was also conducted on the use of the system design (Ogata, Watanabe, Hori, & Ohsuga, 1995b). As this study was presented under a large research design that integrated the marketing and advertisement fields, the authors called the developed system a "Marketing-Advertisement Integrated Support System." The system showed that simple advertisement scenarios can be generated using the author's narrative generation method through the evaluation and verification by an advertisement company.

In the next step, Kayamori and Ogata (2003) and Ogata and Kayamori (2003) presented a new advertisement analysis and a simulation. In particular, this study divides each advertisement into the elements of story and non-

Figure 2. Images of micro and macro in television commercials

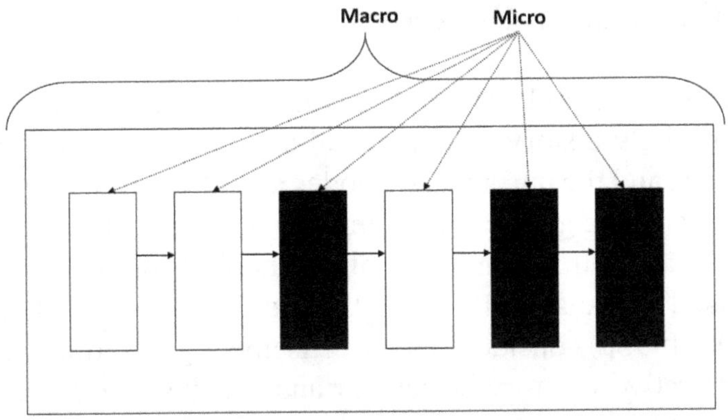

story to analyze the relationships between the brand, a term emphasizing the semantic side of a product, introduced in an advertisement and their elements in the advertisement. Moreover, the author tested the program that generated advertisement scenarios using the results of analysis.

Subsequently, Yoshio and Ogata (2007) more precisely reanalyzed the brand appearance patterns in television commercials as a set of "brand appearance strategies." Shimura and Ogata (2008) considered the methods that generate the brand appearance patterns in single events in advertisement scenarios. If the system framework introduced in the first stage of the advertisement generation study addresses the macro side of advertisement generation, these approaches corresponds to the micro side. A more developed advertisement scenario generation system needs to combine these two sides.

Since 2017, the author has conducted collaborative research with an advertisement production company (Ito, Sasaki, & Ogata, 2018; Ono, Ogata, & Ito, 2019; Ono, Sasaki, Ito, & Ogata, 2018; Ono, Sasaki, & Ogata, 2018, 2019a, 2019b). The work also included movies such as *Manbiki Kazoku* [*Shoplifters*] by Hirokazu Koreeda (1962–) (2018) which received a Palme d'Or award in the 2018 Cannes Film Festival. This study conducted a semantic analysis of many advertisement films to bring their results to advertisement planning and generation, including automatic planning and generation using AI and narrative generation. As regards automatic advertisement scenario generation, the author aims to re-construct and use the studies mentioned in this chapter, including the basic framework of advertisement generation and the brand introduction strategies, in generating advertisement scenarios.

Macro Design of the Advertisement Scenario Generation System

Ogata, Hori, and Ohsuga (1994, 1995, 1996a) defined a hierarchical narrative structure, especially a story's hierarchical structure, in which the terminal nodes are events and the intermediate nodes are relations binding the events, and modeled narrative generation as any applications to this structure. This was explained in detail in Chapter 1 of this book. Moreover, based on this idea, Ogata and Watanabe (1994, 1995) and Ogata, Watanabe, Hori, and Ohsuga (1993, 1995a) consider that advertisement generation can also work on similar narrative or story structures and, in this sense, advertisement scenario generation is a kind of narrative generation.

An important issue in the author's narrative generation framework is how to define the relation system for semantically constructing narrative structures, especially story structures. Further, in the application to the advertisement field, we need to consider the relations unique to the advertisement realm in addition to the more general narrative and story relations. Figure 1 shows a generated example of an advertisement story shown in Ogata, Wanatabe, Hori, and Ohsuga (1995a) and this program description uses various relations, including "contrast," "command-obeying," and "prohibition-violation." Although the latter two relations that use the pair type for story generation are necessarily unique only in the advertisement genre, extremely short stories or instantaneous narratives typically seen in the advertisement genre can leave a strong impression on audiences by using such clear narrative structures to a comparatively higher level than in story tree structures. The other relations, including "contrast" and "continuation," are not only possible in the advertisement genre and, in this sense, advertisement narratives or stories, or scenarios, are also common in other general narratives in the largest or the most macro framework. Thus, the role of this advertisement narrative generation system was to provide the framework for advertisement generation as narrative generation.

How Are Brands Positioned?

A point to focus on in advertisement narratives is that they are narratives incorporating brands. For example, in a type of commercial films, a product's image and the name of the product directly appear and, in another type of films, the story, namely characters and actions, places, and so on, symbolically

Figure 3. Inserting a brand into an advertisement story

Process 1;

(**飲む**[drink] ((AGT AGT%**新入社員**[new employee]) (OBJ OBJ%**栄養ドリンク**[nutritional drink]) (TIM TH%**夜中**1[midnight 1]) (LOC OBJ%**会社**1[office 1])))

Process 2;

((**疲れる**[get tired] ((AGT AGT%**新入社員**[new employee])(OBJ NIL)(TIM TH&**夜中**1[midnight 1])(LOC OBJ&**会社**1[office 1])))

(**飲む**[drink] ((AGT AGT%**新入社員**[new employee])(OBJ OBJ%**栄養ドリンク**[nutritional drink])(TIM TH&**夜中**1[midnight 1]) (LOC OBJ&**会社**1[office 1]))))

(**元気になる**[become healthy] ((AGT AGT%**新入社員**[new employee])(OBJ NIL)(TIM TH&**夜中**1[midnic 1])(LOC OBJ&**会社**1[office 1])))

(**やり遂げる**[to accomplish] ((AGT AGT%**新入社員**[new employee])(OBJ OBJ%**仕事**)(TIM TH&**夜中**1[midnight 1]) (LOC OBJ&**会社**1[office 1])

(**ほっとする**[to be relieved] ((AGT AGT%**新入社員**[new employee])(OBJ NIL)(TIM TH&**夜中**1[midnight 1]) (LOC OBJ&**会社**1[office 1]))))

Process 3;

((**疲れる**[get tired] ((AGT AGT%**新入社員[商品]**[new employee[product]])(OBJ NIL)(TIM TH&**夜中**1[midnight 1]) (LOC OBJ&**会社**1[office 1])))

(**飲む**[drink] ((AGT AGT%**新入社員[商品]** [new employee[product]])(OBJ OBJ%**栄養ドリンク**[nutritional drink])(TIM TH&**夜中**1[midnight 1]) (LOC OBJ&**会社**1))))

(**元気になる**[become healthy] ((AGT AGT%**新入社員[商品]** [new employee[product]])(OBJ NIL)(TIM TH&**夜中**1[midnight 1]) (LOC OBJ&**会社**1[office 1])))

(**やり遂げる**[to accomplish] ((AGT AGT%**新入社員[商品]** [new employee[product]])(OBJ OBJ%**仕事**)(TIM TH&**夜中**1[midnight 1]) (LOC OBJ&**会社**1[office 1])

(**ほっとする**[to be relieved] ((AGT AGT%**新入社員[商品]** [new employee[product]])(OBJ NIL)(TIM TH& **夜中**1[midnight 1]) (LOC OBJ&**会社**1[office 1]))))

Process 4;

((**疲れる**[get tired] ((AGT AGT%**新入社員[商品]** [new employee[product]])(OBJ NIL)(TIM TH&**夜中**1[midnight 1]) (LOC OBJ&**会社**1)))

(**飲む**[drink] ((AGT AGT%**新入社員[商品]** [new employee[product]])(OBJ OBJ%**栄養ドリンク**[nutritional drink])(TIM TH&**夜中**1) (LOC OBJ&**会社**1[office 1]))))

(**元気になる**[become healthy] ((AGT AGT%**新入社員[商品]** [new employee[product]])(OBJ NIL)(TIM TH&**夜中**1[midnight 1]) (LOC OBJ&**会社**1[office 1])))

(**やり遂げる**[To accomplish] ((AGT AGT%**新入社員[商品]** [new employee[product]])(OBJ OBJ%**仕事**)(TIM TH&**夜中**1[midnight 1]) (LOC OBJ&**会社**1[office 1])

(**ほっとする**[to be relieved] ((AGT AGT%**新入社員[商品]** [new employee[product]])(OBJ NIL)(TIM TH&

represents the brand using various strategic techniques, such as personification. Thus, advertisement narratives are narratives connected to brands using a variety of methods and techniques. Additionally, brands are divided into "substantial brands," that is, physical products, and "non-substantial brands," that is, shapeless brands, such as services. The techniques and strategies of advertisement narrative generation will also be influenced by these differences.

Many substantial brands' commercial films fuse the brand with the development of a character's action. For instance, the event sequence of a version of "Chūgoku Ryokucha" (Chinese green tea) by Suntory is as follows:

1. A woman drinks the product.
2. The woman dances in front of a waterfall.
3. The woman dances in front of the waterfall.
4. The woman dances in front of the waterfall.
5. The woman dances in front of the waterfall.
6. A drop of water falls into a glass filled with the green tea/"Sarari Yonsen'nen (4000 nen)."
7. The woman drinks the product in front of the waterfall.
8. A close-up of the woman with the product/the name of product ("Chūgoku Ryokucha")/the name of the company ("Suntory").
9. The product/the name of product ("Chūgoku Ryokucha")/the name of company ("Suntory").

In this TV commercial, the product's name and company's name appear separately. At the same time, the brand is shown at the level of the character's events, such as "the woman dances in front of a waterfall." Furthermore, the woman herself may be a semantic symbol or personification of the fresh green tea. Thus, the appearance of the brand is divided into two types: the brand connecting to event development in a story and the brand being an individual in a story.

The relationship between story and brand in advertisement narratives can be seen from the two viewpoints of micro and macro. By the term micro, the author means a viewpoint that focuses on the appearance, or a type of association between an advertisement event and a brand, in a specific place in a story. On the other hand, macro indicates the problem regarding brand appearance in an entire advertisement story. In the above example, from the micro perspective, various brand-related events, such as "A drop of water falls into a glass filled with the green tea filled in a glass," "The woman drinks the product in front of the waterfall," and "A close-up of the woman with the product." From the macro level, in the first half in this commercial film, we can observe events mainly including the character's actions and the appearance of the brand as connected to the actions and, in last half, the story's flow where the individual brand's appearance gradually increases.

Figure 2 shows an image of the relation between events and a brand in an entire commercial work. The white box and black boxes, respectively, correspond to an event without, and an event with the brand. Simply speaking, an advertisement narrative as a sequence of events is structured by the arrangement of the events with the brand and the events without it. The above micro means the relationship of each event with the brand and the macro is equivalent to the construction of an entire narrative. From a generation perspective, micro level processing corresponds to a kind of technique system that controls the brand's appearance in each event and the macro level processing is related to a kind of control strategy for directing the usage of each of the techniques. Based on the order of generation, there are two contrasting methods: the top-down order, where the micro structure is generated according to the macro structure, and the bottom-up order, where the macro structure is constructed after several micro techniques have been applied.

In addition, in a current advertisement project of a company, the planning and generation of a future advertisement scenario can be scheduled in two directions. One is planning and generation that faithfully reflect the result of the advertisement narrative analysis. Although there is room for considering detailed event-level analysis needs, analyzing actual commercial films themselves is required for this direction. On the other hand, the other direction is planning and generation by, so to speak, destruction and reconstruction (Ogata, 2007). In this direction, the author aims to create narratives, advertisement scenarios in this case, beyond the existing narratives. However, also in the latter case, the detailed semantic analysis of commercial works is necessary for the advertisement generation by destruction, reconstruction, and, moreover, compounding and recombination.

Here, the author would like to describe the main terms. First, "story" means the sequence of events developed by any actions of the characters in an advertisement narrative. Events that construct a story are differentiated from events that only show the brand and exit the story at a different level. Moreover, in "story-type brand introduction," a brand appears incorporated in a story. In contrast, in "non-story-type brand introduction," the brand is not directly in the story and appears outside it. In the story-type brand introduction, the brand appears with a direct association to the events constructing the story. On the other hand, in the non-story-type brand introduction, the brand appears independently from the events constructing the story. In many examples, a brand appears in a shot during an event of the story and, at the same time, the brand appears in a different level (for instance, in the credits).

Adding the micro and macro levels to the above brand introduction methods results in the following:

1. **Story-type Brand Introduction/Micro**: The brand is introduced in each event.
2. **Non-story-type Brand Introduction/Micro**: The brand is introduced outside the world of events constructing the story.
3. **Story-type Brand Introduction/Macro**: Events involving the brand and events without it are placed in the story as a sequence of events.
4. **Non-Story-Type Brand Introduction/Macro**: The brand introduction parts outside the story are placed in an entire advertisement narrative.

Although these become the basic standards of advertisement narrative analysis, they are the chief elements of advertisement scenario generation.

The above considerations were created by Kayamori and Ogata (2003) and Ogata and Kayamori (2003). Depending on the desired result, the author creates an experimental advertisement scenario generation mechanism. This mechanism builds an entire narrative tree for simple advertisement narratives using the narrative tree structure generation mechanism and inserts brand information into specific events. Figure 3 shows examples of the system's input and output; the hierarchical relation tree is not described in this figure. In the figure, the product is a "Nutritional Drink" (the specific brand name is not used in this case). The first simple advertisement event is extended through the following incremental process:

- **Process 1** is the first skeleton.
- **Process 2** extends the above into an entire story without a brand.
- Although **Process 3** and **Process 4** include the brand, the former performs story-type brand introduction whereas the latter process uses non-story-type brand introduction.

Table 2. Categorizing story-type brand introduction strategies

Standard or Defamiliarization	Objects of Rhetorical Strategies	Types of Rhetorical Strategies	Meanings
Regular Rhetorical Strategy or Standard Rhetorical Strategy	Ordinary processes of manufacturing, purchasing, and using	**R1**: Rhetorical Strategy of Manufacturing Process [*1 example*]	In the manufacturing process, the product is manufactured in the typical way.
		R2: Rhetorical Strategy of Purchasing Process [*21 examples*]	In the purchasing process, the product is purchased in the typical way.
		R3: Rhetorical Strategy of Core Usage Method [*465 examples*]	The product is used in the original way, that is, according to the core usage method.
Irregular or Defamiliarization Rhetorical Strategy	Action	**R4**: Rhetorical Strategy of Deviated Use or Defamiliarization Use [*29 examples*]	The product is used in a strange way instead of the typical way.
		R5: Rhetorical Strategy of None Usage [*329 examples*]	The product is not used, such as in the case of only "holding" It.
	Person	**R6**: Rhetorical Strategy of a Character as a Narrator [*94 examples*]	The character directly sends the message to the audience, outside the advertisement story.
		R7: Rhetorical Strategy of Defamiliarization of a Character's State [*85 examples*]	All or parts of the character change to extraordinary or strange ones.
		R8: Rhetorical Strategy of a Character's Defamiliarization of the Product Usage Action [*53 examples*]	The character performs an action deviating from the typical product usage action.
	Product or object	**R9**: Rhetorical Strategy of Defamiliarization of a Product's State [*92 examples*]	All or part of the product changes to another (strange) form.
		R10: Rhetorical Strategy of Defamiliarization of the State of an Object Related to the Product [*5 examples*]	All or part of an object related to the product changes to another (strange) form.
		R11: Rhetorical Strategy of the Product as a Subject [*387 examples*]	The product exists independently, such as simply being in the background, instead of being an object to be used by the character.
	Location	**R12**: Rhetorical Strategy of the Defamiliarization of a Background [*116 examples*]	All or part of the background changes to an extraordinary background or an image background.

Table 3. Rhetorical strategy types of sample commercials (a part)

TVCF Client/Product/Title of CM/Shot no.	R1	R2	R3	R4	R5	R6	R7	R8	R9	R10	R11	R12
Kirin Co., Ltd./Ich-ban shibori/"Kyoto Kamo Nasu"/9	ü											
Asahi Breweries, Ltd./Clear Asahi/"Over 100 Million"/7		ü										
Asahi Breweries, Ltd./Asahi STYLE FREE/"Take a shower"/3			ü									
KUREHA Corp./NEW kurerapu/Part. "Hole prank"/5				ü								
Yakult Honsha Co., Ltd./Kurozu drink [Black vinegar drink]/ Part. "New experience of black vinegar, doubt, FD "/2					ü							
Yakult Honsha Co., Ltd./Yakult 400/-/6					ü	ü						
LOTTE Co., Ltd./Yukimi Daifuku [Snow viewing rice cake]/-/2							ü					ü
Coca-cola (Japan) Co., Ltd./AQUARIUS Freestyle/-/4					ü			ü				
ASAHI Soft Drinks Co., Ltd./CITRULLINE WATER/Part. "Appearance"/1									ü			ü
Kao Corporation/Toilet Magiclean/-/2			ü							ü		
Coca-cola (Japan) Co., Ltd./*GEORGIA* (Canned coffee bland)/ Part. "Grandma in heaven"/5					ü						ü	
Coca-cola (Japan) Co., Ltd./Coca-cola/Part. "Bird's Nest"/3												ü
Calbee, Inc./Jagabee/-/2			ü				ü					
Kao Corporation/Biore sara-sara powder sheet [Biore non-wetness powder sheet]/"Biore in morning"/3			ü									ü
Kao Corporation/Healthya ryokucha [Healthya green tea]/Part. "2018 summer"/2			ü				ü					
SOFT99 Corp./Fukupika/Part. "Fukupika dance"/3			ü					ü				
NEC Corp./ShieldPRO/Part. "Green rockets"/3				ü							ü	
Morinaga & Co., Ltd./Ototo/Part. "Ownership battle"/ 5				ü			ü					
Asahi Breweries, Ltd./Asahi STYLE FREE/"Take a shower"/7					ü	ü						
NISSAN MOTOR CO., LTD./TIIDA/-/7											ü	ü
Dainihon Jochugiku Co., Ltd./Sanpoll/-/2			ü				ü				ü	
Eisai Co., Ltd./Chocola BB Lucent C/Part. "Reverse spin"/2				ü					ü		ü	
Suntory Holdings Ltd./PEPSI NEX/Summer ver. Part. "Forever"/7					ü				ü		ü	
Suntory Holdings Ltd./Premium Malt's/Part. "Hammock"/3			ü					ü			ü	
Suntory Holdings Ltd./Premium Malt's/ Part. "Hammock"/4			ü					ü			ü	ü
Meiji Co., Ltd./ Kinoko no yama [Chocorooms]/-/5				ü			ü		ü			ü

Rhetorical Strategies for Preliminary Categorization of Story-type Brand Introduction

Although Kayamori and Ogata (2003) and Ogata and Kayamori (2003) attempted to analyze the brand introduction methods of story-type and non-story-type considering micro level rhetorical strategies, Yoshio and Ogata (2007) and Shimura and Ogata (2008) reconsidered and recategorized the above attempt as shown in the following.

First, the author divided television advertisement videos into two large categories depending on whether a brand is used in the core usage method or not. The latter was further divided into subcategories. Here, "core usage method" means that a product is used according to the ordinary usage method or order.

(A) The brand is introduced and used according to the core usage method

1. Using 100 television commercial videos as analysis material, the author hierarchically categorized substantial brands (the lower hierarchy corresponds to the brand names) to describe the core usage method of each brand using one or more adequate verbs and the predicate representation including one or more verbs. Examples include "X eats an apple," "X drinks a glass of beer," and "X drives a car." However, the core usage method is frequently a sequence of events (or actions), such as "*He takes the cap off the beer bottle/He holds the beer bottle in his hand/He pours the bottled beer into a cup/He holds the cup in his hand/He raises a toast, he drinks the beer/He holds the cup in his hand.*"
2. The advertisement videos analyzed contained events where only an action different from the central action of a core usage method is represented, such as "*X has a glass of beer.*" The author interpreted whether this action is included as a part of a core usage method.

(B) The brand is introduced and used in a way deviating from the core usage method

1. **The brand is used in a way deviating from the core usage method:** This is the case where a product is used in an extraordinary method different from the core usage method, such as, *a) A woman throws a canned beer, b) A man dances holding a nutritional canned drink, c) A woman lies down holding a canned beer, d) A man smells the coffee's smell, e) A woman walks holding a soft drink, f) A man holds the refreshing drink to the sunlight, and g) A woman pures water over herself a canned alcoholic drink.*
2. **The brand is simply placed:** In this case, the brand is simply put in display, for instance in the background, without it being used. Examples include *a) The seasoning is on a table/A woman gazes at it/She eats the dish/She stares at the seasoning placed on the table, b) A car is stopping, c) A car is stopping/A man sees the car/He talks to a woman/He looks at the scenery, and d) A soft drink is putting on the flore.*
3. **The brand transforms into another object:** For example, *a) A cake disappears, b) A plastic bottle of juice transforms into the doll of a dog, and c) A cake transformes into the doll of a dog.*

Table 4. Usage order in products in sample commercials

Category	Actions
Bleach	She puts the bleach into the washing machine.
Clothing detergent	He holds the clothing detergent in his hand. He puts the clothing detergent into the washing machine.
Kitchen detergent	She applies the kitchen detergent to the cutting board
Eye-drops	He drips eye-drops into the eye. He holds the eye-drops in his hand.
Liquid soap	She takes liquid soap from its container. She washes her hands with the liquid soap. She rinses her hands with water.
Toilet detergent	He picks up the toilet detergent. He applies the toilet detergent to the toilet.
Shampoo	She holds the shampoo in her hand. She washes her hair with the shampoo.
Hair styling product	He applies the hair styling product to his hair.
Facial cleanser	She applies the facial cleanser to her face. She rinses her face.
Hair growth tonic	He applies the hair growth tonic to his scalp.
Antiperspirant spray	She holds the antiperspirant spray in her hand. She sprays the antiperspirant spray onto her body.
Deodorant	He holds the deodorant in his hand. He sprays the deodorant. He holds the deodorant in his hand.
Lotion	She opens the lotion cap. She applies the lotion to a cotton pad. She applies the lotion to her face.
Skin cream	He takes some skin cream on his hand. He applies the skin cream to his face.
Mascara	She applies the mascara to her eyelashes.
Paper car wash wipes	He wipes the car with the paper car wash wipes.
Medicine	She takes out the medicine with her hand.
Ointment	He takes out the ointment with his hand. He applies the ointment to his body.
Writing instrument	She holds the writing instrument in her hand. She writes with the writing instrument.
Sweat-wipe sheet	He takes a sweat-wipe sheet out from the container. He wipes his body with the sweat-wipe sheet. He closes the lid of the sweat-wipe sheet container.
Plastic wrap	She opens the plastic wrap box. She pulls out the plastic wrap. She covers the dish with the plastic wrap.
	He pulls out the plastic wrap. He covers the dish with the plastic wrap, he wraps the dish with the plastic wrap.
Gummy candy	She holds the gummy candy in her hand. She eats the gummy candy. She holds the gummy candy in her hand.
Candy	He holds the candy in his hand. He eats the candy.
Purin	She holds the purin container in her hand. She scoops out the purin with a spoon. She feeds the purin to a person. She eats the purin.
Ice cream	He holds the ice cream in his hand. He eats the ice cream. He holds the ice cream in his hand.
	She holds a cup of ice cream in her hand. She takes a scoop of ice cream with a spoon. She eats the ice cream.

Category	Actions
Gum	He opens the gum's paper wrapper. He chews the gum.
Chocolate	She eats chocolate.
Snack food	He holds the snack food in his hand. He eats the snack food. He holds the snack food in his hand.
Cup ramen	She opens the cup ramen lid. She holds the cup ramen in her hand. She eats the cup ramen. She drinks the cup ramen broth.
Cup yakisoba	He holds the cup yakisoba in his hand. He eats the cup yakisoba.
Instant ramen	She holds the instant ramen in her hand. She pours hot water onto the instant ramen. She eats the instant ramen.
Instant soup	He eats the instant soup. He holds the instant soup in his hand.
Nattō	She mixes the natto.
Dietary supplement (solid)	He eats the dietary supplement. He holds the dietary supplement in his hand.
Retort packaged food	She takes the retort packaged food out from hot water. She puts the retort packaged food onto rice. She holds the rice. She eats the rice.
Canned food	He opens the lid of the canned food. He prepares the canned food. He takes the food with chopsticks. He eats the food.
Frozen food	She heats the frozen food. She holds the frozen food with chopsticks. She smells the scent of the frozen food. She eats the frozen food.
Sōmen noodles	He dips the sōmen noodles into the dipping sauce. He eats the sōmen noodles. He dips the sōmen noodles into the dipping sauce.
Cheese	She eats the cheese.
Vegetable juice	He holds the vegetable juice in his hand. He drinks the vegetable juice. He holds the vegetable juice in his hand.
Green tea (in a plastic bottle)	She opens the green tea cap. She takes the plastic bottle in her hand. She pours the green tea into a cup. She holds the green tea in her hand. She smells the fragrance of the green tea. She drinks the green tea. She holds the cup in her hand.
Black tea (plastic bottle pack)	He holds the black tea in his hand. He drinks the black tea.
Black tea (tea bag)	She makes the black tea. She drinks the black tea.
Oolong tea	He holds the oolong tea in his hand. He drinks the oolong tea.
Other tea	She takes the tea in her hand. She holds the tea in her hand. She drinks the tea. She holds the tea in her hand.
Milk	He holds the milk in his hand. He drinks the milk.

4. An object used by the brand product transforms into another

Category	Actions
Water (plastic bottle)	She takes the water in her hand.
	She holds the water in her hand.
	She drinks the water.
	She holds the water in her hand.
Beer bottle	He takes the cap off the beer bottle.
	He holds the beer bottle in his hand.
	He pours the bottled beer into a cup.
	He holds the cup in his hand.
	He raises a toast, he drinks the beer.
	He holds the cup in his hand.
Canned beer	She takes the canned beer in her hand.
	She opens the canned beer.
	She holds the canned beer in her hand.
	She pours the canned beer into a cup.
	She drinks the beer.
	She holds the cup in her hand.
	He takes the canned beer from the store shelf.
	He puts the canned beer into his basket.
	He scans the canned beer at the checkout counter.
	He carries the canned beer.
	She produces canned beer.
Canned chūhai	He takes the canned chūhai in his hand.
	He opens the canned chūhai.
	He pours the canned chūhai into a cup.
	He holds the cup in his hand.
	He drinks the chūhai.
	He holds the cup in his hand.
Shōchū	She opens the shōchū bottle.
	She pours the shōchū into a cup.
	She drinks the shōchū.
	She holds the cup in her hand.
Whiskey	He pours the whiskey into a glass.
	He drinks the whiskey.
	He holds the glass in his hand.
Soda (plastic bottle)	She pours the soda into a cup.
	She holds the cup in her hand.
	She drinks the soda.
	She holds the cup in her hand.
Sports drink (plastic bottle)	He holds the sports drink in his hand.
	He drinks the sports bottle.
	He holds the sports bottle in his hand.
	He places the sports drink on the table.
Fermented milk drink	She holds the fermented milk drink in her hand.
	She drinks the fermented milk drink.
	She holds the fermented milk drink in her hand.
Juice (plastic bottle)	He drinks the juice.
	He holds the juice in his hand.
Nutrition and health supplement	The store clerk gives the nutrition and health supplement to the customer.
	She holds the nutrition and health supplement in her hand.
	She opens the nutrition and health supplement.
	She drinks the nutrition and health supplement.
	She holds the nutrition and health supplement.
Canned coffee	He opens the canned coffee.
	He holds the canned coffee in his hand.
	He drinks the canned coffee.
	He holds the canned coffee in his hand.
Coffee (plastic bottle pack)	She drinks the coffee.
	She holds the coffee in her hand.

Category	Actions
Instant coffee	He puts the instant coffee into a cup. He holds the cup in his hand. He drinks the instant coffee. He holds the cup in his hand.
Instant coffee	She opens the instant coffee seal. She smells the fragrance of the instant coffee. She puts the instant coffee into a container. She closes the lid of the container.
Instant cocoa	He holds the instant cocoa cup in his hand. He drinks the instant cocoa. He holds the cup in his hand.
Mayonnaise	She holds the mayonnaise in her hand. She adds the mayonnaise to the dish. She eats the dish.
Stock	He adds the stock to the dish.
Salt	She holds the salt (container) in her hand. She adds the salt on the dish.
Ajinomoto	He holds the Ajinomoto (container) in his hand. He adds the Ajinomoto to the dish.
Sauce	She puts the sauce into the dish.
	He adds the sauce to the dish.
Vinegar	She puts the vinegar into the dish.
	He adds the vinegar to the dish.
	She pours the vinegar into the glass.
Rice vinegar	He adds the rice vinegar to the dish.
Ponzu vinegar	She holds the ponzu vinegar in her hand. She adds the ponzu vinegar to the dish.
	He adds the ponzu vinegar to the dish.
	She puts the rice vinegar on the dish.
Soup stock	He holds the soup stock in his hand. He pours the soup stock into a bowl. He puts the soup stock into the dish.
Furikake	She holds the furikake in her hand. She adds the furikake to the rice. She eats the rice.
Green tea rice seasoning	He holds the green tea rice seasoning in his hand. He opens the green tea rice seasoning. He adds the green tea rice seasoning to the rice. He adds hot water to the rice, he eats the rice.
Printer	She turns the power on. She selects the target image. She prints.
	He operates the PC.
PC	She opens it. She looks at the screen. She types.
Washing machine	He loads the laundry. He takes the laundry out.
Television	She watches.
Refrigerator	He opens the door. He puts the food in. He closes the door.
Vacuum cleaner	She cleans
Hot water dispenser	He presses the button to pour hot water.
Digital camera	She holds it in her hand. She zooms. She takes a photograph. She holds it in her hand.
Digital camera	He looks at stored images.

Category	Actions
Air-conditioner	She circulates air.
Electric shaver	He holds it in his hand. He shaves his beard. He holds it in his hand.
Electronic dictionary	She writes with the stylus. She presses the button.
Microwave	He heats the food.
Smartphone	She holds it in her hand. She watched television.
	He holds it in his hand. He makes a call.
Rice cooker	She opens it. She puts rice into a bowl.
Car navigation system	He operates it.
Light bulb	She illuminates.
Component stereo	He inserts a CD.
Car	She gets in, she drives. She gets out.

Table 5. Examples of regular or standard rhetorical strategies [R1, R2, R3]

etoric/Brand	Image
: Rhetorical ategy of nufacturing cess rin Co., Ltd.] *ban Shibori*	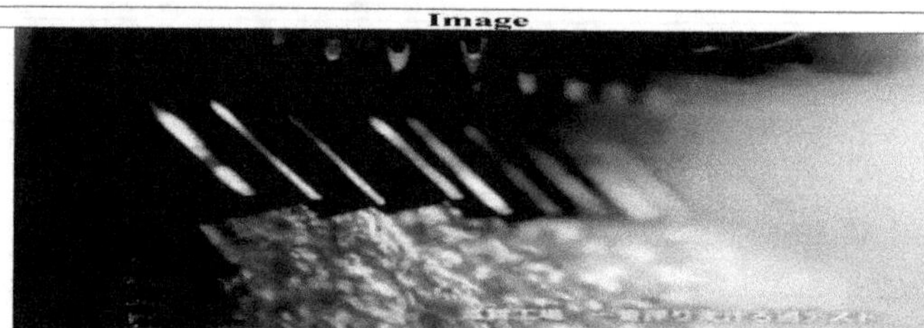
: Rhetorical ategy of chasing cess ahi weries, Ltd.] *ar Asahi*	
: Rhetorical ategy of Core ige Method e Procter & nble npany of an Ltd.] *un Prosonic*	

Table 6. Examples of irregular or defamiliarization rhetorical strategy (action) [R4, R5]

Rhetoric/Brand	Image
R4: Rhetorical Strategy of Deviated Use or Defamiliarization Use [NEC Corporation] *Shield PRO*	
R5: Rhetorical Strategy of None Usage [Suntory Holdings Ltd.] *Kuro uroncha (Black oolong tea)*	

Table 7. Examples of irregular or defamiliarization rhetorical strategy (person) [R6, R7, R8]

Rhetoric/Brand	Image
R6: Rhetorical Strategy of a Character as a Narrator [Dydo Drinco, Inc.] *Yanroncha level care (Yanlong tea level care)*	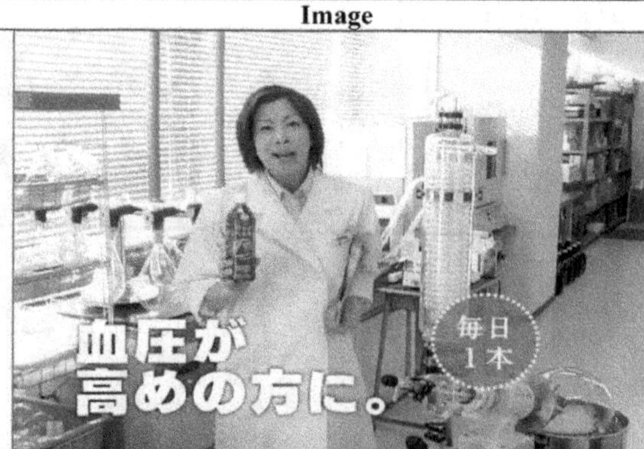
R7: Rhetorical Strategy of Defamiliarization of a Character's State [CASIO COMPUTER Co., Ltd.] *Exword*	
R8: Rhetorical Strategy of a Character's defamiliarization of the Product Usage Action [Coca-cola (Japan) Co., Ltd.] *AQUARIUS Freestyle*	

Table 8. Examples of irregular or dedfamiliarization rhetorical strategy (product or object) [R9, R10, R11]

Rhetoric/Brand	Image
R9: Rhetorical Strategy of Defamiliarization of a Product's State [ASAHI Soft Drinks Co., Ltd.] *Citrulline water*	
R10: Rhetorical Strategy of Defamiliarization of the State of an Object Related to the Product [Kao Corporation] *Toilet Magiclean*	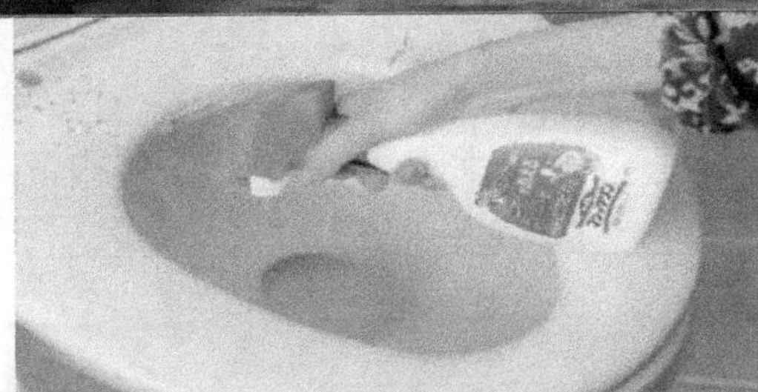
R11: Rhetorical Strategy of the Product as a Subject [Kirin Co., Ltd.] *Kirin Tanrei nama "No. 1"*	

Table 9. Example of irregular or defamiliarization rhetorical strategy (location) [R12]

Rhetoric/Brand	Image
R12: Rhetorical strategy of the de-familiarization of a background [Matsushita Electric Works, Ltd. (Current name: Panasonic Corporation)] *Eco idea*	

Table 10. Mixed example: two rhetorical types

Rhetoric/Brand	Image
Mixed Example: Two Rhetorical Types (1) [Canon Inc.] *EOS kiss X2*	
Mixed Example: Two Rhetorical Types (2) [Hitachi, Ltd.] *Gomi dash cyclone CV-SL10*	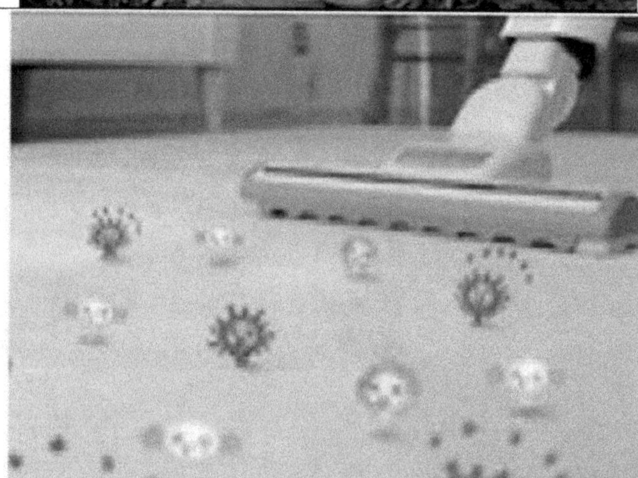

Table 11. Mixed example: three rhetorical types

Rhetoric/Brand	Image
Mixed Example: Three Rhetorical Types (1) [Dainihon Jochugiku Co., Ltd.] *Sanpoll*	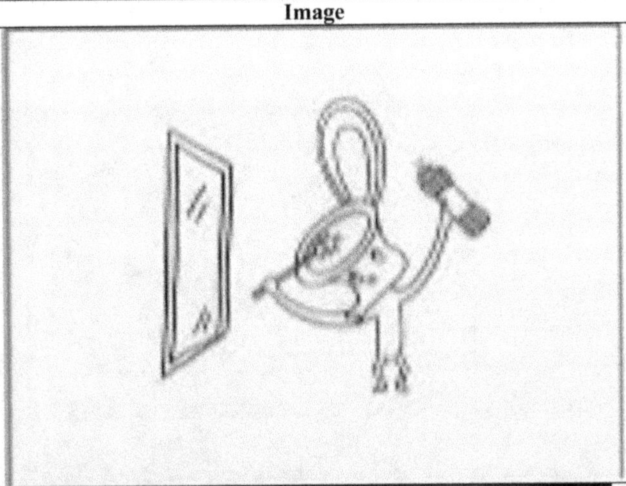
Mixed Example: Three Rhetorical Types (2) [Morinaga & Co., Ltd.] *Ottotto*	

Table 12. Mixed example: four rhetorical types

Rhetoric/Brand	Image
Mixed Example: Four Rhetorical Types (1) [Kao Corporation] *Biore U*	
Mixed Example: Four Rhetorical Types (2) [Kao Corporation] *Biore U*	

Table 13. Mixed examples: not using R3 (core usage method)

Rhetoric/Brand	Image
Mixed Example: None Using R3 (Core Usage Method) (1) [SOFT99 Corporation] *Glaco*	
Mixed Example: None Using R3 (Core Usage Method) (2) [Meiji Co., Ltd.] *Maca no genki*	

Figure 4. Hierarchical system of products (1): food (food)

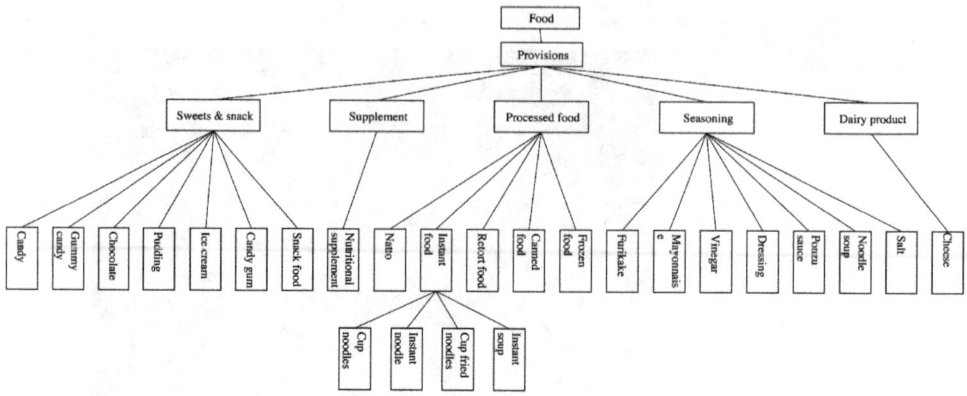

Figure 5. Hierarchical system of products (2): food (drink)

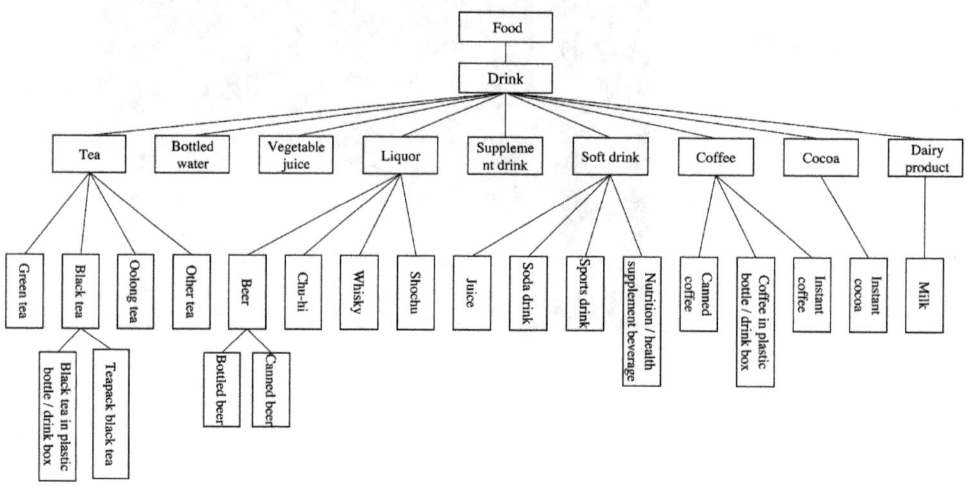

Figure 6. Hierarchical system of products (3): industrial product

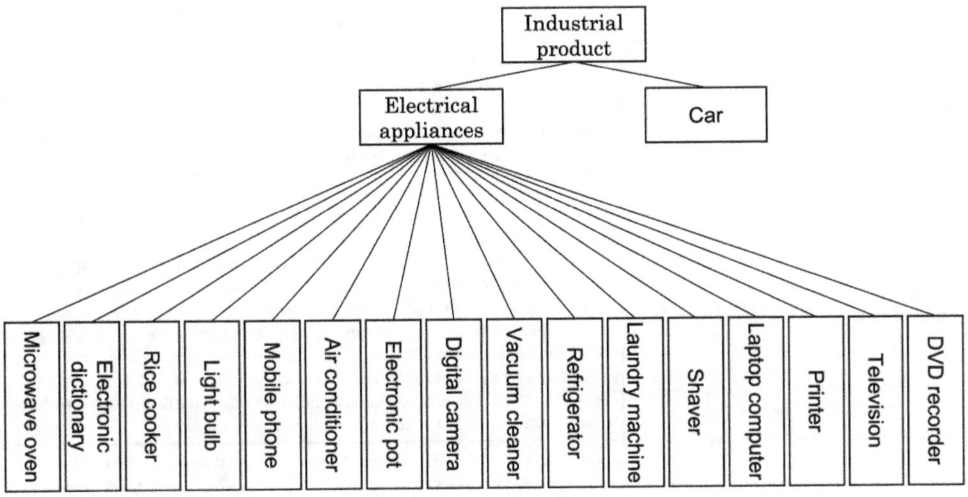

Figure 7. Hierarchical system of products (4): daily commodities

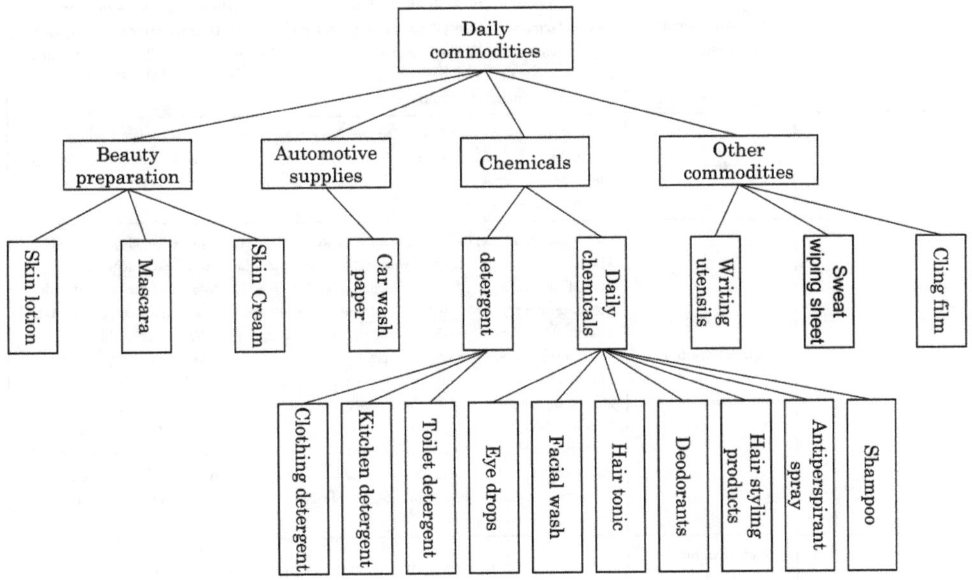

Table 14. Types and examples of characters appearing in the sample commercials

Types of Characters	Types of Change	Examples
Character (Human)	<No Change>	Middle-aged man, Teenage woman, Young Man, Middle-aged farmer, Male construction worker, Male soccer player, Male dancer, Male carpenter, Salesman, Male office worker, Boy, Girl, Young man, Young woman, Adult man, Adult woman, Male teacher, Female teacher, Male singer, Female singer, Primary schoolboy, Primary schoolgirl, Office lady, Housewife, Female inn employee, Female cheerleader
Character (Metamorphose s)	Brand Character	Qoo, Character whose body is light blue, Kyoro-chan, Character like a bird who has a body that symbolizes the product and has legs and a beak, Kakukakusikajika, Character who has a deer's head and a human's body with eyeglasses, Biore mama, Mother character in animation style, Biore mama's children, Gomizu, Character who has a simple body and head that symbolizes garbage, Hiyoko-chan, Symbol character of Nissin Chicken Ramen, Karl uncle and his friends, Karl's symbol character, Poporo-kun, Character whose body is made of pasta and has a simple face, hands, and legs, Shirokuma (white bear)-kun, White bear with fever, Kichinto-san, Character that looks like a housewife who dresses in red overalls with her hair tied back, Kinoyama-san, Character whose simple face is drawn on "Kinoko no Yama" with hands, legs, and a guitar, Character whose head is like the product (cheese) with a simple face, hands, legs, and body, Poteko, Hi-chew king
	Fictional Character	Sakura Momoko's family, Sakura Momoko, Sakura Sakiko (Momoko's older sister), Sakura Hiroshi (Momoko's father), Sakura Sumire (Momoko's mother), Sakura Tomozou (Momoko's grandfather), Sakura Kotake (Momoko's grandmother), Characters of *Kyojin no Hoshi* (*A Star of the Giants*), Hoshi Hyūma, Hoshi Ittetsu (Hyūma's father), Hoshi Akiko (Hyūma's older sisther), Samon Hosaku and his children (three boys and two girls), and a woman, Characters of *Dokonjo-gaeru* (*The Gutsy Frog*)], Pyonkichi (a yellow frog), Hiroshi, Yoshizawa Kyoko, Boys and girls, Gachapin-mukku, Hello Kitty, Ultraman
	CG or Animation Style Animals	Dinosaur in CG style, Object like a yellow bird, Blue elephant who is smaller than the product, Yellow green elephant who is smaller than the product, Character with a simple face on "pupurun (gummy)" and has simple hands and legs, Deer who sometimes acts and poses like a human, White rabbit who can act like a human, Red fish, Characters of a tabby dog, a raccoon dog, a squirrel, and a deer with heads as big as their bodies, Characters becoming animated characters
	Real Animal	Squirrel, Lion who can act like a human, Penguin who can act like a human, Bulldog who can act like a human), Black dog, Dog wearing a hat cap, Curly-haired white dog, Cat that can write its name with a pen, White and black dog, White dog that has brown ears and brown forehead
	Defamiliarization of Costume and Appearance	Child wearing mouse costumes, Family wearing mouse costumes, Alien, Hayami Imaichi, Man of the Heian period style, Man of primitive style, Man dressed like a schoolgirl, Man of samurai style, Man of nobleman style, Man who is wearing an ordinary suit with *chonmage* hairstyle, Man who is semitransparent and has an aura with a strange head size, Woman who takes off *shishimai* (lion dance) costume and has the costume, Children in angel costumes, Woman who is as big as the product and dressed in an angel costume, Children who are dressed in tights like wrapping paper of Calpis, Five Men who are dressed in costumes in which the product is drawn on the head, Man with a panda's head, Man who became a giant by eating the product, Child who dressed in tights and put on headgear, Man who put a doll in his hand, Woman with eyeglasses who dressed in the pink shirt and has a radio cassette, Milk cow in a costume, People who dressed in costumes of a refrigerator, a telephone, and an aquarium, Fox like a costume that has red hair, Man who dressed in a robot costume
	Personification of Object (Product Itself)	Dancing product, Running product like a car, Character with a body and face that are the product and has hands and legs, Corn (product) which has a simple face
	Personification of Object (Other Than Product)	Character personifying a toilet, Character personifying the stuffed doll of a blue bear, Speaking back of hand

Figure 8. Hierarchical system of place (stage, background)

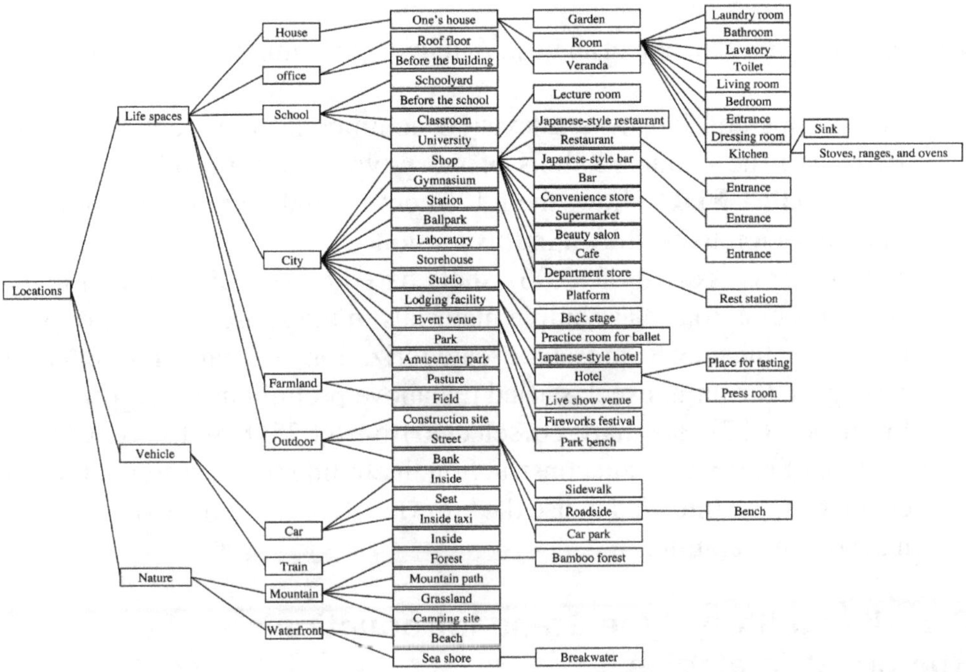

object: For example, *A boy sprinkles the salt (= brand) to anthropomorphic rice ball with hands and feet.*

5. **The brand is used in an extraordinary place:** In this case, a brand is used at the place where the brand is typically not used and this case corresponds to the use of extraordinary place information in an event. An example is *A car is running under the sea.*

RHETORICAL STRATEGIES FOR ACTIONS AND OBJECTS: A DETAILED ANALYSIS OF TELEVISION ADVERTISEMENT VIDEOS AND THE EXTRACTION AND CLASSIFICATION OF STORY-TYPE BRAND INTRODUCTION STRATEGIES

Based on the above ideas and results, the author presented the revised and expanded version of the advertisement narrative generation and brand introduction strategies into the advertisement events. This section provides the detailed description of the brand introduction strategies and related topics.

Analysis Method and Order

The flow of the advertisement analysis of television commercials is as follows:

1. The number of Japanese television commercial clips collected was 1,123 and the extracted shots of story-type brand introduction were 5,076. (**APPENDIX** shows the 1,123 commercials based on the product categorization to be shown in this section.)
2. The author classified the 5,067 shots based on the above preliminary categorization to use as sample materials in the subsequent analysis.
3. Based on the newly presented categorization of brand introduction strategies, the author reclassified the above preliminary categories.
4. From the 1,017 shots analyzed, selected from 5,076 shots, the knowledge content of advertisement construction, including brands (products) and characters, are intended to use the knowledge content for advertisement narrative generation.

Classifying Story-Type Brand Introduction's Rhetorical Strategies

This section shows the concluding part of the analysis. Table 2 presents the categories of story-type brand introduction strategies. The meaning of "Types of Rhetorical Strategies," for R1 through R12 will be explained in the following part.

Table 3 shows the correspondence between the shots in the sample commercial clips and the revised strategy types. A shot in a commercial clip does not necessarily correspond to only one strategy type. There are many shots using several rhetorical strategies types. Therefore, the total number of shots (1,677) corresponding to the categorized rhetorical strategy types is larger than the total number of sample shots. This indicates the concept of complexity of brand introduction strategies. The number of strategy types that a shot includes was as follows:

- 1 rhetorical strategy type used in a shot: 553
- 2 rhetorical strategy types used in a shot: 307
- 3 rhetorical strategy types used in a shot: 118
- 4 rhetorical strategy types used in a shot: 39

Figure 9. Action knowledge base (a part)

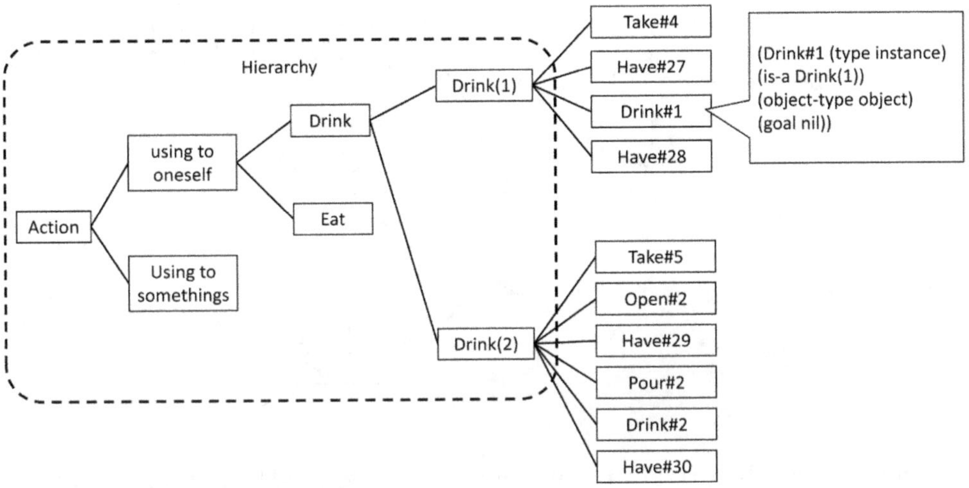

Figure 10. Object knowledge base (a part)

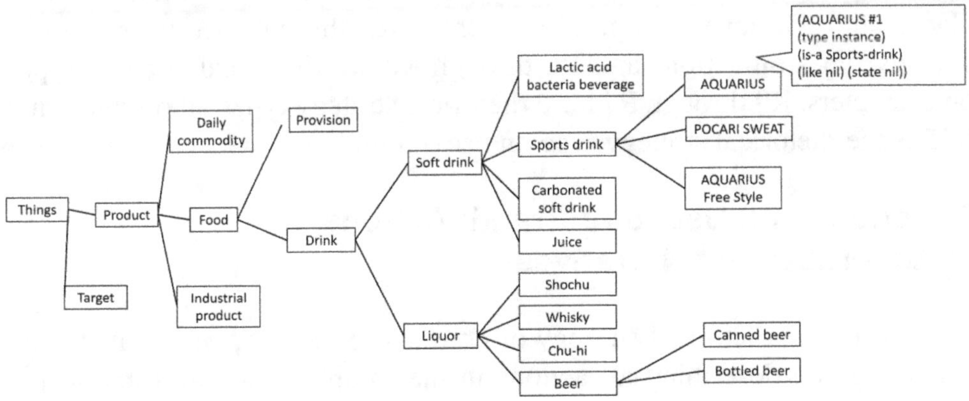

Figure 11. Place knowledge base (a part)

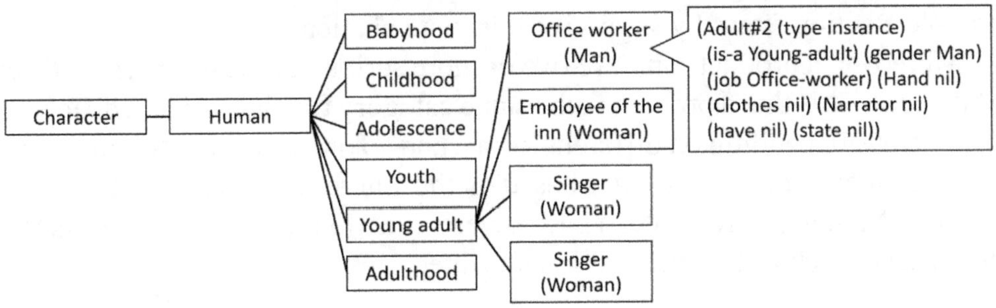

Figure 12. Character knowledge base (a part)

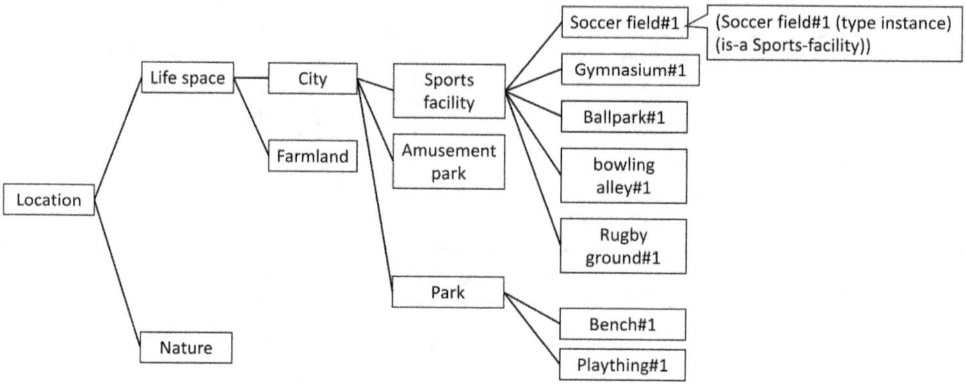

Finally, these rhetorical strategy types can be defined by regular, standard, or ordinary usage or irregular, extraordinary, deviating, or defamiliarization usage of the elements included in or constructing each event, namely actions, objects (a product and other related objects), characters, backgrounds (mainly places and, moreover, temporal elements). R3 through R5 are rhetorical strategy types on actions and R6 through R8 are rhetorical strategy types on characters. R9 through R11 are rhetorical strategy types on products and R12 is the rhetorical strategy type on backgrounds.

Functions and Usage of Brand: Actions in Advertisement Narratives

R3, that is, the rhetorical strategy of core usage method, was analyzed by extracting and describing all actions in the sample shots. In Table 4, the author describes each action in the extracted core usage method according to the action's order. Each action order corresponds to a kind of script (Schank and Abelson, 1977) of the flow of a product's usage. From the viewpoint of scripts for story generation (Ono & Ogata, 2018), we can acquire more complete action flows by supplying adequate elements.

Moreover, a product can have two or more scripts for the usage flow. One of the reasons is that a product in the same category frequently has a different usage flow, depending on different forms, and so on, such as different forms of ice creams. Another reason is based on the ambiguity in a product's usage method. Namely, a product can have several functions and several scripts for core usage, such as a multifunctional printer.

Table 15. Generation examples of the defamiliarized rhetoric

Rhetorical Strategy	Program Code
R4: Rhetorical Strategy of Deviated Use or Defamiliarization Use	(peek#1 (agent adult#2) (counter-agent nil) (instrument nil) (object aquarius#1) (location soccer-ground#1) (goal nil) (from nil) (time noon#1) (narration nil) (caption nil))
R5: Rhetorical Strategy of None Usage	(adult#5 (type instance) (is-a adult) (have kirin-lager-beer#1)) (touch#1 (agent adult#5) (object pleo#1) (location garage#1))
R6: Rhetorical Strategy of a Character as a Narrator	(adult#5 (type instance) (is-a adult) (narrator t))
R7: Rhetorical Strategy of Defamiliarization of a Character's State	(adult#5 (type instance) (is-a adult) (state huge-size#1)) (wash#1 (agent deer#1) (instrument merit#1) (object pleo#1))
R8: Rhetorical Strategy of a Character's Defamiliarization of the Product Usage Action	(adult#5 (type instance) (is-a adult) (gender women) (state dance#1))
R9: Rhetorical Strategy of Defamiliarization of a Product's State	(merit#1 (type instance) (is-a shampoo) (like beer#1) (state nil)) (merit#1 (type instance) (is-a shampoo) (like nil) (state floating-object#1))
R10: Rhetorical Strategy of Defamiliarization of the State of an Object Related to the Product	(pleo#5 (type instance) (is-a target) (like sho-chu#1) (state nil)) (pleo#5 (type instance) (is-a target) (like nil) (state floating-object#1))
R11: Rhetorical Strategy of the Product as a Subject	(exist#1 (agent fukupika#1) (location road#1))
R12: Rhetorical Strategy of the Defamiliarization of a Background	(wipe#1 (agent adult#6) (instrument fukupika#1) (object cube#1) (location roof-floor#1))

Table 16. Examples of gradual generation (1) (partial sentences are included)

Brand = *Nissan Cube* *(Cube is a car brand name, and Chibi-Maruko is a Japanese animeation character*	
Rhetorical Strategy	**Program Code**
1) R3: Rhetorical Strategy of Core Usage Method	(drive#1 (agent adult#10) (counter-agent nil) (instrument nil) (object cube#1) (location road#1) (goal nil) (from nil) (time night#1) (narration nil) (caption nil)) "An adult man drives a Cube in a road at night."
2) R4: Rhetorical Strategy of Deviated Use or Defamiliarization Use	An adult man eats a Cube in a road at night.
3) R7: Rhetorical Strategy of Defamiliarization of a Character's State	Chibi-Maruko eats a Cube in a road at night.
4) R9: Rhetorical Strategy of Defamiliarization of a Product's State	Chibi-Maruko eats a personifying Cube in a road at night.
5) R12) Rhetorical Strategy of the Defamiliarization of a Background	Chibi-Maruko eats a personifying Cube in a living room at night.

Table 17. Examples of gradual generation (2)

Brand = *Merit* *(Merit is a shampoo brand name, and Chibi-Maruko is a Japanese animeation character)*	
1) R3: Rhetorical Strategy of Core Usage Method	(wash#1 (agent adult#8) (counter-agent nil) (instrument merit#1) (object head#1) (location bathroom#1) (goal nil) (from nil) (time night#1) (narration nil) (caption nil))
2) **i) R7) Rhetorical Strategy of Defamiliarization of a Character's State** **ii) R8) Rhetorical Strategy of a Character's Defamiliarization of the Product Usage Action** **iii) R9: Rhetorical Strategy of Defamiliarization of a Product's State** **iv) R12) Rhetorical Strategy of the Defamiliarization of a Background**	(wash#1 (agent deer#1) (counter-agent nil) (instrument merit#1) (object head#1) (location living-room#1) (goal nil) (from nil) (time night#1) (narration nil) (caption nil)) (merit#1 (type instance) (is-a shampoo) (like beer) (state nil)) (deert#1 (type instance) (is-a animal) (hand nil) (clothes nil) (narrator nil) (have nil) (state dance))

Examples of Advertisement Rhetoric

This section introduces actual examples of the R1 to R12 rhetorical strategies and mixed patterns of them.

Regular or Standard Rhetorical Strategies [R1, R2, R3]

R1 in Table 5 shows the scene of "filtaration of work" in beer manufacturing process. In the scene of R2 in Table 5, a shopper is checking the price of the product, a beer, on the machine. R3 in Table 5 is a scene where a business man is using the product in the typical way, electric razor.

Irregular or Defamiliarization Rhetorical Strategy (Action) [R4, R5]

In the scene of R4 in Table 6, a man puts a personal computer on the ground like the goal of rugby football. This is extraordinary, abnormal usage, extremely different from ordinary usage and, semantically, considered appealing based on the attribute of "robust and durable." An attribute differs from a function and means the merit and characteristics in a product aiming to appeal to someone. Thus, intentionally emphasizing and representing an attribute of a product to make it appealing is a method of a deviated and defamiliarizing rhetorical strategy type. On the other hand, R5 in Table 6 represents a man who is only holding the product, black oolong tea, without drinking it.

Irregular or Defamiliarization Rhetorical Strategy (Person) [R6, R7, R8)

The woman in R6 in Table 7 is directly showing the appealing effects of the product, *Yanroncha*, to the audience outside the story. Next, in R7 in Table 7, two extremely small elephants are the users of the product, an electronic dictionary. Specifically, we can interpret this as follows: the characters in the advertisement story are changing to extraordinary states. However, another interpretation can be considered, namely, that the electric dictionary was extremely large. Finally, in R8 in Table 7, a human character, who has a plastic bottle of the product, Coca-Cola, is rotating his body out of doors. This example can be interpreted as follows: the character simply has the product, namely R5, Rhetorical Strategy of None Using, is at play, or we have a mixture of rhetorical strategies R8 and R5.

Irregular or Defamiliarization Rhetorical Strategy (Product or Object) [R9, R10, R11]

First, in R9 in Table 8, an enormous plastic bottle of the product, a refreshing drink, is thrust into a waterfall. Next, in R10 in Table 8, after a human character uses the product, a toilet cleaner, a figure simulating the image of the smell of the water in the toilet is the object for the product. In this case, although the toilet bowl and the water, the usage objects for the product do not change, the related object does. Finally, the product, a beer, in R11 in Table 8, is independently standing on a hill or it has been put there. If we regard a human character as the subject, this scene may be interpreted as follows: the character put the canned beer on the hill. The author interpreted this scene as "the product, as the subject, exists at a place."

Irregular or Defamiliarization Rhetorical Strategy (Location) [R12]

In R12 in Table 9, the product, an automatic washer, is standing on the grass field in a residential area. Because automatic washers are not normally placed at such a place, this shot is classified as using the strategy of defamiliarization of a background.

Mixed Example: Two Rhetorical Types

Next, the author introduces mixed rhetorical strategy examples, where one or more rhetorical strategy types are added to R3, the core usage method.

First, (1) in Table 10 shows a scene employing R7, the Strategy of Defamiliarization of a Character's State: an abnormal character (user), is taking a picture in the ordinary way. On the other hand, (2) in Table 10, although the product, a cleaner, is used normally, personified specs of dust appear from the carpet, which is the cleaner's usage object, and absorbed by the cleaner. This example is interpreted as R10, the rhetorical strategy of defamiliarization of the state of an object related to the product.

Mixed Example: Three Strategy Types

In Table 11, in (1), although the product, a toilet cleaner, is used to clean a toilet bowl, the character using the cleaner is the toilet bowl itself; this is an application of R7, the rhetorical strategy of defamiliarization of a character's state. Further, the background is extremely simple, corresponding to R12, the strategy of defamiliarization of a background. In the shot of (2) in Table 11, ordinary persons are eating the product, a snack, although they are a little theatrical. However, as a cat is also eating the snack, this is interpreted as R7, the Rhetorical Strategy of Defamiliarization of a Character's State. Additionally, the box of the snack is standing on the table (R11, the Rhetorical Strategy of the Product as a Subject).

Mixed Example: Four Rhetorical Strategy Types

Although the foreground in (1) in Table 12 is a figure in which two animation characters are using the product, a milky lotion for skin care (R7, the Rhetorical Strategy of Defamiliarization of a Character's State) and the background is also a figure of a musical image (R12, the Rhetorical Strategy of Defamiliarization of a Background). Furthermore, the characters are singing a song (R8, the Rhetorical Strategy of a Character's Defamiliarization of the Product Usage Action). On the other hand, a character in (2) of Table 12 is washing the hands using the product, a soap solution. At the same time, a part of the soap solution is changing to a cloud (R9, the Rhetorical Strategy of Defamiliarization of a Product's State) and three small animation characters are riding on the cloud (R7, the Rhetorical Strategy of Defamiliarization of a Character's State).

Mixed Examples: None Using R3 (Core Usage Method)

The following two examples use several rhetorical strategy types without the core usage method.

First, although the product in (1) of Table 13 is a water repellent for cars, and so on, this commercial is extremely symbolic. In particular, the water repellent is flying in the air (R11, the Rhetorical Strategy of the Product as a Subject) by changing into a missile (R9, the Rhetorical Strategy of Defamiliarization of a Product's State). Moreover, a character is riding on the missile of water repellent (R4, the Rhetorical Strategy of Deviated Use or Defamiliarization Use). On the other hand, in (2) of Table 13, the product, a canned drink, is placed (R11, the Rhetorical Strategy of the Product as a Subject) on the lane of a bowling alley (R12, the Rhetorical Strategy of the Defamiliarization of a Background). Furthermore, the canned drink is used as a bowling pin (R4, the Rhetorical Strategy of Deviated Use or Defamiliarization Use) and the bowling ball, the object related to the product as a pin, is destroyed and transformed by the collision into the product, the bowling pin.

Knowledge Contents Appearing in TV Commercials

Although the purpose of the above advertisement analysis is to implement the rhetorical strategy types and their combination as a mechanism of advertisement generation, it requires using the knowledge content. As stated above, the main strategic techniques can be defined by both the regular and standard usage of the elements in an event and also their irregular and deviated usage. The regular or standard rhetorical strategy types and irregular or deviated types can be defined operationally from the perspective of the usage method of content knowledge. This section shows the content knowledge collected from the sample commercials. The following shows the content knowledge of products, characters, and backgrounds.

Products

The author classified the foods appearing in the sample products under the sub-categories of "food" (Figure 4), "drink" (Figure 5), "industrial product" (Figure 6), and "daily commodities" (Figure 7).

Characters

Characters are divided into the category of normal, realistically ordinary humans in the sense of marketing targets and diverse characters in the broader sense, including animals, who ordinarily do not use the products, and more strange and abnormal creatures. Table 14 describes the characters in all the analyzed shots under the above two broad categories.

Places (Stages, Backgrounds)

Although the main stage in television commercials is "the space of daily life," other "vehicle" and "natural world" can also be the stages. In addition to the spatial element, the temporal element, such as seasons and the temporal division of a day, also becomes a part of the narrative stages. However, it was difficult to check the temporal division in the stories and scenes of the analyzed clips more than the spatial characteristics, although there are commercials in which the temporal element, such as the season, is an important point of appeal. Figure 8 shows the spaces acquired by the sampled commercials and the shots of the Story-type Brand Introduction Strategies.

FROM ANALYSIS TO GENERATION

Many academic advertisement investigations have the problem of reception of advertisement works, such as advertising effects; however, few studies have dealt with advertising generation, production, and creation. Notably, advertising semiotic investigations such as Fukuda (1990) and Zaltman (2003) have been presented on advertisement studies based on organic analyses of advertising structures. The support of advertisement creation processes was proposed by Honiden, Yoshioka, & Yuri (2006). Further, Kawamura (2019) proposed oriented modeling of an advertisement production process for production support.

Although rhetoric is generally a conscious technique for relaying effects and surprises to the receivers in sentences that emphasize the value of language as in literature, the author uses rhetoric by extending this concept to image narratives and the consideration that the entire product introduction rhetoric is performed based on techniques that rely on a specific strategy via advertising narratives is not necessarily included in a figurative representation genre. Rhetoric is related to the degree from relatively simple to complex.

Ogata (2007) presented the idea of "deconstruction and reconstruction" as a basic concept for the author's narrative generation system study. This refers to an idea that uses existing narrative texts or narratology and literary theories through deconstruction to create new possibilities of narratives based on the reconstruction of the elements. It is also associated with intertextuality in existing narrative parts (Nakashima & Ogata, 2006; Ogata & Nakashima, 2006). This research on advertisement generation is related to the framework of deconstruction and reconstruction in that, using existing television advertisement examples, the aim is to realize simulation generation possibilities in the analysis range and the combinatorial generation possibilities beyond the aforementioned method. A similar approach was shown by Hori (2007) in the context of a creativity aid using AI instead of narrative generation.

Next, the author classified the rhetoric and implemented a system for the rhetoric and their combinations. In particular, this system selects a product or brand as the input information and generates the conceptual representation of a single event as the output information. In an example of the name "ABC," of a car, the system the following conceptual representation, (運転する#1[drive#1] (agent 成人#10[man#10]) (counter-agent nil) (instrument nil) (object ABC#1) (location 道路#1[road#1]) (goal nil) (from nil) (time 夜#1[night#1]) (narration nil) (caption nil)). The meaning is "A man drives ABC on the road at night."

The processing's flow is as follows:

1. The user selects a product or brand.
2. The user selects a type of rhetoric in three types of standard rhetorical types and the system generates a single event according to the selected type of rhetoric. The system can finish the processing here.
3. Continuously, the user selects a type of defamiliarized rhetoric and transforms the aforementioned standard type into another form. This processing can then be repeated.

In particular, the system initially selects an event based on a standard rhetorical type and transforms the event into the defamiliarized event.

In the standard rhetoric, a single event is generated. This event is standard in all the events such as actions and places related to a product or brand. The "standard" means the ordinarily used methods for a product or brand and the places where the product or brand can be used. The author explains the case in which the user selects the product of "Merit," a brand of shampoo, and the rhetoric of the core usage method in the product introduction.

The information for the product, "Merit," selected by the user is stored in the object knowledge base and the standard range of relating elements is described as the properties of the upper class of the level in which "Merit" is included. The upper class, in this case, is "shampoo" and the core usage method (washing hair), place (bathroom), character (nil), time (nil) are described in the class. According to this information, the system hierarchically searches and acquires the corresponding information for each category. Further, in the case of "nil," the system acquires all of the values in the lowest nodes in the corresponding knowledge bases and then selects an element from all of the acquired categories at random to synthesize a single event. For instance, the system generates a conceptual representation that means, "a woman washes her hair in the bathroom at night," (洗う#1[wash#1] (agent 成人女性#2[woman#2]) (counter-agent nil) (instrument メリット#1[Merit#1]) (object 髪#1[hair#1]) (location 風呂場#1[bathroom#1]) (goal nil) (from nil) (time 夜#1[night#1]) (narration nil) (caption nil)).

Regarding the definition of core usage methods, the author collects the corresponded actions for all product categories of approximately 90 types from the shots in the samples of the rhetorical types and temporarily arranges them. Each core usage method is a sequence of actions or a kind of script. The acquired scripts of the core usage methods were totally 251.

The core usage methods are stored in the action knowledge base shown in Figure 9. Different script definitions are provided for the same action in different product or brand categories.

Figure 10, 11, and 12 show the object, place, and character knowledge bases respectively.

As shown in this example, the defamiliarization rhetoric is basically conducted by exploring elements such as actions, characters, objects (products), and places (backgrounds) outside the standard ranges using knowledge bases. We could utilize another method for defamiliarization usage rhetoric that uses the core usage method of a product for another object (product) that has the same form as the former product.

Table 15 shows generation examples of the defamiliarized rhetoric.

Furthermore, the gradual or repeated application of rhetorical techniques facilitates the generation of complicated rhetoric as shown by the examples in Table 16 and 17.

In this chapter, the author defined twelve types of products or brand introduction rhetoric by analyzing TV commercials and developed an experimental system that simulates the analysis results. This experimental system simulates the techniques using a knowledge base constructed using the analyzed commercials. However, this system has a mechanism that generates a simple event.

As the author described earlier on, this study does directly apply to TV commercials. Instead, the objective is to create a new form of advertisement with automatic generation function beyond existing advertisement forms. Further, although advertising narratives are pragmatic and the production and distribution processes are organized in many cases, the numerous narrative contents are mythic or folkloric. Advertising narratives are not simply pragmatic narratives for the consumer. They have implications and realistic potential applications to rhetorical techniques and can assist in the generation of strategies for general narrative.

FUTURE RESEARCH DIRECTIONS

This section focuses on future technological problems and perspectives from a broad perspective.

1. **Revision and expansion as basic rhetorical mechanisms**: The author has conducted several attempts of program development of advertising rhetorical methods in this chapter using narrative generation systems, including INGS. The author will synthetically revise and expand the previous attempts.
2. **Cooperation with current advertising scenario generation systems**: The author has proceeded with an integrated study of advertising analysis and advertising scenario generation. The defamiliarization rhetoric in this chapter will be particularly applicable and useful in automatic advertising scenario generation of advertising scenarios with rhetorical features.

3. **Expansion and utilization as the rhetoric system for general narrative generation**: Advertising rhetorical methods analyzed in this chapter are not necessarily unique to advertising. Therefore, the author aims to expand to apply and use the results as methods of general narrative generation. In particular, they can be integrated in the narrative discourse mechanism in INGS.
4. **Beyond advertisement (fusion with general narrative generation mechanisms)**: Although advertising narratives are very pragmatic and the production and distribution processes are also highly organized in many cases, the described narrative contents in the advertising works have mythic and folkloric characteristics. As this chapter showed, advertising narratives include diverse defamiliarized, deviated, and non-realistic stories and descriptions. However, people can naturally accept strange worlds and stories, in which characters have many abnormal actions and natural objects are transformed and metamorphosed, because they are merely advertisement narratives. Similarly, we accept strange, abnormal, and extraordinary narratives in the framework of dreams and folklore. In this sense, advertising narratives are a kind of ideal or theoretical narratives or narrative generation. In particular, in advertising narrative generation, each narrative and story is generated to transcend the limitation of the naturalness of events. In particular, like dream narratives, advertising narratives intend to complete advertising goals instead of representing reality. In this point, advertising narratives are not merely pragmatic narratives. They will give important indications and actual applicable possibilities for general narrative generation and narrative rhetorical strategies.

CONCLUSION

This chapter focused on an advertising analysis in previous narrative analyses of the author. The second sub-section of **BACKGROUND** showed that the advertising analysis is based on the idea of "advertisement as a narrative." Dependent on the basic idea, the objective of this chapter was to analyze how a brand, i.e., a product or a service, was introduced into the events in story-like television commercial films using real commercials. **APPROACHES TO ADVERTISEMENT GENERATION AS NARRATIVE GENERATION** introduced the previous fundamental attempts for an advertising narrative generation system study that was planned as an application of narrative generation study and has been continued in advertising analyses and the design and implementation of experimental systems from two viewpoints of the macro-framework and the methods introducing brands into advertising works. Thus, the advertisement analyses have been conducted for the introduction of advertising narrative generation systems. These systems are positioned as the application systems of the Integrated Narrative Generation System (INGS) in Chapter 1 and a type of content generation mechanisms in the *Geinō* Information System (GIS) in Chapter 2 (the first sub-section of **BACKGROUND** comprehensively showed the author's previous narrative analyses for diverse narrative genre for the development of application systems).

Next, the central part of this chapter, **RHETORICAL STRATEGIES FOR ACTIONS AND OBJECTS: A DETAILED ANALYSIS OF TELEVISION ADVERTISEMENT VIDEOS AND THE EXTRACTION AND CLASSIFICATION OF STORY-TYPE BRAND INTRODUCTION STRATEGIES** described in detail the rhetorical analysis of television commercials dependent on three regular or ordinary rhetoric types and nine irregular or extraordinary rhetoric types. In particular, the author presented the following categorization of the rhetorical strategies:

- **R1: Rhetorical strategy of manufacturing Process**
- **R2: Rhetorical strategy of purchasing process**
- **R3: Rhetorical strategy of core usage method**
- **R4: Rhetorical strategy of deviated use or de-familiarization use**
- **R5: Rhetorical strategy of no usage**
- **R6: Rhetorical strategy of a character as a narrator**
- **R7: Rhetorical strategy of de-familiarization of a character's state**
- **R8: Rhetorical strategy of a character's de-familiarization of the product usage action**
- **R9: Rhetorical strategy of de-familiarization of a product's state**
- **R10: Rhetorical strategy of de-familiarization of the state of an object related to the product**
- **R11: Rhetorical strategy of the product as a subject**
- **R12: Rhetorical strategy of the de-familiarization of a background**

The author explained all of the rhetorical types and the relations using actual examples. Moreover, in the last half of this section, the author systematically showed knowledge bases acquired through the advertising narrative analysis, including characters, places, and objects. This basic idea is that diverse advertisement narratives are generated based on the gradual complication from a simple ordinary event through the application of rhetorical techniques.

Furthermore, as this study has been applied to the development of advertising narrative generation techniques, the last part in this chapter, **FROM ANALYSIS TO GENERATION**, mentioned the systematization's topic. Although this section stated examples that generate fragmental and short advertising events using the acquired knowledge and techniques, the author currently conducts newer advertising analysis and system development (Ono, Sasaki, & Ogata, 2019a, 2019b). However, the proposed study in this chapter is utilized in the new development in the viewpoint of precise modeling of advertising events to be generated. From a different perspective, this advertising narrative generation study considers deconstruction and reconstruction in the point that it aims to develop the combinatorial generation possibilities beyond the simulative generation possibilities based on existing texts. In this sense, it is similar to the method of intertextual narrative generation that is a foundation of the author's narrative generation idea. Intertextual narrative generation is described in Chapter 1 (Ogata, 2019c) in Ogata (2019b).

ACKNOWLEDGMENT

This chapter's research was supported by JSPS KAKENHI Grant Number18K18509, Kayamori Foundation of Information Science Advancement Research Grant, The Telecommunication Advancement Foundation Research Grant, and AOI TYO Holdings Inc.

REFERENCES

Abe, H., Ogata, T., & Hanada, K. (2008). Televi kōkoku no shot ni okeru brand dōnyū rhetoric no bunseki—Kōkoku no monogatari seisei system no ichi kikō no tameni [An analysis of brand introduction rhetoric in a shot in television commercials: For a mechanism of an advertisement narrative generation system]. In *Proceedings of the 16th Meeting of Literature, Cognition, and Computer Research Group II* (16G-05). Tokyo, Japan: Japanese Cognitive Science Society.

Akaishi, M. (2006). A dynamic decomposition/recomposition framework for documents based on narrative structure model. *Transactions of the Japanese Society for Artificial Intelligence*, *21*(5), 428–438. doi:10.1527/tjsai.21.428

Eisenstein, S. (Director) (1925). *Battleship Potemkin* [Motion picture]. Москвá, Russian Federation: Mos Film.

Endo, Y., & Ogata, T. (2003a). Hyper-comic system as consideration of rhetoric. In *Proceedings of the 4th International Conference on Cognitive Science* (pp. 111–116). Oakbrook Terrace, IL: Cognitive Science Society.

Endo, Y., & Ogata, T. (2003b). Narratology of manga and application to hyper-comic. In *Proceedings of the 17th Annual Conference of the Japanese Society for Artificial Intelligence* (2G2-03). Tokyo, Japan: The Japanese Society for Artificial Intelligence.

Endo, Y., & Ogata, T. (2004a). A rhetorical analysis of a Japanese comic for hyper-comic system. In *Proceedings of the 18th Congress of the International Association of Empirical Aesthetics* (pp. 502–508). Rome, Italy: University of Rome Tre.

Endo, Y., & Ogata, T. (2004b). Multilayered discourse in hyper-comic. In *Proceedings of the 9th International Symposium on Artificial Life and Robotics*, Vol. 1 (pp. 49–52). Oita, Japan: International Society of Artificial Life and Robotics.

Fukuda, T. (1990). *Monogatari marketing* [Story marketing]. Tokyo, Japan: Takeuchishoten-shinsha.

Genette, G. (1972). *Discours du récit, essai de méthode, figures III.* Paris: Seuil.

Honiden, S., Yoshioka, N., & Yuri, S. (2006). *Kangaeru contents "smartive"* [Thinking content "smartive"]. Tokyo, Japan: Maruzen.

Hori, K. (2007). *Sōzō katsudō shien system no riron to ōyō* [Theories and applications of creative activity support systems]. Tokyo, Japan: Ohmsha.

Ishii, K., & Ogata, T. (1998). A study on the methodology of hypertext novels based on a prototyping. In *Proceedings of the 12th Annual Conference of the Japanese Society for Artificial Intelligence* (pp.701–702). Tokyo, Japan: The Japanese Society for Artificial Intelligence.

Ito, T., Sasaki, A., & Ogata, T. (2018). Kōkoku ni kansuru Creative Genome wo mochiita CM concept *haiku* no seisei [The generation of CM concept *haikus* using the Creative Genome regarding advertisements]. In *Proceedings of the 35th Annual Conference of the Japan Cognitive Science Society* (pp. 516–517). Tokyo, Japan: Japan Cognitive Science Society.

Kamada, M., & Ogata, T. (2013a). Analysis and imitation of the Japanese letter notation in narrative texts. In *Proceeding of the 27th Annual Conference of the Japanese Society for Artificial Intelligence* (2I4-5in). Tokyo, Japan: The Japanese Society for Artificial Intelligence.

Kamada, M., & Ogata, T. (2013b). Monogatari bunshō ni okeru moji hyōki—Monogatai seisei system ni okeru moji hyōki chōsetsu kikō ni mukete [Character notation in narrative sentences: Toward a character notation adjustment mechanism in narrative generation systems]. In *Proceedings of the 26th Special Interest Group on Language Sense Engineering Research Group in the Japanese Society for Artificial Intelligence* (p. 29). Tokyo, Japan: The Japanese Society for Artificial Intelligence.

Kamada, M., & Ogata, T. (2013c). Monogatari seisei system ni okeru bun seisei kikō to moji hyōki kikō [Sentence generation and character representation mechanisms in a narrative generation system]. In *Proceedings of the 19th Annual Conference of the Association of Natural Language Processing* (pp. 698–701). Tokyo, Japan: Association of Natural Language Processing.

Kanai, A. (2013). Eizō kankyō no monogatari to setsudan ni yoru kihan riron—"Wakarinikusa" no ninchi wo megutte [Narratives in image environments and a norm theory by cutting: Concerning the cognition of "uncertainness"]. In A. Kanai, S. Dobashi, & S. Tsuda (Eds.), *Media kankyō no monogatari to kōkyōken* [Narratives in media environments and public spheres] (pp. 193–229). Tokyo, Japan: Hosei Daigaku Shuppankyoku.

Kanai, A., & Niwa, Y. (2008). *Eizō henshū no riron to jissen* [Theories and practices of image editing]. Tokyo, Japan: Hosei Daigaku Shuppankyoku.

Kanai, A., Ogata, T., & Shinohara, K. (2002). Film rhetoric generation system for intensive cognitive system. In *Proceedings of IEEE International Conference on Systems, Man and Cybernetics (CD-ROM)*. The Institute of Electrical and Electronics Engineers.

Kanai, A., Ogata, T., & Shinohara, K. (2003). Identical and different elements in film rhetoric generation. *Transactions of the Japanese Society for Artificial Intelligence, 18*(2G), 114–121. doi:10.1527/tjsai.18.114

Kawamura, Y. (2019). An attempt of the commercial film production support system based on the image rhetoric of commercial film. In T. Ogata & T. Akimoto (Eds.), *Post-narratology through computational and cognitive approaches* (pp. 292–317). Hershey, PA: IGI Global. doi:10.4018/978-1-5225-7979-3.ch007

Kayamori, O., & Ogata, T. (2003). Analysis and scenario generation of TVCFs based on the relations between story and brand. In *Proceedings of the 17th Annual Conference of the Japanese Society for Artificial Intelligence* (2G2-06). Tokyo, Japan: The Japanese Society for Artificial Intelligence.

Koreeda, H. (Director) (2018). *Manbiki kazoku* [Shoplifters] [Motion picture]. Tokyo, Japan: AOI Pro.

Mishima, Y. (1963). Gogo no eikō. Tokyo, Japan: Kōdansha.

Miyoshi, T. (1993). Mienai fukushū [Revenge invisible]. In *Mienai fukushū* [Revenge invisible]. Tokyo, Japan: Kōdansha.

Mori, Y., & Ogata, T. (2005). Hyper-comic no kōsō to hyper-link oyobi koma gōsei no jidōka [A hyper-comic and the automatization of its hyper-linking and frame synthsizing]. In *Proceedings of the 19th Special Interest Group on Language Sense Processing Engineering* (pp. 1–11). Tokyo, Japan: The Japanese Society for Artificial Intelligence.

Nakashima, M., & Ogata, T. (2006). Narrative generation system and intertextuality: A consideration on the concept and prototype systems. In *Proceedings of the 20th Annual Conference of Japanese Society for Artificial Intelligence* (2E2-2). Tokyo, Japan: The Japanese Society for Artificial Intelligence.

NHK Science & Technology Research Laboratories. (2017). *TVML, TV program making language—TVML toha nanika?* [TVML, TV program making language: What's TVML?]. Retrieved October 5, 2017, 11:00, from http://www.nhk.or.jp/strl/tvml/)

Ogata, T. (2007). To the rhetoric of story from Propp: Centering around the rhetoric of de-composition and re-composition. *Cognitive Studies, 14*(4), 532–558.

Ogata, T. (2016). Kabuki as multiple narrative structures. In T. Ogata & T. Akimoto (Eds.), *Computational and cognitive approaches to narratology* (pp. 400–431). Hershey, PA: IGI Global. doi:10.4018/978-1-5225-0432-0.ch016

Ogata, T. (2018a). An integrated approach to narrative generation: From Mishima and kabuki to narrative generation systems. In T. Ogata & S. Asakawa (Eds.), *Content generation through narrative communication and simulation* (pp. 49–147). Hershey, PA: IGI Global. doi:10.4018/978-1-5225-4775-4.ch002

Ogata, T. (2018b). Kabuki ni mukete (1)—Shiisei to henshūsei no monogatari kara tajūsei to jitsuzon no monogatari he [Toward kabuki (1): From narratives of arbitrariness and editing to narratives of multiplicity and real existence]. In T. Ogata, Y. Kawamura, & A. Kanai (Eds.), *Jōhō monogatariron—Jinkōchinō, ninchi, shakai katei to monogatari seisei* [Informational narratology: Artificial intelligence/cognition/social process and narrative generation] (pp. 187–208). Tokyo, Japan: Hakutō Shobō.

Ogata, T. (2018c). Kabuki ni mukete (2)—Tajū monogatari kōzō no shosō [Toward kabuki (2): Aspects of multiple narrative structures]. In T. Ogata, Y. Kawamura, & A. Kanai (Eds.), *Jōhō monogatariron—Jinkōchinō, ninchi, shakai katei to monogatari seisei* [Informational narratology: Artificial intelligence/cognition/social process and narrative generation] (pp. 209–244). Tokyo, Japan: Hakutō Shobō.

Ogata, T. (2019a). Kabuki as multiple narrative structures and narrative generation. In T. Ogata & T. Akimoto (Eds.), *Post-narratology through computational and cognitive approaches* (pp. 192–275). Hershey, PA: IGI Global. doi:10.4018/978-1-5225-7979-3.ch005

Ogata, T. (2019b). *Toward an integrated approach to narrative generation: Emerging research and opportunities*. Hershey, PA: IGI Global.

Ogata, T. (2019c). What are narrative generation phenomena? In T. Ogata (Ed.), *Toward an integrated approach to narrative generation: Emerging research and opporunities* (pp. 1–58). Hershey, PA: IGI Global.

Ogata, T. (2019d). Areas of narratives or narrative genres. In T. Ogata (Ed.), *Toward an integrated approach to narrative generation: Emerging research and opportunities* (pp. 59–161). Hershey, PA: IGI Global.

Ogata, T. (2019e). Theoretical or philosophical considerations for an integrated narrative generation approach. In T. Ogata (Ed.), *Toward an integrated approach to narrative generation: Emerging research and opportunities* (pp. 315–403). Hershey, PA: IGI Global.

Ogata, T., & Endo, Y. (2004). Narrative discourse theory and a hyper-comic system. In *Proceeding of the 18th Annual Conference of the Japanese Society for Artificial Intelligence* (2D2-01). Tokyo, Japan: The Japanese Society for Artificial Intelligence.

Ogata, T., Hori, K., & Ohsuga, S. (1994). Towards narrative text generation based on narrative techniques and strategies. In *Proceedings of the International Federation for Information and Documentation* (pp. 296–300). Academic Press.

Ogata, T., Hori, K., & Ohsuga, S. (1995). A basic framework of narrative generation system as creative interface. In Y. Anzai, K. Ogata, & H. Mori (Eds.), *Symbiosis of human and artifact (20A)* (pp. 679–684). Amsterdam, The Netherlands: Elsevier. doi:10.1016/S0921-2647(06)80107-3

Ogata, T., Hori, K., & Ohsuga, S. (1996a). A basic framework for narrative conceptual structure generation based on narrative techniques and strategies. *Jinkō Chinō Gakkaishi*, *11*(1), 148–159.

Ogata, T., Hori, K., & Ohsuga, S. (1996b). An analysis of narrative structure and generation process for narrative generation system. *Cognitive Studies*, *3*(1), 72–109.

Ogata, T., & Kanai, A. (2010a). Story to shōhin = brand—Seisei ni muketa kōkoku no bunseki [Stories and products = brands: An analysis of advertisements toward narrative generation]. In T. Ogata & A. Kanai (Eds.), *Monogatariron no jōhōgaku josetsu—Monogatari seisei no shis? to gijutsu wo megutte* [An introduction to informatics of narratology: Around the thoughts and technologies of narrative generation] (pp. 42–115). Tokyo, Japan: Gakubunsha.

Ogata, T., & Kanai, A. (2010b). Monogatari seisei to shite no kōkōku seisei no tame ni—Danpenteki kōsatsu [For advertisement generation as narrative generation: A fragmental consideration]. In T. Ogata & A. Kanai (Eds.), *Monogatariron no jōhōgaku josetsu—Monogatari seisei no shisō to gijutsu wo megutte* [An introduction to informatics of narratology: Around the thoughts and technologies of narrative generation] (pp. 384–447). Tokyo, Japan: Gakubunsha.

Ogata, T., & Kayamori, O. (2003). Automatic TV-CF generation based on the analysis of the relation between story and brand. In *The 4th International Conference on Cognitive Science* (p. 143). Oakbrook Terrace, IL: Cognitive Science Society.

Ogata, T., & Mori, Y. (2005). A systematic view of narrative discourse based on "hyper-comic" system with automatic generation mechanism. In *Procceding of the 19th Annual Conference of the Japanese Society for Artificial Intelligence* (3D3-01). Tokyo, Japan: The Japanese Society for Artificial Intelligence.

Ogata, T., & Nakashima, M. (2006). Monogatari seisei system to kan-text-sei—Taikeiteki kenkyū ni muketa kokoromi [Narrative generation systems and inter-textuality: An attempt to a systematic study]. In *Proceedings of the 23rd Annual Conference of the Cognitive Science Society* (pp. 270–271). Tokyo, Japan: Japanese Cognitive Science Society.

Ogata, T., Tachibana, S., & Tomite, S. (2009). Animated movie generation from narrative conceptual representation and the automatic camerawork—Analysis and simulation of "Tokyo story." In *Proceedings of the 26th Annual Meeting of the Japanese Cognitive Science Society* (P2-32). Tokyo, Japan: Japanese Cognitive Science Society.

Ogata, T., & Watanabe, K. (1994). Jinkō chinō ni yoru kōkoku sakusei shien system no tame no monogatari-gata CF no kōzō bunseki to kōkoku no monogatari seisei katei no kentō [Structural analysis of narrative commercial films for an advertising making support system by artificial intelligence and the consideration of advertising narrative generation processes]. In *Proceeding of the Japan Academy of Advertising* (pp. 101–104). Tokyo, Japan: The Japan Academy of Advertising.

Ogata, T., & Watanabe, K. (1995). Jinkō chinō ni yoru kōkoku sakusei shien system no tame no monogatari-gata CF no kōzō bunseki to kōkoku no monogatari seisei katei no kentō [Structural analysis of narrative commercial films for an advertising making support system by artificial intelligence and the consideration of advertising narrative generation processes]. *Journal of Advertising Science*, *31*, 153–157.

Ogata, T., Watanabe, K., Hori, K., & Ohsuga, S. (1993). Life style marketing he no monogatari seisei system no ōyō [Application of narrative generation systems to life style marketing]. In *Proceedings of the Joint Conference of System Simposium and Intelligent System Simposium of the Society of Instrument and Control Engineers* (pp. 217–224). Tokyo, Japan: The Society of Instrument and Control Engineers.

Ogata, T., Watanabe, K., Hori, K., & Ohsuga, S. (1995a). A basic framework of the application of narrative generation system for integrated support of marketing/advertisement. *Journal of the Japan Society for Management Information*, *4*(1), 19–42.

Ogata, T., Watanabe, K., Hori, K., & Ohsuga, S. (1995b). Monogatari seisei system ni yoru kōkoku sōsaku shien wo mokuteki to shita televi comercial no kōzō bunseki [A structural analysis of television commercial films for advertising creation support by a narrative generation system]. *Journal of Advertising Science*, *30*, 1–22.

Ono, J., & Ogata, T. (2018). Acquiring short scripts and setting a case frame in each acquired script: Toward random story generation. *Journal of Robotics, Networking and Artificial Life*, *5*(3), 212–216. doi:10.2991/jrnal.2018.5.3.15

Ono, J., Ogata, T., & Ito, T. (2019). A basic study for using folk tale's motifs to narrative generation. In *Proceedings of the 33rd Annual Conference of the Japanese Society for Artificial Intelligence* (1F2-NFC-1-05). Tokyo, Japan: The Japanese Society for Artificial Intelligence.

Ono, J., Sasaki, A., Ito, T., & Ogata, T. (2018). Creative Genome no parameter ni motozuku kōkoku story seisei ni mukete [Toward advertising story generation based on parameters of Creative Genome]. In *Proceedings of the 58th Special Interest Group on Language Sense Processing Engineering* (pp. 35–47). Tokyo, Japan: The Japanese Society for Artificial Intelligence.

Ono, J., Sasaki, A., & Ogata, T. (2018). Kōkoku no Creative Genome parameter wo riyō shita shinsō gakusyū kikō no kokoromi—Kōkoku seisei to monogatari seisei no ketsugō ni mukete [Attempt of a deep learning mechanism using the advieriting Creative Genome parameters: Toward the connection between advertisement generation and narrative generation]. In *Proceedings of the 60th Special Interest Group on Language Sense Processing Engineering* (p. 43). Tokyo, Japan: The Japanese Society for Artificial Intelligence.

Ono, J., Sasaki, A., & Ogata, T. (2019a). Kōkoku no Creative Genome parameter wo riyō shita shinsō gakusyū kikō no kokoromi—Kōkoku seisei to monogatari seisei no ketugō ni mukete [A Consideration of a Method for Narrative Generation Based on the Analysis of CMs: Toward the Organic Combination between an Integrated Narrative Generation System and "Creative Genome"]. In *Proceedings of the 61st Special Interest Group on Language Sense Processing Engineering* (pp. 39–46). Tokyo, Japan: The Japanese Society for Artificial Intelligence.

Ono, J., Sasaki, A., & Ogata, T. (2019b). Tōgō monogatari seisei system to "Creative genome" wo riyō shita kōkoku plot seisei syuhō no ichian [An idea of the advertising plot generation method using an integrated narrative generation system and "Creative genome"]. In *Proceedings of the 33rd Annual Conference of the Japanese Society for Artificial Intelligence* (1F2-NFC-1-04). Tokyo, Japan: The Japanese Society for Artificial Intelligence.

Ono, J., Zhang, Y., & Ogata, T. (2012). Defamiliarization rhetoric and scenario generation for events using the constraints in conceptual system. In *Proceedings of the 26th Annual Conference of Japanese Society for Artificial Intelligence* (1N1-OS-1a-1). Tokyo, Japan: The Japanese Society for Artificial Intelligence.

Otogi Zōshi. (1958). In T. Ichiko (Ed.), Nihon koten bungaku taikei, 38 [Japanese classic literature collection, Vol. 38]. Tokyo, Japan: Iwanami Shoten. (Original work published the Muromachi era)

Ozu, Y. (Director), & Yamamoto, T. (Producer) (1953). *Tokyo monogatari* [Tokyo story] [Motion picture]. Japan: Shōchiku.

Parajanov, S. (Director) (1968). *The color of pomegranates* [Motion picture]. Yerevan, Armenia: Armenfilm.

Propp, V. Y. (1968). *Morphology of the folktale* (L. Scott, Trans.). Austin, TX: University of Texas Press. (Original work published 1928)

Schank, R. C., & Abelson, R. P. (1977). *Scripts, plans, goals, and understanding: An inquiry into human knowledge structures*. Lawrence Erlbaum.

Seki, K. (1957). *Mukashibanashi to shōwa* [Folktales and funny stories]. Tokyo, Japan: Iwasaki Shoten.

Seki, K. (1966). *Mukashibanashi no rekishi* [The history of the folktale]. Tokyo, Japan: Shibundō.

Seki, K. (1977). *Nihon no mukashibanashi—Hikaku kenkyū josetsu* [Japanese folktales: An introduction to the comparative study]. Tokyo, Japan: NHK Shuppan.

Seki, K. (1978a). *Nihon mukashibanashi taisei, 2—Honkaku mukashibanashi, 1* [Compilation of Japanese folktales, Vol. 2: Ordinary folktales, No. 1]. Tokyo, Japan: Kadokawa Shoten.

Seki, K. (1978b). *Nihon mukashibanashi taisei, 3—Honkaku mukashibanashi, 2* [Compilation of Japanese folktales, Vol. 3: Ordinary folktales, No. 2]. Tokyo, Japan: Kadokawa Shoten.

Seki, K. (1978c). *Nihon mukashibanashi taisei, 4—Honkaku mukashibanashi, 3* [Compilation of Japanese folktales, Vol. 4: Ordinary folktales, No. 3]. Tokyo, Japan: Kadokawa Shoten.

Seki, K. (1978d). *Nihon mukashibanashi taisei, 5—Honkaku mukashibanashi, 4* [Compilation of Japanese folktales, Vol. 5: Ordinary folktales, No. 4]. Tokyo, Japan: Kadokawa Shoten.

Seki, K. (1978e). *Nihon mukashibanashi taisei, 6—Honkaku mukashibanashi, 5* [Compilation of Japanese folktales, Vol. 6: Ordinary folktales, No. 5]. Tokyo, Japan: Kadokawa Shoten.

Seki, K. (1979a). *Nihon mukashibanashi taisei, 1—Dōbutsu monogatari* [Compilation of Japanese folktales, Vol. 1: Animal tales]. Tokyo, Japan: Kadokawa Shoten.

Seki, K. (1979b). *Nihon mukashibanashi taisei, 7—Honkaku mukashibanashi, 6* [Compilation of Japanese folktales, Vol. 7: Ordinary folktales, No. 6]. Tokyo, Japan: Kadokawa Shoten.

Seki, K. (1979c). *Nihon mukashibanashi taisei, 8—Warai banashi, 1* [Compilation of Japanese folktales, Vol. 8: Jokes and anecdotes, No. 1]. Tokyo, Japan: Kadokawa Shoten.

Seki, K. (1979d). *Nihon mukashibanashi taisei, 9—Warai-banashi (Shōwa), 2* [Compilation of Japanese folktales, Vol. 9: Jokes and anecdotes, No. 2]. Tokyo, Japan: Kadokawa Shoten.

Seki, K. (1980). *Nihon mukashibanashi taisei, 10—Warai-banashi (Shōwa), 3* [Compilation of Japanese folktales, Vol. 10: Jokes and anecdotes, No. 3]. Tokyo, Japan: Kadokawa Shoten.

Seki, K., Nomura, J., & Ōshima, H. (Eds.). (1979). *Nihon mukashibanashi taisei, 12—Kenkyū hen* [Compilation of Japanese folktales, 12: Researches]. Tokyo, Japan: Kadokawa Shoten.

Seki, K., Nomura, J., & Ōshima, H. (Eds.). (1980). *Nihon mukashibanashi taisei, 11—Shiryō hen* [Compilation of Japanese folktales, 11: Data]. Tokyo, Japan: Kadokawa Shoten.

Sekkyōshū. (1977). In Y. Muroki (Ed.), Shinchō nihon koten shūsei [Shinchō Japanese classic collection]. Tokyo, Japan: Shinchōsha. (Sekkyōbushis were narrated from the end of the medieval period to the early modern period in Japan)

Shimura, K., & Ogata, T. (2008). Story kōkōku ni okeru brand shutsugen rhetoric no bunseki to seisei [Analysis and generation of the brand appearance rhetoric in story advertisements]. In *Proceedings of the 70th National Convention of Information Processing Socirty of Japan*, Vol. 4 (pp. 813–814). Tokyo, Japan: Information Processing Society of Japan.

Tachibana, S., & Ogata, T. (2009a). Eizō satsuei ni okeru kisoku no bunseki—"Tokyo monogatari" wo sozai to shite [Extracting rules for the camerawork: Using "Tokyo story" as a material]. In *Proceedings of the 71th National Convention of Information Processing Society of Japan*, Vol. 4 (pp. 759–760). Tokyo, Japan: Information Processing Society of Japan.

Tachibana, S., & Ogata, T. (2009b). A camerawork setting system based on rules and simulation of "Ozu rules." In *Proceedings of the 23th Annual Conference of Japanese Society for Artificial Intelligence* (1J1-OS2-11). Tokyo, Japan: The Japanese Society for Artificial Intelligence.

Takahashi, R. (2003–2006). *Maison ikkoku, Vols. 1–15*. VIZ Media LLC. (Original work published 1980–1987)

Yamamura, M. (1988). Maiko satsujin jiken [A murder of an apprentice geisha]. In *Miyako odori satsujin jiken* [A murder of miyako dance] (pp. 7–53). Tokyo, Japan: Shinchōsha.

Yoshimoto, T. (1968). *Kyōdō gensō ron* [The theory of communal Illusion]. Tokyo, Japan: Kawade Shobō Shinsha.

Yoshio, T., & Ogata, T. (2007). Story to brand no kankei ni chakumoku shita kōkōku bunseki chishiki wo riyōshita kōkoku scenario jidō seisei system [An automatic advertisement scenario generation system using the advertisement analysis knowledge focused on the relationships between a story and a brand]. In *Proceedings of the 69th National Convention of Information Processing Society of Japan*, Vol. 4 (pp. 507–508). Tokyo, Japan: Information Processing Society of Japan.

Zaltman, G. (2003). *How customers think: Essential insights into the mind of the market*. Harvard Business School Press.

Zhang, Y., Ono, J., & Ogata, T. (2011). An advertising rhetorical mechanism for single event combined with conceptual dictionary in narrative generation system. In *Proceedings of 7th International Conference on Natural Language Processing and Knowledge Engineering* (pp. 340–343). The Institute of Electrical and Electronics Engineers. 10.1109/NLPKE.2011.6138221

Zhang, Y., Ono, J., & Ogata, T. (2012). Single event and scenario generation based on advertising rhetorical techniques using the conceptual dictionary in narrative generation system. In *Proceedings of the 4th IEEE International Conference on Digital Game and Intelligent Toy Enhanced Learning* (pp. 162–164). The Institute of Electrical and Electronics Engineers. 10.1109/DIGITEL.2012.46

KEY TERMS AND DEFINITIONS

Advertisement Narrative Generation: Advertising has been one of the most important targets since the first step of the author's narrative generation study. Research into narrative analyses and advertising-specific narrative generation systems has continued since the first design and implementation of the proposed advertising narrative generation system. One of the reasons for this is that it is relatively easy to analyze and generate advertisement works because advertisement narratives compress rich narrative rhetoric and diverse narrative techniques generally in a short story or plot.

Integrated Narrative Generation System (INGS): INGS is a synthetic narrative generation system architecture integrating the previous studies by the author. From the broadest perspective, INGS is divided into two types of part, knowledge and procedure. INGS aims at a kind of narrative synthesizer that integrates a variety of narrative techniques, methods, rhetoric, and knowledge into an organic and dynamic generation framework.

Multiple Narrative Structures Model: This is a conceptual model by which a narrative text is multiply constituted, and the narrative generation process multiply executed. This is not a special narrative phenomenon, but the multiplicity is positioned as an essentially important concept for considering and designing narrative generation mechanisms in the author's narrative generation study.

Rhetorical Strategies of Advertisement Narratives: In the context of the author's study, this means strategies of how a product, including a service and company's image, is inserted into an advertisement. In this sense, each advertisement is a narrative in which a product is strategically introduced or positioned. This chapter presents twelve categories of the rhetorical strategies for product introduction.

APPENDIX

Table 18. List of TV commercial films used as samples

Brand Category	Brand Name		
Clothing detergent	[Kao Corporation] *Attack*	[Kao Corporation] *Widehaiter*	
Kitchen detergent	[Kao Corporation] Kitchen Haiter		
Eye drops	[ROHTO Pharmaceutical Co., Ltd.] *Rohto Cool 40α*	[ROHTO Pharmaceutical Co., Ltd.] *Rohto Yojunsui*	[SANTEN PHARMACEUTICAL CO., LTD.] Sante 40
	[ROHTO Pharmaceutical Co., Ltd.] *V Rohoto*		
Liquid soap	[Kao Corporation] *Biore U*		
Toilet detergent	[Kao Corporation] *Toilet Magiclean*	[Kao Corporation] *Toilet Quickle*	[Dainihon Jochugiku Co., Ltd.] *Sanpoll*
Shampoo	[Kao Corporation] *merit*	[P&G, The Procter&Gamble Company of Japan Ltd.] *h&s*	[ROHTO Pharmaceutical Co.,Ltd.] *OXY SHAMPOO*
Hairdressing agent	[Kao Corporation] *Cape*	[Shiseido Japan Co., Ltd.] *UNO*	
Facial wash	[Kao Corporation] *MEN's Biore*	[ROHTO Pharmaceutical Co., Ltd.] *OXY*	[ROHTO Pharmaceutical Co.,Ltd.] *MENTHOLATUM Acnes*
Hair tonic	[Kao Corporation] *SUCCESS*	[Shiseido Japan Co., Ltd.] *Sengan Senka Perfect Series*	
Antiperspirant spray	[Kao Corporation] *8×4*	[Mandom Corporation] *GATSBY BIOCORE DEODORANT SPRAY*	[Mandom Corporation] *GATSBY Face Care Aqua Dance*
Deodorants	[P&G, The Procter&Gamble Company of Japan Ltd.] *Febreze*	[Kao Corporation] *Resesh*	
Skin lotion	[Shiseido Japan Co., Ltd.] *AQUA LABEL*	[Shiseido Japan Co.,Ltd.] *BENEFIQUE*	[ROHTO Pharmaceutical Co., Ltd.] *50 no Megumi*
Skin cream	[Shiseido Japan Co., Ltd.] *AQUA LABEL*		
Mascara	[P&G, The Procter&Gamble Company of Japan Ltd.] *MAXFACTOR*	[Maybelline New York] *Volume Express*	
Car wash paper	[SOFT99 Corp.] *Fukupika*		
Medicine	[Kowa Co., Ltd.] *Q&P KOWA GOLD*	[SSP Co., Ltd.] *EVE QUICK*	[SSP Co., Ltd.] *CUTINA*
	[SSP Co., Ltd.] *HYTHIOL C*	[Taisho Pharmaceutical Co., Ltd.] *Corakku II*	[Otsuka Pharmaceutical Co., Ltd.] *Nature Made*
Ointment	[Kowa Co., Ltd.] *VANTELIN KOWA*	[Otsuka Pharmaceutical Co., Ltd.] *Oronine H Ointment*	[TAIHO Phamaceutical Co.,Ltd.] *ZENOL*

continued on following page

Table 18. Continued

Brand Category	Brand Name		
	[ROHTO Pharmaceutical Co., Ltd.] *MENTHOLATUM*		
Writing utensils	[ZEBRA Co., Ltd.] *Mckee*	[PILOT CORPORATION] *e-GEL*	[PILOT CORPORATION] *Dr. Grip G SPEC*
	[PILOT CORPORATION] *HI-TEC*	[PILOT CORPORATION *FRIXION*	[PILOT CORPORATION] *petit 1*
	[PILOT CORPORATION] *MULRI BALL*	[PILOT CORPORATION] *Bordmaster*	[PILOT CORPORATION] *Latte*
Sweat wiping sheet	[Kao Corporation] *Biore Sara-sara Powder Seat [Biore non-wetness powder sheet]*		
Cling film	[Kureha Corporation] *New Kure Wrap*		
Gummy candy	[Meiji Co., Ltd.] *Fruits_ gummi*	[Meiji Co., Ltd.] *Qun*	
Candy	[LOTTE Co., Ltd.] *Nodo Ame Zero*	[Kasugai Seika Co., Ltd] *Kuro Ame*	[Meiji Co., Ltd.] *Nodo Ame Fruit*
	[Morinaga & Co., Ltd.] *Chupa Chupa*		
Pudding	[Ezaki Glico Co., Ltd.] *Torori Cream on Pudding*		
Ice cream	[Ezaki Glico Co., Ltd.] *papico*	[LOTTE Co., Ltd.] *Yukimi Daifuku [Snow viewing rice cake]*	[Morinaga & Co., Ltd.] *Choco Monaka Jumbo*
	[Ezaki Glico Co., Ltd.] *Giant Cone*	[Ezaki Glico Co., Ltd.] *palitte*	[LOTTE Co., Ltd.] Soh
	[Eskimo (Morinaga Milk Industry Co.,Ltd.)] *Pino*		
Candy gum	[Ezaki Glico Co., Ltd.] *Kiss Mint*	[LOTTE Co., Ltd.] *ACUO*	[LOTTE Co., Ltd.] *XYLITOL*
	[Meiji Co., Ltd.] *XYLISH*	[Ezaki Glico Co., Ltd.] *POs-Ca*	[Ezaki Glico Co., Ltd.] *BREO*
Chocolate	[Morinaga & Co., Ltd.] *DARS*	[Morinaga & Co., Ltd.] *CHOCOBALL*	[Meiji Co., Ltd.] *Almond Glico*
	[Fujiya Co., Ltd.] *LOOK Chocolate*	[LOTTE Co., Ltd.] *Rich Fruit Chocolate*	[LOTTE Co., Ltd.] *SASHA*
	[Meiji Co., Ltd.] *galbo chips*	[Meiji Co., Ltd.] *Milk Chocolate*	[Meiji Co., Ltd.] *Hokkaido Chocolate & Potate*
Snack food	[Morinaga & Co., Ltd.] *Ottotto*	[Morinaga & Co., Ltd.] *Morinaga Biscuits*	[Meiji Co., Ltd.] *Kinoko no Yama*
	[Calbee, Inc.] *A Ra Poteto*	[Calbee, Inc.] *Kappa ebisen (Shrimp flavored chips)*	[Calbee, Inc.] *Jagabee*
	[Calbee, Inc.] *Kata-age potato*	[Ezaki Glico Co., Ltd.] *Pocky*	[Ezaki Glico Co., Ltd.] *Sozaiha PRETZ*
	[Meiji Co., Ltd.] *Curl*	[Meiji Co., Ltd.] *Kopan*	[Tohato Inc.] *Bokun Habanero*

continued on following page

Table 18. Continued

Brand Category	Brand Name		
	[Tohato Inc.] *Poteko*	[Ezaki Glico Co., Ltd.] *CRATZ*	[HOUSE FOODS CORPORATION] *O'ZACK*
	[HOUSE FOODS CORPORATION] *Tongaricorn*	[Otsuka Foods Co., Ltd.] *Micro Magic*	
Cup noodles	[Acecook Co., Ltd.] *Super Cup*	[NISSIN FOODS HOLDINGS CO., LTD.] *CUPNOODLE*	[NISSIN FOODS HOLDINGS CO., LTD.] *Donbei*
	[Toyo Suisan Kaisha, Ltd.] *Maruchan Geki Men*	[Toyo Suisan Kaisha, Ltd.] *Maruchan Menzukuri*	[SANYO FOODS.Co., Ltd] *Sapporo Ichiban*
	[Toyo Suisan Kaisha, Ltd.] *Bari Uma*	[Toyo Suisan Kaisha, Ltd.] *Maru-chan Mukashi Nagara no Chukasoba*	[Toyo Suisan Kaisha, Ltd.] *Maru-chan Akai Kitsune to Midori no Tanuki*
	[NISSIN FOODS HOLDINGS CO., LTD.] *CUP NOODLE MISO*	[NISSIN FOODS HOLDINGS CO., LTD.] *Gonbuto*	[NISSIN FOODS HOLDINGS CO., LTD.] *Takoyaki Ramen*
	[NISSIN FOODS HOLDINGS CO., LTD.] *RAOH*	[NISSIN FOODS HOLDINGS CO., LTD.] *Menshokunin*	
Cup fried noodles	[Toyo Suisan Kaisha, Ltd.] *Maru-chan Yakisobabento*	[Toyo Suisan Kaisha, Ltd.] *Maru-chan Yakisobabagon*	[MYOJO FOODS CO.,LTD.] *Ippei-chan Yomise no Yakisoba*
	[NISSIN FOODS HOLDINGS CO., LTD.] *U.F.O.*		
Instant noodle	[NISSIN FOODS HOLDINGS CO., LTD.] *Nissin Chicken Ramen*	[HOUSE FOODS CORPORATION] *Umakacchan*	[NISSIN FOODS HOLDINGS CO., LTD.] *Demae_itcho*
	[MYOJO FOODS CO.,LTD.] *Charumera*		
Instant soup	[Ajinomoto Co., Inc.] *Knorr Cup Soup*		
Natto	[Mizkan Holdings Co., Ltd.] *Kin no Tsubu*		
Nutritional supplement	[Otsuka Pharmaceutical Co., Ltd.] *SOYJOY*	[Otsuka Pharmaceutical Co., Ltd.] *Caloriemate*	
Retort food	[Otsuka Pharmaceutical Co., Ltd.] *Bon Curry Gold*		
Canned food	[HAGOROMO FOODS CORPORATION] *Seachicken*		
Frozen food	[Ajinomoto Co., Inc.] *Reito Shokuhin ebi shumai [Frozen food shrimp shumai]*	[Ajinomoto Co., Inc.] *Reito Shokuhin ebi pilaf [Frozen food shrimp pilaf]*	[Ajinomoto Co., Inc.] *Reito Shokuhin Gyoza [Frozen food gyoza]*
	[Ajinomoto Co., Inc.] *Reito Shokuhin Shizen Kaito Obento series [Frozen food obento Series]*		

continued on following page

Table 18. Continued

Brand Category	Brand Name		
Somen	[NAGATANIEN CO., LTD.] *Sou Ramen*		
Cheese	[Megmilk Snow Brand Co., Ltd.] *Yukizirushi Hokkaido 100 Camembert Cheese*		
Vegetable juice	[ITO EN, LTD.] *Jujitsu Yasai*	[KAGOME CO., LTD.] *Yasai Seikatsu 100*	
Green tea	[Kirin Co., Ltd.] *Namacya*	[Suntory Holdings Limited] *Iyemon*	[ITO EN, LTD.] *Oi Ocha*
	[ITO EN, LTD.] *Katekin Ryokucha*	[Coca-cola (Japan) Co., Ltd.] *Ayataka*	[Coca-cola (Japan) Co., Ltd.] *SAORI*
	[Coca-cola (Japan) Co., Ltd.] *Hajime Chaka*		
Black tea in plastic bottle/drink box	[Kirin Co., Ltd.] *Gogo no Kocha*	[Suntory Holdings Limited] *Lipton LIMONE*	
Teapack black tea	[Suntory Holdings Limited] *FLAVOUR TEA assoutment*		
Oolong tea	[Suntory Holdings Limited] *Suntory Oolong Tea*	[Suntory Holdings Limited] *Suntory Black Oolong Tea*	
Other teas	[Coca-cola (Japan) Co., Ltd.] *Sokenbicha*	[DyDo DRINCO, INC.] *Yaroncha Level Care*	[Suntory Holdings Limited] *Suntory Goma Mugicha*
	[Yakult Honsha Co., Ltd.] *Bansoreicha*	[ITO EN, LTD.] *Ten'nen Biko Jasmine Tea*	[Kao Corporation] *Healthya Ryokutya [Healthya green tea]*
	[Kirin Co., Ltd.] *Ururucha*	[Coca-cola (Japan) Co., Ltd.] *Karada Meguricha*	
Milk	[Megmilk Snow Brand Co., Ltd..] *MEGMILK*		
Bottled water	[Kirin Co., Ltd.] *Volvic*	[Suntory Holdings Limited] *Vittel*	[Suntory Holdings Limited] *Ten'nensui Aso*
	[ITO EN, LTD.] *evian*		
Bottled beer	[Kirin Co., Ltd.] *Ichi-ban Shibori*	[Kirin Co., Ltd.] *KIRIN Akiaji*	[Asahi Breweries, Ltd.] *Asahi Premium JUKUSEN*
	[Kirin Co., Ltd.] *KIRIN LAGER BEER*	[Sapporo Breweries Limited] *SAPPORO Nama BEER*	
Canned beer	[Kirin Co., Ltd.] *KIRIN TANREI NAMA*	[Kirin Co., Ltd.] *NODOGOSHI NAMA*	[Asahi Breweries, Ltd.] *Asahi SUPER DRY*
	[Asahi Breweries, Ltd.] *Asahi STYLE FREE*	[Asahi Breweries, Ltd.] *ASAHI'S ORIGINAL BREW HONNAMA AQUA BLUE*	[Asahi Breweries, Ltd.] *ASAHI'S ORIGINAL BREW HONNAMA DRAFT*
	[Asahi Breweries, Ltd.] *Clear Asahi*	[Suntory Holdings Limited] *JOKKI NAMA*	[Suntory Holdings Limited] *Premium Malt's*
	[Sapporo Breweries Limited] *Sapporo Kuro Label*	[Asahi Breweries, Ltd.] *Asahi SUPER DRY*	[Sapporo Breweries Limited] *Mugi to Hop*

continued on following page

Table 18. Continued

Brand Category	Brand Name		
	[Suntory Holdings Limited] *Kin-mugi*	[Sapporo Breweries Limited] *Hokkaido Gentei SAPPORO CLASSIC*	[Sapporo Breweries Limited] *SAPPORO BEER FINE*
	[Kirin Co., Ltd.] Smooth	[Asahi Breweries, Ltd.] *Gubi Nama*	
Chu-hi	[Kirin Co., Ltd.] *Hyoketsu*	[Asahi Breweries, Ltd.] *Asahi Vegesh*	[Asahi Breweries, Ltd.] *Asahi Syunka Shibori Fruttio*
	[Suntory Holdings Limited] *Calori*	[Suntory Holdings Limited] *AWAS*	[Kirin Co., Ltd.] *TWO DOGS*
	[Kirin Co., Ltd.] *KIRIN ZERO*	[Kirin Co., Ltd.] *Sparkling Hop*	
Shochu	[Asahi Breweries, Ltd.] *Kanoka*		
Whiskey	[Asahi Breweries, Ltd.] *BLACK NIKKA*	[Suntory Holdings Limited] *SUNTORY OLD*	
Soda drink	[Kirin Co., Ltd.] NUDA	[Kirin Co., Ltd.] KIRIN LEMON	[Coca-cola (Japan) Co., Ltd.] Coca-cola
	[Coca-cola (Japan) Co., Ltd.] *Coca-cola zero*	[Coca-cola (Japan) Co., Ltd.] *No calorie Coca-cola*	[Coca-cola (Japan) Co., Ltd.] *Fanta*
	[Suntory Holdings Limited] *CC Lemon*	[Suntory Holdings Ltd.] *Pepsi-cola*	[Suntory Holdings Ltd.] PEPSI *NEX*
	[ASAHI Soft Drinks Co., Ltd.] *MITSUYA CIDER*	[ASAHI Soft Drinks Co., Ltd.] *Lemon wo Shibotta MITSUYA CIDER*	[Otsuka Pharmaceutical Co., Ltd.] *Oronamin C Drink*
Sports drink	[Otsuka Pharmaceutical Co., Ltd.] *POCARI SWEAT*	[Coca-cola (Japan) Co., Ltd.] *AQUARIUS Freestyle*	[Coca-cola (Japan) Co., Ltd.] *AQUARIUS*
Lactic acid bacteria beverage	[Calpis Co., Ltd.] *Calpis*	[Yakult Honsha Co., Ltd.] *Pretio*	[Yakult Honsha Co., Ltd.] *Yakult 400*
	[Yakult Honsha Co., Ltd.] *Yakult SHEs*	[KAGOME CO., LTD.] *Syokubutsusei Nyusankin Labre*	
Juice	[Coca-cola (Japan) Co., Ltd.] *Qoo*	[Suntory Holdings Limited] *DAKARA*	[Suntory Holdings Limited] *Vitamin Water*
	[Calpis Co., Ltd.] Welch's	[DyDo DRINCO, INC.] MiU	[ASAHI Soft Drinks Co., Ltd.] CITRULLINE WATER
	[Yakult Honsha Co., Ltd.] *THORPEDO*	[Coca-cola (Japan) Co., Ltd.] *AQUARIUS Vitamin Guard*	[Coca-cola (Japan) Co., Ltd.] *Minute Maid*
	[Japan Sangaria Beverage Co., Ltd.]		
Nutrition/health supplement beverage	[Morinaga & Co., Ltd.] In Jelly]	[Coca-cola (Japan) Co., Ltd.] *AQUARIUS Sports Jelly*	[Kowa Co., Ltd.] *Ekicaba Kowa*
	[Eisai Co., Ltd.] *Chocola BB Royal 2*	[Yakult Honsha Co., Ltd.] *Kurozu Drink [Black vinegar drink]*	[Shiseido Japan Co., Ltd.] *Collagen*

continued on following page

Table 18. Continued

Brand Category	Brand Name		
	[Meiji Co., Ltd.] *Maka no Genki*	[Ajinomoto Co., Inc.] *amino VITAL*	[Ezaki Glico Co., Ltd.] *P ga Mitibiku Collagen*
	[HOUSE FOODS CORPORATION] *UKON NO CHIKARA*	[Meiji Co., Ltd.] *Amino Collagen*	[Taisho Pharmaceutical Co., Ltd.] *Livita Gurukosu Care*
	[Taisho Pharmaceutical Co., Ltd.] *ALFE neo*	[Taisho Pharmaceutical Co., Ltd.] *Lipovitan D*	[Taisho Pharmaceutical Co., Ltd.] *Lipovitan Fine*
	[Otsuka Pharmaceutical Co., Ltd.] *Fibe Mini*	[TAIHO Phamaceutical Co., Ltd.] *solmack*	[TAIHO Phamaceutical Co.,Ltd.] *TIOVITA*
Canned coffee	[Kirin Co., Ltd.] *FIRE*	[Coca-cola (Japan) Co., Ltd.] *GEORGIA*	[Suntory Holdings Limited] *BOSS*
	[DyDo DRINCO, INC.] *D-1*	[DyDo DRINCO, INC.] *DyDo DEMITASSE COFFEE*	[Japan Tobacco Inc.] *Roots*
	[Nestle Japan Limited] *TAKUMI*	[ASAHI Soft Drinks Co., Ltd.] *WONDA*	
Coffee in plastic bottle/drink box]	[Ajinomoto AGF, Inc.] *Blendy*		
Instant coffee	[Ajinomoto AGF, Inc.] *Blendy*	[Nestle Japan Limited] *GOLD BLEND*	[KATAOKA & CO.,LTD.] *ASTORIA*
	[KATAOKA & CO., LTD.] *MON CAFÉ*		
Instant cocoa	[Morinaga & Co., Ltd.] *MORINAGA MILK COCOA*		
Mayonnaise	[Kewpie Corporation] *Kewpie Half*	[Ajinomoto Co., Inc.] *Pure Select Koku Uma*	
Dashi	[Ajinomoto Co., Inc.] *Hondashi*		
Salt	[Ajinomoto Co., Inc.] *Yasashio*		
Ajinomoto	[Ajinomoto Co., Inc.] *Ajinomoto*		
Tare (Sauce)	[Ajinomoto Co., Inc.] *Cook Do*	[Ebara Foods Industry, Inc.] *Kimchi Nabe no Moto*	[Ebara Foods Industry, Inc.] *Shogayaki no Tare*
	[NAGATANIEN CO., LTD.] *Oishiku Dekicyau! Cyahan to Yasaiitame*	[NAGATANIEN CO., LTD.] *Wafu Mabodofu*	
Vinegar	[Mizkan Holdings Co., Ltd.] *Syokusu*		
Kozu source	[Yazuya Co., Ltd.] *Jukusei Yazuya no Kousaku*		
Ponzu sauce	[Mizkan Holdings Co., Ltd.] *Ajipon*	[YAMASA CORPORATION] *Konbupons*	

continued on following page

Table 18. Continued

Brand Category	Brand Name		
Dashitsuyu	[Mizkan Holdings Co., Ltd.] *Oi Gatsuo Tsuyu*	[YAMASA CORPORATION] *Konbu Tsuyu*	
Frikake (Sprinkle)	[NAGATANIEN CO., LTD.] *Otona no Furikake*		
Ochazuke	[NAGATANIEN CO., LTD.] *Ochazuke Nori*	[NAGATANIEN CO., LTD.] *Gokuuma! Soup Chazuke*	[NAGATANIEN CO., LTD.] *Ume-boshi Chazuke*
	[NAGATANIEN CO., LTD.] *Hiyashi Oolingchazuke*		
Printer	[Canon Inc.] *Color image RUNNER*	[Canon Inc.] *PIXUS*	[Canon Inc.] *Satera*
	[Canon Inc.] *SELPHY*		
Laptop computer	[NEC Corp.] *LaVie L*	[NEC Corp.] *ShieldPRO*	[NEC Corp.] *VALUESTAR W*
	[TOSHIBA CORPORATION] *Note PC Qosmio*	[TOSHIBA CORPORATION] *dynabook*	[Sony Corporation] *NEW TYPE Z*
Laundry machine	[Hitachi, Ltd.] *BEAT WASH*	[Matsushita Electric Works, Ltd. (Current name: Panasonic Corp.)] *Eco Idea*	[TOSHIBA CORPORATION] *Heat Pump Air Conditioner Hybrid Drum*
Refrigerator	[Hitachi, Ltd.] *Refrigerator*	[Matsushita Electric Works, Ltd. (Current name: Panasonic Corp.)] *Eco Idea*	[TOSHIBA CORPORATION] *Hoon Sengoko*
	[Mitsubishi Electric Corporation] *Refrigerator*	[SANYO Electric Co., Ltd.] *Hyoon*	
Television	[Panasonic Corporation] *VIERA*	[Mitsubishi Electric Corporation] *Liquid Crystal Television REAL*	[Sharp Corporation] *AQUOS*
	[Victor Company of Japan, Limited] *Hi-Vision Liquid Crystal Television EXE*	[TOSHIBA CORPORATION] *CORE*	[TOSHIBA CORPORATION] *DIGITAL FACE*
	[TOSHIBA CORPORATION] *Stereo Television*	[TOSHIBA CORPORATION] *Liquid Crystal Television REGZA*	
DVD Recorder	[Panasonic Corporation] *DVD Player*		
Vacuum cleaner	[Hitachi, Ltd.] *Gomi Dash Cyclone*	[TOSHIBA CORPORATION] *Cyclone Cleaner*	[TOSHIBA CORPORATION] *Dasutoron*
Electronic pot	[ZOJIRUSHI CORPORATION] *Electronic Pot*		
Digital camera	[Canon Inc.] *EXY DIGITAL*	[Canon Inc.] *PowerShot*	[Panasonic Corporation] *LUMIX*
Video camera	[Sony Corporation] *Handycam*	[Sharp Corporation] *My Video*	[Panasonic Corporation] *Burenbi*

continued on following page

Table 18. Continued

Brand Category	Brand Name		
	[Victor Company of Japan, Limited] *Digital Video Camera*	[SANYO Electric Co., Ltd.] *Digital Movie Xacti*	[Panasonic Corporation] *Digital Video Camera*
	[Sony Corporation] *Digital Still Camera*	[CASIO COMPUTER CO., LTD.] *EXILIM EX*	
Air conditioner	[Hitachi, Ltd.] *PAM Air Condeitioner*	[Hitachi, Ltd.] *Yukaoki Air Conditioner*	[TOSHIBA CORPORATION] *Air Conditioner Daiseikai*
Shaver	[Matsushita Electric Works, Ltd. (Current name: Panasonic Corp.)] *RAMDASH*	[The Procter&Gamble Company of Japan Ltd.] *Braun Prosonic*	
Electronic dictionary	[CASIO COMPUTER CO., LTD.] *EX-word*		
Microwave oven	[Hitachi, Ltd.] *Microwave Oven*	[TOSHIBA CORPORATION] *CALOLIER*	
Mobile phone	[Panasonic Corporation] *Mobile Phone VIERA*	[SANYO Electric Co., Ltd.] *PHS H"*	
Rice cooker	[TOSHIBA CORPORATION] *Hoon Gama*	[ZOJIRUSHI CORPORATION] *Suihan ja (Rice cooker)*	[TOSHIBA CORPORATION] *COOK MORE*
Car navigation system	[Panasonic Corporation] *Strada F Class*	[Sony Corporation] *Car Navigation System*	
Light bulb	[TOSHIBA CORPORATION] *Toshiba Lighting & Technology*		
Component	[Sony Corporation] *Component*	[SANYO Electric Co., Ltd.] *Component*	
Car	[Daihatsu Motor Co., Ltd.] *CONTE*	[Daihatsu Motor Co., Ltd.] *TanTo*	[Daihatsu Motor Co., Ltd.] *TanTo CUSTOM*
	[Daihatsu Motor Co., Ltd.] *MOVE*	[Daihatsu Motor Co., Ltd.] *MOVE CUSTOM*	[MITSUBISHI MOTORS CORPORATION] *ek wagon*
	[MITSUBISHI MOTORS CORPORATION] *COLT*	[NISSAN MOTOR CO., LTD.] *X-TRAIL*	[NISSAN MOTOR CO., LTD.] *ELGRAND*
	[NISSAN MOTOR CO., LTD.] *cube*	[NISSAN MOTOR CO., LTD.] *SERENA*	[NISSAN MOTOR CO., LTD.] *TEANA*
	[NISSAN MOTOR CO., LTD.] *TIIDA*	[NISSAN MOTOR CO., LTD.] *DUALIS*	[NISSAN MOTOR CO., LTD.] *NOTE*
	[NISSAN MOTOR CO., LTD.] *MARCH*	[SUZUKI MOTOR CORPORATION] *SWIFT*	[SUZUKI MOTOR CORPORATION] *LANDY*
	[SUZUKI MOTOR CORPORATION] *WAGON R*	[Toyota Motor Corporation] *ISIS*	[Toyota Motor Corporation] *VANGUARD*
	[Toyota Motor Corporation] *VELLFIRE*	[Toyota Motor Corporation] AURIS	[Toyota Motor Corporation] *SIENTA*

continued on following page

Table 18. Continued

Brand Category	Brand Name		
	[Honda Motor Co., Ltd.] *BEAT*	[Honda Motor Co., Ltd.] *FIT*	[Honda Motor Co., Ltd.] *FREED*
	[Honda Motor Co., Ltd.] *STREAM*	[NISSAN MOTOR CO., LTD.] *OTTI*	[Toyota Motor Corporation] *VITZ*
	[Toyota Motor Corporation] *RAV4*	[Toyota Motor Corporation] *VOXY*	[Toyota Motor Corporation] *RUMION*
	[Toyota Motor Corporation] *CAROLLA FIELDEER*	[Toyota Motor Corporation] *PRIUS*	[Toyota Motor Corporation] *Opa*
	[Honda Motor Co., Ltd.] *ELYSION*	[MAZDA Motor Corporation] *DEMIO*	[MAZDA Motor Corporation] *ATENZA*
	[MAZDA Motor Corporation] *MPV*	[NISSAN MOTOR CO., LTD.] *PRIMERA WAGON*	[NISSAN MOTOR CO., LTD.] *PRIMERA*
	[SUBARU CORPORATION] *R2*	[Toyota Motor Corporation] *ist*	[Toyota Motor Corporation] *WISH*
	[Toyota Motor Corporation] *MARK X*	[Toyota Motor Corporation] *CROWN*	[MITSUBISHI MOTORS CORPORATION] *DELICA*
	[Honda Motor Co., Ltd.] *MOBILIO*	[Honda Motor Co., Ltd.] *That's*	[Honda Motor Co., Ltd.] *STEP WGN*
	[Honda Motor Co., Ltd.] *Odyssey*	[Honda Motor Co., Ltd.] *ACCORD*	[Daihatsu Motor Co., Ltd.] *MIRA*
	[SUBARU CORPORATION] *PLEO*		

Chapter 4
Toward External and Internal Practices of Narrative Generation

ABSTRACT

From relationships to the level of multiple narrative structures, the geinō information system (GIS) described in the previous chapter corresponds to a social-level narrative generation mechanism beyond the narrative generation as a simple substance or as something including many narrative generation as a simple substance. From a wider viewpoint here, as an overview, this chapter attempts to discuss the future of the narrative generation systems. In this chapter, the author discusses the topic from two directions: external narrative generation and internal narrative generation. The "external" is the outer world beyond the "I" and ordinarily refers to a society or societies. Conversely, the "internal" is equal to the world inside the "I." Therefore, although the practice of narrative generation toward the outer direction ordinarily indicates the social development of narrative generation, the practice of narrative generation toward the internal is equal to the direction of narrative generation study for drawing "I" in any style.

DOI: 10.4018/978-1-5225-9943-2.ch004

INTRODUCTION

The *Geinō* Information System (GIS) explained in Chapter 2 corresponds to the narrative generation mechanism at the social level over narrative generation as a simple mechanism, which contains several or many simple narrative generation mechanisms in relation with the level of multiple narrative structures. Therefore, the consideration of social narrative generation was partially discussed previously. In this chapter, the author would like to discuss a future vision of the narrative generation system study by the author. This future vision is not limited merely to a social direction. Rather, it has two directions: "external narrative generation" and "internal narrative direction." These are placed as the practice of narrative generation. Therefore, this chapter discusses the "external practices of narrative generation" and the "internal practices of narrative generation." The "external" is the outside world beyond this "I" and is ordinarily a society; conversely, the "internal" means the inside world for me. Therefore, although practices of narrative generation toward the outside mean narrative generation in the ordinary sense, practices of narrative generation toward the inside are related to the direction for describing this "I" by various methods.

In **BACKGROUND**, the author introduces Takaaki Yoshimoto (1924-2012), who was a poet, critic, and philosopher, with special consideratiosn of Yoshimoto's concept of "communal illusion" (Yoshimoto, 1968a) and his identification of three types of illusions—personal illusion, pair illusion, and communal illusion—which will direct the future development of the author's narrative generation study. As another background, the author describes a unique literary genre called "*watakushi shōsetsu* (I-novel)," which was introduced and practiced in modern Japanese literature. This discussion is related to the plan of "*watakushi monogatari*" that appears in the last half in this chapter.

The next section, **TOWARD EXTERNAL OR SOCIAL PRACTICES OF NARRATIVE GENERATION**, comprehensively considers the external practices of narrative generation and is structured as follows. First, ***Geinō* Information System: The Overall Architecture and Thought** introduces the overall architecture of the GIS and the various components, including sender and receiver mechanisms. GIS offers the design for the basic framework for the external practices of narrative generation. Next, **GIS with INGS and *Geinō* History** describes how GIS is constructed as a synthetic system that involves a narrative production-consumption process containing INGS for a narrative generation-reception process and considers the problem of the *geinō* history generated by GIS. In particular, the author introduces the problem that the development and distribution of narrative generation systems and the production and distribution of the narrative contents generated by them need to be integrated into a single mechanism through the social development of narrative generation systems. Third, in **Multiple Narrative Structures: Relationships Between Narrative Generation-Reception Mechanism in INGS and Narrative Production-Consumption Mechanism in GIS**, the author reconfirms that GIS is a system based on the multiple narrative structures model, which is one of the conceptual backgrounds of the integrated narrative generation study conducted by the author. Business approaches to narrative social development and distribution, particularly more practical and realistic narrative methods, are discussed.

In contrast, **TOWARD INTERNAL OR PERSONAL PRACTICES OF NARRATIVE GENERATION**, the author describes an internal practice of narrative generation, which means the author's personal practice of making personal narratives. The results can also be developed and distributed socially in the linkage with the above social direction. This section focuses on planning materials for internal and personal narratives and the roles and functions of the narrative generation systems, INGS and GIS, for dealing with the materials. In particular, through **Narrative Generation to the Inside or Toward *Watakushi Monogatari***, **Overview of Contents for *Watakushi Monogatari***, and **Narratives as Fluid Process: Personal View**, the author discusses the *watakushi monogatari*, which is an outline of the narrative works or the narrative genre created using the narrative generation systems, INGS and GIS.

Finally, **THE FIELD OF NARRATIVE GENERATION AS SOCIAL AND COLLECTIVE PHENOMENA** discusses the problem of unintentional narrative generation, as well as topics related to social and philosophical narratology. In particular, it examines how the collection of various personal and organizational narrative generation phenomena produces larger narratives and how they are fixed into people's consciousness, influencing social processes. This is the part that is categorized in the genre of "emergent narratives" in the narrative genre system in Chapter 2 (Ogata, 2019c) in Ogata (2019b) and is related to such academic ideas as "narrative economics" by Shiller (2017). In addition, Yoshimoto's philosophy of communal illusion analyzed and interpreted at the macro-level narrative formation process of a nation—

for example, Japan, as a communal illusion related to micro-level narrative generation processes in village communities.

In addition, this chapter is an extended version of Ogata (2018a, 2018b).

BACKGROUND

Takaaki Yoshimoto

The author considers two topics in the context of the external development of narrative generation. The first part introduces Takaaki Yoshimoto. Although not a narratologist in a narrow sense, he was a critic and philosopher in Japan who was interested in the human mind, literature and personal discourses, and social and institutional problems through unique systematic thought. On the other hand, in the context of the internal development of narrative generation, the author engages with a genre of stories in modern Japanese literature called *watakushi shōsetsu* or *shishōsetsu* (a type of autobiographical stories) that aim to represent literary works and narratives drawn from the ordinary and familiar world of each writer.

The "theories on narratives and societies by Takaaki Yoshimoto" here do not mean narrative sociology, literature sociology, or the sociology of literature and narratives. The sociologies of literature and narratives, as seen in Pierre Bourdieu's (1930-2002) (1996) study and a sociological study of *Genji Monogatari* (Ando, 2006), are areas that study the societies that appear in narratives and literature, the mutual relationships among characters, and the historical backgrounds as represented or in their associations with actual historical and social backgrounds. The approaches in fact overlap with traditional literary research that emphasizes literary or narrative writers in many parts.

The author considers narratives and societies from the viewpoint of narrative generation mechanisms. To provide an overview, the approach focuses on how narratives are developed and distributed in a society. According to Sapir's (2014) general understanding of the sociology of literature (2014), such problems have been treated by the flow of sociology of literature proposed by Bourdieu. However, the author's study of narratives and societies corresponds to the level of social narrative production and consumption as applicable in the context of the author's narrative generation system study.

The theory of narratives and societies by the author also has several levels. A level is an area where intentionally created narrative contents that are socially developed and distributed overlap with various narrative businesses. Another area is the one that considers how our societies come into existence as narratives. The latter is not an area that considers how narratives are developed and distributed intentionally and consciously at the social level. Rather, it treats the problem of how our societies are constructed through narratives without clear intentionality and consciousness or how institutional narratives, in the broad sense (the greatest manifestations being nations), are socially developed and distributed.

The above two approaches are equal in terms of the fact that they commonly treat the problem of social development of narratives or social narratives. However, an essential difference between them is as follows. In the former, narratives as substances or as contents exist in any form. In contrast, narratives as substances do not exist in the latter. For instance, although documents for law can exist, they are not narrative contents themselves that are similar to the former, but are rather documents that include narratives. The above difference is related to the difference between intentional and conscious narratives and between unconscious and unintentional narratives. Although the intentions and consciousness for socially realizing specific laws and institutions exist in the latter, they are not different from the personal and organizational intentions and strategies that aim to distribute specific narrative contents themselves. Narratives in this case are things that emerge from the collection of laws and institutions and are more ambiguous than narratives in the former sense. However, the narratives that substantially dominate our consciousness, thoughts, and lives are very strong. For example, in the present world, nations have strong powers that support the narratives in the latter sense, or nations exist in the background of strong narratives, as individually stronger narratives. However, although there are many documents and texts supporting the construction of nations' narratives, we do not call them nations.

The social narratives here correspond to the narrative types that are located most externally in the multiple narrative structures model. Although only few studies have dealt with the narratives themselves in the above sense as their main theme, relevant studies have been conducted by various philosophers and thinkers. Takaaki Yoshimoto was one of such philosopher. He was a critic without any power who continued his discursive activities based on the special contexts of post-war society in Japan and did not belong to any university or research institution. The main media for presenting his achievements were not academic journals, but a magazine that he himself published along with some companions; various magazines and newspapers; books from different publishers; oral presentations at universities and other places; and talks with several writers, critics, philosophers, and researchers. The writing style of his essays was different from the style of academic papers. In particular, he used unique words and a culture-dependent style that contained the unique nuances of Japanese. Although Yoshimoto was in conversation with Michel Foucault (1926-1984) (Yoshimoto, 1980) and Jean Baudrillard (1929-2007), he is not famous in the English-speaking world, as there are no English translations of his writings (His *Kyōdō Gensō Ron* (Yoshimoto, 1968a) was translated to French (Yoshimoto, 2016)). Although there are Wikipedia descriptions of his works in English, French, German, Italian, Russian, and Chinese, they are incomplete. However, this thinker or philosopher, who was the father of Banana Yoshimoto (1964-), a renowned novelist, had a very significant impact on people, along with Yukio Mishima (1925-1970), in the post-war society in Japan. Currently, a collection of several of his writings is published.

In addition, the author would like to provide an additional explanation regarding the positioning of philosophers and thinkers in modern Japan. Philosophical studies, in the narrow sense, in modern Japan were performed by a group of "researchers of a specific (Western) philosopher" or a group of "researchers of a specific (Western) philosophical thought." Researchers here refer to those researchers in universities who professionally study philosophers and philosophical thoughts from the West. The translations and introductions of Western philosophical books and papers have certainly made substantial contributions to thoughts and studies in modern Japan. However, thinking that no original Japanese thought had been produced until then is quite strange. The author has introduced Japanese literary histories by Furuhashi (2010) in the **BACKGROUND** section in Chapter 2 (Ogata, 2019c) in Ogata (2019b) and charted the traditions of essays and critiques in Japan. These demonstrate a Japanese philosophical and critical tradition in the broad sense. With this tradition as a background, philosophers in the broad sense or true sense as "thinking people" were novelists, writers, and critics who are successors of those who wrote essays and critiques in Japanese literature in modern Japan, instead of the researchers in universities. Natsume Sōseki (1867-1916) and Mori Ōgai (1862-1922), who were known as great writers, wrote many critiques and essays and have been regarded as philosophers and thinkers. Further, novelists such as Yukio Mishima and Kenzaburō Ōe (1935-) were also a type of philosophers and thinkers. Although Yoshimoto was not an academic researcher who professionally studied specific philosophers or philosophical thoughts, he was clearly a philosopher and critic in the above sense and has been regarded as one of the leading critics along with Hideo Kobayashi (1902-1983). In addition, philosophers and critics similar to Yoshimoto include, since Kobayashi, Shūichi Katō (1919-2008) (literature, social thought), Shigehiko Hasumi (1936-) (literature, film theory, social thought), Kōjin Karatani (1941-) (literature, social thought, philosophy), etc. Although Katō, Hasumi, and Karatani belonged to universities, they were more critics and philosophers in the broad sense than academicians. However, compared to them, Yoshimoto was rather conspicuous in that he engaged in very broad discoursal activities. In addition, Kumagusu Minakata (1867-1941) was also a great philosopher who did not belong to any university but constructed a unique research system. Minakata produced a wide variety of writings covering literature, religion, folklore, ecology, biology, etc.

Yoshimoto was born (December 1924) almost at the same time as Mishima (January 1925). Both Mishima and Yoshimoto did not participate in the Pacific War. However, their boyhood and early youth overlapped with the period of war. Although Mishima had actually been examined for conscription, he was exempted from being drafted due to his accidental sickness. The war certainly had a heavy impact on the formation of their thoughts and characteristics. Although Mishima died in 1970, 25 years after Japan's defeat with a cry of "*Ten'nō Heika Banzai* (Long live the Emperor!)," Yoshimoto had also been a passionate supporter of the Emperor system in Japan until his early youth. At the time of defeat in the war, he felt suspicious about why several Japanese people did not die from the shock of the defeat.

In contrast to Mishima who was born at *Yamanote* (the hilly sections) of Tokyo, Yoshimoto was born in *Shitamachi* of the same Tokyo. However, they were common city-dwellers. Mishima, however, was an elite who became a bureaucrat from the law department of the University of Tokyo through Gakushūin, and Yoshimoto was an elite engineer who graduated from the Tokyo Institute of Technology through Yonezawa Technical College (the present-day Department of Engineering of Yamagata University). Yoshimoto became a chemical engineer who worked for both companies and at a university. Yoshimoto, who was a passionate supporter of Japan's militarism in his boyhood and early youth, transformed into a young man who believed in leftism and hoped for a future socialistic Japan, similar to many young intellectual people in the period. He directed labor movements in companies and read Karl Marx (1818-1883), Adam Smith (1723-1790), and Georg Wilhelm Friedrich Hegel (1770-1831). Further, he began to write poems, essays, and critiques regarding literature, politics, and societies to became one of the famous writers. Simultaneously, in the period of campaign against the Japan-U.S. Security Treaty in 1960, he participated as one of the theoretical directors, and during this period, his books and essays were treated as symbolic thought and philosophy for this social movement. A primary and important theme of his thought and philosophy in the age was the problem of "war responsibility." This theme was partially influenced by the thoughts, feelings, and emotions from his boyhood when he was a positive supporter of the Pacific War. He exploratively considered why thoughts and emotions were produced in him and several other people to comprehend synthetically the mutual relations among the individual, family, and societal levels. Considering war responsibility laid the groundwork for his later systematic theory of three types of illusions, namely individual, pair, and communal (or collective) illusions.

After the failure of the campaign against the Japan-U.S. Security Treaty, he stopped engaging with direct political movements and began to publish theoretical journals called "*Shiko*" with his two friends, Gan Tanigawa (1923-1995) and Ichirō Murakami (1920-1975). Afterwards, Tanigawa returned to his homeland, Kitakyūshū (north Kyūshū), to continue his political activity. Murakami also left *Shiko* to continue his own writing activity. Murakami hoped for the restoration of the Emperor system in a true sense and killed himself using a Japanese sword after the death of Mishima. *Shiko* became a central medium for Yoshimoto's writing activity, and he published, in addition to long theoretical critiques, topical and timely essays on literature, politics, societies, economics, sciences, engineering, and so on. This can be compared to, for example, Jean-Paul Sartre (1905-1980) continually discussing current events in addition to long philosophical and literary books, papers, and essays. The first theoretical paper by Yoshimoto published in *Shiko* was *Gengo ni totte Bi toha Nanika* (Yoshimoto, 1965). He periodically published *Kyōdō Genso Ron* (Yoshimoto, 1968a), and *Shinteki Genshō Ron Josetsu* (Yoshimoto, 1971a) and *Shinteki Genshō Ron* (*Honron*) (Yoshimoto, 2008). *Shikō* continued till 1997. Other writers and researchers also published many essays and papers in *Shiko*.

Through his long career, Yoshimoto continued to publish *Shiko,* long and short essays and critiques in diverse magazines and newspapers, and many books. His long philosophical and theoretical works include *Shomotsu no Kaitaigaku* (Yoshimoto, 1975), *Genji Monogatari Ron* (Yoshimoto, 1982), *Mass Image Ron* (Yoshimoto, 1984), and *High Image Ron* I, II, III (Yoshimoto, 1989a, 1990, 1994). In addition, Yoshimoto conducted dialogues with many novelists, thinkers, philosophers, researchers, poets, and others to compile them into books. Yoshimoto also made presentations about various themes.

The author listened to his presentation on Asian and African societies at Waseda University when the author was a student of the university. Yoshimoto, who was originally an engineer and chemical researcher, summarized his presentation on large papers, and gave a speech on the papers. His many works have been collected in several kinds of collections. After his death, a new collection was begun to be published (2014-) by Shōbunsha in Tokyo. His works cover the following topics:

- Psychological theories, criticisms, and studies: This includes theories on mental phenomena.
- Linguistic theories, criticisms, and studies: This includes studies and theories on aesthetic linguistics and Ryūkyū (Okinawa) literature.
- Literary theories, criticisms, and studies: This includes studies on novelists' and poets' criticisms and theories, poetics (in the narrow sense), novel theories, classical literature theories, and literary criticisms.
- Political theories, criticisms, and studies: This includes studies and theories on war responsibility and criticisms of thinkers and philosophers.
- Social theories, criticisms, and studies: This includes theories on communal illusions, image theories of high-consumption societies, and theories on the Japanese nation's origin.
- Culture and civilization theories, criticisms, and studies: This includes theories on cultures and civilizations of Japan and Africa.
- Religious theories, criticisms, and studies: This includes theories and criticisms on Shinran (1173-1263) and Jōdo-shinshū, and Jesus (6 or 4 BC-30 AD) and Christianity.

  Yoshimoto also published the following types of papers and books:
- Situational comments and criticisms on diverse social events in his age (Yoshimoto, 1968b).
- Essays, papers, and books on literary works: This includes *Genji Monogagari Ron* (Yoshimoto, 1982).
- Essays, papers, and books on authors and poets: This includes Kenji Miyazawa (1896-1933) (Yoshimoto, 1989b); Minamono no Sanetomo (1192-1219) (Yoshimoto, 1971b); Georges Bataille (1897-1962) in Yoshimoto (1975).
- Essays, papers, and books on philosophers, thinkers, and researchers: This includes Karl Marx (Yoshimoto, 1966) and Masao Maruyama (1914-1996) (Yoshimoto, 1963).

- Dialogues with other people: This includes talk with Jun Etō (1932-1999) (Yoshimoto, 2017). Yoshimoto (1985) and Yutaka Haniya (1909-1997) (1985) state their opinions regarding dialogues between Yoshimoto and Haniya.
- Presentations and lectures: Yoshimoto (2014-2015) collects his many presentations.

In addition to these numerous publications, from his early career, he also successfully explored and provided systematic views about humans and society.

Regarding his chosen situational topics of discourse, a characteristic of Yoshimoto's social discourse was that he sometimes ignored opinions considered as common sense or good sense and presented his opinions from the viewpoint of his systematic thought and philosophy. For example, regarding the Fukushima Daiichi nuclear accident that occurred as a consequence of the Great East Japan Earthquake on March 11, 2011, based on his opinion that the problems of science and technology should be attempted to be solved using science and technology itself, Yoshimoto opposed in principle the decommissioning of the reactor based on political judgements. This opinion was criticized by many authors and researchers. Further back, regarding the Tokyo subway sarin attack by Aum Shinrikyō (March 1995), Yoshimoto expressed a certain appreciation of Shōkō Asahara (1955-2019), who was the founder of Aum Shinrikyō, as a religious man. Further, during the Gulf War (1991), several Japanese authors launched the "antinuclear movement" that objected the nuclear strategy by the USA. In contrast, Yoshimoto criticized it based on his logic that if they object to the USA, they should also simultaneously object to the nuclear strategies of Russia and China. Thus, while similar to but also in a different sense from Yukio Mishima, although Yoshimoto did not negate the acts of the general public in principle, he presented opinions that were sometimes different from common and conventional ideas to continue his long discursive activities.

One of his typical philosophical ideas is the "three illusions theory." The author has sometimes referred to this theory in other places in this book and in Ogata (2019b). Yoshimoto considered that human worlds are full of unique ideal illusions that are divided into three main areas, namely, areas of individual illusions (corresponding to personal worlds), pair illusions (concerning families), and communal illusions (corresponding to nations and societies). Yoshimoto comprehended humans and worlds as the organic synthesis of these areas. Furthermore, Yoshimoto stated that each area has relative autonomy. Therefore, he did not think that the area of communal illusions subsumes the other two areas. According to his theory, for instance, the three illusional areas have equal priority and importance. Therefore, a nation as a communal illusion cannot violate the world of individual illusions of each individual. Although social powers treat NEET (young people Not in Education, Employment, or Training) and people who have withdrawn from society as an anomaly, Yoshimoto respected their existences in principle. However, the three illusions theory is not a theory that unquestioningly respects people's individuality. Yoshimoto considered that the three illusional areas exist as firm frameworks that impose constraints or limit each person. Therefore, although in the world of a person's individual illusions, it is possible and easy for a person to transcend the worlds of communal illusions, it is not easy to actually weaken and disrupt a nation's power.

Watakushi Shōsetsu

In this chapter, the author presents the concept of "*watakushi monogatari* (Autobiographical-narrative or I-Narrative)." This is not merely an academic concept, but a concept that closely related to the author's research plan of narrative generation for bridging to actual narrative creation in the future. The term *watakushi monogatari* reminds Japanese people of *watakushi shōsetsu* (Autobiographical-story or I-Novel)." *Watakushi shōsetsu* is a genre of novels in modern Japanese literature, which has so far influenced artistic and literary novels and stories and has played a greater role than mere entertainment. It is a unique genre that can have a spellbinding effect on readers. The author's objective in referring to *watakushi monogatari* is not necessarily to use it in the same sense as the I-Novel in Japan. However, as *watakushi monogatari* that the author uses is related to the *watakushi shōsetsu* in the sense of narratives related to private and personal topics through various mechanisms, this section provides a simple explanation about the *watakushi shōsetsu* for readers' reference information.

After the Meiji Restoration, although the Japanese were shocked by Western culture, civilization, science, and technology, they took it in their stride under the slogan of "Japanese spirit with Western learning." In the area of literature, the impact of Western literature was huge. However, the Japanese, after the Meiji era, introduced the Western literature in Britain, French, German, and America and Russian literature through energetic translations. After the Meiji era, many Japanese people began considering their own literature and narrative resources as poor and inferior ones. Modern Japanese literary authors felt strongly, and with a sense of inferiority, that though great novelists also existed, such as Murasaki Shikibu (the middle of the Heian era), Santō Kyōden (1761-1816), Ryūtei Tanehiko (1783-1842), and Kyokutei Bakin (1767-1848), Japan's literary tradition was antique-looking compared to Western literature in the same period and that only non-realistic, fanciful, and ludic narratives dominated. Further, many authors thought that Japanese literature should be reformed. (Simultaneously, a particularly important fact in the Meiji era was that many authors were actually intimately aware of Edo literature. This is comparable to a typical type of psychological tendency of intellectual people in modern Japan who rationally considered *kabuki* as something detestable despite being fond of it as an emotional level.)

In addition, the dominance of and adoration for the West was the highest in music. In the new Meiji government, the music education dedicated to traditional Japanese and Asian music was substantially reduced. By contrast, the Meiji government reorganized the music education system drastically by positioning the fundamentals of musical knowledge on the Western music. Traditional Japanese music was stored and inherited only in the musical band belonging to the Imperial Family in the format of traditional performing arts (*geinō*), such as *nō*, *ningyō jōruri*, *kabuki*, and Japanese dances and songs. However, before the end of World War II (or the Pacific War), many people leaned *shamisen*, *yōkyōku* (the narration with melody of *ningyō jōruri*), and Japanese songs including *nagauta* and *kiyomoto*. However, it can be said that in present-day Japan, the destruction strategy of Japanese traditional music that was begun by the Meiji government has become almost completely successful. In particular, modern-day Japanese people do not have an ear for traditional Japanese music. Currently, many people are almost conscious of the fact that Japanese music works, similar to the Western music, were composed and created by musicians and artists. The knowledge of traditional Japanese music is passed down currently through different organizations, such as via the school education in Kokuritsu Gekijō (the National Theatre in Japan) and the *Iemoto* system. However, or at the same time, the vestiges of traditional Japanese music have clearly been introduced, in terms of forms and contents, into modern and contemporary popular music, especially in the music genre called *Enka*. (The author would like to believe that although reviving and reconstructing the tradition that was once eliminated is not easy, the current situation is not ultimately hopeless.)

Now, in the domain of literature in the Meiji era, although literary works in the West were actively translated, direct introduction was effectively linked to the previous Japanese literary tradition, especially in the Edo era.

In the previous literature, particularly in the literature of the preceding Edo era, many long and complex stories and novels with strong narrativity were written. In a sense, the novels were considered to have similar tendencies as contemporary Western novels. The Meiji era began in 1867. In the West and Russia, in the second half of nineteenth century, many famous writers, including Charles Dickens (1812-1870), William Thackeray (1811-1863), Victor Hugo (1802-1885), Gustave Flaubert (1821-1880), Émile Zora (1840-1902), Fyodor Dostoevsky (1821-1881), Ivan Goncharov (1812-1891), and Lev Tolstoy (1828-1910), wrote long and complex novels. In a broad way, it was the period in which the paradigm of literature shifted to realism and then to naturalism from the preceding romanticism. Toward the late nineteenth century, the naturalistic tendency became strong, as illustrated in the writings of Guy de Maupassant (1850-1893) and Anton Chekhov (1860-1904). However, novels by these writers did not necessarily sacrifice of narrativity, and the Japanese writers of the Meiji era understood this fact.

Among the writers in the Meiji era, or more correctly speaking, among writers through the early to middle era of the Meiji rule, several intensively followed the literary tradition of the Edo era (although Kōda Rohan (1867-1947) and Ozaki Kōyō (1868-1903) consciously utilized the literary tradition, Sōseki Natsume, who was a researcher of British literature, also had very rich knowledge of Edo literature) focused on the sentence level in novels, and eventually resulted in the "*Genbun-icchi Undo* (Vernacular Movement)" aimed at colloquial and realistic novels. Through the history of Japanese literature, for example, in the Kamakura era, narrative sentence style described mainly using *hiragana* that was used as spoken language in the Heian era (794-1185 or 1192) was gradually formalized and became a classical norm. In these literary circumstances, writers in the Kamakura era developed the representation style of *Setsuwa Bungaku* (the genre of narrative literature in Medieval Ages in Japan) to reform the literary style. Additionally, in Edo literature, as the language of narratives and literature became gradually separated from the spoken language of the same period, the Vernacular Movement in the Meiji era emerged as a natural phenomenon. Furthermore, instead of dense narrativity and patterns in novels and dramas (*kabuki* works) in the Edo era, many writers in the Meiji era began to hope to realistically represent the realities of humans and societies.

Although many literary works in the Edo era certainly appear to be non-realistic because of the figurative, formal, and ludic language used, through narratives by Bakin, *ningyō jōruri*, and *kabuki*, magnificent histories and politics were selected as the materials (*jidaimono* in *ningyō jōruri* and *kabuki*), and they simultaneously illustrated, the precise descriptive techniques used to represent various human formations and relations (*sewamono* and *jidaimono* in *ningyō jōruri* and *kabuki*). In particular, literature in the Edo era established both macro narratives and micro narratives. Existing theory suggests that, this could have been possible because of the fusion of realistic literary techniques (including narrative techniques in addition to linguistic techniques) from the Western literature and the narrative tradition of the Edo literature. However, the key Japanese writers in the Meiji era positively introduced only the aspects of realistic, and later, naturalistic description. By contrast, they did not introduce the aspects of narrative and literary structure and construction and further reduced them. The narrative structures of Honoré de Balzac (1799-1850), Dostoevsky, Tolstoy, Flaubert, and Zora were rather serially linked to writers in the Edo era, such as Bakin and Tanehiko, and narratives of *ningyō jōruri* and *kabuki*.

Finally, the direction that required both realism in language and narrative began to consider narrative structures as objects to abandon. Thus, based on realism in language and narrative, many writers attempted to describe the situations surrounding them in as real a way as possible using the spoken language that they ordinarily spoke. Natsume Sōseki and his friend Masaoka Shiki (1867-1902) also presented a literary concept called "*shasei-bun* (literary sketch)." Further, the *watakushi shōsetsu*, a novel genre with a narrator who was an actual person who aimed to describe as faithfully and realistically as possible the small world that the narrator could look at and directly experience, was created, and the literary movement of "naturalism" was developed by a group of writers who believed in the concept of the *watakushi shōsetsu* (The naturalism by Zora tried to describe a whole family by introducing the natural concept of "genetic inheritance" into literature with its essential characteristic being describing various social levels and aspects concerning the family as a whole. Therefore, Zora's naturalism was not different from the *watakushi shōsetsu*). On the other hand, writers who did not intend to adopt the format of *watakushi shōsetsu* and naturalism, such as Kōda Rohan and Izumi Kyōka (1873-1939), were not located in the central position in the literary world in their period (however, their literary position is high according to current literary evaluation).

According to Akiyama, Katsumata, and *Watakushi Shōsetsu Kenkyūkai* (2014), although *watakushi shōsetsu* was a concept that was presented by Kōji Uno (1891-1961), the true origin was explored by various writers, critics, and researchers. Although many textbooks describe that the novel *Futon* by Tayama Katai (1872-1930) (1908) was the first *watakushi shōsetsu*, there are different opinions. The term "*Shinkyo-shōsetsu*" was presented as a similar or related concept. *Shinkyo-shōsetsu* is a concept that emphasizes the part of "fact" in a novel and restricts the interference of the author or the author's mental state in the narration of the events. It is an approach that is being increasingly used to study diaries and *zuihitsus* (essays). Regarding the definition of *watakushi shōsetsu*, including *shinkyo shōsetsu*, which is a unique and original novel genre in Japan, Hijiya=Kirschnereit (1981) stated that, in the *watakushi shōsetsu*, a fact (or facts) experienced by a focused character is (are) narrated and faithfully represented, the writer and reader are mutually understood, "a focused character is a single," and the structure of "the character in the work = narrator = author" is followed. In particular, *watakushi shōsetsu* is a novel in which "the narrator who is a real author" "describes faithfully and without any fictions" "the events (that) occurred near the real author." In many cases, the *watakushi shōsetsu* describes relations between a man and a woman, relations between the author and family, and small events related to an ordinary incident. Readers liked the kind of novels where an author ruined himself by becoming obsessed with a woman and alcoholism. In the case of the author of a typical *watakushi shōsetsu*, linking the author's works would automatically generate the author's autobiography. The tradition of *watakushi shōsetsu* has continued until now, and in developing *watakushi shōsetsu*, novels were divided into "*Jun-bungaku* (Pure literature)" and "*Taishū-bungaku* (Popular literature)." Many readers have a sense that *Jun-bungaku* is equivalent to *watakushi shōsetsu*.

Several *watakushi shōsetsu* criticisms have been offered for a long time. Although the description of the author of this book also contains critical parts, the author does not necessarily criticize *watakushi shōsetsu* thoroughly because it was a literary practice or a rare practice with unique characteristics, which aimed to eliminate the vestiges of literature and *geinō* from the Edo era. Another reason is that though it may be a contradiction, the author considers that the role of *Watakushi shōsetsu* was not merely reducing the effects of previous tradition (if *watakushi shōsetsu* aimed at a complete separation from the tradition of Japanese literature, it has failed). First, *watakushi shōsetsu* has a deep relationship with "*nikki-bungaku* (diary literature)" that existed as a separate and unique literary genre in Japan. Although *nikki* (diary) is a type of discourse in which a writer describes individual events, it has led to the creation of a literary genre in Japan. In modern literature, *Romazi Nikki* by Ishikawa Takuboku (1886-1912) (1977) and *Danchōtei-nichijō* by Nagai Kafū (1879-1959) (2001a, 2001b, 2001c, 2001d, 2002a, 2002b, 2002c) were masterpieces. Many politicians, and even Emperors, also wrote *nikkis*.

According to Furuhashi (2010), although the *Nikki-bungaku* in the Heian era had both directions approaching *Monogatari* (Narrative), including fictional elements and complete description of facts, the author considers that *watakushi shōsetsu* also had both directions or possibilities. In addition, a *nikki*'s description is divided into the description of facts or events and the description of thought concerning the facts or events. *Watakushi shōsetsu* also has possibilities for various directions. Through the lens of the concept of *watakushi shōsetsu*, works by writers, such as Jun'ichiro Tanizaki (1886-1965) and Yukio Mishima, who are considered to be far from *watakushi shōsetsu*, can also be perceived as manifestation of the development of the respective authors' "I (*watakushi*)." In case of Tanizaki, his early short stories include many stories similar to *watakushi shōsetsu,* and in his long novel, *Sasameyuki* (Tanizaki, 1949), his life in Kansai and his actual experiences are described fictionally. *Kamen no Kokuhaku*, an early work by Mishima (1949), was a *watakushi shōsetsu* constructed intentionally as a fiction. In contrast to Tanizaki and Mishima, although Ryūnosuke Akutagawa (1892-1927), in his early career, continued to write highly fictionalized short stories, he wrote genuine *watakushi shōsetsu*s, including *Haguruma* (Akutagawa, 1969c) in his last period. Although Akutagawa-prize is known as the most authoritative literary prize for *Jun-bungaku* (Pure literature), especially for *watakushi shōsetsu* and similar works, this phenomenon is a kind of irony from Akutagawa's literary career. Kōbō Abe (1924-1993) received the Akutagawa-prize. However, Mishima and Haruki Murakami (1949-) did not receive the Akutagawa-prize.

Furthermore, *watakushi shōsetsu* also offers magnificent examples. Shimazaki Tōson (1872-1943) first wrote very aesthetic poems as a poet of romanticism and a fictional novel called *Hakai* (Shimazaki, 2006). However, thereafter, Shimazaki continued to mainly write long novels of *watakushi shōsetsu* and was evaluated as a typical novelist of *watakushi shōsetsu.* However, although his last long novel, *Yoakemae* (Shimazaki, 1966a, 1966b), was certainly a kind of *watakushi shōsetsu*, it was also a magnificent historical novel. The model of the hero of this novel is the author's real father and the story develops according to his father's life. Therefore, the novel's world contains Shimazaki Tōson himself. The world where this novel is developed is a small town in the mountains of the Nagano prefecture in Japan, and the protagonist or hero (Tōson's father) continues to live in this town except for studying abroad in his youth. The viewpoint of narrative is always limited to the protagonist's narrow world, and the tendencies of the world around him are informed to him through only letters, newspapers, and other rumors.

This novel was able to successfully describe concretely the situation in Japan before the Meiji Restoration through the viewpoint of a common person who lives in a small country town. In particular, in this novel, the tradition of *watakushi shōsetsu* achieved an organic synthesis of both every individual person and the wider society to describe a society and period as a whole. Moreover, although the main character of *Aru On'na* by Takeo Arishima (1878-1923) (1995) was a woman and the construction of this novel was not similar to ordinary *watakushi shōsetsu*, the character with divided reason and sensibility was a manifestation of the author (Arishima) himself. *An'ya Koro* by Naoya Shiga (1883-1971) (1973) was more closer to *watakushi shōsetsu*; it was an example of *watakushi shōsetsu* having achieved synthetic narrativity.

Furthermore, author Yūko Tsushima (1947-2016) (2014), the daughter of Osamu Dazai (1909-1948), positioned *watakushi shōsetsu* in the line of *Katarimono-bungei* (Narration-literature) in Japan (Akiyama, Katsumata, and *Watakushi Shōsetsu Kenkyūkai*, 2014). The *Katarimono-bungei* was a large movement in Japanese literary history and Japanese narrative history that began from *Heike Monogatari* (1991, 1993) and *Sekkyōbush* (Sekkyōshū, 1977) in the medieval ages to reach *geinō*s such as *ningyō jōruri* and *kabuki*. This movement is considered to be the most distinct from *watakushi shōsetsu*. However, Tsushima, based on her intuition as a narrator, considered that the tradition of *katarimono-bungei* originally began from the narration of "*watakushi* (I)" by shrine maidens and that the "*watakushi monogatari* (I-Narrative)" by shrine maidens was actually multiple narrations including different "*watakushi*s" in a narration. Her conceptualization that connects *watakushi shōsetsu* to the multiple "*watakushi*s" is an interesting opinion that is overlapped with the author's multiple narrative structures model.

When describing *watakushi monogatari* in this chapter, the author does not intend to imitate *watakushi shōsetsu* in a narrow sense. If the tradition *of watakushi shōsetsu* in Japan developed to have wider and more free directions and possibilities, considering *watakushi monogatari* proposed by the author in the relation with *watakushi shōsetsu* is useful, which is why the author has introduced *watakushi shōsetsu* here.

TOWARD EXTERNAL OR SOCIAL PRACTICES OF NARRATIVE GENERATION

This section describes the developmental plan for both the external (social) directions and the internal (personal and individual) directions of the author's narrative generation study. These directions are related to the bridge from research to content creation. This bridging may influence the development of narrative generation systems (including the Integrated Narrative Generation System: INGS).

The entire narrative generation system (as technology) is a large framework containing "the narrative generation system as a single system" that is equivalent to INGS. It refers to a comprehensive narrative production-consumption model or system in contrast to the narrative generation-reception model or system of INGS. Their overlapping is a concrete form of multiple narrative structures model.

The external development = social development of INGS corresponds to connecting the INGS and the narrative contents produced by INGS, ranging from the collaborative production forms with humans based on supporting methods, to the societies as the external area of the systems, which more concretely expand to the directions of social distribution and consumption mechanisms. The area is positioned in the environment that occupies the outermost area in the multiple narrative structures model. Although a typical form is business, it can include new possibilities such as the virtual *geinō* production company. In addition, trials of pure artistic and literary experiments will also be possible.

Originally, as the author's narrative generation system study itself was an extremely individual and personal attempt, the development will approach the internal, personal, and individual fields in contrast to the above external directions. Simultaneously, internal development will also be conducted using INGS and the narrative production-consumption mechanisms including INGS. The external and social development requires more technological, mechanical, and formal discussion. By contrast, as the internal and individual development requires the author's "personal" knowledge as content knowledge, it needs and stimulates the discussion that is more related to the aspect of narrative content.

Here, the author presents *Geinō* Information System (GIS) as a larger system framework with INGS. Both developments of the external = social and the internal = personal and individual of INGS will be realized through

GIS. In the whole study by the author, GIS has originally been planned as the system for realizing the social development of narrative generation through the multiple and complex usage of the narrative generation system as a single system, namely INGS. At the same time, GIS was a plan that aimed to incorporate the author's individual and personal narrative contents into the mechanism based on various *geinō* genres, especially *kabuki* and *ningyō jōruri*. Thus, the GIS is a system framework for developing and expanding toward both external and internal directions. Although different from INGS, as GIS is still in the stage of fragmental implementation, the author explains GIS according to the vision for execution.

Geinō Information System: The Overall Architecture and Thought

This section introduces GIS and discusses the basic concept for the development from INGS to GIS. The system architecture is currently fluid and will be expanded and revised through future concrete design and implementation. In addition, "why *geinō*" is an important point. For the author, narrative was not literature and arts, but *geinō*. In the narrative genre system (Chapter 2 (Ogata, 2019c) in Ogata (2019b)), narrative as literature and arts was only one part. Further, this is also related to the fact that the author's narrative generation study aims or what the author ultimately aims for through the narrative generation study is not what is advanced and avant-garde, but what is "commonplace" and "non-unique." According to the author, narrative is the farthest from anything represented by words such as "advanced," "avant-garde," and "innovative."

Now, a conceptual composition of the GIS is shown in Figure 1. Although it is based on an original design by Kawamura and Ogata (2000a), the INGS mechanism is included as a part. The architecture will be revised and changed in future progressions. Regarding previous attempts, refer to papers such as Kawamura and Ogata (2000b) and Amino, Kawamura, and Ogata (2002). GIS corresponds to a narrative production-consumption mechanism that includes the multiple use of INGS as a narrative generation-reception mechanism. The system is, at the highest level, composed of a sender mechanism, a receiver mechanism, and a *geinō* history.

Simply speaking, GIS can be summarized as a system that continuously generates *geinō* contents through the interaction between the sender mechanism and receiver mechanism. In the sender mechanism, a variety of

Figure 1. Draft architecture of the GIS

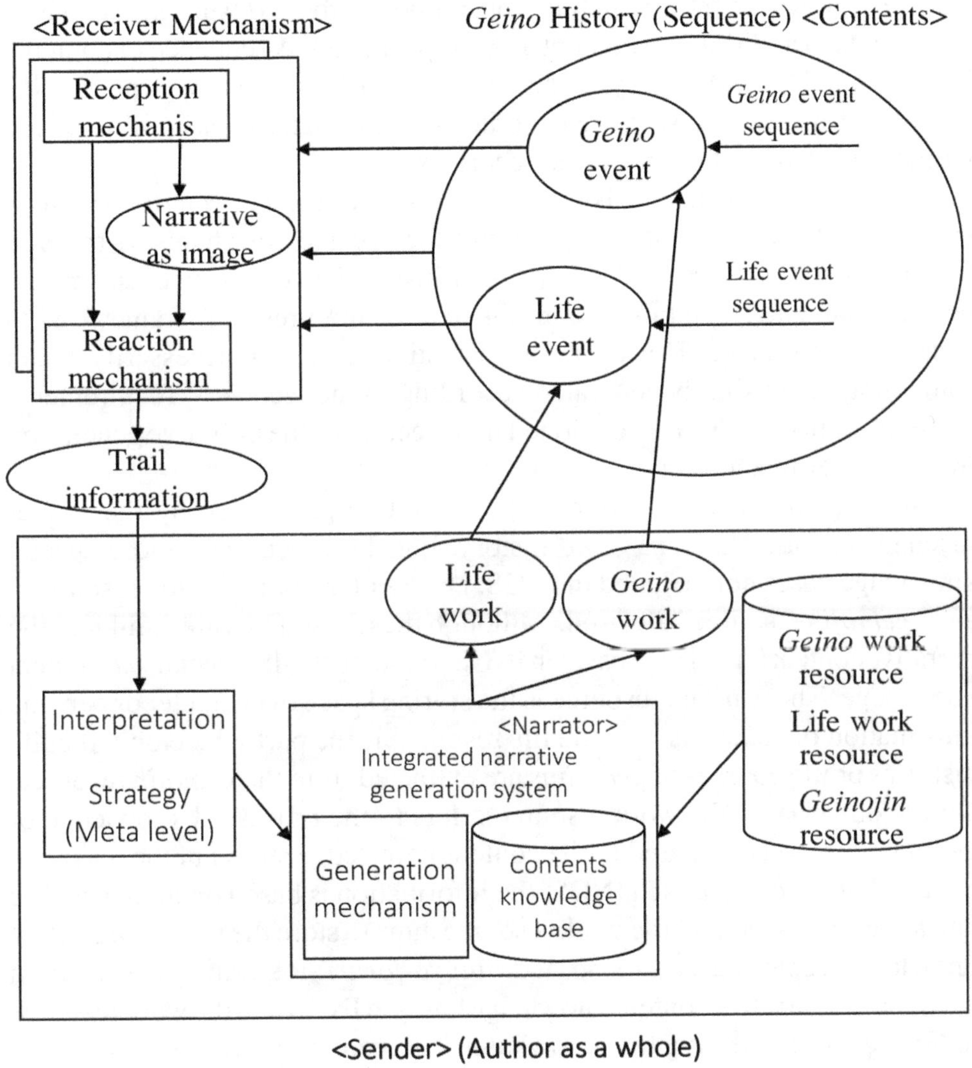

methods coexist and, at the highest level, it divides into two types of sender mechanism: *geinō* works and *geinōjins*. Thus, the GIS is also a system that creates contents relating to the subjects of *geinō* works called *geinōjins* in addition to simply creating contents as *geinō* works. Further, as mentioned in the above overview of informational narratology, receivers of a narrative frequently realize the level at which they themselves are the senders, beyond their positioning as simply pure reception subjects. In this sense, the receivers can also be regarded as the subjects that generate narratives. GIS produces various types of content through the above mutuality between sender and receivers. However, the production-consumption process results in a sequence of contents in addition to a content as a simple object. A sequence of contents is here called "*geinō* history." Namely, from a more macro viewpoint, GIS can be defined as a system that produces *geinō* histories according to the mutuality between the sender and receivers.

First, the sender mechanism receives the external information (various traces of information output by mainly the receiver mechanism) through the "interpretation strategy" mechanism inside the sender mechanism and performs an interpretation according to the interpretation framework to generate a narrative. However, the narrative does not necessarily mean something being simply generated according to the receivers' requirements and expectations. INGS is positioned in the center of the narrative generation by the sender mechanism.

The sequence and reception of the narrative generation by the sender mechanism form the next *geinō* history. For the processing, more special knowledge bases not included in INGS, i.e. "work resource," "life resource," and "*geinōjin* resource" are used. Although these resources are similar to the narrative content knowledge bases in INGS, these store GIS-specialized content knowledge. The *geinōjin* resource is the storing space of character or persona information of *geinōjin*s, and the life resource is the part where only the life histories or life courses of *geinōjins* are extracted from the *geinōjin* resource. On the other hand, the work resource refers to the processed knowledge of *geinō* works, and compared to the content knowledge stored in the narrative content knowledge bases in INGS, the information is based on more specific narratives. In particular, the work resource aims to store the narrative content knowledge centering on *kabuki* and *ningyō jōruri*. Regarding the division between the narrative content knowledge bases in INGS and the work resource in GIS, *geinō*-related content knowledge is stored in the work resource.

The receiver mechanism includes any mechanical reception subjects in addition to the human receiver. These output any reception results based on the reception experiences of an individual work, a sequence of works, and the life history of a *geinōjin*. A *geinō* history is sequentially formed through the circular and repetitive interaction between the sender and receiver mechanisms. The content of a *geinō* history is classified into both a *geinō* event and the sequence and a life event and the sequence. The former means each generated work, and the sequence and the latter correspond to the flow of life of the actor in a *geinōjin*. Works here also include scandals focused on an actor's life itself.

Regarding the division of the content knowledge of INGS and GIS, the knowledge of INGS is more general, and further, each dictionary corresponds to the most general knowledge group. GIS's knowledge as more special knowledge is also divided into large two groups in the axis of generality/specialty. The most special group in GIS is the author's private and individual knowledge, and the various knowledge resources are prepared as relatively more general knowledge.

In addition, although the *kabuki* knowledge mechanisms described in Chapter 2 in this book are also related to INGS, regarding the relations with GIS, many elements are currently included in the work resources, such as "fact (event)," "work," "genre," "material," "story or plot," "place (stage)," "time (era)," "*geinō* style," "text," "production," "direction," and "staging." However, works in GIS do not necessarily indicate *geinō* works and artistic litserary works in the narrow sense. They include scandals about and the life stories of *geinōjin*s. They are rather related to *geinōjin* resources and life resources. On the other hand, only a minor element was related to the *geinō* history and reception mechanism. Precisely speaking, the *geinō* history is related to comparatively many elements such as the historical features of works, characters, and actors, and staging. In the future, surveying and analyzing *kabuki* focusing on the dynamical aspects of narrative production and consumption will be required to extend and revise more concretely the knowledge resources in GIS. Thus, GIS modeling will help design the survey and analysis of *kabuki* more organically.

The following sub-sections describe in detail GIS architecture that is structured based on the sender and receiver mechanisms and *geinō* history. The author also discusses several special topics.

Sender Mechanism

The following three parts explain the design of the main mechanisms of which GIS is composed. The first sender mechanism receives the external information, which mainly includes traces of information that are outputted by the reception mechanism in GIS through the GIS mechanism of "interpretation strategy" to select an interpretation according to an interpretation framework. Lastly, the sender mechanism generates the information based on this interpretation. However, this process does not necessarily mean that GIS simply generates narratives according to the desire and expectation of the receivers. INGS is placed in the central position of narrative generation by the sender mechanism.

The sequence and repetition of generation by the sender mechanism forms "*geinō* histories." In order to continuously generate *geinō* histories, GIS uses special knowledge bases called "work resource," "life resource," and "*geinōjin* resource," which are not included in INGS. Although these are similar to the narrative content knowledge bases in INGS, the unique content knowledge is contained in GIS.

The *geinōjin* resource is equivalent to a place where *geinōjin* information is stored. The *geinōjin* information includes the information on *geinō* performers and actors/actresses as well as their personal information. In particular, the life resource is the part from which only the life histories are extracted in the *geinōjin* resource. The work resource means the knowledge used to process *geinō* works using various methods. In particular, it comprises information based on more specific narratives in comparison with the content knowledge to be stored in the narrative content knowledge bases in INGS. Further, in this mechanism, the author aims to store the narrative content knowledge with a focus on *kabuki* and *ningyō jōruri.* The life resource comprises the information that analyzes and processes the information relating to *geinō* works using various methods.

The work resource in GIS is prepared to store more specific and concrete narrative information compared to the knowledge representation of the narrative content in the narrative content knowledge bases in INGS. For example, analyzing and studying *kabuki* aims at the storage of the specific narrative content knowledge, including various information and knowledge relating to a *kabuki* work. In particular, knowledge in INGS is more general and several of the dictionaries in INGS are placed into the most general knowledge group. The knowledge of GIS is also categorized into two major groups on the axis of generality/specialty. Although the most special knowledge is private and personal knowledge of the author in Ogata's (2018b) research plan, various knowledge resources are prepared in GIS as general knowledge more than the above knowledge type.

Although the contents generated by the sender mechanism of GIS include contents of life works, in the wide sense, such as scandals (documents) of *geinōjins* relating to *geinō* works, in addition to the narrative contents in the narrow sense, this discussion focuses on *geinō* works in the narrow sense, such as dramas. To give an overview for the production, first, GIS mainly decides the following elements: "authors" (single or several), "performers" or "actors/actresses" (ordinarily several), and "chronicle (the part or parts)." One or more authors decided here first create the "story" of a narrative and, for this story, the "characters" and "places, stages" (of which there are usually several) are selected and decided upon. Based on the above, INGS in GIS exercises the entire narrative generation process from the overview stage to the detailed stage. Both INGS and GIS enact the entire process by performing their unique respective tasks and through their mutual interaction. In addition, of course, in case of such narratives as dramas, the stage performance occurs after the scenario is produced. If INGS can enact this process, the possibilities to be considered may include the CG and robot representation, mainly through the narrative representation mechanism. However, here, with the limitation to the generation at the level of scenario, the author would like to describe the various concepts that appear in the sender mechanisms.

First, GIS has several authors similar to authors working in the theatre in *kabuki*. In this sense, the author her/himself is also a kind of *geinōjin* and has her/his life story. In the case of a dramatic or theatrical narrative, the author, for specific performers, chooses and sets particular places = stages and the historical context to produce a narrative content where specific characters are actively performing. A GIS system is akin to a theatrical group and also holds a collection of limited numbers of performers (actors, actresses, or *geinōjins*). Each performer also holds the stories of their respective life histories. The performers are related to the *geinōjin* resources and life resources in Figure 1. Technologically, the main body of an author is equivalent to INGS, which generates each work utilizing the knowledge resources involved in GIS in addition to unique knowledge.

The knowledge resources also include the "characters" extracted from existing *kabuki*, *ningyō jōruri*, modern and contemporary novels, et cetera. The characters in *kabuki* and *ningyō jōruri* are already fluidized as they commonly appear among many works in ordinary cases. By contrast, in modern and contemporary novels, although there are examples of a specific character appearing in many works by the same author, they do not reach such sufficient fluidity as do *kabuki* and *ningyō jōruri* narratives in which a specific character freely plays an active part in the different works written by different authors, although this phenomenon may, from a larger perspective, be considered to be the fixation of a character. An experimental direction will also be possible to intentionally create such a narrative situation. This is related to the work resource in Figure 1, in addition to the next topics of places = stages and chronicle. GIS also stocks or organizes the typical places and stages that have appeared in previous narrative works. Furthermore, they can be designed as a knowledge base based on the reference of the works (work resource) from *ningyō jōruri* and *kabuki* to modern and contemporary novels.

Kabuki and *ningyō jōruri* have prepared typical epochs, such as the eras of Sugawara no Michizane (845-903) and *Heike Monogatari* (1991-1993), particularly in their association with places = stages. These correspond to main components along with characters and places = stages. "Worlds" similar to them in *kabuki* are not explicitly defined in modern and contemporary narratives. However, in fact, many narratives are produced according to the framework of each age divided by the Meiji Restoration (1868), the Pacific War (1941-1945), the age of high economic growth (1954-1973), and so on, although, as in the following description, they still do not seem to reach the stage that is established as a kind of historical or chronic common illusion. One research (very distant) possibility is to inductively construct a chronicle in each epoch from existing narrative works to be utilized by GIS and INGS.

Chronicle and Narrator

Here, the author would like to mention the problem with respect to chronicle and narrator. In the following description, the narrative representation mechanism is placed in the narrative discourse mechanism in INGS. The content, as the thing narrated, which means a narrative discourse structure in the wide sense, is divided into the structural and deep level and the representative and surface level. The latter level also includes the generation of music and image in addition to language generation. A basic method is to process all surface representation mechanisms in the semiotic level similar to language. The generation subject in the narrative representation mechanism is also called a narrator. Thus, the narrator in INGS is a mechanism covering both the narrative discourse mechanism and narrative representation mechanism. Additionally, the former is called the narrative discourse mechanism in the wide sense and a whole mixing both the former and latter is called the narrative discourse mechanism in a broad sense.

On the other hand, the story generation mechanism is a qualitatively different mechanism. In the current INGS, the author clearly calls the generation subject in the narrative discourse mechanism "narrator" (in addition, Hasumi (2014), perhaps dependent on his idea that "novel" is different from "narrative," uses the word "speaker" instead of "narrator"). However, the generation subject in the story generation mechanism is not named clearly, although clearly a different mechanism from that of a narrator needs to be used for the story generation mechanism.

Compared to the "historical narratology" by Noe (1996, 2007), Kan (2010), etc., the above narrator means a subject who concretely narrates a historical material. The thing that corresponds to this "historical material" is a story or narrative content in the present components of INGS. Who produces, composes, and prepares this content? This problem is not necessarily clear in the case of a real history. However, in the historical case, it will be possible to differentiate the thing called a "chronicle" from the layer corresponding to a story. In this case, the layer of chronicle + story is set on the stage before the narrative discourse by the narrator. Further, a chronicle or story must be constructed by subjects who differ from the narrator in the case of narrative discourse. The author regards a chronicle or story as one subject here. Who created them? A chronicle means a discourse as itself in which events that occurred are described within a specific temporal range. Although a history uses past events as its materials, it needs to be edited and organized in chronicle form because people cannot witness the real occurrence themselves from the materials. The subject of such a chronicle must differ from the narrator who, in a sense, freely narrates the materials. It is straightforward to say, the name of the "chronicle author" or simply "author" may be given to the existence. Namely, it also includes the meaning of an author of real events or materials. However, the materials of a chronicle are themselves frequently a fiction, in the case of a fictional narrative. Simultaneously, from the viewpoint of the narrator of a narrative discourse, a group of events for the materials will be relatively able to be interpreted as real events. Thus, the following shows the possible division: the generation subject of the story generation mechanism (including a chronicle author) is simply called an "author"; the generation subject in the narrative discourse mechanism is called a "narrator."

The author would like to consider the above problem using the example of the story composition of, so to speak, *jidaimono* in *ningyō jōroru* and *kabuki*. Tetsuya Imao (1931-2013) (2010) discussed the classification between *jidaimono* and *sewamono*, which is in contrast to the *jidaimono*. Generally, a *jidaimono* story is produced dependent on a history that preliminarily exists as a so-called communal illusion or collective illusion (Yoshimoto, 1968). Such a history corresponds to the chronicle prior to the story and the narrative discourse (in the broad and wide senses). The story of *jidaimono* arranges and rearranges, as if it slips through in the chronicle, many historical facts, including characters, places = stages, and times in an attempt to create the reverse story, which is hidden in the lower layers of the story, according to the framework of necessary chronological and historical facts. Concurrently, in this case, a better or appropriate direction is to define the composition of "chronicle/story/narrative discourse" rather than the above composition of "chronicle+story/narrative discourse." In the method of narrative production based on such composition, the chronicle layer in the three stages of chronicle/story/narrative discourse is simply an existing fact that is too difficult to shift. In this case, the author may be able instead to regard both story and narrative discourse as the level corresponding to the narrative discourse in INGS, such as "chronicle/story+narrative discourse." When this *kabuki* method is applied, it will be able to treat the level of a chronicle as that which corresponds to a part of the narrative content knowledge stored in the knowledge bases. Thus, the present story generation mechanism and narrative discourse mechanism are equivalent to the narrator who alters and edits those structures and contents. This is a possible method to be considered.

On the other hand, however, Japanese history from modern to contemporary, which will be dealt with in Ogata's narrative study (particularly in the phase of future actual narrative production), has not still been holding, so to speak, provisions and determinism as seen in histories that authors of *kabuki* and *ningyō jōruri* had previously treated as their objectives or materials. In current society, people are in a whirlpool, a fluid situation, where histories, which are not limited, are being generated. This means that we are existing in the way that chronicles themselves are produced, defined, and determined. Although a minimum "chronicle as a historical chronology" exists, it has not been established in the case of modern and contemporary ages as a communal or collective illusion of people as it was in previous ages. Alternatively, although there may be various possibilities of fixation, a fluid situation that does not result in the ending is continuing. This is because, through the temporal flow from modern to contemporary, many events for which the evaluation and positioning have not been established are proliferating and the world of historical events has not been boiled down to a common illusion. At the same time, there have many attempts to bring out a chronicle or chronicles from prewar days to after the war via media, literature, and cinema in present-day Japan. The present situation must be a step wherein the hot composition process of a history or histories, a chronicle or chronicles as a collective illusion is just progressing. Concurrently, in this situation, the evaluation of what has really occurred is also shaking in a complex manner. The basis itself of story or narrative generation, which is dependent on a determined chronicle or chronicles that believe in a common illusion, as past *kabuki* writers did, has not been established in the above fluid situation.

Therefore, even in the case of treating history, a chronicle itself does not exist as the proposition and background of the story. At the practical level, a possible method to be considered is to spin stories to fill the gap between existing historical facts (which exist in a huge number). In this case, a chronicle is a kind of framework for the mere background situation or information for establishing a story. A chronicle is also defined as a type of macro framework at a similar level to characters and places = stages. Concurrently, because story generation continues according to a chronicle as a framework that has neither efficient meaning nor basis, and based on recognizing its fact, the chronicle itself frequently changes. Alternatively, an essential issue regarding the present narrative is in the work of creating a chronicle, and further in the acts of not creating some chronicle.

An essential problem regards what narrative information is to be given preliminarily. This means the starting point of the narration by the narrator in the narrative discourse. In particular, should a chronicle be given or not? In the first case, the author (as a tentative term) of a story creates a story on the basis of this chronicle and the narrator transforms it into a narrative discourse. On the other hand, the second case requires the author for the chronicle. Further, based on the chronicle, both the author of the story and the narrator are required for a narrative discourse. Another idea of the conceptual division is to instead integrate the author of the story and the narrator of the narrative discourse into the same category described as "narrator A/narrator B" to indicate the difference of levels. Alternatively, the chronicle and story could be given the name of the author to describe them as "author A/author B." The latter, author B, is a naming method or concept definition for a relatively non-deterministic chronicle. In contrast, the former, author A, is set for a relatively deterministic chronicle. These refer to a method or model for defining the different features of authors according to the characteristics of the chronicles. Another method or model is, dependent on a preliminarily given chronicle, to define both the author who is the subject that composes and edits the chronicle and the narrator who is the subject of the narrative discourse.

According to the above discussion, from the perspective of the relationships between INGS and GIS, what relationships exist among chronicle, story, and narrative discourse? If chronicles are previously stored in the knowledge base, the level of an author corresponds to that of a subject who generates various stories using those chronicles. On the other hand, a narrator corresponds to a subject who narrates the generated story using various methods, namely narrative discourse techniques. However, there are several levels in chronicles. If a chronicle description itself is preliminarily given to GIS as an already completed object, it is not necessary to especially set the level of the authors of chronicles. By contrast, if anyone must describe a chronicle by a considerable effort, the descriptive subject must be added to a part of the subject of narration as an independent author of the chronicle. Thus, as a possibility, GIS includes three types of subject: the author of the chronicle, the author (or author of the story), and the narrator. Most fundamentally, although the chronicle's author is a unique subject for GIS, the author of the story and the narrator are the subjects of narration in INGS. On the other hand, as a more special form, there is the author of the story at the level of *geinō* history. A *geinō* history is, of course, a story composed as a result and the author cannot create it completely intentionally. However, if story generation at the level of planning also exists, the story generation of a *geinō* history can also exist, at the level of planning at least. Furthermore, the narrative discourse corresponding to a story at this level will also be able to exist efficiently. The subjects of narration for these narratives, including the story as a *geinō* history and the narrative discourse, are considered to be contained in GIS.

Finally, in this study, the definition of INGS and GIS may not be narrative generation systems at a personal level. They are, rather, narrative generation systems that strongly consider the narratological, structural, and communal levels. If structural and narratological ideas are inserted into the idea of communal illusion by Yoshimoto (1968), the characteristic as communal illusion is the degree issue and the narrative discourse techniques can also be interpreted at the communal level. Of course, the layer of chronicle is the communal level. In terms of the degree issue, the narrator of a story (and chronicle) is a narrator at the more communal and structural levels, while the narrator of a narrative discourse is a narrator at the more personal level. Concurrently, it is necessary to prepare mechanisms to properly utilize the level of "narrator" as a large framework in accordance with the stages of narrative generation or to change the degree of usage.

Receiver Mechanism

A receiver mechanism outputs, through various methods, the reactive information based on the reception experiences of contents including a work, a sequence of works, and the life history of a *geinōjin*. The receiver for GIS includes humans and, further, virtual existences that are programmed, such as artificial readers and artificial audiences. The subject of a narrator also includes both virtual existence or an artificial author and a real existence or human, from the viewpoints of functions or agents. From the level of

GIS, one of the narrative generation-reception functions, such as INGS, can also be human. Designing GIS for an entire narrative generation system by a human-machine system is one of the real and actual methods used. This does not necessarily mean that humans compromise the processing for which complete automation is impossible by a machine. GIS sequentially forms a *geinō* history through the iterative and circulative interaction between the sender and receiver mechanisms. Of course, the receiver mechanism can be comprised entirely of reception subjects where each of the subjects is a machinery mechanism.

The receiver mechanism evaluates, by some method, generated narratives, including single works, and further their sequence, to represent them as responses in some form. In particular, the receiver mechanism outputs the response information based on the reception experiences for a *geinō* work, a sequence of *geinō* works, the life histories of *geinōjin*s, and so on. In the architecture of GIS in Figure 1, the term "trace information" means fragmental information in various forms of the evaluations and responses from receivers. The trace information, in the content phenomena of real societies, includes providing direct information to the sender using language, such as discourse, e-mail, blogs, critiques, and *geinō*-related news as more indirect methods, as well as rumors, reputations, word-of-mouth communication, scandals, and so on. The trace information in GIS is equivalent to various expressions of the reception information by both real and virtual receivers and is regarded as a parameter referred to in various senses by the sender mechanism.

Although literary theories, including reception theory and reader-centric criticism, are certainly newly focused on the side of receivers in literary works, there are no clearly systematized studies that efficiently enable the application and usage in narrative generation systems in these areas. Evaluating a narrative is always very subjective and vague. Although literary histories are considered to be described and constructed based on objective evaluation of literary works as far as possible, it is also clear that each literary history is a very ideological institution or social system. Further, the reception theories of Hans Robert Jauss (1921-1997) (1970) and Walfgang Iser (1926-2007) (1976) clearly demonstrated the historical variability of literary evaluations. GIS utilizes the basic idea of Akimoto and Ogata (2013) that attempted a very simple experimentation of the evaluation and generation cycle using virtual or artificial receivers, in the narrative discourse mechanism in INGS.

Narrative and literary evaluation is also very difficult with respect to psychological meaning. For example, people often use language representations such as "interesting (面白い)" in reading a literary or narrative work. This word, "interesting" or "interestingness," is frequently used for research and is a complex and subtle concept that synthesizes a variety of senses and nuances in this case. The "interestingness" of a narrative is a more general word, although the synthetic and complex characteristics of narratives are not different from those of research. However, the evaluation criteria for "interestingness" are not so simple, even though or perhaps because people frequently and ordinarily use them. There are also historical differences in terms of what people find interesting. For instance, in the local situation in contemporary Japan, in the personal opinion of Ogata, from approximately the 1970s to 1980s, very readable and understandable texts were lionized, excessively distributed, and flourished. The trend of this age was linked to the dislike for unreadable and incomprehensible texts. The trend of hating such discourse as "difficult works are interesting" has recently risen to an extreme and, for instance, some people, who name "authors of novels," have published a book criticizing "difficult literary works that they have not read" as if boasting of this fact. This may, however, be a symptom of returning to an age in which people find "difficult literary works" interesting. On the other hand, from personal experience, the feeling of reading James Joyce's (1882-1941) *Ulysses* (1922) and finding it extremely interesting can also be regarded as a fact. Although it is, perhaps, influenced by the tide of liberal principles in the current age, what is interesting or what people ought to find interesting is a kind of institutional norm and there are also people who efficiently feel the interestingness of such works that are generally recognized to be difficult. Additionally, as a related topic, the narrative generation game system by Ono and Ogata (2018) conducts a study analyzing the functions and techniques of "surprise" in stories and narratives.

Another problem is related to the point that the evaluation by the receiver in GIS is directed to various objects. It may be directed to a single work, the life contents and events of *geinōjin*s, and the sequences of the respective contents. The evaluation and response from, so to speak, a fixed point for a single work form the complicated information that makes up the traces of narrative reception, in the relationships with the evaluation and response through a temporal flow for the history and sequence of works that include the first work. To process this kind of trace information within a system, it will be necessary to classify and analyze the objects to be evaluated. Further, because a closed evaluation within a system is not actually useful, experiments will also be necessary in the form opened to business-like narrative distribution processes. In this case, if only evaluations by receivers and consumers are privileged, more specific public works will be evaluated as good. However, this kind of majority rule is only one of the possibilities of narrative and literary evaluations. Moreover, evaluating one's own works from the sender side will be able to be an element to be considered. In particular, a work that has not even one reader has been read by the author through the writing process. Therefore, the author can also be a receiver of the work at a minimum. For example, the self-satisfaction by an author herself or himself may also be a possible evaluation criterion. This standpoint emphasizes self-satisfaction rather than selling or being read by many readers. In such cases, a perverted logic will also exist, such as that not selling proves the high evaluation.

In addition, this problem of multiple or plural evaluations is related to a merit in the multiple narrative structures model. In particular, if, in this model, the narrative evaluations can be diverse and plural, the narrative generation and distribution can be operated in a direction where diverse and plural evaluations are retrieved from the receivers. This may be a merit in considering narrative social distribution.

Additionally, in *kabuki* and *ningyō jōruri*, complex stories were sometimes created. "*Naimaze*" (Miura, 2000) is a narrative technique for complex narrative creation. For instance, Osamu Hashimoto (1948-2019) (2012) criticizes the extremely complex story, as if rejecting the readers' understanding, of the scenario of *ningyō jōruri* of *Honchō Nijū-shikō* by Chikamatsu Hanji (1725-1783) and others (1766). The scene reaching the *seppuku* of Hayano Kanpei in *Kanadehon Chūshingura* (1948) uses a narrative technique in which the character, Kanpei, does not know what the audiences know. In the episode of Kojirō, the sacrifice of Taira no Atsumori (1169-1184) in *Ichinotani Futaba Gunki* by Namiki Sōsuke (1695-1751) and others (1751) uses a narrative technique in which the other characters and audiences do not know of the event Kojirō and Atsumori know. Such techniques concerning mutual recognition between characters and asudiences (Oikawa & Ogata, 2012; Ogata, 2018c) are sometimes used for complex stories and narrative developments. Such complicity of stories is not necessarily associated with how interesting stories are. By contrast, it is sometimes discussed in relation with the poor quality of a narrative work. This problem is also related to the multiple narrative structures. However, the problem of abnormally complex stories may be a consequence of a single author's individual narrative techniques and creation strategies. In particular, the problem may be the narrative techniques in an institutional sense.

GIS WITH INGS AND *GEINŌ* HISTORY

The concrete mechanism of GIS is designed as a complex mechanism in which, ordinarily, several INGSs are contained. As stated above, although a *geinō* history is gradually extended through the mutual association between the sender and receiver mechanisms, the author will consider this problem based on the viewpoints of the following several narrative generation concepts covered by Ogata's (2016, 2019a, 2019d) narrative generation research: (1) circular narrative control, (2) narrative fluidity and fixation, (3) narrative norm-deviation, and (4) narrative generation-reception and production-consumption. Further, the relationships between INGS and GIS are also included within this perspective.

Figure 2 shows the execution principle of GIS (including INGS) that has the above elements. This figure indicates that, in a narrative generation process as a generation-reception or single form, a narrative generation is performed iteratively and cyclically. This narrative state, seen as a synthetic whole, corresponds to a fluid state in narrative generation. In this iterative and cyclical narrative generation process as a fluid state, the receiver side, comprising one or more humans or one or more artificial subjects created virtually within a system, functions as a kind of control mechanism for the narrative generation. A content as a thing that is relatively fixed is frequently pushed to the outside of the cyclical process through the interaction between the narrator side and receiver side. In particular, in the fluid, iterative, and cyclical generation process, a fixed narrative content frequently jumps to the outside from the ring of the circulation. This external or outside world is, for instance, the world of narrative distribution. This theoretical framework calls the level of generating externalized and fixed narrative contents themselves the narrative production-consumption process or system. Concurrently, the level of narrative generation in the above fluid situation is called the narrative generation-reception process or system. A kind of communication as generation control is also performed among the receivers who form a part of this narrative production-consumption process. Additionally, the receiver, in the production-consumption process in addition to the narrative generation-reception process, can be seen as the receiver as a single or collective object. Furthermore, in the narrative production-consumption process, the interaction, which is similar to the association between the sender and receiver in the narrative generation-reception process, is performed to form an iterative and cyclical process.

Figure 2. The execution principle of GIS in the relationship with INGS

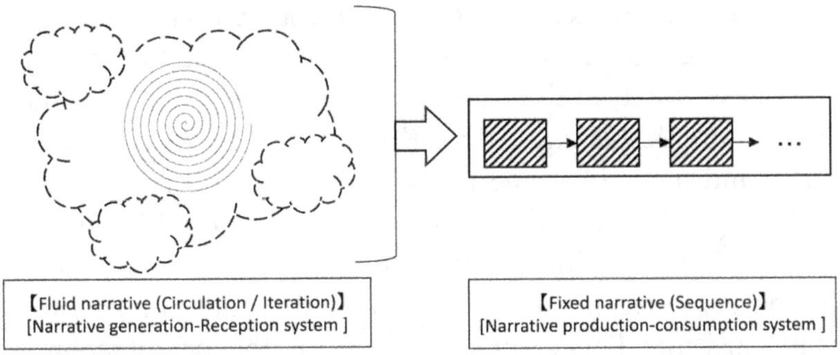

As described above, by *geinō* history in the framework of GIS, the author means a sequence of contents at various levels that the sender and receiver mechanisms gradually create through their mutual relationships. This repeated and circulative narrative generation is a core part of GIS.

However, this iterative and cyclical characteristic is also formed, in several respects, in the generation-reception process of INGS as a single system. One is the circulative and iterative generation between music and a conceptual narrative, and INGS enables the generation order from music to a conceptual structure in addition to music from a conceptual structure (Kobayashi & Ogata, 2004a, 2004b; Ogata & Akimoto, 2007; Akimoto, Endo, & Ogata, 2013; Akimoto & Ogata, 2014). The circulative and repeated characteristic here also means the non-specification of the processing order. INGS mechanism aims to realize the circulation and repetition in such a sense in all sides of generation. Similarly, the circular narrative control is also used in the narrative discourse control mechanism that introduces the theory by Jause mentioned above to perform the circular and iterative processing of narrative discourse that theoretically goes on ad infinitum, through the interaction between the narrator and narratee who are virtually set in the system. This is a mechanism that models a narrative generation process as the response and intentional deviation by the narrator (sender) of the requirement and expectation by the narratee (receiver) to repeat narrative discourse generation based on their relationship. As explained above, the circulation and repetition in INGS system means, at first, free and flexible control that does not determine the order of processing among the modules and representation forms, and, secondly, is concretized as narrative discourse control based on the informational feedback from the receiver.

GIS including INGS need, in addition to the above, to have unique circulation and repetition. First, the idea of narrative discourse control using reception theory (Akimoto & Ogata, 2012, 2013) can be extended to the production-reception process. Next, in the framework of the informational narratology (Ogata, Kawamura & Kanai, 2018), Kawamura (2018) and Kanai (2018) rather understand narrative generation processing from the perspective of reception. The large framework is that the receiver side subjectively proposes the parameters for generation and production, while, on the other hand, the sender side purely performs the generation and production according to the receiver's proposition. This reversal is, in itself, not special technologically. At the level of GIS strategy, this type of mechanism means multiplization whereby sender- and receiver-oriented strategies are treated equally. The flexibility of the processing order is also related to the narrative's plural characteristic. An important point is how narrative production-reception can be multiplexed and an entire narrative diversified, including the generation-reception. In the future, discussing narrative multiplicity and diversity using *kabuki* as material will be concretely linked to the design and development of such system architecture.

Next, the author would like to think from the perspective of the fluidity and fixation of narrative generation. INGS is divided into the following large level generation phases of story, narrative discourse, and narrative representation, and this division shows that each side of a narrative is generated by respective generation phases. Finally, "a whole narrative is formed" can be considered to indicate that diverse narrative possibilities are hidden in each step of the generation, namely the fluid state, and they are finally fixed into a narrative. In GIS, as stated above, INGS is placed in the mechanism to exercise a relatively more fluid narrative generation and the specific parts in GIS in the narrow sense, except for the part of INGS, are positioned in the same way in the mechanism to ensure more fixed narrative generation or production.

However, to create a more essential mechanism, the processing of fluidity and fixation needs to be performed recursively. The above description comprised a model in which the narrative generation process as a single substance is considered an undifferentiated and fluid processing, and each content that was fixed in the fluid situation forms a narrative sequence in the production-consumption. Moreover, there can be a situation where this sequence itself becomes fluid. Next, the iterative process, in which the fluid state in this next step again results in a fixed state, recursively occurs and the movement of the re-fixation regarding a sequence of fixed narrative contents as a fluid situation occurs once more. Ultimately, such circulation and repetition continues ad infinitum. A possibility of a method for implementing such a model in GIS is the processing through which fixed narrative contents by GIS or the sequence are returned to INGS once more as some knowledge elements. In addition, because the fact that such processing and process are themselves fixed is clearly a problem, it is also necessary to make such situations fluid at the meta-level or from different perspectives. When an entire society is seized as a large narrative production-consumption apparatus, each narrative generation and reception within the society also becomes a fluid process that aims at fixation on a higher level. Although its scale is relatively small, GIS is a mechanism that can be considered more technologically and explicitly and is a system, viewed as a whole, that multiply stores the fluid narrative generation-consumption processes. When the author considers the concepts of fluidity and fixation through the recursive feature, the framework itself will be able to be regarded as fluid.

Multiple Narrative Structures: Relationships Between Narrative Generation-Reception Mechanism in INGS and Narrative Production-Consumption Mechanism in GIS

The narrative production-consumption model is a model that includes several narrative generation-reception models to perform the transformation, ranging from a single model, such as a news-story generation system to social model such as a television company as the collection of a single mechanism. The multiple narrative structures model which is one of the basic concepts in the author's narrative generation study deals with a narrative generation phenomenon as a multiple and hierarchical object, and the multiplicity and hierarchical feature are related to the division of story/narrative-discourse/narrative-representation in the generation-reception mechanism. Each generation-reception mechanism in a narrative production-consumption mechanism is considered as a state based on fluidity. In particular, each narrative generation mechanism becomes an element in the narrative production mechanism through the process in which the fixation in a timing of the fluid and continuous generation and circular and repetitive generation becomes an element in a larger narrative generation process. From a more macro viewpoint, in the situation in which a variety of narratives is always generated through a circular mechanism, only one narrative is fixed in a timing to generate a large narrative through the collection. For example, in a television company, single-level narrative generations are always conducted by several journalists, reporters, and writers (they are accompanied with internal reception acts). A sequence of narrative production-consumption is produced through the externalization and fixation of a part in the narrative generation process. GIS fixes a part of the narratives that are continuously generated by the inside narrative generation-reception mechanism (INGS) to produce larger narratives and change the corresponding *geinō* history. These results influence the parameters for the generation and reception of INGS to change the fluid state. Simultaneously, from a more macro perspective, GIS is also in a fluid state ordinarily. Where does the fixation in this case exist? To consider the problem, we will need to return to the problem of narrative generation society or a large narrative generation mechanism of a society (this will be an important theme for the next book).

Dependent on the above consideration, the author will design and implement INGS in association with the development of INGS. The entire narrative generation system with INGS and GIS is conducted using generation control based on main concepts including circularity and fluidity-fixation. Further, it is multiply realized through the narrative generation-reception process as a single body and its complicated narrative production-consumption process. Although INGS and GIS respectively conduct fluid narrative generation and fixed narrative generation comparatively, all stages and levels of narrative generation, in their inside and the relations with their outside, essentially hold the characteristics and mechanisms of fluidity and fixation. Moreover, INGS corresponds to the narrative generation-reception process as a single body, and GIS plays narrative production-consumption. However, both processes have mutual relationships beyond a simple division, such as the production-consumption process has circularly returned to the generation-reception process.

Abstractly speaking, the author's objectives in the development of external narrative generation are the social development and distribution of the narrative generation systems themselves and the contents generated by the narrative generation systems. The social development of narrative generation systems means that the narrative generation systems themselves are socially developed and distributed through any other system. In the process, various users use the narrative generation systems. In the second social development and distribution of the contents generated by the narrative generation systems, the users buy the contents, for example, as books, to read them. The social development and distribution of narrative generation for the author means constructing the mechanisms and social systems that contain both of the above aspects, and they are organically associated each other.

The first development and distribution of narrative generation systems, provided by the system developer, i.e. provided by the author, mean that generated or generating narrative generation systems are incorporated in the social system for the users' usage. On the other hand, the end user uses the narrative generation systems for any practices of narrative generation. The forms of narrative generation systems include INGS as a single system, INGS with GIS or GIS with INGS, and specific application systems (including INGS as a single system or GIS with INGS). In particular:

- The user makes a narrative in collaboration with INGS, GIS, or the application systems. For instance, the collaborations include the following possibilities:
    - INGS generates the plot or story, and the user makes the detailed parts to complete the entire narrative.
    - By contrast, the user makes the plot, and INGS generates the detailed parts to complete the entire narrative.
    - Mixed form of both the user and INGS.
    - GIS creates a character's story, and INGS and the user generate the narrative using the character.
    - GIS and INGS generate several narratives and a sequence of narratives at a larger level, and the user enters into the narrative.
- Another possibility is that INGS and GIS generate narratives for which looking and appreciating are more the main objectives than reading:
    - For example, there are completely automated experimental spaces similar to theatres where INGS and GIS automatically generate a narrative with automated generated music and CG image, and further, the associating robots and automated puppets play automatically.
- Improvisational narrative and literature can be realized.

On the other hand, in the second style, the narrative contents generated and produced by narrative generation systems, i.e. INGS, GIS, and various further application systems, are socially developed and distributed. The system developer's objective is to develop such higher-level systems. On the other hand, the users (appreciators and readers) receive the narrative contents generated through narrative generation systems using any system. The user buys specific narrative content by selecting from the many prepared and diverse contents. A possibility of narrative generation systems is the mass production of narrative contents. An opinion about narrative mass production concerns the difficulty of evaluation. As GIS prepares the evaluation mechanism of generated narratives, the evaluation mechanisms for the generated narratives need to be incorporated into the narrative generation systems. The narrative evaluation can be conducted according to both the human's standards and mechanical methods.

Implementing the above integrated system as a business system is a rational method in the current society. Regarding the relations of narrative with business and management, various studies have been conducted. One type is the research on the production, distribution, and reception of contents at various levels. Moreover, one approach analyzes from the narratological viewpoint various business themes, including organizational cultures, human behaviors, and management strategies. In Japan, Kanai, Takai, Nakanishi, and Morioka (2009) analyze the multiplicity of human behaviors in organizations according to the multiple narrative structure of *Rashōmon* by Ryuūosuke Akutagawa (1969a) and Akira Kurosawa (1910-1998) & Itō (1950), and Kusunoki (2014) analyzes management strategies as a type of problem caused by story generation. Recently, many attempts have been made to engage in conscious story generation and narrative generation from the perspective of management strategies (for instance, Baddilo, Donovan, & Trevarthen, 2016). By contrast, the author's idea is not directly related to the narratological tendency of management and business practices. Many existing content studies do not also incorporate the level of automated generation of contents themselves into the framework. The author will progress the problem of narrative production-consumption; however, it will adopt a management science perspective based on the structure of entire narrative generation study, i.e., in the positioning as a partial and organic mechanism in the structure.

TOWARD INTERNAL OR PERSONAL PRACTICES OF NARRATIVE GENERATION

This section discusses considerations for internal narrative generation or, in other words, actual narrative creation. It also presents some basic data for narrative creation.

Narrative Generation to the Inside or Toward *Watakushi Monogatari*

The description of the social development of narrative generation systems through the *Geinō* Information System (GIS) in the first half of this chapter was fundamentally neutral and general in that it elucidated the element of the author of narrative contents. If we understand that system to be a tool, then the system creators become tool suppliers, normally sharply differentiated from those who used that tool to produce specific contents. However, one of the goals of the author's narrative generation research is the production of a personal work that in some sense is the author own, although not simplistically linked to so-called autobiographies or *watakushi shōsetsu*s (I-Novels). Furthermore, going beyond the level of tools and their usage, the author is also aiming for the form of a tool that is at once a work. The words "Narrative Generation to the Inside" in this section's title signifies the author's narrative generation on the personal and individual level, meaning a narrative work or narrative contents created through a process of private narrative or *watakushi monogatari* (I-Narrative).

The author has previously had a variety of opportunities to talk about *Watakushi monogatari* and the idea of private research for the purpose of creating just that (Ogata, 2000, 2011a, 2011b, 2011c, 2011d, 2011e). The *watakushi monogatari* referred to here signifies a narrative based on the author's private themes, subjects, and topics. In the context of narrative generation systems, it has to do with the use of the author's private and personal knowledge. This corresponds to the inclusion of private and personal elements as knowledge, prepared and acquired in advance, for a GIS equipped with an INGS.

There is a lot activity on the research front on life histories (life courses) (Saito & Honda, 2001), autobiographical memory (Sato, Ochi, & Shimojima, 2008), and oral history (Mikuriya, 2007). A sociological study by Ito (2009) discusses the restoration and generation of alcoholics' self-narrative. Narrative therapy (Morioka, 2008), which is done in the field of clinical psychology as a development from psychoanalysis, has traditionally dealt with issues relating to the composition, recomposition, and reorganization of an individual's private history, but Kato (2003) and Eguchi, Saito, and Nomura (2006) bring together issues of personal narratives, including clinical psychological approaches, within the broader framework of psychiatry.

However, as mentioned above, the narrative generation to the inside or *watakushi monogatari* that the author is thinking of does not necessarily narrowly signify the author's purely personal narrative. The word *watakushi monogatari* reminds us of the *watakushi shōsetsu*, and to the extent that an *watakushi shōsetsu* takes the form of a narrative that intermingles fiction on a basis of private and personal information, the *watakushi monogatari* discussed here resembles the *watakushi shōsetsu*. Yet in actual fact, the *watakushi shōsetsu* narrowly defined is likely something else, which means that the *watakushi monogatari* discussed here differs from the *watakushi shōsetsu* (narrowly defined). As one grows older, there is a tendency to want to write down autobiographical text or investigate and record the achievements of ancestors. Very few of these are normally distributed through commercial publishing routes, but make up one of the principal genres of the world of self-publishing together with hobby novels, poetry, and essays. It would seem that the author himself is also no exception to falling into that trap of mediocrity.

Nonetheless, as exemplified by Futabatei Shimei's (1864-1909) *Heibon* [*Mediocrity*] (Futabatei, 1949), Robert Musil's (1880-1942) (1930, 1932) *The Man without Qualities*, Ōgai Mori's historical biographies like *Shibue Chūsai* (Mori, 1999) and *Izawa Ranken* (Mori, 1973), James Joyce's (1922) *Ulysses*, and Yukio Mishima's (1971) *Ten'nin Gosui*, the original meaning of the *novel*, as representative of modern narrative, is the unheroic narration of fragments of the lives and ways of life of the perfectly ordinary person, the mediocre human being that anonymously walks the street. In fact, what the author is aiming for (or what the author aspires to) is perhaps this kind of novel in the original sense or a variation thereof. Moreover, the "topic of I or myself as a topic," which the author will discuss below, is nothing more than a topic for show, selected by chance—it is not Ryūnosuke Akutagawa's (1969b) *Kappa*, but we cannot ourselves decide our greatest fate, which is our own birth—to express such mediocrity, and in that sense its private nature is perhaps rather a general one, and taking it further, it is perhaps what can be termed "an other one." Being *without qualities* is a quality in itself, so summarizing the above, we should maybe call it "I/other narrative without qualities," but that doesn't quite roll off the tongue, so the author decids to compress all those meanings and call it *watakushi monogatari*.

The *watakushi monogatari* is created by using GIS or INGS in a special way. The systems consist of an independent, general organization, based on which it imports what can be called personal and private knowledge contents. Potentially, the "I" that exists external to the systems can perceive the narratives created and expressed there as "fluid" and can create a narrative on another

Figure 3. Image of watakushi monogatari generation/production based on fluid and fixed

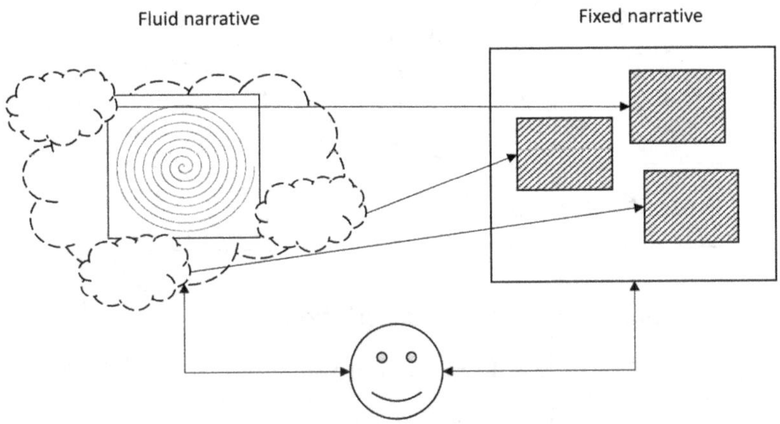

new level—a *watakushi monogatari* that is fixed. Alternatively, the cycle or repetition of narrative generation by the GIS equipped with INGS may aggregate all fluid narratives created at some point to create a "*watakushi monogatari* that has been fixed." GIS can be thought of as a mechanism for setting up the "external I" of INGS within the system. It is possible to imagine a variety of other possibilities. In the end, the aim is the building of an organization that brings together multiple concrete mechanisms.

Figure 3 shows in abstract form a narrative concretization mechanism based on the concepts of fluid and fixed. Documents on the personal and private level are correlated with narrative contents knowledge bases on various INGS levels as well as GIS locations. What the author wants to focus on in this chapter is not the aspect of narrative generation mechanisms, but the personal and private knowledge contents itself, on the foundation of which the mechanism for creating the *watakushi monogatari* is built. This knowledge contents makes up the most personal and private knowledge contents and knowledge base used by the GIS with INGS. The GIS as a whole, including the INGS, continuously and repetitively generates/gives birth to fluid narratives through its interface. The fluid narratives are cyclically, repetitively, and continuously generated using the GIS's troupe (including actors) and stage equipment. In this fluid narrative generation, the process itself can be identified as a work, where the receiver can observe various aspects of that process at any time using the interface. This expression can happen in a variety of forms. At present, this generation image is not especially distinguished from the

Figure 4. Overview of knowledge contents needed for watakushi monogatari

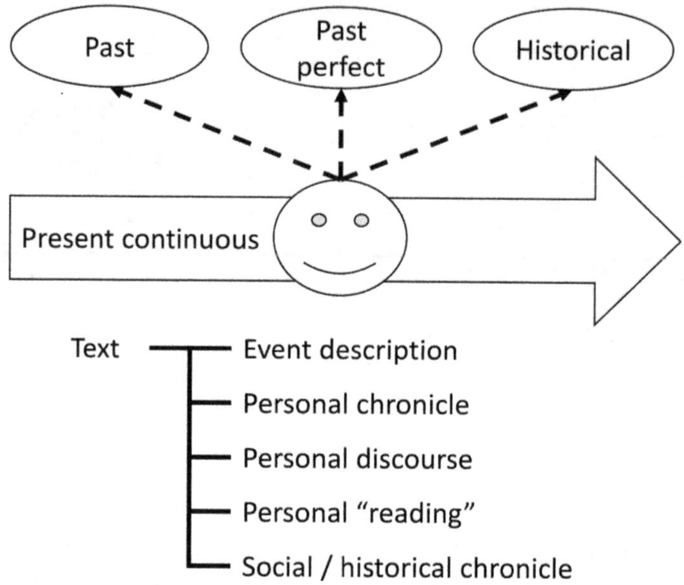

general GIS narrative generation image. The work to reach a conclusion on this point is still ahead of us.

Furthermore, the descriptive text making up the knowledge contents needed for the *watakushi monogatari* as discussed above can also be seen as an independent unit by itself or as several partitioned units of writing. This is the knowledge used to create narrative contents through the narrative generation system, but at the same time, it in itself makes up a text that appears like narrative contents in various senses. As such, it is not necessary for the narrative contents to be used exclusively inside the system, but it is possible to try to publish them in books and magazines, on paper and online, as multiple writings appropriately partitioned and ordered in some form or contents. We can imagine a series has the words *The Narrative Generation of XXX* in all its titles (in this case, "narrative generation" simultaneously includes the meaning "anti-/non-narrative generation" (Ogata & Kanai, 2010). Although this is also connection with the "categorization of personal texts" discussed in the next section, they make up a unit belonging to the category criticism and thought (critiques and essays), contrasting with the academic papers that have been at the center of the author's writing so far. Moreover, the fluid form of the *watakushi monogatari* (operated in computers or online) and its fixed state (expressed in the forms of books, be they paper-based or electronic) make up another group called narrative works and narrative contents.

Overview of Contents for *Watakushi Monogatari*

Figure 4 summarizes the overall private and personal knowledge contents for creating *watakushi monogatari* through the use of GIS with INGS. The "temporal divisions" signify the extent of time needed to cover the materials prepared, while the "text" is the types of text to be concretely collected or created and edited within that timeframe.

Temporal Divisions

As shown in Table 1, the basic structure of the private and personal knowledge contents is texts correlated within four types of temporal division.

Text Categorization

In accordance with the abovementioned temporal divisions, the author constructs a chronological account of events considered as facts as represented by the following list. However, the contents of "personal texts" and "personal readings" should not be written anew, but should be edited and organized.

- **Event descriptions**
- **Personal chronologies**
- **Social and historical chronologies**
- **Personal texts**
- **Personal readings**

Event Descriptions

Firstly, there is a category called "event descriptions," which is a temporal and chronological list of past matters thought to have happened to the author—or which "I myself" think have happened—as well as a somewhat detailed explanation thereof. Outside of that, there are descriptions of matters directly

Table 1. Four types of temporal division [the author uses "I" instead of "the author" in this table]

Temporal Division	Overview	Concrete Description
Present Continuous	The first temporal division of **present continuous** refers to a temporal division that has been proceeding since March 11, 2011 and that will undoubtedly continue to proceed for the time being.	It started with the day on which occurred the so-called Great East Japan Earthquake, and me, the period indicated by this temporal division is one from which I can survey the past and construct the future, meaning an extremely dynamic period. At least, the private and personal knowledge contents of the *watakushi monogatari* is composed centering on the main axis of this present continuous temporal division. In that sense, this temporal division is likely to act as a kind of perspective (base) for surveying the whole.
Past	The next temporal division of **past** is the memories that demarcate *my myself*, which I define as the period from November 25, 1970 to the abovementioned March 11, 2011. The first day is the day Yukio Mishima died, and subjectively, that was the day my conscious and self-aware *days of life* started. I lived through days at school (elementary school, middle school, high school, university, graduate school) and at work, like most people do. In my life course, which tends to turn our atypical probably due to personality issues, I also had a few years of drifting unemployment. On April 1, 1997, my home as a working professional became the university and it has remained so to this day. If the above is a general framework for matters of a shared illusory past, then there must also exist anti-illusory matters and it should be possible to somehow determine individual illusory matters as well. Furthermore, since there ought to be a correlation between the consciousness that does not allow matters in the form of such facts to surface and the atypical act of narrative creation via narrative generation systems, we will basically avoid describe this kind of matters directly. In this sense, the *watakushi monogatari* differs from the *watakushi shōsetsu* (I-Novel) or autobiographical records narrowly defined.	I decide that this basically covers the period from around 1925 to November 25, 1970. Of these years, the segment from 1958 to 1970 make up my days of unconscious autobiography. I refer to the segment from 1925 to the end of this period as the past perfect. By contrast, the temporal division from November 25, 1970 to March 11, 2011 corresponds to my days of conscious autobiography. This is how I distinguish the two. I can trace memories of matters relating to myself that happened during this past perfect division until November 25, 1970 back to a certain point. I remember my first day of elementary school (April 1964) and I also remember watching the opening ceremony of the Tokyo Olympics on TV at home on October 10 of the same year. I also remember starting to watch what was probably the first episode of the TV cartoon *Obake no Qtarō* (1965-1967), based on a work by Fujiko Fujio (Hiroshi Fujimoto (1933-1996) and Motoo Abiko (1934-)), at some place like my father's company office (while eating sweets I got from a female employee), and going home in the middle while crying. My father worked at a big iron company next to the sea in Kawasaki, and I was born close to there, but we moved to a hilltop house in the outskirts of Yokohama with a direct train connection to the Kawasaki company while I was still very young, so I have almost no clear memories of living in Kawasaki. Yet I have numerous memory fragments in my head, such as a sky dark with smoke, orange flames bursting from the factory chimneys, a landscape of the sea and gigantic buildings, passenger airplanes passing right overhead after departing Haneda Airport, and people chatting in a narrow back street (yet these might include many scenes and narratives that I made up later on). This past perfect transcends myself and connects with a sphere of relations with others—and especially with those whom I have a blood connection. The work of clarifying this goes beyond my own memories and demands various objective labors such as listening to stories and exploring records, making it a kind of restoration work, where the work will perhaps lean more either to memories or to records. Moreover, the line between objective and subjective is far from clear in this work.
Past Perfect	I define a third temporal division that I call **past perfect.**	This refers to a temporal division where *I* become indeterminate or where I don't exist objectively, but people directly connected with me, such as my parents, do.
Historical	I also define a fourth temporal division that I dub **historical.** This refers to a temporal division where I myself don't exist of course, but the existence of those connected with me is also indeterminate and there is a vast historical space transcending the private where time intermingles without distinction.	Yet it is impossible deal with the entirety of that vast realm, so I have no choice but to extract certain historical scenes based on some criteria that relate to myself.

related but not directly experienced by the author, making up what could be thought of as an outer edge of the author's experiences. Specifically, this should center on the narratives of the author's father and mother and other surrounding narratives. Furthermore, the author can add matters in the past space—time that are unrelated to the author, but in reality selected because of certain criteria or reasons. In this way, the author describes and organizes matters in accordance with three temporal divisions—present continuous, past, past perfect.

We can imagine a variety of possibility for their use. For example, it is possible to have an organization that places present continuous event descriptions starting from March 11, 2011 at the center, from which past event descriptions and past perfect event descriptions are derived. In concrete terms, we have present continuous event descriptions in the form of something like a diary at the center, and in those descriptions, there appear pointers that connect with past and past perfect event descriptions and it is from those pointers that we can go back to the past and the past perfect. The aggregate of those is past event descriptions and past perfect event descriptions. If so, looking from the perspective of the present continuous event descriptions or from a dynamic perspective, the only thing that exists is always just the event descriptions of each moment, so that the past and past perfect event descriptions are gradually formed in step with the progression of the present continuous event descriptions. We can also see the organization of the descriptions as each text having been created and edited independently, but that the abovementioned arrangement is made in the form of meta-level organization.

As for the way the descriptions themselves are written, they are not necessarily so structured or formalized, but can also be written as regular text. If so, those descriptions themselves become narratives that can be read as themselves. The reason for making independent the personal chronologies described next is to facilitate such event descriptions as regular text.

This is a type of hypertext. The private and personal knowledge contents needed for the *watakushi monogatari* considered in this paper has an overall structure where each entry follows a temporal framework as well as links with a chronological order. It is edited as a hypertext that takes as its units all possible levels, including the word level, the level of partial or whole sentences, the level of some kind of textual extent, and the level of whole works or texts. The primary aim of regular hypertexts is receiver convenience, but here the main goal of the hypertext structure is sender use. In this way, the knowledge contents texts edited in a hypertext format are also used as material for data analysis as well as utilized in INGS and GIS after undergoing various types of treatment. The knowledge is not necessarily used directly, but the natural language could be divided into simple and incomplete conceptual expressions or conceptual units, and then be processed for storage in the INGS narrative knowledge base. It is possible to acquire and variously make use of frequency and co-occurrence information for the divided words.

Personal Chronologies

Seen from a bottom-up perspective, we especially refer to the chronological information of the individual-level temporal divisions (past perfect, past, present contunuous), made up of the abovementioned aggregate of event descriptions, as "personal chronologies." This corresponds to the organized time management of the event descriptions, excluding the detailed explanations. Conversely, from a top-down perspective, it becomes an indicator for event descriptions on the individual level, a kind of index for reaching personal and detailed event descriptions. It is structural indicative information that can be used when creating event descriptions. Past perfect event descriptions are compatible with the social and historical chronologies described next.

Social and Historical Chronologies

"Social and historical chronologies" are general chronological descriptions that do not overlap with any account of "my own life" as the author and can very broadly be divided into three areas——one is a chronology of Japan during and after the Meji era, another is a chronology of Japan before that, and the last one is a chronology of other relevant regions that are not Japan, meaning the world. When actually trying to describe this, it is necessary to have some kind of foundation that relates to myself as the author in some way. Rather than something based on objective records, it might be better to call it something approaching the realm of imagination and illusion. Moreover, since Japanese and world history are texts and narratives in their own right, this also has to do with the issue of what historical narrative "I myself" as the author choose.

Table 2. Types of the author's texts

Categories	Overview	Sab-categories
Published academic contents	Includes papers and essays in academic journals (paper-based and electronic) and books.	(1) Academic papers (published in journals) (2) Academic papers (published in conference abstract books, technical reports, bulletins, reports, etc.) (3) Academic papers (published in books) (4) Essays and miscellaneous texts (published in journals, abstract books, etc.) (5) Degree papers (bachelor's thesis, master's thesis, doctoral dissertation)
Unpublished contents related to academic activities	Unpublished texts related mainly to the abovementioned academic papers and other research activities.	(1) Papers rejected for publication (2) Applications for grants, etc. (3) Unfinished or early drafts of papers, etc.
Materials from conferences and classes	Projection material, handouts (resumes), etc. related to talks and lessons.	(1) Projections and handouts from conference presentations, etc. (2) Projections and handouts for class (3) Records from talks and lessons (photos, videos, audio, etc.)
Contents that was not published	Contents stored privately without it having been presented or published anywhere.	(1) Novels (or something similar) (2) Poems (or something similar) (3) Critiques and essays (or something similar) (4) *Tanka* (a type of Japanese short poems) poems (or something similar) (5) Diaries, journals, records, notes, etc. (6) Letters (paper) (7) E-mails, discussions done electronically, etc. (8) Photos, videos, etc.
Lesson records and large volumes of notes from self-study, research gatherings, and other occasions written for record or contemplative reasons		

Personal Texts

This is a chain of many statements written by the author in the past and in the present continuous (Table 2). The list is a kind of discography, but the list does not necessarily contain only presented or published statements, but it is a more comprehensive and inclusive or personal and private list that also includes unpublished contents. Ideally, it is an aggregation of all texts written by the author to date, including texts that the author simply did not publish anywhere as well as texts that were not approved by the system (including documents rejected including papers and even research plans). It becomes an enormous project if the author wants to include remnants of manuscript revisions, but it goes without saying that the author does not think to save them because they are excellent, but the author saves them simply because they exist, doing nothing but saving them as a record of a man without qualities. In that sense, it might be more meaningful to save things that are not excellent. As such, these are private things, but they are also *other* things that are normally dissolved in statistics, and the *I/other narrative without qualities* includes such things as well. The author categorizes the author's personal texts from 1970 to 2011 into the following four types. When it comes to the description format, it is perhaps more helpful to use a purely chronological order, as attempted in Shinchōsha's 30 volumes of *The Complete Works of Kōbō Abe* (1997-2009).

Moreover, the author obviously wrote texts also prior to 1970, but most of those have been disposed of. Besides compositions, research projects, and other texts that the author wrote at kindergarten and elementary school, the author spontaneously wrote quite a lot of texts for "my own amusement" for the author, such as the *imagined and fantasized timetable* imaginary baseball teams, homemade comics, something akin to novels, and other things the author does not quite remember. In particular, the imagined and fantasized timetable came about when the author made up an imaginary railway company, imagined and made by the author station names, the structures of major stations, designs of surrounding town areas, a railway-centered route map, and train designs and names. Based on that, the author created a train timetable for one day across several notebooks, which to the author as an elementary school student was a big project taking several years. As such, the author read thick timetables and different kinds of railway magazines as well as went beyond timetables for readers to start making train diagrams (the railway magazines frequently included examples of diagrams from railway companies), but that is around when the author's passion subsided. The author's memory is indistinct, but the author seems to recall that the author wrote most of it from the author's third to fifth year of elementary school. This is unrselated, *Garden, Ashes* by the Yugoslavian writer Danilo Kiš (1935-1989) (1975) tells the story of a person (the narrating boy's father) who keeps writing an enlarged version of an encyclopedic timetable and finally disappears.

The category of personal texts from 2011 to the present includes all new publications that the author keeps producing as well as texts and records that the author writes on the side, but at the center lies the *diaries* or *journals*. As the author mentioned, amid their brief swelling, they organically connect with all kinds of texts and records discussed until this point (this suggests a gigantic and mediocre narrative) and all of that then flows into the *watakushi monogatari* created using GIS including INGS.

Personal Readings

Table 3 includes all or main content types that the author has read until this point. However, this is figurative, because it actually also includes everything that the author has *heard* and *seen*, yet if the author were to summarize it all and call it *appreciated*, that would give it a strange nuance, so the author will say *read* instead. This is of course likely an expression of the author's disposition to think centering on reading. Specifically, this includes novels, music, pictures, movies, TV series, radio, newspapers, magazines, and plays. It also includes things that the author is reading in present continuous. The author considers the following literary texts as the principal subjects of reading, investigation, analysis, and dissected use as they are deeply related to the contents side of narrative generation systems as well as their existential side——since it is a list not formed by research necessity but personal interest and curiosity. The categories mixes ones where genres and authors are identified with others where that is not the case. Categories like *kabuki*, *ningyō jōruri*, and Yukio Mishima carry the meaning of subjects that should be treated or that the author wants to treat with exceptional care

Furthermore, in Ogata (2010) of Ogata and Kanai (2010), the author attempted a comprehensive list of all books the author had read until a given point, but that was actually a preparation for the work of systematizing private and personal knowledge that the author is talking about here. Although that list was not exhaustive, the following section includes part of it (1973-1974) as a concrete example and shows in Figure 5 a conceptual image of a mechanism for generating *watakushi monogatari* that connects with other knowledge. In terms of private and personal stories for the author, the research and development of INGS and GIS is conducted to realize this kind of image.

Narratives as Fluid Process: Personal View

This section is based on Ogata (2010) but has been partially revised and given a table for better readability. In addition, this section uses the first person, "I, my, me," etc., instead of the third person, "the author," etc. to indicate the author himself.

Let us assume that the fact that a *novel* takes a certain form means the existence of fixed contents. By contrast, for example when a person reads it, that *novel* changes from something fixed to something fluid in their head. Moreover, it is made fluid even by imaging the process of it being written

Table 3. Contents read or received by the author

Categories	Overview
Japanese literature from antiquity to the present (works: narratives, novels, poems, waka, haiku, *etc.)*	This includes literary texts belonging to Japanese (language) literary history in a broad sense, containing not only literary works narrowly defined but also ideological and philosophical texts. In this sense only, it is the kind of *Japanese literature* prescribed by Shūichi Katō (1975, 1980). Yet here there is a division between works and commentaries. Firstly, the "works" category covers narratives, novels, and poems. The category also includes detective stories and popular literature from Rampo Edogawa (1894-1965) and Seichō Matsumoto (1909-1992) to the present, reading done indiscriminately without emphasizing or being so conscious of the reading itself, thus differing from the reading of novels that the author has read consciously. As such, it is difficult to precisely define and exhaust this category.
World literature from antiquity to the present (works: narratives, novels, poems, plays, etc.)	This becomes centered on the West, but it includes literary works from different countries from past to present and including the present continuous.
Japanese literature from antiquity to the present (commentaries, critiques, miscellaneous writings, etc.)	Miscellaneous writings and thoughts make up a distinct genre in Japanese literature, a tradition that connects with modern essays and critiques. Moreover, this also goes for publications like monthly reports, expositions of commentaries, and commentaries and book reviews in newspapers and magazines. This is the same for **world literature from antiquity to the present (commentaries, critiques, miscellaneous writings, etc.)**, so that it includes commentaries, ideology, critiques, and philosophy in a broad sense. Furthermore, it also contains the category of research papers and materials, which includes academic research papers that are not commentaries and reviews in an even narrower sense.
Music (especially Western classical music and jazz, but perhaps the truly important are the other genres)	Besides records, CDs, DVDs, and other standalone media, this also includes live houses and other texts in the form of one-time experiences, just like with *kabuki* and *ningyō jōruri*. As for the timeframe, this centers on the mainly Western classical music that the author listened to from around 12 to my early 30s and the jazz that the author has passionately listened to after that and especially in the 30s, but there are also other listening experiences that creep in on a large scale and on an everyday basis without the author actually having any special intention to listen to them, such as Shōwa-period pop songs, *enka*, pop, folk, new music, rock, chanson, and other forms of modern popular music (although these are also used in the TV programs and movies listed below) as well as music from advertisments and on the street (stations, shops). It is extremely difficult to define and exhaust these. Moreover, *kabuki* and *ningyō jōruri* are naturally very musical as well, so they cannot be separated from musical experience.
Video works like movies and TV	They can be stored as videos and other forms of fixed media, but watching things at cinemas or at home are one-time experiences just like the above, meaning that they require some form of record to be made into text. The experience of watching videos through the super-everyday medium of TV is especially important. It includes a rich variety of categories like drama series, news, and entertainment as well as a rich variety of subgenres.
Kabuki *and ningyō jōruri and texts (research and commentaries)*	This includes texts on the level of literary work such as scripts (for *kabuki* and *ningyō jōruri*) as well as texts that research and comment on those, all texts as letters written on paper, mainly in the form of books. Another important group are those that the author watched (in reality, this also includes *listened*), which primarily means performances of *kabuki* and *ningyō jōruri* that the author watched at theaters, but also includes those that the author saw through media like images and video. Of the latter, performances recorded as videos on a personal level can be stored like books, but watching a stage is a one-time experience, where auxiliary materials like programs and performance information can be stored as forms of text, but the experience of watching itself does not become a text unless the author records it in some way. As such, this part has a strong connection to event descriptions of the author's actions. Even so, any kind of reading possesses the quality of direct and singular experience. What matters here is that the actual experience of watching a performance cannot be stored as a text unless there is an indirect record made by the author (although this does not obstruct the existence of a reading record), which distinguishes it for example from reading experiences of books that in themselves can be stored.
Works by Yukio Mishima (and related research and commentaries)	As with *kabuki* and *ningyō jōruri*, the author creates this category as independent from the others. Besides texts written by himself, this also includes research and commentaries relating to Mishima. Moreover, the author hs read a considerable number of the works in this category twice or more, so there needs to exist some kind of marker relating to such event descriptions.
Other miscellaneous narratives	The author has not seen enough these to warrant an independent category, but the author has experiences of intensively watching theater other than *kabuki* and *ningyō jōruri*, especially during the author's university days and the years thereafter. Likewise, the author has also appreciated comics, art galleries, and museums, so this category contains anything that falls outside of the categories above.

Figure 5. Conceptual image of watakushi monogatari generation

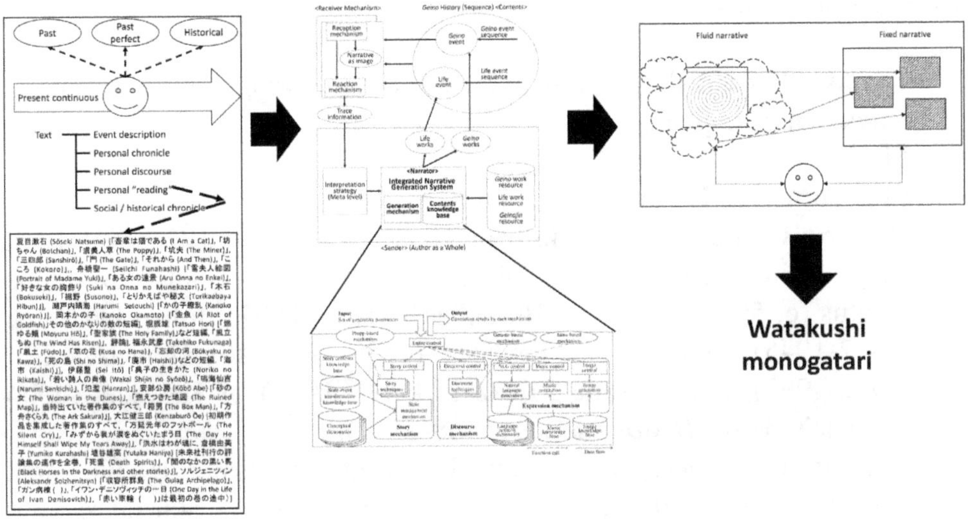

or generated. This means that it exists as the kind of *text* that is discussed in narratology, and in that sense, we can think of it as fluidly opcn to a varicty of possibilities. The *novel* is open and fluid, yet also closed and fixed. I frame the *novel* as something possessing such a duality, governed by such a sense. The topic here is about approaching the simultaneity (simultaneous existence) or contradiction of the fluid and the fixed in the *novel* from a certain perspective that is not analytical, but speculative or methodological.

Speaking of which, this concept of *novel* that accompanies such an awareness and sense is something experienced by me in a limited timeframe, formed through extremely limited volume and quality of reading. Tracing back old memories, my reading until 1970 was largely unconscious, but become conscious after that (the dividing line is somewhere after Jules Verne (1828-1905) and around Sakae Tsuboi's (1899-1967) (1952) *Nijūshi no Hitomi*). During the twelve years from 1971 to 1983, I read quite a lot of *novels*. Nonetheless, it seemed that there were those who read more *novels* among my fellow middle school, high school, and university students (yet toward the end of this period, the proportion of my reading taken up by *novels* was gradually decreasing), and it also seemed that there were those who consciously, analytically, and critically read *novels* as objects of research and criticism. This is nothing but my subjective *novel* experience or perhaps simply a confirmation of the fact that my notion of *novel* came out of somewhere.

Moreover, my reading during that period did not just include *novels*, but also the following:

- **Poems**: *Waka*, ballads, *tanka*, Japanese and foreign modern poetry, *haiku*, Chinese poetry, etc.
- **Plays**
- **Philosophy and thought**
- **History**: Includes everything from Herodotus (c. 484 BC-c. 425 BC), Gaius Iulius Caesar (100 BC-44 BC), Publius (or Gaius) Cornelius Tacitus (c. 56-c. 120), Carl Burckhardt (1818-1897), Johan Huizinga (1872-1945), and Arnold Toynbee (1889-1975) to Kinji Imanishi (1902-1992) and Ichisada Miyazaki (1901-1995) of Kawade Shobō Shinsha's *World History* series (1968-1972).
- **Art commentaries**: I first received a strong impression by Jun Etō's (1988) *Sakka ha Kōdōsuru* [*Writers Act*], and early on started reading numerous studies on and introductions of writers, literary critiques, and art commentaries either written in the form of *expositions* of *novels* or published in art magazines or newspapers. Seiichi Yoshida (1908-1984), Yukio Miyoshi (1926-1990), Shun Akiyama (1930-2013), Miyoko Tanaka (1936-), etc. In this sense, I likewise read quite a lot of records, music magazines, music commentaries in newspapers, art books, and art commentaries at exhibitions.
- **Psychological and mental analyses**
- **Natural science books**

This includes both *literature* narrowly defined and *literature* broadly defined. Yet here, as a way to isolate the preparatory work for writing a methodological introduction to the (super private) *novel*, I will limit my account to tracing my initial experience of the *novel*.

I should mention that if we switch our attention from the private to the social, as will be discussed later on, this period came right before the time of literary diffusion, so the *novels* I actually read were many canonized classical literature, modern literature, and contemporary literature. Of course, when I was reading the *novels*, it was not my intention in the slightest to do so for the sake of cultivation, but in this sense, when seen through the eyes of someone who knows what happened afterward, that reading can be evaluated as having been *traditional and honorable* as well as culturally refining. Considering that it was untouched by the circumstances of post-1980 Japan, this knowledge about and these memories of *novels* are perhaps fossilized. The thickest trunk of my memories are the *novels* that make up the reality of *cultivation* and thoroughly contributed to depreciating their predominance, taking literary discourse as their accomplice.

If I were to espouse an extreme generational argument, the disassembly of cultural refinement was promoted by the collusion of *babyboomers* and *nerds*, and if I were to term this assertively, this fossilized reading suggests a *rejection* of both (passive and active, respectively). The *novels* extracted from the memories below are generally of that nature, but the ones listed here are not all those that I read in that period. Since I am relying exclusively on memory without referencing any diaries or reading records from that time, there ought to be quite a few books that either I do not remember or left no impressions. In fact, the following were included as well, but I took them out of the actual list.

- **Detective *novels* and mysteries for my amusement**: Includes Rampo Edogawa, Seichō Matsumoto, and Seishi Yokomizo (1902-1982) as well as Misa Yamamura (1934-1996), Kyōtarō Nishimura (1930-), Jirō Akagawa (1948-), and Yasuo Uchida (1934-2018). I read quite a lot, but with a few exceptions, I did not subjectively consider this the reading of *novels*. Moreover, it was before and after this period that I really read a lot.
- **Atlases, field guides, and encyclopedia**: A representative work is the *Heibonsha World Encyclopedia* (Hayashi, 1955-1960).
- **Railway timetables and railway magazines**: Includes railway magazines such as *Tetsudo Journal* and *Tetsudo Pictrocial.* More specifically, this category was on par with or more significant than *novel*-like and narrative texts during my period of unconscious hobby reading.
- **Comics**: Mitsuteru Yokoyama (1934-2004), Osamu Tezuka (1928-1989), Fujio Akatsuka (1935-2008), Sampei Shirato (1932-), Shigeru Mizuki (1922-2015), Kazuo Umezu (1936-), Gō Nagai (1945-), etc.
- **TV**:
    - A vast amount of monster series, starting with *Ultra Q* (1966) and *Ultraman* (1966-1967).
    - Suspense dramas, detective dramas, drama serials, and one-off dramas.
    - Comedic light performances for TV, including *Tenamo'nya Sandogasa* by Makoto Fujita (1933-2010) and Minoru Shiraki (1934-), Conte 55gō by Jirō Sakagami (1934-2011) and Kin'ichi

Hagimoto (1941-), the Drifters including Chōsuke Ikariya (1931-2004) and Cha Katō (1943-), Ryūtarō Kamioka (1942-), Beat Takeshi (Takeshi Kitano) (1947-) and Beat Kiyoshi (1946-), Ken Shimura (1950-), and Hitoshi Matsumoto (1963-), etc.

- ◦ A vast amount of TV animation, including *Tetsujin 28gō* (1960) by Mitsuteru Yokoyama, *Tetsuwan Atom* [*Astro Boy*] (1963-1966) by Osamu Tezuka, *Obake no Qtarō* (1965-1967) by Fujiko Fujio, *Mahōtsukai Sally* [*Sally the Witch*] (1966-1968) by Mitsuteru Yokoyama, *Kyojin no Hoshi* [*Star of the Giants*] (1968-1971) by Noboru Kawasaki (1941-) and Ikki Kajiwara (1936-1987), *Gegege no Kitarō* (1968-1969) by Shigeru Mizuki, *Lupin III* (1971-1972) by Monkey Punch (1937-2019), and *Tensai Bakabon* (1971-1972) by Fujio Akatsuka.
- ◦ A vast amount of TV commercials.

- **Movies and videos**: Especially, European movies, including avant-garde, by Luchino Visconti (1906-1976) and others as well as numerous Japanese works, including ones by Tatsumi Kumashiro (1927-1995), Chūsei Sone (1937-2014), Toshiya Fujita (1932-2097), Akio Jissōji (1937-2006), Kijū Yoshida (1933-), Seijun Suzuki (1923-2017), and Shūji Terayama (1935-1983). Also pornos and so-called adult videos by Nikkatsu and Ōkura created many unique directors.
- **Theater and stage performances**: *Kabuki* at Kabukiza and other theaters, small theater performances in the 1980s.
- **Music**: Until my 20s, I went to classical music concerts (especially Gustav Mahler (1860-1911) and Johann Sebastian Bach (1685-1750)). Later, it has been live jazz at live houses (especially in Tokyo, Yokohama, Kōfu, and other Japanese cities). With regard to my listening to music,
  - ◦ I started with contemporary Shōwa pop (including Mieko Hirota (1947-) and Yukari Itō (1947-)).
  - ◦ *Group Sounds* (including Blue Comets and Tigers of Kenji Sawada (1948-)).
  - ◦ *New Music* (including Yumi Arai (Matsutōya) (1954-) and *Elaine* and many songs by Miyuki Nakajima (1952-)), all of which I listened to largely unconsciously.
  - ◦ And then went on to listen to a lot of Western classical music (taking the occasional detour through Japanese classical music), especially privileging Mahler (all works starting with Symphony No. 1 and especially James Levine's (1943-) No. 5), Robert Schumann (1810-1856) (from his intermezzi to violin

Table 4. List of novels that I read consciously [1971-1978]

Reading Period	Works (Mainly Novels)
1971-1974	Takeo Arishima: *Umareizuru Nayami* (*The Agony of Coming into The World*); *Cain no Matsuei* (*The Descendents of Cain*); *Chiisaki Mono he*; *Aru On'na* (*A Certain Woman*). Ernest Hemingway: *A Farewell to Arms*; *The Sun Also Rises; For Whom the Bell Tolls*; *The Old Man and the Sea*; *The Killers*, *The Snows of Kilimanjaro*, and other short stories. Franz Kafka: *The Metamorphosis*; *The Judgment* and other short stories; *The Trial; The Castle*; *America*. André Malraux: *Man's Fate*. John Steinbeck: *The Grapes of Wrath*. William Faulkner: *The Sound and the Fury*; *Light in August*. Henry Miller: *Tropic of Capricorn*; *Tropic of Cancer*. Nathaniel Hawthorne: *The Scarlet Letter*. Ryūnosuke Akutagawa: *Rashōmon*; *Hana* (*The Nose*); *Kappa*; *Haguruma* (*Spinning Gears*); Other short stories. Ichiyō Higuchi: *Takekurabe* (*Child's Play*); *Nigorie* (*Troubled Waters*); *Jusan'ya* (*The Thirteenth Night*); other short stories. Naoya Shiga: *Kinosaki nite* (*At Cape Kinosaki*); *Kozō no Kamisama*; other short stories; *An'ya Koro* (*A Dark Night's Passing*). William Shakespeare: *Hamlet*; *Macbeth*; *Othello*; *King Lear*. Albert Camus: *The Outsider*; *The Plague*; *The Fall; Exile and the Kingdom*. Yūzō Yamamoto: *Robō no Ishi* (*A Stone by the Roadside*); *Shinjitsu Ichiro* (*Towards the Truth*). Yōjirō Ishizaka: *Wakai Hito*. Tatsuzō Ishikawa: *Watakushi Hitori no Watakushi*. Margaret Mitchell: *Gone with the Wind*. Erich Remarque: *Arch of Triumph*. Emily Brontë: *Wuthering Heights*. [**Poems**] Mikhail Lermontov; Fyodor Tyutchev; Nikolay Nekrasov; Vladimir Mayakovsky; Sergei Yesenin; Toson Shimazaki; Hakushu Kitahara; Sakutaro Hagiwara; Tatsuji Miyoshi; Haruo Sato; Shizuo Ito; Paul Valéry; Other many poets. Lev Tolstoy: *Resurrection*; *Anna Karenina; War and Peace*; *Childhood*; *Boyhood*; *Youth*. Fyodor Dostoevsky: *Poor Folk*; *Crime and Punishment*; *The Idiot*; *Demons*; *The Brothers Karamazov*; *The Raw Youth*; *The House of the Dead*; *Notes from Underground*; *Humiliated and Insulted*; *The Village of Stepanchikovo*. Ivan Turgenev: *First Love*; *Fathers and Sons*; *Rudin*; *On the Eve*; *A Nest of Gentlefolk*; *Smoke*; *Virgin Soil*. *Alexander Sergeyevich Pushkin*: *The Captain's Daughter*; *Eugene Onegin*; *The Queen of Spades*. Maxim Gorky: *Childhood*; *The Life of a Useless Man*; *My University*; *The Mother*; *The Lower Depths*. Nikolai Alexeevich Ostrovsky: *How the Steel Was Tempered*. [**Short Stories**] Leonid Andreyev; Vsevolod Garshin; Ivan Bunin. Saneatsu Mushanokōji: *Yujo* (*Friendship*); *Ai to Shi* (*Love and Death*). Sachio Itō: *Nogiku no Haka* (*The Wild Daisy*). Theodor Storm: *Immensee*; *Viola Tricolor*. Johann Wolfgang von Goethe: *The Sorrows of Young Werther*; *Faust*; Poems. Jacques-Henri Bernardin de Saint-Pierre: *Paul and Virginia*. Abbé Prévost: *Manon Lescaut*. Madame de La Fayette: *La Princesse de Clèves*. George Sand: *Little Fadette, A Domestic Story*. Colette: *Green Wheat*. Françoise Sagan: *Hello Sadness*. Alain-Fournier: *Le Grand Meaulne*. James Hilton: *Goodbye, Mr. Chips*. Hans Christian Andersen: *E no Nai Ehon* (*Billedbog uden Billeder*). Antoine de Saint-Exupéry: *The Little Prince*; *Night Flight*. Charles Lindbergh: *The Spirit of St. Louis*. Charles Dickens: *A Christmas Carol*. Anne Frank: *The Diary of a Young Girl*. Kazumi Nishikawa: *Hikyō Saiiki Hachinen no Senkō*. Daisaku Kishigami: *Mouhitotsu no Ishihyōji*. Norio Nagayama: *Muchi no Namida* (*Tears of Ignorance*). Etsuko Takano: *Hatachi no Genten*. Hermann Hesse: *Beneath the Wheel*; *Demian, The Story of Emil Sinclair's Youth*; *Steppenwolf*; *Narcissus and Goldmund*; poems. Thomas Mann: *Tonio Kröger*; *Death in Venice*; *Buddenbrooks*; *The Magic Mountain*; *Royal Highness*; *Doctor Faustus; The Holy Sinner*; *Lotte in Weimar, The Beloved Returns; Confessions of Felix Krull*; *Joseph and His Brothers Vol.1. The Stories of Jacob*; *Little Mr Friedemann*; *Mario and the Magician*. Friedrich Nietzsche: *Thus Spoke Zarathustra. A Book for All and None*. Rainer Maria Rilke; *The Notebooks of Malte Laurids Brigge*. Hans Carossa: *Childhood; Utsukushiki Madoi no Toshi* (*Das Jahr der schönen Täuschungen*). Jean-Paul Sartre: *Nausea*; *The Wall*; *The Flies*. André Gide: *Strait is the Gate*; *La Symphonie pastorale*; *The Fruits of the Earth*; *Si le grain ne meurt; Les Caves du Vatican*; *The Counterfeiters*; Other short stories. Jean Genet: *Our Lady of the Flowers*; *Miracle of the Rose*; *The Thief's Journal*; *The Maids*. Guy de Maupassant: *Pierre et Jean*; *Une vie*; *Dumpling*; Other short stories. Gustave Flaubert: *Madame Bovary*; *Sentimental Education*; *Bouvard et Pécuchet*. Stendhal: *The Charterhouse of Parma*; *The Red and the Black*; *Vanina Vanini*. Choderlos de Laclos: *Dangerous Liaisons*. David Harvard Lawrence: *Sons and Lovers*; *Women in Love*; *The Rainbow*; *Lady Chatterley's Lover*; *The White Peacock*; *St Mawr*; Other short stories. Mikhail Sholokhov: *The Quiet Don*; *Virgin Soil Upturned*; Short storyes. Romain Rolland: *Jean-Christophe*. Motojirō Kajii: *Lemon*; Other short stories; Haruo Sato: *Den'en no Yuutsu* (*Melancholy in the Country*); *Tokai no Yūutsu* (*Melancholy in the City*). Atsushi Nakajima: *Sangetsuki*. Ango Sakaguchi: *Hakuchi* (*The Idiot*); Other short stories. Yoshikichi Furui: *Yukigakure*. Yasunari Kawabata: *Izu no Odoriko* (*The Dancing Girl of Izu*); *Hana no Warutsu*; *Yukiguni* (*Snow Country*); *Koto* (*The Old Capital*); *Senbazuru* (*Thousand Cranes*); *Tanagokoro no Shōsetsu* (*Palm-of-the-Hand Stories*); *Kinju* (*Of Birds and Beasts*); *Jojoka* (*Lyric poem*); *Aisuru Hitotachi*; *Hi mo Tsuki mo*; *Meijin* (*The Master of Go); Kawa no Aru Shitamachi no Hanashi; Utsukushisa to Kanashimi to* (*Beauty and Sadness*); *Nemureru Bijo* (*The House of the Sleeping Beauties*); *Mizuumi* (*The Lake*); *Kataude* (*One Arm*); *Tanpopo* (*Dandelions*); *TōKyō no Hito*; *On'na de Aru Koto*. Yukio Mishima: *Utsukushii Hoshi* (*A Beautiful Star*); *Tōzoku*; *Kamen no Kokuhaku* (*Confessions of a Mask*); *Junpaku no Yoru*; *Ai no Kawaki* (*Thirst for Love*); *Ao no Jidai*; *Kinjiki* (*Forbidden Colors*); *Shiosai* (*The Sound of Waves*); *Sizumeru Taki*; *Kinkakuji* (*The Temple of the Golden Pavilion*); *Bitoku no Yoromeki*; *Kokyo no Ie* (*Kyoko's House*); *Utage no Ato* (*After the Banquet*); *Kemono no Tawamure* (*The Frolic of the Beasts*); *Gogo no Eiko* (*The Sailor Who Fell from Grace with the Sea*); *Ongaku*; *Kinu to Meisatsu* (*Silk and Insight*; *Tōnorikai*; *Manatsu no Shi* (*Death in Midsummer*); *Tamago*; *Yukoku* (*Patriotism*); *Mikumano Mode* (*Acts of Worship*); *Hashizukushi* (*The Seven Bridges*); *Rokumeikan*; *Raio no Terrace* (*The Terrace of The Leper King*); *Hojo no Umi* (*The Sea of Fertility tetralogy*) (Haru no Yuki (*Spring Snow*); *Honba* (*Runaway Horses*); *Akatsuki no Tera* (*The Temple of Dawn*); *Ten'nin Gogui* (*The Decay of the Angel*); *Taiyo to Tetsu* (Sun and Steal); Other short stories and essays. Raymond Radiguet: *The Devil in the Flesh*; *Le Bal du comte d'Orgel*. Oscar Wilde: *The Picture of Dorian Gray*; *Salomé*. Ōgai Mori: *Maihime* (*The Dancing Girl*); *Vita Sexualis*; *Seinen* (*Young Men*); *Gan* (*The Wild Geese*); *Shibue Chusai*; *Takasebune* (*The Boat on the Takase River*. Kyōka Izumi: *Uta-andon*; (*A Song by Lantern Light*); *Koya Hijiri* (*The Holy Man of Mount Kōya*); *On'na Keizu*; *Nihonbashi*. [**Text Reader**] Yukio Mishima; Jun'ichirō Tanizaki; Yasunari Kawabata; Hiroshi Noma. Kōyō Ozaki: *Konjiki Yasha* (*The Golden Demon*. Osamu Dazai: *Shayo* (*The Setting Sun*); *Ban'nen* (*The Late Years*); *Shinchaku Shokoku-banashi* (*A new version of countries' tales*); *Tsugaru*; *Ningen Shikkaku* (*No Longer Human*; *Villon no Tsuma* (*Villon's Wife*); Other short stories. [**Philosophy**] René Descartes; Dōgen; Zeami Motokiyo; Shinran; *The Old Testament*; *The New Testamen*. [**Japanese Classical Literature**] *Kodai Kayōsyū* (*Collection of Ancient Songs*); *Chūsei Kinseu Kayōsyū* (*Collection of Medeival and Edo Songs*); *Man'yoshu* (*Collection of Ten Thousand Leaves*); *Kokin Wakashu* (*Collection of Japanese Short Poems of Ancient and Modern Times*); *Shinkokin Wakashu* (*New Collection of Short Poems Ancient and Modern Times*); *Kinkai Wakasyū*; (*Collection of Short Poems by Minamoto-no Sanetomo*) *Izumi Shikibu Kasyū*; (*Collection of Short Poems by Izumi Shikibu*); *Gikeiki* (*The Chronicle of Yoshitsune*); *Sagoromo Monogatari*; *Yoru no Nezame* (*Wakefulness at Midnight*); *Heike Monogatari* (*The Tale of the Heike*); *Genji Monogatari* (*The Tale of Genji*); Akiko Yosano: *Genji Monogatari* (*The Tale of Genji*); Ueda Akinari: *Ugetsu Monogatari* (*Tales of Rain and the Moon*); *Harusame Monogatari* (*Tales of Spring Rain*); Shikitei Samba: *Ukiyoburo*; *Ukiyodoko*. Ihara Saikaku: *Koshoku Ichidai-otoko* (*The Life of an Amorous Man*); *Koshoku Gonin-on'na* (*Five Women Who Loved Love*); *Nippon Eitaigura* (*The Eternal Storehouse of Japan*); *Seken Munezan'nyo* (*Reckonings that Carry Men Through the World or This Scheming World*); *Saikakuoridome*. [**Heiancyō Joryū Nikki (Women's Diaries)**] *Kagero Nikki* (*The Mayfly Diary*); *Murasaki Shikibu Nikki* (*The Diary of Lady Murasaki*); *Izumi Shikibu Nikki*; *Sarashina Nikki* (*The Sarashina Diary*); *Sanukinosuke Nikki*. [Japanese Classical Essays] *Makura no Soshi* (*The Pillow Book*); *Tsurezuregusa* (*The Harvest of Leisure*); *Hōjōki*. *Tamenaga Shunsui*: *Shunshoku Umegoyomi* (*Colors of Spring: The Plum Calendar*); *Genshin*: *Ojo Yoshu* (*Essentials of Birth in the Pure Land*).

continued on following page

Table 4. Continued

Reading Period	Works (Mainly Novels)
1974-1978	Sōseki Natsume: *Wagahai ha Neko dearu* (*I Am a Cat*); *Botchan*; *Gubijinso* (*The Poppy*); *Kofu* (*The Miner*); *Sanshirō*; *Mon* (*The Gate*); *Sorekara* (*And Then*); *Kojin* (*The Wayfarer*); *Kokoro*; *Michikusa* (*Grass on the Wayside*); *Higangusi Made* (*To the Spring Equinox and Beyond*); *Meian* (*Light and Darkness*); Other short stories and essays. Kafū Nagai: *Amerika Monogatari* (*American Stories*); *France Monogatari* (*French Stories*); *Danchōtei Nichijō*; *Sumidagawa* (*Sumida River*); *Udekurabe* (*Geisha in Rivalry*); *Hiyorigeta*; *Okamezasa*; *Tsuyu no Atosaki*; *Hikage no Hana*; *Bokuto Kitan* (*A Strange Tale from East of the River*); *Odoriko*; *Ame Syōsyō*. Shimei Futabatei: *Ukigumo* (*The Drifting Cloud*); *Sono Omokage* (*An Adopted Husband*); *Heibon*. Shūsei Tokuda: *Shin Setai*; *Ashiato*; *Kabi*; *Tadare*; *Arakure; Kasō Jinbutsu*; *Syukuzu*. Tōson Shimazaki: *Hakai* (*The Broken Commandment*); Haru (*Spring*); *Ie* (*The Family*); *Shinsei*; *Yoakemae* (*Before the Dawn*). Jun'ichirō Tanizaki: *Okuni to Gohei* (*Okuni and Gohei*); *Otsuyagoroshi*; *Jōtarō*; *Kyōfu Jidai*; *Chijin no Ai* (*Naomi*); *Sasameyuki* (*The Makioka Sisters*); Other short stories and essays. Seiichi Funahashi: *Yuki-fujin Ezu* (*Portrait of Madame Yuki*); *Aru On'na no Enkei*; *Suki na Onna no Munekazari*; *Bokuseki*; *Susono*; *Torikaebaya Hibun*; Other short stories. Harumi Setouchi: *Kanoko Ryōran*. Kanoko Okamoto: *Kingyo* (*A Riot of Goldfish*); Other shorts stories. Tatsuo Hori: *Moyuru Hō*; *Sei Kazoku* (*The Holy Family*); *Kazetachinu* (*The Wind Has Risen*; Other short stories and essays. Takehiko Fukunaga; *Fūdo*: *Kusa no Hana*; *Bōkyaku no Kawa*; *Shi no Shima*; *Haishi*; *Kaishi*; Other short stories. Sei Itō: *Noriko no Ikikata*; *Wakai Shijin no Syōzō*: *Narumi Senkichi*; *Hanran*. Kōbō Abe: *Suna no On'na* (*The Woman in the Dunes*); *Moetsukita Chizu* (*The Ruined Map*); *Tanin no Kao* (*The Face of Another*); All early novels, short stories, and essays; *Hako-otoko* (*The Box Man*); *Hakobune Sakuramaru* (*The Ark Sakura*). Kenzaburō Ōe: All early novels, short stories, and essays; *Man'en Gan'nen no Football* (*The Silent Cry*); *Mizukar Waga Namida wo Nuguitamae Hi* (*The Day He Himself Shall Wipe My Tears Away*); *Kozui ha Waga Tamashii ni Oyobi* (*The Flood Invades My Spirit*). Kazumi Takahashi: *Hi no Utsuwa* (*Vessel of Sadness*); *Sutego Monogatari*; *Waga Kokoro ha Ishi ni Arazu*; *Yutsu naru Toha* (*A Melancholy Faction*); *Jashumon* (*Heretical Faith*); *Nihon no Akuryō*; *Daraku*; *Sange*; *Shiroku Nuritaru Haka*; *Tasogare no Hashi*; Many essays and papres (*Bungaku no Sekinin*; *Waga Kaitai*). Nobuhiko Matsugi: *Hikaru koe*. Syō Shibata: *Saredo Waga Hibi*; *Okuru Kotoba*. Yumiko Kurahashi: Many early novels, short stories, and essays (*Watashi no Naka no Kare he*; *Sumiyakist Q no Boken* (*Adventures of Sumiyakist Q*); *Amanon-koku Tanboki* (*Journey to Amanon*); Mieko Kanai: *Ai no Seikatsu*; *Yume no Zikan*; *Kishibe no Nai Umi*; *Yoru ni Natte mo Asobitsudukero*. Makoto Oda: *America*. Kenji Maruyama: *Natsu no Nagare* (*Summer's Passage*); *Mahiru Nari*. Jun Ishikawa: *Hakutōgin*; *Yakeato no Jesus* (*Jesus in the Ashes*); Ryū Murakami: *Kagirinaku Tomei ni Chikai Blue* (*Almost Transparent Blue*). Hidetoshi Sotooka: *Hokkikō*. Taruho Inagaki: *Issen Ichibyo Monogatari* (*One Thousand and One Second Stories*); Chutaro Oguri: *Kokushikan Satsujin Jiken* (*Black Death Hall Murders*). Hiroshi Noma: *Kurai E*; *Hōkai Kankaku*; Other short stories; *Shinku Chitai* (*Zone of Emptiness*); *Waga Tō ha Soko ni Tatsu*; *Saikoro no Sora*; *Seinen no Wa, All volumes of Vol.1 to Vol.5*. Taijun Takeda: *Fūbaika*; *Kizoku no Kaidan*; *Mori to Mizuumi no Matsuri*; *Fuji*; *Keraku*; *Shibasen*; *Hikarigoke*; Other short stories and essays. Toshio Shimao: *Syuppatsu ha Tsui ni Otozurezu*; Other short stories; *Shi no Toge* (*The Sting of Death*). Kenji Nakagami: *Karekinada* (*The Sea of Withered Trees*); *Chi no Hate Shijo no Toki* (*Supreme Time at the End of the Earth*); *Sen'nen no Yuraku* (*A Thousand Years of Pleasure*); *Kumanosyu*; *Misaki* (*The Cape*); Other novels, short stories, and essays. James Joyce: *Ulysses*; *Dubliners*; *A Portrait of the Artist as a Young Man*; *Finnegans Wake*. Virginia Woolf: *To the Lighthouse*; *Mrs. Dalloway*. *Henry James*: *The Bostonians*; *The American*; *The Wings of the Dove*; *Roderick Hudson*; *The Turn of the Screw*; Other short stories. Émile Zola: *Thérèse Raquin*; *L'Assommoir*; *Nana*; *La joie de vivre*; *Germinal*; *La Terre*. Honoré de Balzac; *The Lily of the Valley*; *Eugénie Grandet*; *The Quest of the Absolute*; *Sarrasine*. Ivan Alexandrovich Goncharov: *Oblomov*. Marquis de Sade: *Justine, or The Misfortunes of Virtue*; *Juliette*; *The 120 Days of Sodom, or the School of Libertinage; Philosophy in the Boudoir*; *Eugénie de Franval*. Georges Bataille: *Madame Edwarda*; *Story of the Eye*; *Histoire de l'œil*; *L'Abbé*; *L'Erotisme*; Essys and papers. Guillaume Apollinaire: *The Eleven Thousand Rods*. Syozō Numa: *Kachikujin Yapū*. John Cleland: *Fanny Hill*. Jean-Jacques Rousseau: *Émile, ou De l'éducation*; *Julie ou la Nouvelle Héloïse*; *Les Confessions*; *Les Rêveries du promeneur solitaire*; Other essays and papers. Michel Eyquem de Montaigne: *The Essays*. Niccolò di Bernardo dei Machiavelli: *The Prince*. Georg Wilhelm Friedrich Hegel: *The Phenomenology of Spirit*. Vladimir Nabokov: *Lolita*; *The Real Life of Sebastian Knight*. Nobuo Kojima: *Hoyo Kazoku* (*Embracing Family*); Shima. Homer: *Iliad*. François Rabelais: *The Life of Gargantua and of Pantagruel*. Victor Hugo: *The Hunchback of Notre-Dame*. Gottfried Keller: *Green Henry*. Aldous Huxley: *Point Counter Point*. Roger Martin du Gard: *The Thibaults*. Samuel Beckett: *Malone Dies*. Andrei Bely: *The Silver Dove*. Louis-Ferdinand Céline: *Journey to the End of the Night*. Cesare Pavese: *The Fine Summer*. Günter Grass: *The Tin Drum*. Saul Bellow: *Dangling Man*. Jerome Salinger: *Nine Stories*. Jean-Marie Gustave Le Clézio: *Terra Amata*; *The Interrogation*. Philippe Sollers: *Le Parc*. Drieu La Rochelle: *Gilles*. Mario Vargas Llosa: *Conversation in the Cathedral*. José Donoso: *The Obscene Bird of Night*. Julio Cortázar: *Hopscotch*. Manuel Puig: *Kiss of the Spider Woman*. García Márquez: *The Autumn of the Patriarch*. Shintaro Ishihara: *Kaseki no Mori* (*Petrified Forest*). Gen'ichiro Takahashi: *John Lennon Tai Kaseijin* (*John Lennon vs the Martians*. Yutaka Haniya: *Shirei* (*Death Spirits*); *Yami no Naka no Kuroi Uma* (*Black Horses in the Darkness and other stories*); Other short stories, and many essays and papers. Aleksandr Solzhenitsyn: *The Gulag Archipelago*; *Cancer Ward*; *One Day in the Life of Ivan Denisovich*; *The Red Wheel Vol.1*. **[Essays and Papers]** Bunzō Hashikawa; Kōichi Isoda; Takaaki Yoshimoto; Masao Maruyama; Taijun Takeda; Hiroshi Noma; Yukio Mishima; Kanji Nishio; Naoki Komuro; Kiyoshi Kasai; Kinji Imanishi; Tadao Umesao; Takeo Kuwabara); and others.

concertos, especially *Fantasie*, Vradimir Horowitz's (1903-1989) *Kreisleriana*, all symphonies, the cello concertos, and all those small and peculiar works), and Bach (especially a number of his passions).

- Furthermore, jazz starting with Roland Hanna (1932-2002) (whose record I first purchased) (countless, varied, and diverse pieces of music including Japanese contemporary jazz, Bill Evans (1932-2002) (from his *Undercurrent* with Jim Hall (1930-2013)), Miles Davis (1926-1991), Hank Mobley (1930-1986), Lee

Morgan (1938-1972), Blue Mitchell (1930-1979), Ann Burton (1933-1989), Junko Onishi (1967-), Chihiro Yamanak (1976-), Keiko Lee (1965-), and Diana Krall (1964-)); that is my history of listening to music (with the exception of Elvis Presley (1935-1977) and Ringo Starr (1940-) in my early days, I have not listened to much in that tradition, and almost no rock).

- **Paintings**: From books and art exhibitions. Including Giorgio de Chirico (1888-1978), Paul Delvaux (1897-1994), Hieronymus Bosch (c. 1450-1516), and Matazō Kayama (1927-2004).

If we were to consider what truly has had an effect on my brain, then we cannot underestimate the power of these non-*novels*, and in a sense, they probably serve a kind of underlay for the *novels*, hidden yet true.

Table 4 shows a record of my conscious reading of *novels* from about 1971 to 1978.

From about 1978 to 1983, I read fewer *novels* and started reading more from the genres below, until the volume of *novels* read finally become a very small part of my total reading.

- **Literary critiques and commentaries**: As I started university, I began reading the works of Takaaki Yoshimoto. After that, I read a lot of Yutaka Haniya, Shūichi Katō, Taijun Takeda (1912-1976), and Hiroshi Noma (1915-1991). After graduating, it was Kōjin Karatani, Kenji Nakagami (1946-1992), Shin'ichirō Kurimoto (1941-), and Shigehiko Hasumi. I regularly read Seigō Matsuoka (1944-) while standing in the Hōrindō Bookstore in Takadanobaba in Tokyo.
- **Philosophy and thought**: What I read the most around this time was Kiyoshi Miki (1897-1945), but I also read quite a bit of other philosophical texts in the narrow sense, like Plato (427 BC-347 BC), Aristotle (384 BC-332 BC), Hegel, Friedrich Nietzsche (1844-1900), Edmund Husserl (1859-1938), and Sigmund Freud (1856-1939) (My favorite read in high school was Otoya Miyagi (1908-2005), Masaki Shimazaki (1831-1886), and other introductions to psychological and mental analysis.)
- **Social sciences**: Especially Adam Smith (1723-1790) (also *The Theory of Moral Sentiments* (Smith, 1759)), Karl Marx, Max Weber (1864-

1920), research and expositions on them (Hisao Ōtsuka 1907-1996, etc.), Masao Maruyama, Ikutarō Shimizu (1907-1988), works related to methodology of social science, economics (back when principles of Marxist economics was still used in classrooms), law (criminal law and things that interested me), political science (especially comparative political science and polimetrics), sociology (especially the sociology of law and the sociology of the family of Kizaemon Ariga (1897-1979) and Takeyoshi Kawashima (1909-1992)), and a mix of other studies, introductions, expositions, and textbooks as well as the kind of leftist and progressive texts that were flooding Iwanami Shoten at the time. Furthermore, I read almost no business administration or marketing theory and especially no advertising theory as I did not believe them to be (proper) disciplines.

At the center of my reading from about 1983 onward became computer manuals as well as textbooks, expositions, and research papers on artificial intelligence, cognitive science, information science, and engineering, which I needed for my job. What I kept reading repeatedly the most were textbooks, expositions, and manuals on programming languages and especially Lisp, which is a programming language for artificial intelligence, and I was pretty much reading them indiscriminately. I vividly remember that after some time, when I had started reading texts related to artificial intelligence, I was astounded by the simplicity of the view on humanity and worldview there. What I was reading in terms of literature was been little more than *novels* in the form of a small number of literary and ideological critiques and commentaries. Yet at the same time, my reading of ideological works in literary theory, narratology, structuralism, and semiotics, as well as philosophical studies and narrative-related academic studies (also folklore and entertainment, *geinō* in Japanese, history) increased dramatically. It was from that direction that I tried to reconnect with literature. In a most direct sense, the computer-related books and books on narratology, literary theory, structuralism, and semiotics that I read back then are what is supporting my work life today. Of course, what in turn supports that are the aforementioned *novels* and the numerous books related to them. As for *novels* (*novel*-like texts), I read Riichi Yokomitsu (1898-1947), Yasunari Kawabata (1899-1972), Saiichi Maruya (1925-2012), Yumiko Kurahashi (1935-2005), Tsutomu Minakami (1919-2004), Yukio Mishima, Shūsaku Endō (1923-1996), contemporary urban legends, Yasutaka Tsutsui (1934), Natsuo Kirino (1951-), Seichō Matsumoto, Seishi Yokomizo, Jirō Akagawa, Misa Yamamura, Kyōtarō Nishimura, Yasuo Uchida, Shizuko

Natsuki (1938-2016), Nobuhiko Kobayashi (1932-), Fumiko Enchi (1905-1986), Yaeko Nogami (1885-1985), Kaoru Takamura (1953-), Sōji Shimada (1948-), Liza Dalby (1950-), Denis Diderot (1913-1784), Eiji Ōsita (1958-), Alexander Afanasyev (1826-1871), Hakuchō Masamune (1879-1962), Masuji Ibuse (1898-1993), Shintarō Ishihara (1932-), André Breton (1896-1966), David Harbert Lawrence (1985-1930), Earnest Hemingway (1899-1961), *Sekkyōbushi* (Sekkyōshū, 1977), Tanehiko Ryūtei, Ango Sakaguchi (1906-1955), Richard Powers (1957-), Kenji Miyazawa, Jonathan Swift (1667-1745), Charles Dickens, Ikuo Kameyama (1949-), Italo Calvino (1923-1985), Danilo Kiš, Victor Hugo, and others, mixing light and heavy, hard and soft (including also a lot of silly things). Excepting that small number, the core of my reading was for a time the translated *Genji Monogatari*. I reread Akiko Yosano's translation and read through the ones by Fumiko Enchi and Jakuchō Setouchi (1922-). Subsequently, although not *novels* in the narrow sense, I read and still read a lot of scripts for *ningyō jōruri* and *kabuki*.

Furthermore, as I was buying books for about 100,000 yen a month for a while, the bookstore was of prime importance to me. It developed in such a way that I had during my middle and high school days the Yūrindō in Isezakichō, Yokohama, the Yūrindō and Eishōdō by Yokohama Station, the various new and used bookstores at Gumyōji Temple (737-), Yokohama, and Tokyo Station Yaesu Book Center. During my university days, I added the Hōrindō in Takadanobaba, Tokyo, the used bookstores along Waseda-dōri in Tokyo, the Waseda University Co-op Store, ad Kinokuniya in Shinjuku, Tokyo. Later I added the Yokohama Station Lumine Yūrindō, the Taiseidō in Shibuya, Tokyo, the Maruzen in Nihonbashi, Tokyo, the Kinokuniya in Kamiōoka, Yokohama, and the Yaesu Book Center. Going all the way to the present, I have extended my scope of exploration to Junkudō and Kinokuniya across the country (especially Sapporo, Morioka, Sendai, etc.), Yaesu Book Center at Tokyo Station, the many bookstores around Ginza, Tokyo replete with *kabuki* books, and used bookstores in Morioka and elsewhere, but because of the advance and development of online bookstores, my opportunities for bookstore exploration have become fewer than before.

In this way, my focused reading of *novels* finished around 1978 or 1983 at the latest. After that, I recommenced a project of systematically reading *novels* (toward certain aims) in early 2007.

During this time, my *classical (fossilized) memories* of *novels* have intermingled with generative activities and have been disassembled and reconstructed as an extension of that genre. Moreover, amid interventions and the mixing of other genres, they were disassembled and made fluid. This movement of fluid and fixed has generate a large number of peculiar seeds, and the process is still ongoing.

If I make a definite linguistic expression, I feel that it might lose an important sensation, but the reason for the budding of my feeling of "Why does it have to be that way?" was not necessarily purely through my experiences of reading, but it might have been affected by various other personal circumstances. When connecting some experience to ideology, the sensational part of the experience becomes the subject and you must describe something comprehensive and pluralistic as logical and explicit, which tends to require that some things are cut off and that some awkward bits remain. Poetry and then literature might exist as a method for ideologizing without needles abstraction. Also in such cases, the feeling of "Why does it have to be that way?" does not originate from or apply only to these *novels*, but it might very well have budded from all the *novels* that came later or when observing (personal or social) landscapes.

(If I were to make a slight addition, this sensation of "Why does it have to be that way?" was not accompanied by a sensation of open possibility that an alternative is readily available, but it was accompanied by a strong sensation of some necessity that it did not have to be that way but it turned out that way as a result. I expressed that sensation using a variety of words in my notes from back then, but at least it is certain that it did strongly carry any meaning of openness to possibilities. As such, it was sensationally true that fluidity had an immediate flipside of fixedness. Moreover, it was not necessarily so that this sensation mature through the process of reading these *novels* or through a relationship with the *novels* and nothing else. Rather, when considering it from a certain consciousness now, it is perhaps more accurate to say that it is projected unto the world of *novels*. Nonetheless, it is a truth of mine that the *novel* was an overwhelming existence.)

Conversely, these *novels* perhaps rather existed strictly as such. Moreover, when I started reading *novels* consciously was also when I started writing *novels* (*novel*-like things for the sake of *novels*) consciously, meaning that it was when I started moving the direction of the act of *production*, although imperfect. That is, those two events were synchronized. In any event, it is required of the professional that they complete their tasks, but here the *novel* professional (mistakenly?) started moving in the opposite direction, but this really signified that you step out endlessly toward incompleteness and openness when it comes to both reading and that which resembles production. Even if those *novels* themselves, upon being read, did not start moving directly toward liberation from sensationally having to exist as they are, it is certain that there was movement in that direction as a whole.

If I were to employ a slightly figurative wording, that was perhaps the direction of a kind of death. Death is fluid and liberation itself. These circumstances that were nothing but wholly personal made me feel a sense of resistance toward the *novel* (the *novel* in general but also these *novels* that possess a cohesion that likely removes them from death the most) and their completeness and fixedness. Here, the *novels* are dissolved, made fluid. While each of the *novels* was completed and fixed because of the strength of their vitality, acquiring the character of unyielding individual units, they were now constantly visited by a sense of incomprehension and boredom that resists exactly that. This might just be a reflection of a scattered mental state (death in a figurative sense), but then conversely, I felt that completeness and fixedness are the lowest point of vitality that also do not mean full death, created only at a location that you randomly reach on account of an extremely half-baked force, which does not allow for your own volition but is entirely passively inevitable. In Aristotelian thought, *fate* is linked with unease, but I felt the fate of something fixed that is accompanied by next to no sense of unease. In other words, as I mentioned in the beginning, it is a state where there is more or less no reason for the "Why does it have to be that way?" This kind of melting is fluid and indeterminate. Rather than the phenomenon itself, it is the emptiness on its flipside and a kind of death. Taking that to its extreme, we go beyond human principles and reach a world that is just completely covered in constantly boundless phenomena, where nothing takes any shape. At some point I started to think that, in a way, that can somehow be reproduced with mechanical lack of intention and a fluid logic. That is, the mechanical lacks volition and intention and is indiscriminate. Its operations pursue internal objectives, not organically, but it is done single-mindedly without any reflection as it was determined and conferred from without. Even if some fixed and determinate conclusion is reached in form, that is nothing but a result of chance. Such operations are extremely tough, but that toughness is accomplished only when it is connected with something extremely exhausting. I felt that mechanical morphological operations can actually be a means for capturing melting fluid phenomena only just.

Going via this slightly complicated logic and sensation, this traction towards fluidity and accompanying exhausting orientation toward fluidity are at the same time the reverse of traction and orientation toward fixedness, so the power of the *novel* works was perceived through that double meaning. That is, fluidity and fixedness very much were or became cohabitants in a relationship like front and back. Thus, these *novels* that appeared as fixed melted and changed into something fluid, and I was tempted by the question of why it had to be so, but at the same time, I felt that the basis for that question would easily disperse if I were to face those *novels* in earnest. One possible interpretation is that what had melted was in the end not the *novels*, but my own personal stance as it had not been able to withstand the force of the fixed *novels* and was overcome by the temptation of wanting to be pushed away toward fluidity. In any case, regardless of my personal stance, I am certain of the existence of the pressure of fluidity and the mental tendency to consider the *novels* as what melted. If also feels that it is certain that this chaos and disorder originates from the overwhelming experience of the *novels*. However, as I mentioned above, that becomes possible only when you have the context of reading or receiving texts (in a broad sense) of the various other genres that surround the *novels*. Nevertheless, at that time, *novels* were the world to me.

If such an exploration is possible, you should divert your memories of *novels* as they become fluid in the direction of production and look for methods of production to make *novels* fluid. Furthermore, taking the matter one step further, you should look for methods to anchor *novels* and to cause a movement that intertwines the two.

Now, if we think of the phenomenon of *novel* fluidization as a more conscious effort in the context of *novel* production or the *novel* as a work, then we can imagine a *novel* series where a number of *novel* works melt and blend with each other. We have Balzac's *The Human Comedy*, Emile Zola's *Les Rougon-Macquart*, William Faulkner's (1897-1962) "Yoknapatawpha County," Kenji Miyazawa's "Ihatov," and Kenji Nakagami's "Kumano and Kishū," which are all works set in a shared location. Then there are works that contain multiple stories skewered on an axis of time rather than space, such as *Hōjō no Umi* [*Thes Sea of Fertility*] by Mishima, *Genji Monogatari* [*The Tale of Genji*] that at first glance seems like an accumulation of short *novels*, and the works of Yasunari Kawabata that appear like a methodological succession embodying *Genji Monogatari*, which are all *novels* that contain fluidity from the outset. These are *novel* phenomena where the characters and other elements bleed out and flow around multiple works. That is, this type of melting refers to a situation where the elements of a single work do not remain closed but become open, so that the elements of various works are used and reused fluidly, meaning a fluidization phenomenon on a different level from that of individual *novel* works. Also here, the exhaustion of this situation can shatter the orientation toward fixedness and unity maintained in the case of the abovementioned *novels*, thereby necessitating a greater degree of sinking, a phenomenon of melting and flowing, within my consciousness and memories or within the process of writing. At the same time, we still have to consider the fixedness that nonetheless remains or the possibility of refixation. This means thinking about what kind of fixedness might possibly exist that differs from the rigor of works like these.

If it is so that the *novels* make up a kind of *novel* series inside my head according to some personal standard, then the work of restructuring, which likewise involves letting characters, locations, and other elements escape the framework of the fixed work and facilitates a variety of different combinations, becomes straightforwardly possible, and it conjures up an extremely pragmatic image of the mediocre tasks of disassembly and restructuring. In other words, that is literally business, business in its purest form. People normally try to emphasize the separation or distance from such pragmatism by setting up various membranes of originality. However, the mutual operation of fluidity and fixedness do not constitute anything ascending or dialectical, nor is it something you would want to constitute in such a way, because the ideal is rather its realization on a pragmatic level. The issue at hand is how we can do that. In that sense, this text becomes a method-oriented study. Instead of trying to elevate a mediocre task to the extraordinary (that is exactly the story of narratives in their currently popularized sense), we are asked to disassemble and analyze the mediocre task with even greater seriousness. The transparent pragmatism that remains after the elimination of all sensational description, the unadulterated pragmatic being of something, is the ideal we strive for. We might as well say that the study of the techniques of joining together narrative generation and production is this hidden theme.

Now, as I mentioned above, I was not only addicted to the intensive reading of mainly *novels* during a limited period of time, but at the same time, I also wrote a large volume (converting it into the kind of 400-character sheets that people used to write on, it was likely as many as 10,000 pages) of things that sometimes took the shape of *novels* (something *novel*-like) and other times the shape of poems, *tanka*, diaries, critiques, and a variety of other things. This was probably (meaning that I did not have any rational perception of what on earth I was doing) a type of method for *novels* (although I was not clearly conscious of this, the word *novel* was indispensable, but because of that I all but stopped using the word after that). As such, my memories of reading mixed with my memories of production, so that the two finally made fluid my memories of *novels*, along with my consciousness thereof. This promoted the melting and fluidization of the things I gained from reading on a daily basis, but the themes that I wrote on (themes that I was pretty clearly conscious of even then, but that were also rather high-level comprehensive themes) were about the difficulties of melting and becoming fluid as well as the issue of the fateful yet never idealized fixation. I think this was the start of my awareness relating to the concept of the *system*. Since this was an extremely contradictory logic, back then, my intellectual work and activities were physically quite tough as well.

Firstly, looking at the one extreme of the issue of fluidization, by which I here refer to the work of fluidization, I want to think a bit about fluidization in the process of generation and production. In conjunction with this, we also have fluidization (and fixation) as a work, which is the issue of in what way a certain work arranges (or does not arrange) a fluid state within itself, but I will just briefly address this toward the end of the paper (the issue of fluidization as a work will probably be about how a generation and production process are somehow made to exist inside a work to some extent). If we think about this without complicating the matter, it goes without saying that the *novel* generation process is fluid. When a writer writes their *novel*, they follow a fluid process both inside their head and in their physical work. That is, they explore various possibilities through trial and error within an intellectual space or workspace that is open to various possibilities and directions, or they repeat manuscript revisions over time to create a number of different manuscripts, thereby experiencing a fluid state.

Alternatively, and this might complicate the discussion, but I can give the following example. It is an example of not *novel*, but music generation and production (I also want to show the *novel* process and the music process alongside each other). In some of his work when making records, Miles Davis (he is just a representative, but also someone who went beyond that) would first perform on stage, record that, and then edit it to produce the work. It is unlike both the way of simply editing studio performance and the way of remixing based on fixed disassembly and disassembly fixation (that is the fixated method). I can feel the desire to have the fluidity of the generation and production process overlap with the fluidity of the work. To begin with, most jazz songs borrow something from earlier songs, such as title, melody, or rhythm, and the performance is not a reproduction of fixed songs, but the performance itself overlaps with songwriting. Moreover, the performance is not a process that leads to a complete or final version. In the case of music like this, the production process is made up of a variety of production processes piled on each other, and rather than making up a complete process, they are open to the outside (of course, this means that they are naively open on an unconscious level, as fused music is naturally repulsive). The performance itself is a single work and any edited version of it is a work as well, so that this state of incomplete work production never ends even as it is conducted by different people. As such, the work at any one focal point is temporarily fixed and simultaneously open toward fluidity. The genre system itself possesses this kind of fluidity. Again, this is a case where we cannot really separate the fluidity of the generation and production process from fluidization as a work. There exists the fixed strength called *Miles Davis*. The existence of this strength has two meanings. It is disgusting, but I also feel that it is necessary. Yet it also makes me want to chase it into indiscrimination for its meaninglessness. Alternatively, I am tempted to consider whether there exists a method for retaining indiscriminate strength.

In any case, we are tasked with exploring the issue of fluidity (and fixedness) in *novels* as one method. The reading of *novels* leads to an exhausting fluid state, but regardless of whether that was a good thing or not, it had a value in absolute terms, which is why I thought this has to be made into a method of some sort. The notions that appeared at this point were indiscrimination, unintentionality, and involuntariness. More sensationally, it had to do with the abovementioned exhaustion. This signifies a movement devoid of spontaneous movement toward anything, which does nothing outside of what happens to happen by chance. In other words, it is a fluid state and a state of trickling melting (yet as I mentioned previously, this in itself is a form of fate and fixation). Even if there is at least some external goal or will that can control this movement, that did not originate spontaneously from within the movement, and in this sense, the strength of the movement is endlessly small and had become fluidly indistinguishable from other movements. Nonetheless, we can imagine a variety of models also for this kind of movement. This issue immediately links to the issue of the methodology for *novel* generation and production, or it might even be that issue itself. This is what occurred to me.

In concrete terms, the task that I had to engage in was exploring methods for increasing the fluidity of a work's generation process as well as the converse, methods for increasing its fixedness in parallel. In particular, the nature of fluidity and fluidization remains somewhat unclear. However, I feel that the introduction of machines (*narrative generation systems*) contributes to amplified fluidity. This feeling is that the limit of fluidity (the fluid-state *novel*) is the format of a machine constantly generating (fragments of) *novels* and that itself becoming a work. At the same time, the concepts of fixedness and fixation likewise remain somewhat unclear, but it needs to be at once a format where fluidization is just like fixation and a format where that is not fixed as a relationship. Also taking into account degrees, there are a number of possible concrete methods and ways of being, making it possible to try out a variety of formats. Moreover, the trying out of various methods has itself the potential of becoming a way of producing *novels* in the present.

The following line of thought is also budding inside me. I want to think about the correlation between fluidity in the *novel* generation process and formality as well as about the opposite correlation between fixedness and contents. Being formal means allowing for automatic movement by mechanical operation, so it is a force unconstrained by semantic and content-related alignment and choice. It can be seen as a kind of indiscriminate force. In other words, the fluidity of the *novel* generation process can be raised by increasing the strength of its formal character while its fixedness can be raised by increasing its content-related character. I then add to this schema the interventions of machinery or mechanical character and of humanity or human character. The formality/content and this machinery/humanity are mutually independent concepts on logically different levels. For example, it is also possible to advance formality through human elements and content through mechanical elements. Yet they are related to each other in a practical sense. In other words, machinery/humanity functions as a kind of amplification mechanism for formality/contents, where machinery is effective for amplifying formality and has the characteristic tendency of accelerating it, while conversely, humanity has the characteristic tendency of effectively amplifying contents. Taking an opposite view, if we, for example, introduce human elements when promoting formality, it might easily retrogress, while if we introduce mechanical elements when promoting contents, it might easily retrogress as well. In this way, they can function as a form of tuning valve control.

As a trial, let us schematize the flow of a *novel* from fluidity to fixedness based on these ideas——we have the formal generation formula that has a strong tendency toward fluidization from the outset, so the *narrative generation system* acts as an amplification mechanism for it. Next, we add the content principle, and considering the context for the use of the *narrative generation system*, it will work together with the human principle. Moreover, its character as a content generation formula will be strengthened as the power of *my* principle become stronger.

This kind of process is possible even without the interposition of the *narrative generation system*, but we can assume that it will be a mechanism that is wholly focused on the formal method for *novels*. Of course, humans also possess a mathematical operating capability, meaning a formal capability, for example, without serving something (such capabilities can also be found among certain literature-related capabilities), but the machine is a powerful bearer of formality, and I am looking at it within the context of manifesting a mechanical processing device for *novels*. However, while *machine = non-human*, the artificial intelligence on which the *narrative generation system* is based is a machine that aims to humanize (it aims to imitate human functions and configure a human cognitive model), and in the meantime, there will be contradictions and area infringements. Yet in the end, just like photos and movies are the (pseudo-) humanization of images, the humanization of artificial intelligence is a pseudo-phenomenon and is every bit as false. In other words, it does not fall outside the general framework of machine = non-human. Moreover, this *machine = non-human* is fluid in the sense that it lacks the powers of focus, centrality, and selection. Its very character is fluidity. The *novel*'s fluidity is amplified by machine = non-human = *narrative generation system*, and it is my intention to introduce here fixedness as a work and to do this by eliminating all human, contents character direct use. Nevertheless, if this is done as a story of fluidization ® fixation, then it becomes simply a human process that can be expressed as *disassembly and restructuring*, so a crucial question is how to introduce mechanical principle = non-human principle in this process (especially from fluid to fixed). The *narrative generation system* has to be explored as a tool in such a context. That means we have to be conscious of its relation to what the *novel* seeks or what is sought of the *novel*.

I think of machine = *narrative generation system* in the same way as photos and movies relate to images and instruments to music. The notion of machine *transparency* has to do with why it is a *narrative generation system* and not a word processor. For example, photos and movies have the function of indiscriminately depicting whatever image is thought to correspond to the phenomenon, regardless of our thoughts or intentions, and in the sense that our thinking is not directly projected onto that image, they are *transparent*. A word processor depicts the sentences we write not on paper but on the screen of a machine, in the sense that our intention is directly reflected there, it is even more transparent (only relatively so, as the word processor may be seen as displaying letter-like shapes of its own accord based on its own independent principles and mechanisms, which we then happen to simply interpret as being no different from anything produced by our own hands). What happens through the *narrative generation system* is that the machine becomes less transparent, if only relatively speaking, meaning that it itself creates an image (in this case words) of some subject without going through our intentions. When it comes to how it creates, this, of course, happens in accordance with a composition designed by a human designer, which is the same for photos and movies. As such, there is no room for any real opacity, but in a relative sense, there can be said to exist a great opacity. That is the reason the *narrative generation system* comes to hold one meaning (significance). From the viewpoint of architecture, the expressions artificial intelligence and human information processing are simply superficial figures of speech, for it requires the realization of a machine that lacks human discriminatory capabilities.

Now, I would like to think about fluid states in *novel* generation and production when overlapping with the *narrative generation system* as a machine like this. As a trial, let us suppose three models of *novel* generation:

1. One-way (definite) generation
2. Parallel possibility generation
3. Composite possibility generation

Among these, model (1) is the most fixed, and it is then fluidized as either (2) or (3). The fluidized format then becomes the premise, and it returns to a process of fixation.

Firstly, the *novel* is constantly open toward a variety of possibilities (this, of course, applies not only to the story aspect, but also to the narrative discursive aspect, including what concrete expressions to pick, and various other aspects) at every point in time during its generation, which is normally followed by some bundling and narrowing down of the latent possibilities. This may be done (it is usually so) decisively and fixedly by human will (although a composite of diverse elements), but it may also be done more exhaustingly and more accidentally. Exhaustingly means a pattern where something happens in a certain way despite no necessity for it. It happens as if by chance and is partly related to the word *arbitrary*, which is frequently used in other contexts, but they differ greatly since arbitrary has a stronger meaning of post-event decisiveness and fixedness. The use of exhausting here does not imply anything absolute or fateful about the result. As such, even when it comes to one-way generation, it is possible to have exhausting and fluid generation. Thus, fluidity is not necessarily the equivalent of decision, but can be said to happen according to an essentially different principle. When seen as a phenomenon, unconscious fluidity and exhaustion is realized for many (most) *novels*. Yet, simplifying this further, I will discuss it in a pragmatic fashion.

One possible method for developing this one-way situation toward further fluidity is the parallel arrangement, rather than bundling in a certain direction, of the openness toward various latent possibilities, regardless of whether it is intentional, accidental, or arbitrary. If we think of bundling as happening through the existence of a unified force (also supposing non-intentional forces), regardless of the situation (whether it is intentional or not), then there are also situations with no unified force and with patterns of forces staying put or scattering. In other words, we can suppose a situation where numerous possibilities are developed in parallel toward the realization of various possibilities. Realistically, this leads to the realization of an almost endless number of *novels*, but the thought experiment should be possible as a theoretical problem. This does not only imagine the simultaneous and parallel creation of numerous *novels* but can also be seen imagining the entirety of a situation that encompasses numerous latent possibilities. This kind of situation currently only looks chaotic, but untangling the circumstances, keeping detailed records, and making it a reality can become a realistic task.

Furthermore, and conversely, let us consider the insertion of some kind of non-fluid factor into this fluid and indefinite (what I refer to as exhausting here) generation. That is, we take the most fluid situations, such as one-way, definite generation that simply happens by chance or generation that realizes possibilities in parallel while removing even arbitrariness, and this time around, we study the addition of non-fluid elements there. In the case of one-way generation, when going back to the initial description to select and decide on a single possibility, exhausting elements are removed as much as possible, and unless the circumstances or reasons for the selection are described, it will look the same, but it is different in essence (information about the circumstances and reasons for some selection are added in the background). From the reader's perspective, it is possible to adopt an attitude of "*katte-yomi* (convenient reading)" that is "reception-theoretical and reader-centric critical" (Iwagaki, 2008), which says that they are the same in the end, but it is also possible to intentionally think the opposite and deny that they are the same. It is also possible to hold that even though their appearances as objects or words are the same, the properties included in the objects (words) are likely different. Yet what I want to consider now is not this issue, but it is processing and mechanisms that generate fluid states that exhaustively house various *novel* possibilities while also compositing them using some controlling force. For example, we can imagine an approach that is like music, where the processing consists of simply physically lining up various *novels* generated in parallel. In the case of music, rows (melodies and rhythms) overlapping contrapuntally or according to the law of harmony have an overall effect of physical sound characteristics harmonizing or not, and this also results in some kind of semiotics. Yet in the case of *novels*, since their central characteristics lie not in the sounds of language but in its meanings, or the meanings are not perceived as derived from the sounds but the two rather come at one, it is more difficult than with music for such a physical composite to acquire some kind of overall integrity. However, it should be possible to realize this kind of state experimentally. Also in the case of images, it is difficult to realize harmonious integrity by physically duplicating multiple texts like music, but it is possible to create such a situation experimentally, for example, in the form of some kind of visual art, and it should be possible to create an experiment where a *novel* imitates that as well. An anagram might suggest one direction. It might be possible to think about a small part of the will to control in the most exhausting sense by inserting this kind of metaphor.

In this way, the barrier that stands in the way of the will to achieve composite control is the issue of meaning in *novels*, and more concretely, in their words. Establishing this while bundling exhausting parallel rows, giving rise to a unified whole, is a task anticipated for the next step. Necessary here are the methods for overlapping meaning. One of the basic principles is the overall meaning of the parts overlapping, so that the whole has a single meaning. Alternatively, if this way of speaking excludes the pluralism of meaning, then the whole has to become a single organic cluster. Methods like these are open to a variety of operating possibilities, including the one-way, copy-and-paste reediting of various *novels* generated in parallel and lining up *novels* horizontally, but having all of it make sense when reading them vertically with all kinds of variations. These are methods that approach the *editing* of movies. Characteristic of movies, differing from the digital image technologies that look similar, is that they use the physical object of the film (projected using the camera, subject to powerful physical constraints) as a material and that they require physical editing as an important element of their production, so that the editing principles of arranging the physical material of the film (both in cases when the controlling will to edit plays a major role in advance and when it does not) has such a major power (although it goes without saying that photos are the forerunners). The theme here is the *novel* and not the movie or the photo, but also in the case of the *novel*, the compositional principles are important, and it is perfectly acceptable to imitate presentation techniques as well as imitate things and be influenced (by the abovementioned music, images, etc.) in the generation and production processes. Yet at the same time, this is expressed through language, and in that way, it differs from music and images. While they are methodologically mutually open, they are also mutually closed.

As a plan for having the machine = *narrative generation system* intervenes in the *novel*'s formal generation processing to introduce a fixedness premised on even more fluidized processing, the amplification of content through human elements appears as a theme. Here, we must consider not the elevation of fluidity to fixedness, but fixation while maintaining fluidity. How should we think about the human elements mentioned here? For example, even if wanted to insert human elements into a sequence, selection, and arrangement of photos, there is a major difference between introducing a story that is meaningful in the most meaningless way and introducing some expression that possesses different properties (not in the sense of which one is good or bad, but simply in terms of different properties). In particular, the introduction of expression mentioned here has to be different from control or apperception in any simple sense. Moreover, even if a machine can at least superficially imitate the methods of human production, including human elements there means inserting individuality, which is a movement toward a production methodology for fixation, individualization, and work-making. This kind of directional rejection and refusal is what made me explore the direction of *novel* fluidization and the introduction of the *narrative generation system*, but I am now heading toward its repeated denial. As is an intermediate state, it is a state of fusing mechanical and human principles and having them work together, but I cannot deny that the expressions are pulled toward the bureaucratic terminology of information processing. Alternatively, it has completely fit into that classification system. We could call it collusion, but it would not change anything. On the themes of photos as machines and movies as machines, people do not usually use words like cooperation or collaboration between machines and people. The machines that are computers, especially in the context of artificial intelligence, are approaching the discourse of humans too much, which is why I think such expressions have just become automatic. I think it might be good to draw analogies between the *narrative generation system* of a computer and machine used when writing a *novel* and the machines or devices of photos and movies, and to think about their relationship to human roles, functions, and work. However, photos and movies retain a level of transparency vis-à-vis humans on account of the real visual images that are their supposed objects, which makes them different from the *narrative generation system*. Nonetheless, I believe that it is necessary to think about the relationship between the *narrative generation system* and humans outside of the sense of cooperation, collaboration, and fusion.

In any case, it can only sound clichéd when said, but if I were to start by summarizing the discussion below, it will probably be a search for a path toward fixation (work-making) by introducing principles of humanization (individualization) in the fluid generation that interposed the machine = *narrative generation system*.

When simply dividing the portions of the *novel* generation process into those handled by the machine and those handled by the human, this should show a variety of ways to unify the tasks of the two. One way is perhaps to divide that generation process into components. An example of an extremely simple big-picture formula (and nothing but an example) is dividing the components into narrator and characters, applying the assignment models listed below:

1. ***Narrative generation system* = narrator/human = characters**: The *narration generation system* narrates and the human handles the characters.
2. **Human = narrator/*narration generation system* = characters**: The human narrates and the *narrative generation system* handles the characters.
3. **Narrator = *narrative generation system*/human**: The *narrative generation system* and the human share the narration.
4. **Characters = *narrative generation system*/human**: The *narrative generation system* and the human share the characters.

In these cases, it is possible for even more detailed categorization by deciding whether other elements should be assigned to the *narrative generation system* or to the human, or whether the two should be combined for those elements as well. Although the *novel*'s narrating subject is just a narrator (a comprehensive mechanism) seen in the grand scheme of things, the characters can also be considered a relatively independent element in the generation of the *novel*.

The narrator is probably the most central (yet this does not have to be completely or even considerably visible, and the narrator is not necessarily obvious or explicit about their own centrality) and comprehensive mechanism in the *novel* (if we consider the *novel* as a form of narrative), and depending on your point of view, even the characters can be seen as being under the narrator's control. Yet let us think about the narrators that do not fit this picture, that is, think of the narrator and the characters as relatively independent existences. Then it becomes possible to assign tasks so that the story generation and production is handled by the human, while the discourse is handled by the *narrative generation system* (or the other way around), and the narrative's world setting and building is handled by the human, while other tasks are handled by the *narrative generation system* (or the other way around). This is in case of the abovementioned (3), but the (4) format makes the characters appearing in the *novel* relatively independent and assigns them to either the machine or the human. One of the principal functions of the characters is to perform actions and those actions might be easily mistaken for the progress and development of the abovementioned narrative contents (or rather, they might be thought of as equivalents), but when we separate the characters from the narrator to become independent mechanisms, it immediately becomes possible to separate the two by considering the story generation based on the narrative's structural principles as the sphere of the narrator and the action development prompted by the characters' will or environmental necessity as the sphere of the characters. When assigning tasks within the characters, we can imagine, for example, that the *narrative generation system* handles the operation of external actions, while the human handles the construction of various attribute elements and psychological details (or the other way around). (1) and (2) completely separate the tasks relating to the narrator and the characters. Firstly, in case of human = characters, *narrative generation system* = narrator, we can imagine an organization where the human will take the characters' place and perform actions suitable for different situations, upon which the machine will perform the actions of other characters to create the story or conduct other linguistic actions. Conversely, in case of human = narrator, *narrative generation system* = characters, it is the machine that imitates the characters' actions.

In the above cases, we just roughly partitioned the *novel* generation into the narrator's parts and the characters' parts, but this is based on the idea that the narrative of the *novel* can be organizationally disassembled into components and that those components correspond to mechanisms in the generation process. The *novel*'s organization and generation process are modular. It is possible to separate them into relatively independent components as modules, and if the modular separation is done skillfully, each of them can be realized as relatively simple generation mechanisms and the whole is made up of various relationships and recombinations between the modules. The potential of this modular processing is the basis for the compositional potential of the *narrative generation system*, but rather than task assignment, the truly content-related issue that is up for examination is what should be handled by the human for it to have an effect on the *novel*'s fixedness. In this sense, the problem is not simply how to divide the work between the human and the *narrative generation system*, or how to use the *narrative generation system* as a tool for *creative activity support*. Moreover, when I talk about the human here, it is not simply about the human working. This becomes clear from the fact that the *novel*'s fluidity and fixedness do not purely correlate with machinery and humanity.

Now, even in the case of a mechanism that prioritizes the formal, it has to include some content element for it to be a *novel*. The content of a *novel* includes a concrete theme, characters, setting (time and space), a variety of concrete modeling, ideology and thinking expressed by the narrator and characters (it can be either directly visible and expressed or not directly visible or expressed). The content elements of the *novel* are one with the formal mechanism that composes and arranges them, so it is of course possible for them to be selected and used coincidentally and by chance. Moreover, it is not necessarily so that the content mechanism is equivalent to a human mechanism or that they are linked. Yet as the initiative is transferred to the content side (although the two are like two sides of the same coin, making it only a matter of degree) due to the expediency of content-related formal initiative or arbitrary use, I assume that the strength of the *novel*'s fixation increases. The same thing applies to the formal aspect as the introduction of a human element serves to amplify its power further.

Speaking of which, as I mentioned at the beginning of this section, the sense of frank boredom toward a *novel* (conversely, it can also be called a sense of interestingness in the most boring sense) also has quite a lot to do with the content aspect. The sense of no basis for a story progressing and advancing in this way instead of that or the sense of fortuity has to do with the formal aspect, and since form cannot be selected on the same formal plane at that location, there is an orientation toward parallel generation, duplicating generation, and other kinds of exhaustions. Moreover, the content aspect, including character and world modeling as well as ideology and ways of thinking, can also stir the *novel* receiver's feelings toward other possibilities and hidden realizations, just like with the sense of story fortuity, and play the role of making them aware of the light basis for the possibilities selected and constructed (I digress, but when reading Mishima's *novels*, like *Shizumeru Taki* [*The Sinking Waterfall*] (Mishima, 1955) and *Kinkakuji* [*The Temple of the Golden Pavilion*] (Mishima, 1956), I get the feeling that the narrative content is completely enclosed by pan-exhausting possibilities, while choices and decisions are rather done as if by a seemingly mechanical force and a kind of baseless force). This resonates with the sense of boredom towards the form, which might suggest that the form does not function as a container for the meaningfulness of the content or that the content did not have enough destructive force to produce something interesting enough to suit the form. The reason why I was unable to participate in the *novels* without criticizing them was the sense that the great number of *novels* that exists (and continue to multiply) in a surface corner of our world, modern Japan, if it is modern Japan (already inconspicuous), possesses great content strength but have stopped aspiring to appropriate form. Alternatively, it was perhaps because of the sense that the world of *novels* no longer had the intellectual content to develop the strength of *novel* forms. In the end, form and content are linked, and even if we were to accept the idea that the start of the form is the *novel*'s lifeline, in the end, that is supported by the strength of the content aspect or they must be in a relationship of accomplices. If there were any formal reform, it must be premised on and it must result in content reform. (It naturally goes without saying that reform here has nothing to do with politics.) However, this process is not simple, or perhaps I should say that it cannot be simplified. We have to maintain the *novel*'s formal fluidity without resetting it or amplify it further, while composing an ideology that can grasp its totality, adding content to the *novel* in a fluid state, and then pushing that content toward fluidization as well, while also constructing content for an ideology that overlooks that even bigger totality. The formal and content-related inevitability of the *novel* must be supported by fluidity and that fluid support must be inevitable. (The *novel*'s content does not only include the narrative content expressed there. If it were so, then our discussion here would be dangerous in an extremely clichéd way. The content of a *novel* also includes an ideology that gives a basis for the form and content. This ideology supports content that can only be expressed through a certain form as well as supports forms that are necessary for the expression of certain content.)

The period from about 1971 to 1983, during which I was obsessed with reading *novels* as well as other literary genres like philosophy, ideology, history, music in addition to the humanities and the social sciences, was a transition period for the social phenomenon of the *novel*. *Novels* were read mainly as *tankōbon* books, but also as the following books:

- ***Bunko* paperbacks**: Shinchō bunko, (now gone) Ōbunsha bunko, Kadokawa bunko, Iwanami bunko, Kōdansha bunko, Chūkō bunko, etc.
- ***Shinsho* pockets**: At the time, Iwanami Shinsho was the most popular. The increase in *shinsho* types was a phenomenon that happened later.

- **Literary collections**: World literary collections, Japanese literary collections, and Russian literary collections by Kawade shobō shinsha, Chūō kōronsha, Shinchōsha, Shūeisha, and other publishers (while not *novels*, there were also a number of collections on Japanese and world history.)
- **Individual collections**: Quite a few individual literary collections of active writers like Kenzaburō Ōe, Jun'ichirō Tanizaki, Kōbō Abe, Kazumi Takahashi (1931-1971), Yumiko Kurahashi, and Toshio Shimao (1917-1986) were published and sold relatively cheaply, making them easy to acquire.

Since then, the types and volume of *novels* included in *bunko* paperbacks have increased. They used to contain only conventional classics and recognized works of modern and contemporary literature, contemporary writers on the frontline of so-called pure literature, and the big names of popular literature. but the bunko-ization has now been incorporated into the medium mixing and flow of the *novel* as our day's consumer good as well as its sales strategy (this strategy can be said to have started already with Yukio Mishima on a small-scale, personal level, but it is my impression that it was launched on a grand scale by Kadokawa bunko [the Kadokawa bookstore of Haruki Kadokawa (1942-)]). They rapidly started containing more and more entertainment *novels* and contemporary works that cannot be evaluated with any finality. Moreover, the literary collection projects have disappeared, and collections of currently active contemporary writers are not published so much anymore. Some of these *novels* as consumer goods evolved their quality or carved out an independent area for themselves, becoming the subject of later thought and criticism (Kiyoshi Kasai (1948-), Eiji Ōtsuka (1958-), Hiroki Azuma (1971-), etc. Of course, Takaaki Yoshimoto's genre-indiscriminate critiques, based on his perceptions of the data body of contemporary society, also contributed to this trend). Since then and until now, the *novel* has likely expanded in terms of volume and has unconsciously undergone repetitions of standard form and standard contents. Looking at this boring and dark transition period, it is not that there are no sure signs of the combined formula of formal reform and content reform being on the way. Mechanical methods were introduced as a limit of fluidization that can deprive the *novel* of its fixedness. Envisioning the purest, meaning the most superficial, formalization and thoroughly pursuing that direction means depriving the *novel* of close to all content and individuality, but what awaits beyond that is the acquisition of some form of content element that appears at once arbitrary and inevitable, and I think that can be a turning point for the *novel*.

However, if the content elements of the *novel* have to support mechanical processing, then we have no choice but to try to regress the debate on the possibilities of exhausting mechanisms for *novel* content. For example, consider the phenomenon of various *novels* that consist exclusively of quotations (this is the most blatant form of intertextuality embodied, but by relaxing the criteria in various ways, it should be possible to create some variations). Looking at the end product, quotations are selections of specific narrative contents, and quotations can be given consistency by continuously developing a specific contents theme, thereby giving rise to a totality (purely or simply, it is possible to have formal methods for controlling the whole by following rhetorical flourishes such as anagrams). They are exhausting as a mechanism, but the insertion of human elements makes it possible to increase selection and inevitability, meaning the power of the *novel*'s fixation.

Speaking of which, contrasting with content in the sense of concrete information about characters, setting, other kinds of content, and story development (looking at the Vladimir Propp (1895-1970), for example, the level of the *function* of *violence* corresponds to the narrative's formal aspect, and the method for concretely realizing this *function* corresponds to the content aspect) is content in the sense of ideology. Of course, and as I will discuss further down, the ideology of the *novel* should normally be expressed indirectly through events, story sequences, or way of speaking, so the distinction here is nothing but relative. Ideology is the way of thinking expressed in the *novel*. According to the *Poetics* of Aristotle, it is not the most important element of narrative artworks, but if we expand our perspective, we can see that it is the principle that gives shape to the *novel* as a whole. That is, ideology can be defined broadly and it can be defined narrowly. The latter is less an ideology that encompasses the basic principle making up the *novel* and more of an ideology in the sense of a specific way of thinking confessed or expressed in the *novel*. In order to think about this in the most exhausting way and imagine whether it is possible to cut-and-paste compose just like above, let us consider the case of quoting just one wrsiter for the sake of preserving content consistency. For example, the ideology of Yukio Mishima is very directly expressed in his critique *Bunka Bōei Ron* [*The Defense of Culture*] (Mishima, 1969) and is likewise (although philosophically) very indirectly (or symbolically, poetically) expressed in his critique *Taiyō to Tetsu* [*Sun and Steel*] (Mishima, 1968). If we extract a part as an aphorism, then it should become some form of ideological quotation. Looking not only at his critiques, but also his *novels*, *Kinkakuji*, *Utsukushii Hoshi* [*A Beautiful Star*] (Mishima, 1962), *Akatsuki no Tera* [*The Temple of Dawn*] (Mishima, 1970), or *Eirei no Koe* [*Voices of the Heroic Spirit*] (Mishima, 1966) can all be called ideological *novels*, so it should be possible to extract aphorisms or ideological confessions. Yet, considering a *novel* like *Yukōku* [*Patriotism*] (Mishima, 1973), it is undoubtedly a pivotal confession of the ideology and philosophy of Yukio Mishima the writer, but there is almost no place in that *novel* where that ideology is directly described. Most of the *novel* is actions, depictions of psychology, and accompanying explanations, so that the ideology is indeed expressed only indirectly, so it does not allow for the exhausting expression of ideology through direct quotation or extraction. This kind of format is the most *novel*-like expression of ideology. In other words, ideology here is symbolized by plot, story, and narrative discourse as a whole. A part can only express that ideology indirectly and symbolically when it occupies a certain spot in that totality. As such, ideology in this sense also functions as a principle that regulates the resolution of the story, corresponding to what I earlier referred to as the broad definition. In the end, ideology in the *novel* operates as a force to impede the formal exhausting format and as a force to select one possibility among many. However, it serves no purpose to have this as our conclusion. The extremity of the formal exhausting style must have an impact on the contents of this ideology that flourishes in form. In turn, the ideology must promote such form. Ideology in this sense should perhaps now be called un-ideology or anti-ideology or non-ideology. If we take this state of un-ideology literally, then both the ideology directly confessed somewhere in the *novel* and the ideology expressed through the totality of the *novel* have the potential to be exhaustingly composed of quotations and copy-and-past, continuously avoiding the fixation of the content aspect as much as possible. At present, I do not know how that correlates with the *novel*'s form, but I cannot be sure that a fragment of ideology in the narrow sense will appear from there (the existence of *I* might correlate with that appearance).

Perhaps it is formed by nothing but various quotations (having mentioned Yukio Mishima, there is also Kazumi Takahashi, Dostoyevsky, Riichi Yokomitsu, Ikki Kita (1883-1937), Sōseki Natsume, etc.), but for example, this *I* am dedicating at least some thought to systems, and especially the system of the nation, and more specifically the system of Japan. On this issue, I feel that there exists a sensation that collides with exhaustion and pluralism. The topics of the nation and Japan are multilayered, and are issues of both self-identity and self-division that overlaps with the diversity of the narrators and characters as well as touches on the issues of external cultural pluralism. Yet they are also accompanied by something that collides with multilayeredness, pluralism, and diversity. This means that there also exists a point where extremity of exhaustion switches to total denial. What is important here is that it is *total denial*. In other words, it has to be different from the work of selecting one possibility from many. That total denial has to do with a position that dislocates *me* from being voluntary and universal to being individual. This dislocation does not happen through subjective work but should perhaps rather be thought of as happening through the organization of certain circumstances. Nonetheless, it goes without saying that I do not praise this kind of direction as an active value, but I anticipate that some exhausting fluid state will infiltrate this point in some way. That is, there is a latent direction toward a movement to once more dissolve it into a fluid state.

Furthermore, the abovementioned thematic content is clichéd in itself and is superficial in that it is clichéd. It may be that that *novel* or literary theme is stimulating something inside *me* to form the nucleus of some contents, but that is notional and so it cannot be that it has no relation to any actions or events surrounding *me*. Notions do not cause actions, but notions exist as the background for actions and events, or they exist in unison. If we think in this way, then *my* ideology or *my* contents should rather be hidden in sceneries and situations that are totally familiar to *me*. They are not hidden but are always present, so they are not to be discovered through excavation, but it is better to say that they should be discovered through perception. The issue is from what angle and with what strength the notions will pierce the discovered actions and events of *mine*. Or rather, this design that also mobilizes the unconscious is a task that is all about *my* content.

What I have discussed above was an exploration of paths for *novel* fluidization and fixation, a pursuit of the integration of the two and then a notional examination of their methods, carried out by the *narrative generation system*, the human, and *myself*. Yet it is also possible to approach these issues from the perspective of the *novel* as a work. That is an issue of how fluid elements and fixed elements are arranged within a single *novel* work. The creation of this image is best pursued in connection with a study of the *novel*'s generation and production process, but we did not reach any integrated descriptions here, so I will finally just add an opinion on the *novel* as work, in the sense of a suggestion for future study. Expressed roughly, a fluid work signifies a state where a *narrative generation system* or a human (narrative generation mechanism) is directly moving inside the *novel*, while a fixed work is an already generated text that is still and preserved within the *novel*. In the former case, the generation process itself is part of the *novel* as a work, while in the latter case, the work is merely a result of the generation process. In this way, just like we distinguish between a fluid state and a fixed state in the generation process, the format of a work can be distinguished in the same way. If the fluidity/fixedness of the generation process are mutually independent from those of the work, then we should see a variety of combinations between them. For example, imagine a work that is overall a fluid *novel*, meaning that it is a *novel* constantly in a generation process. In other words, this is a format where the generation process itself makes up the work. Conversely, we can only clearly distinguish between the generation process and the work when the *novel* as a work is completely fixed, regardless of the state of the generation process. Any *novel* produced using the *narrative generation system* already possesses a fluid character, but by directly adding fluid elements to the format of the work itself, we get a multiplex fluid work.

This is nothing but an extremely simple thought experiment, but if we mechanically divide the generation process into the three categories of *fluid*, *fixed*, and *intermediate*, add the styles of *mechanical generation*, *human generation*, and *intermediate generation*, and combine these, then we can make nine types, such as a generation process that does mechanical generation completely fluidly. Moreover, by dividing the intermediate state in *generation process: intermediate* into *fluidity>fixedness* (the fluid character is greater than the fixed character), *fluidity=fixedness* (the two characters are of equal strength), and *fluidity<fixedness* (the fixed character is greater than the fluid character) as well as divide the mechanical and human intermediate states into the three categories of *mechanical>human*, *mechanical=human*, and *mechanical<human*, then the number of combinations increases to 25:

1. Fluid+human
2. Fluid+mechanical
3. Fluid+(human>mechanical)
4. Fluid+(human=mechanical)
5. Fluid+(human<mechanical)
6. (Fluid>fixed)+mechanical
7. (Fluid>fixed)+human
8. (Fluid>fixed)+(human>mechanical)
9. (Fluid>fixed)+(human=mechanical)
10. (Fluid>fixed)+(human<mechanical)
11. (Fluid=fixed)+mechanical

12. (Fluid=fixed)+human
13. (Fluid=fixed)+(human>mechanical)
14. (Fluid=fixed)+(human=mechanical)
15. (Fluid=fixed)+(human<mechanical)
16. (Fluid<fixed)+mechanical
17. (Fluid<fixed)+human
18. (Fluid<fixed)+(human>mechanical)
19. (Fluid<fixed)+(human=mechanical)
20. (Fluid<fixed)+(human<mechanical)
21. Fixed+human+human
22. Fixed+mechanical
23. Fixed+(human>mechanical)
24. Fixed+(human=mechanical)
25. Fixed+(human<mechanical)

At the same time, if we divide the intermediate states of fluidity/fixedness as a work into three categories as well, we get the following five:

1. Fluid
2. Mechanical
3. Intermediate (fluid>fixed)
4. Intermediate (fluid=fixed)
5. Intermediate (fluid<fixed)

As such, by combining the generation process states (25) with the work states (5), there exist a total of 125 states.

This kind of subdividing shows that even in cases where a work is completely fluid, meaning that the work is a generation process itself, the concepts of fluidity/fixedness in the generation process are separate from those of the work, so that the generation process itself can be either fluid or fixed. In the former case, the generation is indefinite, non-selective, and exhaustive, which is a word I have used frequently. In the latter case, it is the opposite as the generation is definite and highly selective. If a machine (*narrative generation system*) is added to the fluid generation, that will further increase the exhaustion.

At the same time, also in the extreme case of a completely fixed work, the generation process can be either fluid or fixed. In particular, if we were to fixate the most fluid generation, then the generation process will not include any unambiguous decisions on plot, story, or other elements, and the generation itself will not have the power to fixate, meaning that the fixation of the work might happen in the manner of flash freezing. A single instance of the work in a fluid state is extracted. If the generation process is completely fluid, then the result of this extraction becomes completely random. Also in the case of the previously mentioned generation process as a series of possibilities and composite generation process, we can imagine a correlation with how this freezing is done (especially the method for fixation as a work in the case of fluid generation). For example, if multiple generations are done in parallel, we can achieve the fixation of all of it by fixating and freezing each of them and somehow bundling each of them into a work. This can be considered a convergent application of the methods above.

Lastly, and this is redundant, but I am interested in the issue of *denial* in literary descriptions and discourse (more broadly, it connects with issues of words and world perception), for example as demonstrated in the *waka* poetry of Fujiwara no Teika (1162-1241), when he writes 見渡せば花も紅葉もなかりけり浦の苫屋の秋の夕暮れ (*Miwataseba Hana mo Momiji mo Nakarikeri Ura no Tomaya no Aki no Yūgure* [*Looking out there are neither blossoms nor crimson leaves*])." The question of why it has to be that way demands a description that it is so as a premise and preceding stage. That it is so is affirmative, but what if the preceding stage contained a description of denial? Taking the above example, we can imagine the impossible situation of a simultaneous scene (location) where there also are blossoms (cherry), and there also are crimson leaves, and then both situations are denied altogether. In the case of affirmation, either of the situations is selected and focused on, which is sure to bring up the question of why it has to be so. Yet in the case of denial, then the decision that it has to be so is really discarded, and it gives rise to movements of fluidization toward other possibilities. However, at the same time, that fictitious image also establishes one instance of fixation. Not only that, but in the case of denial, the contradictory matters and events described there will be able to find a basis without losing their descriptive consistency. Conversely, in the case of affirmation, the consistency of the contradictory images is violated. For example, if we write an affirmation that "there are blossoms (in one place) and crimson leaves (in the same place)," then we get two images that cannot merge. Yet if we write "there no blossoms (in one place) and no crimson leaves (in the same place)," then the two events are denied and can exist in parallel without the need for any merging on the overall level. Right now, the two images contradict each other, but they are fixed as images and hold validity at the same time. Is it not possible that denial has the potential of realizing processing for the integrated expression of diverse narrative descriptions? The work for standardizing the use of denial might be a potent method for fixating fluidity. This is a hypothetical idea on my part.

THE FIELD OF NARRATIVE GENERATION AS SOCIAL AND COLLECTIVE PHENOMENA

From the viewpoint of this study, all the worlds concerned with humans are constructed as narratives. The author uses the three types of illusions theory by Takashi Yoshimoto. This theory divides human worlds (or narrative worlds) into communal or collective illusions, pair illusions, and self illusions (or individual illusions or personal illusions). We can replace the word "illusion" with "narrative: and rename the three types of illusions, communal narratives, pair narratives, and self narratives.

In addition, the narrative genre system by the author can be considered in association with the three illusions theory by Yoshimoto. Among the main books by Yoshimoto, the level of self illusions was discussed in *Shinteki Genshoron Josetsu* (Yoshimoto, 1971a) and its main part (*Shinteki Genshoron Honron* (Yoshimoto, 2008), and *Gengo ni totte Bi toha Nannika* (Yoshimoto, 1965). Next, a part of *Kyōdō Gensō Ron* (Yoshimoto, 1968), a part of *Mass Image Ron* (Yoshitmo, 1984), and a part of *High Image Ron I, II, III* (Yoshimoto, 1989a, 1990, 1994) discussed the level of pair illusions. The level of communal illusions was addressed through parts of *Kyodo Gensoron*, *Mass Image Ron*, and *High Image Ron*. In comparison with the author's narrative genre system, the following genre areas correspond with the narrative genre of self illusions: (5) "Narratives as human physiological and psychological natural phenomena or human physiological and psychological natural phenomena in which narratives appear (or narratives are included)", (1) "Narratives as works in the narrow sense or works in the narrow sense in which narratives appear (or narratives are included)", and (2) "Narratives as works in the wide sense or works in the wide sense in which narratives appear (or narratives are included)". The range through (2), (3) "Narratives as social and emergent phenomena or social and emergent phenomena in which narratives appear (or narratives are included)", and (4) "Narratives that invade real phenomena or real phenomena in which narratives appear (or narratives are included)" can be understood as communal illusions. If the existence of language is social, (1) and (2) are also in the area that bridges self illusions to communal illusions.

When the author considers the three illusional theory in relation with the narrative genre system, pair narratives (illusions) are difficult to position. Although the communal illusions or -narratives and self illusions (narratives) can easily be positioned in the narrative genre system, pair illusional narratives are considered to correspond to the characteristics and contents rather than the category of narrative genres. Narratives at the level of novels and poems may be narratives at the level of pair illusions in contrast to self illusional narratives such as narratives of dreams and delusions or narratives generated in the human consciousness and the communal illusional narratives such as politics and ceremonies. If the category classification of the narrative genre system indicates a type of narrative's narrator's type, the narratives such as movie genre are communal illusional narratives that include many narrators. However, the classification of the narrative genre system is in fact based on the points of the social functions of a narrative genre and whether the narrative generation is intentionally conducted, particularly regarding whether the narrative is socially emerged or is generated by clear individual intentional consciousness. Although the latter is related to the narrator's type in a sense, the narrator is not the type in the above meaning. Simply, the author calls the intentional narratives by individuals or organizations the self illusional narratives and non-intentional narratives that emerge through social groups as communal illusional narratives. In this case, pair illusional narratives are the narratives created by intentional pairs. However, do such narratives exist?

Concluding, the author considers that all narratives include self illusional, communal illusional, and pair illusional elements to some extent and the self illusional and communal illusional narratives have features that overlap with the large category classification in the narrative genre system. The pair illusional narratives sometimes have the same feature. However, they instead have the characteristic that exists commonly in all of the narratives. Pair illusions take a position of bridging self illusions and the communal illusions. Simultaneously, self illusions are a rather common concept through all narrative genres that exist in relation with the central theme in all narrative genres.

Now, a theme that the author would like to consider in the final level in narratives is another meta-level of narratives that emerged through the entirety of narrative generation at various levels. In particular, self illusional and pair illusional narratives are generated in each mechanism, and collective illusional narratives are also generated through considerably intentional mechanisms, such as political narratives and the narratives of legal systems. Furthermore, at the narrative level of collective illusions, narratives at various levels are collected and formed.

Therefore, collective illusional narratives are divided into collective illusional narratives for creating concrete social and collective institution creation and a more meta-level's collective illusional narratives. The narratives in various narrative genres' levels and the narratives as a whole generated through those narratives should be comprehended as distinct. This final narrative level, particularly narrative as a whole, is one of the final topics to be considered in the author's narrative generation study.

For the author, the word "institution" in a broad sense appears at this level. In particular, an entire society based on both the synchronic level and diachronic level (history) is a system of discourse through narratives. A world's image and social image is formed for us through narratives. The narratives in this case concretely contain various narrative genres shown comprehensively by the narrative genre system. A narrative is in fact realized as concrete content through the rhetoric and techniques from macro to micro in each genre in the narrative genre system. Actual world image and social image are influenced by the narrative genres in the narrative genre system that are dominant in society. Although this figure does not describe the origin of a narrative, particularly the origin on which narrative generation is performed, the origin is essentially information that is retrospectively formed from the world image and social image in a society. The above institution corresponds to the "narrative" in this case. In contrast to the concrete institutions such as political and educational institutions, the institution here is more a conceptual institution. However, the author calls this level's institution the first-level institution, and there are several more concrete social institutions in the second-level institution. Another viewpoint is that although the first-level institution can be decided and changed by the characteristic of usage of an entire narrative genre in a society, self illusional narratives such as literature play an important role for the change and result in a kind of evolution in these processes. Literary or artistic narratives are phenomena in which the levels of self illusions and pair illusions break the level of communal illusions. This process's collection changes and innovates the characteristics of the institution of an entire narrative.

FUTURE RESEARCH DIRECTIONS

1. ***Geinō* information system (GIS)**: The author plans to design and develop a system with INGS. An attempt for the design and development is to incorporate the experimental system based on the survey and analysis

of *kabuki* in Chapter 3 in this book as a mechanism of GIS with INGS more consciously. Further, the author also aims to design the virtual construction of a troupe or a *geinō* production company (entertainment agency), which is an actual form of the GIS, and to prepare *geinō* characters in various components in the GIS.

2. **Developing *watakushi monogataris***: Based on the informational framework the author described in this chapter, the author will consider and concretize the various information, construction, and generation methods in relation with the conceptual dictionaries in INGS and GIS.
3. **Philosophical considerations of institutional narratives in the broadest sense**: As described in the previous section, considering social narratives and collective narratives in the broadest sense, this research plan considers narratives from an institutional viewpoint. This research area has an essential connection with the theory of communal illusion by Yoshimoto (1968a) and historical narratology by Haydon White (1928-2018) (1973). Moreover, this area also has a close relationship with philosophical considerations regarding narrative generation. Philosophical considerations of the institutional narrative in the broadest sense will be considered in the future as a core research area that supports the author's wider narrative generation studies.

CONCLUSION

From relationships to the level of multiple narrative structures, the *Geinō* Information System (GIS), described in Chapter 2 corresponds to a social-level narrative generation mechanism beyond the narrative generation as a simple substance or as something including the generation of many narratives as a simple substance. From a broader perspective here, this chapter aims to discuss "the future of the narrative generation systems or the narrative generation study." The author discusses the topic from two directions: "external narrative generation" and "internal narrative generation." The "external" is the outer world beyond the "I" and ordinarily refers to a society or societies. Conversely, the "internal" is equal to the world inside the "I." Therefore, although the practice of narrative generation toward the outer direction ordinarily indicates the social development of narrative generation, the practice of narrative generation toward the internal is equal to the direction of narrative generation study for drawing the "I" in any style. In this chapter, the external direction of narrative generation is closely associated with GIS. In particular, GIS is a type of social narrative production-consumption mechanism that continues to produce cyclically narrative works to generate narrative contents in the larger level beyond each narrative work. On the other hand, the internal direction is related to the aspect of narrative content, especially the author's own story's or narrative's level. In contrast to the external narrative generation through INGS and GIS that pertains to the formal side of narrative generation, the internal narrative generation focuses on the side of contents of narrative works actually created according to the formal side. From a broad viewpoint, the description regarding the internal narrative generation may act as a pointer for a narrative work using narrative generation systems by the author. Therefore, this chapter also dealt with individual and personal information, including histories, novels, and writings regarding the author.

ACKNOWLEDGMENT

This chapter's research was supported by JSPS KAKENHI Grant Number18K18509, Kayamori Foundation of Information Science Advancement Research Grant, The Telecommunication Advancement Foundation Research Grant, and AOI TYO Holdings Inc.

REFERENCES

Akimoto, T., Endo, J., & Ogata, T. (2013). The expansion of paths in the mutual transformation mechanism of music and narrative. *International Journal of Cognitive Informatics and Natural Intelligence*, *7*(4), 44–63. doi:10.4018/ijcini.2013100103

Akimoto, T., & Ogata, T. (2012). A narratological approach for narrative discourse: Implementation and evaluation of the system based on Genette and Jauss. In *Proceedings of the 34th Annual Conference of the Cognitive Science Society* (pp. 1272–1277). Tokyo, Japan: Japanese Cognitive Science Society.

Akimoto, T., & Ogata, T. (2013). Discourse mechanism in narrative generation system: Proposal of a system introducing narrative discourse theory and reception theory. *Cognitive Studies*, *20*(4), 396–420.

Akimoto, T., & Ogata, T. (2014). Circulative narrative generation based on the mutual transformation between narrative conceptual structures and music in the integrated narrative generation system. *Journal of Robotics. Networking and Artificial Life*, *1*(3), 198–202. doi:10.2991/jrnal.2014.1.3.6

Akugtagawa, R. (1969a). Rashōmon. In Akutagawa Ryūnosuke (Nihon bungaku zenshū, 10) [Akutagawa Ryūnosuke (Japanese literature collection, Vol. 10)] (pp. 7–13). Tokyo, Japan: Shinchōsha. (Original work pubulished 1917)

Akutagawa, R. (1969b). Kappa. In Akutagawa Ryūnosuke (Nihon bungaku zenshū, 10) [Akutagawa Ryūnosuke (Japanese literature collection, Vol. 10)] (pp. 389–435). Tokyo, Japan: Shinchōsha. (Original work pubulished 1927)

Akutagawa, R. (1969c). Haguruma [Gear]. In Akutagawa Ryūnosuke (Nihon bungaku zenshū, 10) [Akutagawa Ryūnosuke (Japanese literature collection, Vol. 10)] (pp. 437–467). Tokyo, Japan: Shinchōsha. (Original work pubulished 1927)

Amino, T., Kawamura, Y., & Ogata, T. (2002). Hierarchical generation of geinō-idol stories—Toward geinō information system and narrative marketing. In *Proceedings of 17th Congress of the International Association of Empirical Aesthetics* (pp. 549–552). Rome, Italy: University of Rome Tre.

Ando, T. (2006). *Genji monogatari to monogatari shakai* [The tale of Genji and narrative society]. Tokyo, Japan: Shinwasha.

Arishima, T. (1995). Aru on'na. Tokyo, Japan: Shinchōsha.

Baddilo, A., Donovan, T., & Trevarthen, T. (2016). *Narrative generation: Why narrative will become your most valuable asset in the next 5 years (English edition)*. Scotts Valley, CA: Createspace Independent Publishing.

Bourdieu, P. (1996). *Rules of art: Genesis and structure of the literary field.* Stanford, CA: Stanford University Press. (Original work published 1992)

Eguchi, S., Saito, S., & Nomura, N. (Eds.). (2006). *Narrative to iryō* [Narratives and medicine]. Tokyo, Japan: Kongō Shuppan.

Etō, J. (1988). *Sakka ha kōdōsuru* [Writers act]. Tokyo, Japan: Kawade Shobō Shinsha.

Furuhashi, N. (2010). *Nihon bungaku no nagare* [The flow of Japanese literature]. Tokyo, Japan: Iwanami Shoten.

Futabatei, S. (1949). Heibon. Tokyo, Japan: Shinchōsha. (Original work publised 1908)

Haniya, Y. (1985). Seiji to bungaku to, hosoku—Yoshimoto Takaaki he no saigo no tegami [Politics and literature, supplement: The last letter to Yoshimoto Takaaki]. *Kaien*, *4*(4), 148–163.

Hashimoto, O. (2012). *Jōruri wo yomō* [Let's read jōruri]. Tokyo, Japan: Shinchōsha.

Hasumi, S. (2014). *"Bovary fujin" ron* [A theory of "Madam Bovary"]. Tokyo, Japan: Chikuma Shobō.

Hayashi, T. (Eds.). (1955–1960). Sekai dai hyakka jiten, 1–32 [Heibonsha world encyclopedia, Vols. 1–32]. Tokyo, Japan: Heibonsha.

Heike Monogatari. (1991). In M. Kajiwara & H. Yamashita (Eds.), Shin nihon koten bungaku taikei, 44 [New Japanese classic literature collection, Vol. 44]. Tokyo, Japan: Iwanami Shoten. (Original work published the Kamakura era)

Kenkyūkai, W. S. (Ed.). (2014). Watakushi shōsetsu handbook [I-Novels handbook]. Tokyo, Japan: Bensei Shuppan.

Heike Monogatari. (1993). In M. Kajiwara & H. Yamashita (Eds.), Shin nihon koten bungaku taikei, 45 [New Japanese classic literature collection, Vol. 45]. Tokyo, Japan: Iwanami Shoten. (Original work published the Kamakura era)

Hijiya-Kirschnereit, I. (1981). *Selbstentblößungsrituale: Zur Theorie und geschichte der autobiographischen gattung shishosetsu in der modernen japanischen literatur*. Wiesbaden, Germany: Harrassowitz Verlag.

Honchō Nijūshikō. (1766). [Japan's 24 paragons of filial piety] [Ningyō jōruri]. Chikamatsu Hanji, Miyoshi Shōraku, et al. (Scenario). Osaka, Japan: Takemotoza (First performance).

Ichinotani Futaba Gunki. (1751). [Chronicle of the battle of Ichinotani] [Ningyō jōruri]. Namiki Sōsuke, Asada Icchō, et al. (Scenario). Osaka, Japan: Toyotakeza (First performance).

Imao, T. (2010). *Kabuki—<Tsūsetsu> no kenshō* [Kabuki: The verification of 'common views']. Tokyo, Japan: Hosei Daigaku Shuppankyoku.

Iser, W. (1976). *Der akt des lesens*. Munchen, Germany: Wilhelm Fink Verlag.

Ishikawa, T. (1977). Takuboku ROMAZI NIKKI (T. Kuwabara, Ed. & Trans.). Tokyo, Japan: Iwanami Shoten. (Original work written 1909)

Ito, T. (2009). *Self-help group no jiko monogatariron—Alcohlism to shibetsu taiken wo rei ni* [Self-narratology of self-help groups: Using examples of alcohlism and bereavement experiences]. Tokyo, Japan: Harvestsha.

Iwagaki, M. (2008). *Katte yomi to iu koto* [On convenient reading]. In *Proceedings of the Japanese Cognitive Science Society 15th Workshop of Literature in Cognition and Computer II* (15W-01). Tokyo, Japan: Japanese Cognitive Science Society.

Jauss, H. R. (1970). Literaturgeschichte als provokation. Frankfurt am Main, Germany: Suhrkamp Verlag.

Joyce, J. (1922). *Ulysses*. Paris: Sylvia Beach.

Kan, S. (2010). *Rekishi no tetsugaku—Monogatari wo koete* [The philosophy of history: Beyond narratives]. Tokyo, Japan: Keisō Shobō.

Kanadehon Chūshingura. (1748). [The treasury of loyal retainers] [Ningyō jōruri]. Namiki Sōsuke, Takeda Izumo II, & Miyoshi Shōraku (Scenario). Osaka, Japan: Takemotoza (First performance).

Kanai, A. (2018). Eizō kara story wo ninchisuru koto/shinai koto—Ninchiteki reality no hassei yōuin [Recognizing/unrecognizing stories from the film: The generating factor of cognitive realities]. In T. Ogata, Y. Kawamura, & A. Kanai (Eds.), *Jōhō monogatariron—Jinkōchinō, ninchi, shakai katei to monogatari seisei* [Informational narratology: Artificial intelligence/cognition/social process and narrative generation] (pp. 141–153). Tokyo, Japan: Hakutō Shobō.

Kanai, T., Takai, T., Nakanishi, M., & Morioka, M. (Eds.). (2009). *Katari to katari no aida—Rashōmon-teki genjitsu to ningen no responsibility* [Between narrative and imposture]. Kyoto, Japan: Nakanishiya Shuppan.

Katō, S. (1997) A history of Japanese literature: From the Man'yōshū to modern times (D. Sanderson, Trans. & Ed.). Tokyo, Japan: Kōdansha International. (Original works published 1975, 1980)

Kato, S. (Ed.). (2003). *Katari to chōshu* [Narrating and listning]. Tokyo, Japan: Nakayama Shoten.

Kawamura, Y. (2018). Kōkoku eizō no image, han'no bunseki [The images of advertising films, the response analyses]. In T. Ogata, Y. Kawamura, & A. Kanai (Eds.), *Jōhō monogatariron—Jinkōchinō, ninchi, shakai katei to monogatari seisei* [Informational narratology: Artificial intelligence/cognition/social process and narrative generation] (pp. 141–153). Tokyo, Japan: Hakutō Shobō.

Kawamura, Y., & Ogata, T. (2000a). A conceptual framework on geinō organization model for artificial geinō performers with lives. In *Proceedings of the 5th International Symposium on Artificial Life and Robotics* (pp. 793–796). Oita, Japan: International Society of Artificial Life and Robotics.

Kawamura, Y., & Ogata, T. (2000b). Geinōjin image marketing system no tameno geinōjin homepage no bunseki [Analysis of geinōjins' homepages for a geinōjin image marketing system]. *Journal of Advertising Science*, *40*, 181–192.

Kiš, D. (1975). *Garden, ashes* (W. J. Hannaher, Trans.). San Diego, CA: Harcourt. (Original work published 1965)

Kobayashi, F., & Ogata, T. (2004a). Narrative and music as variation: Transformation of musical structure based on narrative discourse theory. In *Proceedings of the 9th International Symposium on Artificial Life and Robotics*, Vol. 1 (pp. 170–173). Oita, Japan: International Society of Artificial Life and Robotics.

Kobayashi, F., & Ogata, T. (2004b). A musical variation system by the structural correspondence between music and narrative. In *Proceedings of the 18th Congress of the International Association of Empirical Aesthetics* (pp. 555–559). Rome, Italy: University of Rome Tre.

Kurosawa, A. (Director), & Itō, M. (Producer) (1950). *Rashōmon* [Motion picture]. Tokyo, Japan: Tōhō.

Kusunoki, K. (2010). *Story senryaku tosite no kyōsō senryaku—Sugureta senryaku no jōken* [Competitive strategies as story strategies: Conditions of excellent strategies]. Tokyo, Japan: Tōyō Keizai Shinpōsha.

Mikuriya, T. (Ed.). (2007). *Oral history nyūmon* [Introduction to oral histories]. Tokyo, Japan: Iwanami Shoten.

Mishima, Y. (1949). Kamen no kokuhaku. Tokyo, Japan: Kawade Shobō.

Mishima, Y. (1955). *Shizumeru taki* [The sinking waterfall]. Tokyo, Japan: Chūōkōronsha.

Mishima, Y. (1956). Kinkakuji. Tokyo, Japan: Shinchōsha.

Mishima, Y. (1962). *Utsukushii hoshi* [A beautiful star]. Tokyo, Japan: Shinchōsha.

Mishima, Y. (1966). *Eirei no koe* [Voices of the heroic spirit]. Tokyo, Japan: Kawade Shobō Shinsha.

Mishima, Y. (1968). Taiyō to tetsu. Tokyo, Japan: Kōdansha.

Mishima, Y. (1969). *Bunka bōei ron* [The defence of culture]. Tokyo, Japan: Shinchōsha.

Mishima, Y. (1970). Akatsuki no tera. Tokyo, Japan: Shinchōsha.

Mishima, Y. (1971). Ten'nin gosui. Tokyo, Japan: Shinchōsha.

Mishima, Y. (1973). Yūkoku. In Mishima Yukio zenshū, 13 [Mishima Yukio collection, Vol. 13] (pp. 221–249). Tokyo, Japan: Shinchōsha. (Original work published 1961)

Miura, H. (2000). Naimaze. In Y. Hattori, T. Tomita, & T. Hirosue (Eds.), Shinpan kabuki jiten [New edition kabuki dictionry] (p. 310). Tokyo, Japan: Heibonsha.

Mori, O. (1973). Izawa Ranken, 1, 2. In Mori Ōgai zensyū, 16, 17 [Complete collection of Mori Ōgai, Vols. 16, 17]. Tokyo, Japan: Iwanami Shoten. (Original work published 1916–1917)

Mori, O. (1999). *Shibue Chūsai*. Tokyo, Japan: Iwanami Shoten. (Original work published 1916)

Morioka, M. (Ed.). (2008). *Narrative to shinri ryōhō* [Narratives and psychotherapy]. Tokyo, Japan: Kongō Shuppan.

Musil, R. (1930). *Der mann ohne eigenschaften* (Vol. 1). Reinbek, Germany: Rowohlt Verlag.

Musil, R. (1932). *Der mann ohne eigenschaften* (Vol. 2). Reinbek, Germany: Rowohlt Verlag.

Nagai, K. (2001a). Shinpan Danchōtei-nichijō, 1 [New edition Danchōtei diary, Vol. 1]. Tokyo, Japan: Iwanami Shoten. (Original work written 1917–1957)

Nagai, K. (2001b). Shinpan Danchōtei-nichijō, 2 [New edition Danchōtei diary, Vol. 2]. Tokyo, Japan: Iwanami Shoten. (Original works written 1917–1957)

Nagai, K. (2001c). Shinpan Danchōtei-nichijō, 3 [New edition Danchōtei diary, Vol. 3]. Tokyo, Japan: Iwanami Shoten. (Original works written 1917–1957)

Nagai, K. (2001d). Shinpan Danchōtei-nichijō, 4 [New edition Danchōtei diary, Vol. 4]. Tokyo, Japan: Iwanami Shoten. (Original work written 1917–1957)

Nagai, K. (2002a). Shinpan Danchōtei-nichijō, 5 [New edition Danchōtei diary, Vol. 5]. Tokyo, Japan: Iwanami Shoten. (Original works written 1917–1957)

Nagai, K. (2002b). Shinpan Danchōtei-nichijō, 6 [New edition Danchōtei diary, Vol. 6]. Tokyo, Japan: Iwanami Shoten. (Original work written 1917–1957)

Nagai, K. (2002c). Shinpan Danchōtei-nichijō, 7 [New edition Danchōtei diary, Vol. 7]. Tokyo, Japan: Iwanami Shoten. (Original works written 1917–1957)

Noe, K. (1996). *Monogatari no tetsugaku—Yanagita Kunio to rekishi no hakken* [The philosophy of narrative: Yanagita Kunio and the invention of histories]. Tokyo, Japan: Iwanami Shoten.

Noe, K. (2007). *Rekishi wo tetsugaku suru* [Practicing the philosophy of histories]. Tokyo, Japan: Iwanami Shoten.

Ogata, T. (2000). Tajū monogatari kōzō no macro model—Simulation toshite no monogatari josetsu [The macro model of multiple narrative structures: An introduction to narratives as simulation]. *Simulation & Gaming*, *10*(1), 35–46.

Ogata, T. (2010). Shōsetsu—Ryūdō to kotei, Sakuhin no hō he [Novels: Fluidity and fixation, toward works]. In T. Ogata & A. Kanai (Eds.), *Monogatariron no jōhōgaku josetsu—Monogatari seisei no shisō to gijutsu wo megutte* [An introduction to informatics of narratology: Around the thoughts and technologies of narrative generation] (pp. 130–169). Tokyo, Japan: Gakubunsha.

Ogata, T. (2011a). Monogatari seisei no naiyōteki tōgō ni mukete—Monogatari/bungaku/geinō, kabukis/Mishima/「watakushi」, shōsetsu: Ryūdō to kotei (12) [Toward integrating the contents of narrative generation—Narrative/literature/geinō, kabuki/Mishima/"I," novels: Fluidty and fixation (12)]. In *Proceedings of the Japanese Cognitive Science Society 24th Workshop of Literature in Cognition and Computer II* (24G-02). Tokyo, Japan: Japanese Cognitive Science Society.

Ogata, T. (2011b). Monogatari seisei system ha ikani kinō suruka: Shakai no monogatai seisei to watakushi no monogatari seisei he no jo—Ryūdō to kotei (13) [How narrative generation systems function? Introduction to social narrative generation and my narrative generation: Fluidity and fixation (13)]. In *Proceedings of the Japanese Cognitive Science Society 25th Workshop of Literature in Cognition and Computer II* (25G-02). Tokyo, Japan: Japanese Cognitive Science Society.

Ogata, T. (2011c). "Monogatariron no jōhōgaku" no jissen to shite no monogatari seisei system—Ryūdō to kotei (14) [Narrative generation systems as the practice of "informatics of narratology": Fluidity and fixation (14)]. In *Proceedings of the Japanese Cognitive Science Society 26th Workshop of Literature in Cognition and Computer II* (26W-02). Tokyo, Japan: Japanese Cognitive Science Society.

Ogata, T. (2011d). Tsuiho—*Watakushi* kenkyū kara *watakushi* monogatari he [Suppliment: From *watakushi* researchs to *watakushi* narratives]. *Journal of Japanese Society for Artificial Intelligence, 25*(6), 494–513.

Ogata, T. (2011e). Narrative generation system as the practice of "informatics of narratology.". *Journal of Japan Society for Fazzy Theory and Intelligent Informatics*, *23*(5), 14–24.

Ogata, T. (2016). Computational and cognitive approaches to narratology from the perspective of narrative generation. In T. Ogata & T. Akimoto (Eds.), *Computational and cognitive approaches to narratology* (pp. 1–73). Hershey, PA: IGI Global. doi:10.4018/978-1-5225-0432-0.ch001

Ogata, T. (2018a). Gaibu he no monogatari seisei mata ha geinō jōhō system ni mukete [Narrative generation to the outside or toward a geino information system]. In T. Ogata, Y. Kawamura, & A. Kanai (Eds.), *Jōhō monogatariron—Jinkōchinō, ninchi, shakai katei to monogatari seisei* [Informational narratology: Artificial intelligence/cognition/social process and narrative generation] (pp. 327–353). Tokyo, Japan: Hakutō Shobō.

Ogata, T. (2018b). Naibu he no monogatari seisei mata ha watakushi monogatari ni mukete [Narrative generation to the inside or toward I narratives]. In T. Ogata, Y. Kawamura, & A. Kanai (Eds.), *Jōhō monogatariron—Jinkōchinō, ninchi, shakai katei to monogatari seisei* [Informational narratology: Artificial intelligence/cognition/social process and narrative generation] (pp. 355–370). Tokyo, Japan: Hakutō Shobō.

Ogata, T. (2018c). An integrated approach to narrative generation: From Mishima and kabuki to narrative generation systems. In T. Ogata & S. Asakawa (Eds.), *Content generation through narrative communication and simulation* (pp. 49–147). Hershey, PA: IGI Global. doi:10.4018/978-1-5225-4775-4.ch002

Ogata, T. (2019a). A computational, cognitive, and narratological approach to narrative generation. In T. Ogata & T. Akimoto (Eds.), *Post-narratology through computational and cognitive approaches* (pp. 1–84). Hershey, PA: IGI Global. doi:10.4018/978-1-5225-7979-3.ch001

Ogata, T. (2019b). *Toward an integrated approach to narrative generation: Emerging research and opportunities*. Hershey, PA: IGI Global.

Ogata, T. (2019c). Areas of narratives or narrative genres. In T. Ogata (Ed.), *Toward an integrated approach to narrative generation: Emerging research and opporunities* (pp. 59–161). Hershey, PA: IGI Global.

Ogata, T. (2019d). Theoretical or philosophical considerations for an integrated narrative generation approach. In T. Ogata (Ed.), *Toward an integrated approach to narrative generation: Emerging research and opportunities* (pp. 315–403). Hershey, PA: IGI Global.

Ogata, T., & Akimoto, T. (2007). Towards the circular narrative generation based on the correspondences between language-based narrative and music: Systems development based on narrative rhetoric and fundamental considerations about the systems and the circular narrative generation. *Cognitive Studies*, *14*(3), 355–379.

Ogata, T., & Kanai, A. (2010). *Monogatariron no jōhōgaku josetsu—Monogatari seisei no shisō to gijutsu wo megutte* [An introduction to informatics of narratology: Around the thoughts and technologies of narrative generation]. Tokyo, Japan: Gakubunsha.

Ogata, T., Kawamura, Y., & Kanai, A. (2018). *Jōhō monogatariron—Jinkōchinō, ninchi, shakai katei to monogatari seisei* [Informational narratology: Artificial intelligence/cognition/social process and narrative generation]. Tokyo, Japan: Hakutō Shobō.

Oikawa, H., & Ogata, T. (2012). Tōjōjinbutsu no sōgo ninshiki kikō toshite no monogatari seisei no kōsatsu [On the narrative generation as a mutual cognition mechanism among characters]. In *Proceedings of the 29th Annual Meeting of the Japanese Cognitive Science Society* (pp. 540–549), Tokyo, Japan: Japanese Cognitive Science Society.

Ono, J., & Ogata, T. (2018). Surprise-based narrative generation in an automatic narrative generation game. In T. Ogata & S. Asakawa (Eds.), *Content generation through narrative communication and simulation* (pp. 162–185). Hershey, PA: IGI Global. doi:10.4018/978-1-5225-4775-4.ch004

Saito, K., & Honda, T. (2001). *Life cource no shinrigaku* [The psychology of life course]. Tokyo, Japan: Kaneko Shobō.

Sapir, G. (2014). *La sociologie de la littérature*. Paris: La Découverte.

Sato, K., Ochi, K., & Shimojima, H. (2008). *Jidenteki kioku no shinrigaku* [The psychology of autobiographical memory]. Kyoto, Japan: Kitaōji Shobō.

Sekkyōshū. (1977). In Y. Muroki (Ed.), Shinchō nihon koten shūsei [Shinchō Japanese classic literature collection]. Tokyo, Japan: Shinchōsha. (Sekyōbushis were narrated from the end of the medieval period to the early modern period in Japan)

Shiga, N. (1973). An'ya kōro. In Shiga Naoya zenshū, 5 [Shiga Naoya collection, Vol. 5]. Tokyo, Japan: Iwanami Shoten. (Original work published 1922–1937)

Shiller, R. J. (2017). Narrative economics. *Cowles Foundation Discussion Paper, 2069.*

Shimazaki, T. (1966a). Yoakemae, 1. In Tōson zenshū, 11 [Shimazaki Tōson collection, Vol. 11]. Tokyo, Japan: Chikuma Shobō. (Original work published 1929–1935)

Shimazaki, T. (1966b). Yoakemae, 2. In Tōson zenshū, 12 [Shimazaki Tōson collection, Vol. 12]. Tokyo, Japan: Chikuma Shobō. (Original work published 1929–1935)

Shimazaki, T. (2006). Hakai. Tokyo, Japan: Iwanami Shoten. (Original work published 1906)

Smith, A. (1759). *Theory of moral sentiments*. London, UK: Andrew Milla. doi:10.1093/oseo/instance.00042831

Tanizaki, J. (1949). Sasameyuki. Tokyo, Japan: Chūōkōronsha. (Original work published 1943–1948)

Tayama, K. (1908). Futon. In Katai syū [Tayama Katai collection] (pp. 1–102). Tokyo, Japan: Ekifūsha. (Original work published 1907)

Tsuboi, S. (1952). Nijūshi no hitomi. Tokyo: Japan: Kōbunsha.

Tsushima, Y. (2014). "Monogatari" to "uta" no kanjusei [Sensitivity of "narative" and "song and narration"]. In Watakushi shōsetsu handbook [I-Novels handbook] (pp. 193–206). Tokyo, Japan: Bensei Shuppan.

White, H. (1973). *Metahistory: The historical imagination in nineteenth-century Europe*. Johns Hopkins University Press.

Yoshimoto, T. (1963). *Maruyama Masao ron* [On Maruyama Masao]. Tokyo, Japan: Hitotsubashi Shinbunbu.

Yoshimoto, T. (1965). *Gengo ni totte bi toha nanika* [What is beauty for language?]. Tokyo, Japan: Keisō Shobō.

Yoshimoto, T. (1966). *Karl Marx*. Aizuwakamatsu, Japan: Shikō Shuppan.

Yoshimoto, T. (1968a). *Kyōdō gensō ron* [The theory of communal Illusion]. Tokyo, Japan: Kawade Shobō Shinsha.

Yoshimoto, T. (1968b). *Jōkyō he no hatsugen* [Comments to situations]. Tokyo, Japan: Tokuma Shoten.

Yoshimoto, T. (1971a). *Shinteki genshōron josetsu* [A theory of mental phenomena: Introduction]. Tokyo, Japan: Hokuyōsha.

Yoshimoto, T. (1971b). *Minamoto no Sanetome*. Tokyo, Japan: Hokuyōsha.

Yoshimoto, T. (1975). *Shomotsu no kaitaigaku* [Books' anatomy]. Tokyo, Japan: Chūōkōronsha.

Yoshimoto, T. (1980). *Sekai ninshiki no hōhō* [A method of world recognition]. Tokyo, Japan: Chūōkōronsha.

Yoshimoto, T. (1982). *Genji monogatari ron* [Essays on The tale of Genji]. Tokyo, Japan: Yamato Shobō.

Yoshimoto, T. (1984). *Mass image ron* [Essays on mass images]. Tokyo, Japan: Fukutake Shoten.

Yoshimoto, T. (1985). Seiji nante mono ha nai [Politics does not exist]. In *Jūsōteki na hikettei he* [Toward multi-tiered nondetermination] (pp. 8–45). Tokyo, Japan: Yamato Shobō.

Yoshimoto, T. (1989a). *High image ron, 1* [Essays on high images, Vol. 1]. Tokyo, Japan: Fukutake Shoten.

Yoshimoto, T. (1989b). *Miyazawa Kenji*. Tokyo, Japan: Chikuma Shobō.

Yoshimoto, T. (1990). *High image ron, 2* [Essays on high images, Vol. 2]. Tokyo, Japan: Fukutake Shoten.

Yoshimoto, T. (1994). *High image ron, 3* [Essays on high images, Vol. 3]. Tokyo, Japan: Fukutake Shoten.

Yoshimoto, T. (2008). *Shinteki genshōron honron* [A theory of mental phenomena: Main issues]. Tokyo, Japan: Bunka Kagaku Kōtō Kenkyūin Shuppankyoku.

Yoshimoto, T. (2016). *L'Illusion commune* (H. Nakata, Trans.). Nagoya, Japan: Digitalestate.

Yoshimoto, T. (2017). *Yoshimoto Takaaki Etō Jun zen taiwa* [All talks of Yoshimoto Takaaki and Etō Jun]. Tokyo, Japan: Chuōkōronsha.

Yoshimoto, T. (2014–2015). *Yoshimoto Takaaki kōensyū, 1–12* [Yoshimoto Takaaki lecture collection, Vols. 1–12]. Tokyo, Japan: Chikuma Shobō.

KEY TERMS AND DEFINITIONS

External Narrative Generation: The author's external narrative generation means social development and distribution of the narrative generation systems themselves, mainly including "Integrated Narrative Generation System: INGS" and "*Geinō* Information System: GIS," and the contents generated by the narrative generation systems, such as computer games and automatic generation stories and novels. The first objective of external narrative generation is to develop a computational environment for using the above systems.

***Geinō* Information System (GIS):** This means a system model of *geinō* production and consumption. GIS has been considered a framework in which various levels of narrative generation processes are driven by writers, receivers, characters, and actors. Furthermore, GIS is a model for circulative narrative generation that repeats a narrative generation process. In contrast, in the author's narrative generation study, the "Integrated Narrative Generation System: INGS" corresponds to a single-level narrative generation system.

Integrated Narrative Generation System (INGS): INGS is a synthetic narrative generation system architecture integrating the previous studies by the author. From the broadest perspective, INGS is divided into two types of part, knowledge and procedure. INGS aims at a kind of narrative synthesizer that integrates a variety of narrative techniques, methods, rhetoric, and knowledge into an organic and dynamic generation framework.

Internal Narrative Generation: This concept focuses on the contents of narrative works created according to external narrative generation. In particular, through the concept of internal narrative generation, the author aims to create novels and stories of himself. These narrative contents are socially distributed through external narrative generation's mechanisms, mainly including "Integrated Narrative Generation System: INGS," "*Geinō* Information System: GIS)," and their integrated narrative environments.

Multiple Narrative Structures Model: and **Social and Collective Narrative Generation:** This is a conceptual model by which a narrative text is multiply constituted, and the narrative generation process multiply executed. This is not a special narrative phenomenon, but the multiplicity is positioned as an essentially important concept for considering and designing narrative generation mechanisms in the author's narrative generation study. The outermost area of the multiple narrative structures corresponds to the level of "social and collective narrative generation" regarding narrative contents' creation and their social distribution.

Takaaki Yoshimoto (1924–2012): He is a critic and philosopher in Japan who is interested in the human mind, literature and personal discourses, and social and institutional problems through unique systematic thought. Yoshimoto considered human worlds full of unique ideal illusions that are divided into three areas: Areas of individual illusions corresponding to personal worlds; Pair illusions concerning families; and Communal illusions corresponding to nations and societies. He seized humans and worlds as the organic synthesis of the three illusions.

***Watakushi Monogatari*:** *Watakushi Monogatari* ("My Story" or "My Narrative," or "I-Story" or "I-Narrative") means a new narrative genre that should be practiced using narrative generation systems in the future. Although it is a type of story or narrative genre in which narrative worlds and narrations are pursued and represented through "I," this does not necessarily mean that the narrative world to be generated should be narrow and the field related only to "I"; rather, it means that the first-person world gradually opens to the second-person and third-person worlds to construct an entire narrative dependent on their organic relationships.

Conclusion

This section is the final part of this book. It comprehensively summarizes the contents and results and shows future directions and visions. In the recent book, *Toward an Integrated Approach to Narrative Generation: Emerging Research and Opportunities* (Ogata, 2019b), the author mainly treated the basic thoughts and concepts, including philosophical concepts, for the integrated approach. By the contrast, this book mainly showed the analytical and technological approaches to the author's narrative generation study. In particular, simply speaking, in more artistic and literary meanings, the most important future topic is how to actually create narrative works. Thus, the author's narrative generation study will gradually move in the direction of the practice of narrative generation. In this book, narratives did not necessarily include only literary or artistic narratives but covered more *geinō*-like and folktale-like narratives. However, in the future, the creation of an interesting narrative in very broad sense will be a goal of narrative generation studies. Furthermore, problems and themes of literary and artistic quality will emerge as an important topic at this stage.

This book described the following themes in its five chapters. In this introduction, the author stated that books are an excellent medium for writing and systematically and consistently presenting a series of studies; subsequently, the author presented overviews of four long papers and five books that the author has published, aiming to develop a systematic and consistent research story. Further, this section reviewed themes treated in the previous paper and book series and introduces the chapters in this book. Chapter 1 presented and discussed various studies concerning the design and implementation of the systems. The central sections in this chapter explained a narrative generation systems called "Integrated Narrative Generation System (INGS)." In summary, this chapter surveyed and discussed interdisciplinary approaches to primarily artificial intelligence (AI)-based computational narrative or story generation systems by way of introducing cognitive science, narratology, and related literary theories. Chapter 2 undertook a comprehensive survey and analysis of *kabuki*, aiming to explore a narrative generation-reception and a narrative production-consumption model of *kabuki* from the viewpoint of an information system and, in particular, a narrative generation system (the framework of GIS with INGS). In Chapter 3, the author focused on an advertising analysis based on the idea of "advertisement as narrative." This chapter described in detail the rhetorical analysis of television commercial films dependent on three regular or ordinary rhetoric types and nine irregular or extraordinary rhetoric types. This study has been applied to the development of advertising narrative generation techniques in the subsequent study. In Chapter 4, from relationships to the level of multiple narrative structures, the description of GIS described corresponded to a social-level narrative generation mechanism beyond narrative generation as a simple substance or as something including many narrative generations as a simple substance. This chapter attempted to discuss "the future of narrative generation systems" through various descriptions regarding "*watakushi monogatari*," which is a plan or vision that will be created with the narrative generation systems INGS and GIS.

As stated above, in contrast to Ogata's (2019b) book that mainly contains fundamental and philosophical concepts and a research framework of the integrated approach to narrative generation or the post-narratology, this book mainly presented the topics of systematization, including chief systems, INGS and GIS, advertisement narrative analysis for application systems, and plans for future narrative creations as well as a vision of external and internal developments based on all of the studies.

This final section summarizes the main topics in this book and simply presented an overview of the introductory section and four chapters (in **A SUMMARY OF THE BOOK**). This section also provides future research and development directions and a supplementary consideration of several chapters (in **FUTURE RESEARCH DIRECTIONS IN THE BOOK: FROM THEORIES, TECHNOLOGIES, AND SYSTEMS TO THE CREATION OF NARRATIVE WORKS**). Although this book mainly treats the topics of technologies and systems, the most essential element in the future direction of integrated narrative generation studies or post-narratology is the problem of narrative creation, and in particular the creation of actual narrative works.

A SUMMARY OF THE BOOK

As the author preliminarily stated in the introduction, this book mainly dealt with the following topics:

- **The consideration of narrative generation mechanisms through narrative analyses**
- **A fundamental narrative generation system**
- **Applications and the social development and distribution of narrative generation**
- **The production of literary and artistic works or literary and artistic organizations**

Throughout this book, conceptual considerations that were treated in Ogata (2019b) were bridged to the technological and applicable directions of narrative generation. In particular, the book introduced Integrated Narrative Generation System (INGS), *kabuki* analyses and *Geinōo* Information System (GIS), advertisement analyses and application systems, and internal and external approaches to narrative generation. This section summarizes all of previous chapters in the book.

In the first **Introduction, TWO BOOKS THAT INTEGRATE A SERIES OF STUDIES** section stated that books are an excellent medium for writing as well as systematically and consistently showing a series of studies by an author. Further, this introduction section stated the main themes in the recent book by Ogata (2019b) and this book, which together compose a series. This section simply summarizes the chapters in Ogata (2019b) to show the characteristics of this book. Based on the author's opinion regarding the book medium, the next section, **SYNTHETIC APPROACHES TO NARRATIVE GENERATION THROUGH SYSTEM-ORIENTED BOOKS**, presented overviews of four long papers and five books that the author has published, aiming at a systematic and consistent research story. This book and the recent book (Ogata, 2019b) correspond to a current synthesis and expansion of these previous papers and books. Finally, the section titled **THEMES OF THE BOOK SERIES FROM A SUMMARY OF THE BOOKS AND PAPERS AND CONFIGURATION OF THE BOOK** reviewed themes treated in the previous paper and book series and introduced the chapters in this book.

Next, the Integrated Narrative Generation System, INGS, introduced in **Chapter 1 [An Integrated Narrative Generation System: Synthesis and Expansion]** was one of the central parts in this study that aims at an integrated approach to narrative generation. This book presented interdisciplinary and cross-disciplinary approaches to narrative and narrative generation as well as many related fields and concepts. Moreover, this book introduced philosophical concepts supporting this study and describes narrative analyses of *kabuki* and advertisements. However, the objectives of this book did not stop with the research of the receptive aspects such as consideration, survey, and analysis regarding narratives and narrative generation.

In this narrative generation study, all parts were considered as components that served the extreme generative and productive goals. The first and the most important element of the goals corresponded to the design and development of INGS. Further, the *Geinō* Information System, discussed in Chapter 2 and again in more detail in Chapter 4, was related to the system's framework that includes an INGS or many INGSs and is executed with an INGS or INGSs. Ogata and Akimoto (2016), Ogata and Asakawa (2018) and Ogata and Akimoto (2019), which was an enhanced version of Ogata and Akimoto (2016), provided the systematic and comprehensive descriptions of INGS. Although this chapter overlapped with previous papers, with respect to INGS as project, the system has been gradually developed and this new chapter shows the entire system, including newly revised and expanded parts. While the entire structure of the description was based on Ogata and Asakawa (2018), the content was extended and supported with new examples of narrative generation.

The following part described the chapter's structure. First, **BACKGROUND: COMMON LISP** covers Common Lisp, the main programming language for our system development, including INGS and other systems. Although, in previous papers, the author did not explain in detail such purely technological problems, the author's discussion here revealed that the Lisp programming language, being a nonprocedural and functional language with flexible scalability, as the most adequate programming language for INGS and has been expanded gradually. Next, **GENERATION PROCESS OF THE INTEGRATED NARRATIVE GENERATION SYSTEM** described the research and development process of the author itself used as a kind of narrative generation process, referring to many related papers. As in Ogata and Asakawa (2018), this chapter divided INGS into two parts: mechanisms and contents. Since the literary criticism of Russian Formalism (1910s–1930s), narratology divided a narrative or narrative work into form and content and, especially, emphasized the formal side of a narrative. This had significance or meaning as a kind of antithesis to other literary schools of thought (e.g., Social Realism) that emphasized the contents described in narratives and literary works and consider "what is described" to be the core of values of narratives and literary works. In Japan, many theoretical literary studies, such as Natsume Sōseki (1867–1916) (2007) and Takaaki Yoshimoto (1924–2012) (1965), considered the formal aspect in narratives and literature works to be important. The narrative study of the author also divided a narrative generation process into formal aspects and content aspects. However, when the author aimed at actual narrative generation, both formal and content aspects were respectively important. In particular, for performing narrative generation, an unavoidable problem was how the author richly gave content information to a generating narrative. The chief objective of previous narrative generation studies was to present and develop the mechanisms or forms of narrative generation. Beyond such a research-oriented approach, such as how to produce excellent narrative and literary works using narrative generation systems, arose such important problems as substantial discussions and technological developments regarding the side of narrative contents that emerge in a central position of narrative generation. Accordingly, the author presented a paper's construction in which the formal mechanism and contents in narrative generation are clearly divided.

The theme of **Chapter 2 [*Kabuki* as a Synthetic Narrative: Synthesis and Expansion]** is *kabuki*. The study regarding the narrative generation of *kabuki* has been frequently tackled. The author's previous paper on *kabuki* (Ogata, 2019a), the author undertook a comprehensive survey and analysis, aiming to explore a narrative generation-reception and a narrative production-consumption model of *kabuki* from the viewpoint of an information system and, in particular, a narrative generation system. A fundamental concept of the modeling was "multiplicity," or "multiple narrative structures." In addition, the author associated this model with the concept of the *Geinō* Information System (GIS), representing a system model in which multiple narrative generations and production mechanisms or processes are included. In addition, this paper showed the results of concrete analyses of *kabuki*'s elements, including "person," "story," "*tsukushi*," and "*naimaze*." Building on the paper, this chapter revised and extended the part of *kabuki*'s multiple elements using many actual examples and presents an experimental system design and implementation in which several analyzed elements were organically combined with each other. Thus, the *kabuki* study in this book corresponded to a concrete example of the multiple narrative structures model in the author's integrated narrative generation study. A series of *kabuki* studies sought to bridge a comprehensive survey and analysis of *kabuki* based on the concept of multiplicity or multiple narrative structures with a more generalized conceptual model, including both narrative generation-reception and production-consumption processes. The synthesized narrative generation architecture that the author has been designing and developing—that is, the Integrated Narrative Generation System, INGS, shown in detail in Chapter 1, formed the background of this study. Although INGS is a basic, fundamental system that has a narrative generation process by a single subject operating on a basis similar to that of an individual human being, the ultimate goal of the author's narrative generation project, including the development of INGS, was to involve one or more INGSs to realize various narrative generation application systems and social distribution mechanisms using automatic narrative generation functions. In particular, the conceptual model referred to as the *Geinō* Information System, GIS, was associated with INGS in this plan. GIS will be also described in Chapter 4. GIS was designed to be a system model in which multiple narrative generation and production mechanisms or processes perform a social level task using one or more INGSs. It was a framework in which various levels of narrative generation processes can be driven by the authors, receivers, characters, and actors in the system. In short, INGS and GIS were designed, respectively, to correspond to narrative generation-reception and narrative production-consumption as an entire system. A *kabuki* play is a collection of multiple narratives built around original scenarios, authentic and related histories, and actors' private scenes. It was intended for its construction to be connected with plans for a future narrative generation system comprising several other narrative generation systems.

This chapter expanded the narrative generation study of *kabuki* to the above purposes. In particular, in the first section, **BACKGROUND**, the author briefly described the overview of *kabuki* itself and re-introduced the multiple elements of *kabuki*. The former part, dependent on Ogata and Akimoto (2016, 2019) and Ogata and Asakawa (2018), gave the overall information regarding *kabuki*, including the history, characteristics, and basic mechanisms of the stage to the readers. The latter multiplicity approach of *kabuki* analyzed the entire phenomena of *kabuki* to consider multiple structures in each element and the multiple structures among several elements. Although *kabuki*'s multiple elements currently contain fifteen basic elements and two technical elements, the theoretical structure and the elements' number will change according to the study's development in the future. The first component of the central part in this chapter, ***KABUKI*'S MULTIPLE NARRATIVE STRUCTURES SEEN IN CONCRETE EXAMPLE**, expanded the previous description of *kabuki*'s multiplicity by adding actual examples and supplementary explanations. The actual examples will contribute to bridging the future systematic and concrete collection and classification of narrative generation techniques, rhetoric, and strategies and directions to the design and implementation by INGS and GIS. In the next main section, **TOWARD INTRODUCING *KABUKI* KNOWLEDGE INTO GIS WITH INGS: FROM ANALYSIS TO GENERATION**, the author showed the design and implementation of experimental programs for concretizing *kabuki*'s narrative generation. Although Ogata (2018, 2019a) presented the system methods and implementations for analyzed elements in previous works—"person," "story," "*naimaze*," and "*tsukushi* or *zukushi*"—, these programs were connected to each other. In contrast, using program modules of "person," "story," "*naimaze*," and "*tsukushi* or *zukushi*," this chapter presented the simulation of the process that gradually generates more complex narratives according to the mutual relationships of the above modules. The attempts in this chapter will provide a more developed starting point to connect the *kabuki* study to the system design and implementation using INGS and GIS.

The objective of **Chapter 3 [An Analysis of Advertisements for Narrative Generation]** was to analyze how a brand—i.e., a product or a service—was introduced into the events in story-like television commercial films using real commercials. An event in a story-like commercial means an event that is developed through a character's actions. Each brand corresponds to a symbol showing a product or a service and focusing on the semantic aspect of the product or service. An event is simply a set piece including action(s) and the necessary related elements, and it is different from a shot, which is a basic unit in moving films. Events are frequently constructed using several shots. However, shots can also contain several events. The direct objective of this analysis is to invent the algorithms and develop the knowledge systems that introduce a brand into an event in the advertising narrative generation system. These systems are used in the mechanism that constructs the advertising flow. Although this study can be applied directly to the systems that support advertising planning and production, such analysis may also be able to be applied to other forms of advertising generation in the future. Advertisement narratives seen in television commercials have mythic or folkloric characteristics beyond realistic narratives. A mixture of rhetoric regarding the relationships between an action and an object, and the rhetoric of metamorphoses of objects and characters, are also interesting for the mechanism of general narrative generation. Moreover, in the study of a story generation system based on the method of Propp's (1968) narrative morphology, Ogata (2007) presented the concept of "narrative generation by destruction and reconstruction." In this concept, ideas destroy and reconstruct existing narrative texts or narratology and literary theories to create new possibilities not seen in existing texts. This idea is also related to the concept of intertextuality in the sense of the usage of existing texts. This advertising narrative generation was also an example of the studies regarding the destruction and reconstruction in the point that aims to develop the combinatorial generation possibilities beyond the simulative generation possibilities based on existing texts, though it uses existing advertising texts. In addition, a similar plan was seen in the idea of "liquidation and crystallization" presented by Akaishi (2006) and Hori (2007) in the context of creative aid in AI. The philosophical concept of fluidity and fixation by the author is a broader concept that includes the above concepts.

The chapter's configuration was as follows. First, **BACKGROUND** introduced many of the previous narrative analysis research by the author and described the concept of "advertisement as narrative" that regarded all advertisement works as contemporary, mythic, or folkloric narratives. This idea extended the range of advertising approaches of the author to the directions to broader general narratives. Next, in **APPROACHES TO ADVERTISEMENT GENERATION AS NARRATIVE GENERATION**, from two viewpoints of the macro framework and the brand introduction methods into advertising narratives, the author introduced a series of studies of advertising narrative generation in which advertising narrative analysis, system design, and implementation were continued in the positioning of an application of narrative generation. **RHETORICAL STRATEGIES FOR ACTIONS AND OBJECTS: A DETAILED ANALYSIS OF TELEVISION ADVERTISEMENT VIDEOS AND THE EXTRACTION AND CLASSIFICATION OF STORY-TYPE BRAND INTRODUCTION STRATEGIES** was the central portion of this chapter and described the developed advertising narrative analyses of the above methods in detail. Finally, **FROM ANALYSIS TO GENERATION** connects the above advertisement analysis to the design and implementation of an advertisement narrative generation mechanism.

The *Geinō* Information System, GIS, dealt with in **Chapter 4 [Toward External and Internal Practices of Narrative Generation]** corresponds to the narrative generation mechanism at the social level over narrative generation as a simple mechanism, which contains several or many simple narrative generation mechanisms in relation with the level of multiple narrative structures. Therefore, the consideration of social narrative generation was partially discussed previously. In this chapter, the author discussed a future vision of the narrative generation system study. This future vision was not limited merely to a social direction. Rather, it had two directions: "external narrative generation" and "internal narrative direction" (this book's title is based on these two concepts). These were placed as the practice of narrative generation. Therefore, this chapter discussed the "external practices of narrative generation" and the "internal practices of narrative generation." The "external" was the outside world beyond this "I" and was ordinarily a society; conversely, the "internal" meant the inside world for the author. Therefore, although practices of narrative generation toward the outside meant narrative generation in the ordinary sense, practices of narrative generation toward the inside were related to the direction for describing this "I" by various methods.

In **BACKGROUND**, first, the author introduced Takaaki Yoshimoto, who was a poet, critic, and philosopher, with special consideration of Yoshimoto's concept of communal illusion (Yoshimoto, 1968) and his identification of three types of illusions—personal illusion, pair illusion, and communal or collective illusion—which will direct the future development of the author's narrative generation study. As another background, the author described a unique literary genre called "*watakushi shosetsu*," which was introduced and practiced in modern Japanese literature. This discussion was related to the plan of "*watakushi monogatari*" that appears in the last half in this chapter. The next section, in **TOWARD EXTERNAL OR SOCIAL PRACTICES OF NARRATIVE GENERATION**, the author described the developmental plan for both the external (social) directions and the internal (personal and individual) directions of the author's narrative generation study. These directions are related to the bridge from research to content creation. This section dealt with the direction to the external or social practices of narrative generation. In particular, the author introduced the problem that the development and distribution of narrative generation systems and the production and distribution of the narrative contents generated by narrative generation systems need to be integrated into a single mechanism through the social development of narrative generation systems. From the narrative generation systems, this problem corresponds to the theme of relationships between the INGS and GIS. By contrast, the next section, in **TOWARD INTERNAL OR PERSONAL PRACTICES OF NARRATIVE GENERATION**, the author described an internal practice of narrative generation, which meant the author's personal practice of making personal narratives. The results could also be developed and distributed socially in the linkage with the above social direction. This section focused on planning materials for internal and personal narratives and the roles and functions of the narrative generation systems, INGS and GIS, for dealing with the materials. The section, **Narratives as Fluid Process:**

Personal View, provides a philosophical discussion. Finally, **THE FIELD OF NARRATIVE GENERATION AS SOCIAL AND COLLECTIVE PHENOMENA** discussed the problem of unintentional narrative generation, as well as topics related to social and philosophical narratology. In particular, it examined how the collection of various personal and organizational narrative generation phenomena produces larger narratives and how they are fixed into people's consciousness, influencing social processes. This was the part that was categorized in the genre of "emergent narratives" in the narrative genre system in Chapter 2 (Ogata, 2019c) in Ogata (2019b) and was related to such academic ideas as "narrative economics" by Shiller (2017). In addition, Yoshimoto's (1968) philosophy of communal illusion analyzed and interpreted at the macro-level narrative formation process of a nation—for example, Japan, as a communal illusion related to micro-level narrative generation processes in village communities.

FUTURE DIRECTIONS OF THE ENTIRE BOOK: FROM THEORIES, TECHNOLOGIES, AND SYSTEMS TO THE CREATION OF NARRATIVE WORKS

Ogata (2019b) described mainly conceptual topics in the integrated approach to narrative generation (e.g., frameworks of a narrative generation–reception process and a narrative production–consumption process, a narrative genre system) and introduced a new narratology, called post-narratology, and philosophical and thinking considerations regarding the narrative generation study. Further, based on the above considerations, the author stated the design and development of the systems related to narrative generation–reception and narrative production–consumption, narrative analyses of *kabuki* and advertisement related to the above systems, and external or social and internal or personal approaches to narrative generation based on the post-narratology.

Contingent on the above description, the author is considering the future perspectives and directions of the research from the following two aspects.

The first is the integration of the above presented systems, namely, the Integrated Narrative Generation System (INGS) and *Geinō* Information System (GIS), into the design and development of mechanisms that are connected to actual societies or worlds. This area is overlapped with the GIS in the above study, and is related to the direction of the development of business of the author's narrative generation study. In particular, this integration refers to the problem of the development of platforms for socially distributing and sending narratives that are automatically generated and generated through the collaboration between humans and machines. Therefore, this direction is positioned as an extension of GIS research and development that includes INGS. Simultaneously, this direction needs research on the framework of social business.

The second is the practice of narrative generation through a social platform positioned as an extension of concretizing GIS or through other mechanisms, media, and spaces. This practice has the two directions of artistic and literary practice. Artistic practice means practicing narrative generation, or the production and representation, as a mainly fluid act. As stated above, these artistic and literary practices are closely related to the concepts of fluidity and fixation of narrative generation, referred to in detail in Ogata (2019b). Artistic practice is mainly situated as the method for representing the fluidity of narrative generation. By contrast, literary practice is positioned as the method for representing the fixation. In particular, the author considers a method in which fluid narrative generation is experimentally conducted through an artistic practice and a fixed narrative or fixed narratives are created from the above process.

In summary, in more artistic and literary meanings, the most important future topic is the actual creation of narrative works. Thus, the author's narrative generation study will gradually move in the direction of the practice of narrative generation. In this book (and the recent one by Ogata (2019b)), narratives did not necessarily include only literary or artistic narratives but covered more *geinō*- and folktale-like narratives. However, in future, the creation of a narrative that is interesting in a very broad sense will be a goal of narrative generation studies. Problems and themes of literary and artistic quality will emerge as an important topic at this stage. In particular, an important future direction is "creating narratives from the theories and technologies of narrative generation." Here, narrative creation means the problem of how narratives are created as literature and arts. Hence, an integrated study to narrative generation as post-narratology by the author will gradually move toward internal and external narrative creation. Figure 1 shows the future directions to external and internal practices of post-narratology and the creation and social development (including distribution as a main element) of *watakushi monogatari*.

As stated above, in **A SUMMARY OF THE BOOK**, the author summarized each chapter and discussed important concepts and future directions. The recent work by Ogata (2019b) and this book are a series that shows a systematic narrative generation study based on post-narratology. Ogata

Figure 1. From future practices of post-narratology to watakushi monogatari

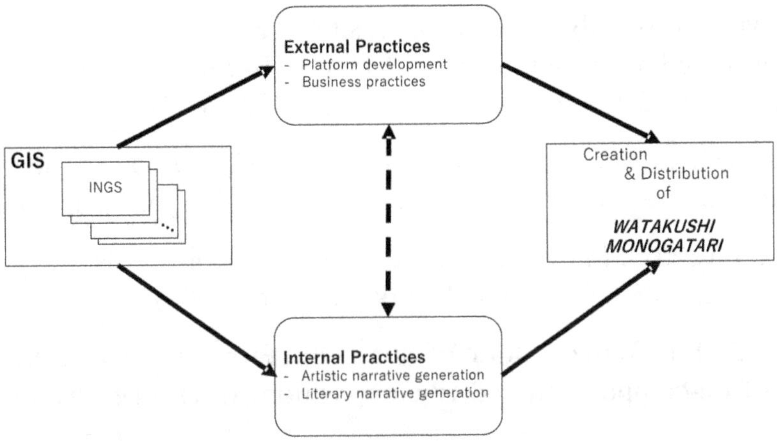

(2019b) mainly contained fundamental and philosophical concepts, as well as a research framework of the integrated approach to narrative generation or post-narratology. This book mainly presented the topics of systematization through core systems, including Integrated Narrative Generation System (INGS), *Geinō* Information System (GIS), and advertising application systems based on the advertisement narrative analysis. Another important theme in this book was the development of narrative generation study through external or social and internal or personal practices and considerations of narrative generation. This chapter also presented vision for further work and future directions in the section **FUTURE DIRECTIONS IN THE ENTIRE BOOK: FROM THEORIES AND TECHNOLOGIES TO NARRATIVE CREATIONS**. In particular, the author presented visions for future narrative creations as well as a plan of external and internal developments based on all of the studies thus far.

ACKNOWLEDGMENT

This chapter's research was supported by JSPS KAKENHI Grant Number18K18509, Kayamori Foundation of Information Science Advancement Research Grant, The Telecommunication Advancement Foundation Research Grant, and AOI TYO Holdings Inc.

REFERENCES

Akaishi, M. (2006). A dynamic decomposition/recomposition framework for documents based on narrative structure model. *Transactions of the Japanese Society for Artificial Intelligence, 21*(5), 428–438. doi:10.1527/tjsai.21.428

Hori, K. (2007). *Sōzō katsudō shien system no riron to ōyō* [Theories and applications of creative activity support systems]. Tokyo, Japan: Ohmsha.

Natsume, S. (2007). *Bungaku ron, 1,2* [Literary theory, Vols. 1, 2]. Tokyo, Japan: Iwanami Shoten. (Original work published 1907)

Ogata, T. (2007). To the rhetoric of story from Propp: Centering around the rhetoric of de-composition and re-composition. *Cognitive Studies, 14*(4), 532–558.

Ogata, T. (2018). An integrated approach to narrative generation: From Mishima and kabuki to narrative generation systems. In T. Ogata & S. Asakawa (Eds.), *Content generation through narrative communication and simulation* (pp. 49–147). Hershey, PA: IGI Global. doi:10.4018/978-1-5225-4775-4.ch002

Ogata, T. (2019a). Kabuki as multiple narrative structures and narrative generation. In T. Ogata & T. Akimoto (Eds.), *Post-narratology through computational and cognitive approaches* (pp. 192–275). Hershey, PA: IGI Global. doi:10.4018/978-1-5225-7979-3.ch005

Ogata, T. (2019b). *Toward an integrated approach to narrative generation: Emerging research and opportunities*. Hershey, PA: IGI Global.

Ogata, T. (2019c). Areas of narratives or narrative genres. In T. Ogata (Ed.), *Toward an integrated approach to narrative generation: Emerging research and opportunities* (pp. 59–161). Hershey, PA: IGI Global.

Ogata, T., & Akimoto, T. (Eds.). (2016). *Computational and cognitive approaches to narratology*. Hershey, PA: IGI Global. doi:10.4018/978-1-5225-0432-0

Ogata, T., & Akimoto, T. (Eds.). (2019). *Post-narratology through computational and cognitive approaches*. Hershey, PA: IGI Global. doi:10.4018/978-1-5225-7979-3

Ogata, T., & Asakawa, S. (Eds.). (2018). *Content generation through narrative communication and simulation*. Hershey, PA: IGI Global. doi:10.4018/978-1-5225-4775-4

Propp, V. Y. (1968). *Morphology of the folktale* (L. Scott, Trans.). Austin, TX: USA University of Texas Press. (Original work published 1928)

Shiller, R. J. (2017). Narrative economics. *Cowles Foundation Discussion Paper, 2069.*

Yoshimoto, T. (1965). *Gengo ni totte bi toha nanika* [What is beauty for language?]. Tokyo, Japan: Keisō Shobō.

Yoshimoto, T. (1968). *Kyōdō gensō ron* [The theory of communal illusion]. Tokyo, Japan: Kawade Shobō Shinsha.

About the Author

Takashi Ogata received his Bachelors in Social Sciences from Waseda University in 1983. Then he received his M.S. from Tsukuba University in 1992 and his Ph.D. from the University of Tokyo in 1995. He has attained industrial experience since 1983 at software development companies. Having been an Associate Professor in the faculty of Engineering at Yamanashi University since 1997, he is now a Professor in the faculty of Software and Information Science at Iwate Prefectural University since 2005. He is a member of the Japanese Society for Artificial Intelligence (JSAI), the Japanese Cognitive Science Society (JCSS), and the Japanese Association for Natural Language Processing. He was also the primary manager of the Literature, Cognition, and Computer (LCC) research group at the JCSS. He received the JSAI best paper award (1996), best paper award from the Japan Academy of Advertising (1996), and other academic awards. In addition to many papers, He has published academic books, including *Monogatariron no Jōhogaku Josetu* (*An Introduction to the Informatics of Narratology*) (T. Ogata & A. Kanai, 2010, Tokyo: Gakubunsha), *Computational and Cognitive Approaches to Narratology* (T. Ogata & T. Akimoto, 2016, IGI Global), *Content Generation Through Narrative Communication and Simulation* (T. Ogata & S. Asakawa, 2017, IGI Global), *Jōho Monogatariron* (*Informational Narratology*) (T. Ogata, Y. Kawamura, & A. Kanai, 2018, Tokyo: Hakutō Shobō), *Post-Narratology Through Computational and Cognitive Approaches* (T. Ogata & T. Akimoto, 2019, IGI Global), and *Toward an Integrated Approaches to Narrative Generation: Emerging Research and Opportunities* (T. Ogata, 2019, IGI Global). Furthermore, he is currently writing and editing three books related to his concept of post-narratology by Japanese and English. His research interests include the following broad areas: (1) Computer science and information technologies: narrative (story) generation (system), artificial intelligence, cognitive science, natural language processing and generation, ontology and conceptual dictionary, etc. (2) Humanities: narratology, literary theories, semiotics, semantics, discourse theory, philosophy, *kabuki* and Japanese arts and entertainments, Japanese narratologies and literary theories, etc. (3) Social sciences: content business, marketing and advertising studies, sociology, social philosophy, etc.

Index

N

P

R

S

T

W

www.ingramcontent.com/pod-product-compliance
Lightning Source LLC
LaVergne TN
LVHW080613300926
849162LV00005B/1140